Lecture Notes in Control and Information Sciences

Volume 475

This series aims to report new developments in the fields of control and information sciences—quickly, informally and at a high level. The type of material considered for publication includes:

1. Preliminary drafts of monographs and advanced textbooks
2. Lectures on a new field, or presenting a new angle on a classical field
3. Research reports
4. Reports of meetings, provided they are
   (a) of exceptional interest and
   (b) devoted to a specific topic. The timeliness of subject material is very important.

More information about this series at http://www.springer.com/series/642

Sophie Tarbouriech · Antoine Girard
Laurentiu Hetel
Editors

Control Subject to Computational and Communication Constraints

Current Challenges

Springer

Editors
Sophie Tarbouriech
LAAS-CNRS
Université de Toulouse
Toulouse
France

Antoine Girard
Laboratoire des Signaux et Systèmes
CNRS, CentraleSupélec, Université Paris-Sud, Université Paris-Saclay
Gif-sur-Yvette
France

Laurentiu Hetel
CNRS, Centrale Lille, UMR 9189—Centre de Recherche en Informatique Signal et Automatique de Lille (CRIStAL)
Université Lille
Lille
France

ISSN 0170-8643 ISSN 1610-7411 (electronic)
Lecture Notes in Control and Information Sciences
ISBN 978-3-030-08701-2 ISBN 978-3-319-78449-6 (eBook)
https://doi.org/10.1007/978-3-319-78449-6

Printed on acid-free paper

This Springer imprint is published by the registered company Springer International Publishing AG part of Springer Nature
The registered company address is: Gewerbestrasse 11, 6330 Cham, Switzerland

Preface

In modern industrial applications and other scientific domains, control theory plays a fundamental role to achieve high performances, optimal production levels, product and service quality, energy savings, and last but not least operational safety. However, in order to achieve these generic goals, different issues due to the heterogeneity of the phenomena affecting the controlled systems have to be taken into account: nonlinearities, time-varying parameters, uncertainties, time delays, signal discontinuities, disturbances, and limitation of the available information and computational resources, among others. Neglecting these effects can lead to economic losses or even disastrous effects (e.g., airplane crashes, plant explosions, etc.). Although in the last years the modern control theory has addressed many of the features above, there are still many theoretical and practical open problems. Also, an important challenge is the development of methods considering the simultaneous presence of these features in the systems. Hence, new analysis and design methods to cope with the complexity of these phenomena, aiming at achieving robust stability and performance of the feedback loops, are of major relevance both in scientific and technological contexts. Control systems are nowadays implemented over communication networks, in the so-called networked control paradigm. Networked control systems (NCSs), or more generally cyber-physical systems (CPS), are spatially distributed systems for which the communication between sensors, actuators, and controllers is supported by a shared communication network. CPSs offer many advantages such as low installation and maintenance costs, reduced system wiring, and increased flexibility of the system. However, their rapid development also led to new technical problems: asynchronism (due to limited computational resource, concurrency, and multitask scheduling algorithms), complex energy management (where switching among several functioning modes must be considered), communication latency and data loss (due to the use of networks), etc. The overall system reliability is a major requirement, since in many applicative domains embedded controllers are involved in safety-critical activities.

It is important to mention that, in general, the conception of the control systems addresses separately the problem related to the design of the control loop and that one related to its implementation using some computational resources, possibly affected by imprecise communications. An integrated approach, in which the constraints due to computational resources and limited information are taken into account, allows the development of control systems with certification guarantees of stability, robustness, and performance. The implicit goal is then to optimize the use of available resources. These separated designs can be very suboptimal, incurring some severe performance losses and significant additional costs. It turns out that these losses and costs become more and more critical for modern (and large scale) applications such as the use of numerous sensors in cars, sensor networks, airspace applications, and the deployment of cellular and internet networks.

Moreover, systems to be controlled usually have a nonlinear behavior, and under network delays, asynchronous samplings, or computational resources constraints, controlling them becomes more and more difficult, especially with a prescribed performance requirement. In particular, the hybrid nature of the system (mixing a continuous dynamics and discrete control sampling) should be taken into account.

Combining the field of control theory, communication theory, and computational resources management, the main goal of the current book is to present different aspects and solutions in order to take into account the complex phenomena and issues mentioned above. In particular, this book pertains to gather several facets of systems subject to information and computational constraints. Then, this book constitutes a large overview of results and techniques with respect to the recent literature, including hybrid dynamical systems, switched systems, Event-Triggered architecture, sampled-data systems, and distributed systems for CPS.

At this aim, this book is organized as follows:

- Part I is devoted to switched and sampled-data systems and consists of Chaps. 1–6.
- Part II is devoted to Event-Triggered architecture and consists of Chaps. 7–11.
- Part III is devoted to distributed control of CPS and consists of Chaps. 12–17.

Note that this partition is somewhat arbitrary as most of the chapters are interconnected, and mainly reflects the editors' biases and interests.

The idea of this book inherits from the organization of the international workshop CO4 organized on the topics in Toulouse in October 26–28, 2016, with the support of ANR.[1] Hence, we would like to thank the contributors of this book because their constant encouragement, enthusiasm, and patience have been essential for the realization of this book.

[1]The French National Research Agency, public body providing funding for project-based research under the authority of the French Ministry of research.

We hope that this volume will help in claiming many of the problems for controls researchers, and to alert graduate students to the many interesting ideas at the frontier between control theory, information theory, and computational theory. There are, of course, many areas which are not represented through a chapter, and therefore we would like to apologize to those whose areas are not profiled.

Toulouse, France
December 2017

Sophie Tarbouriech
Antoine Girard
Laurentiu Hetel

Contents

Contributors

A. Adimoolam VERIMAG, Université Grenoble-Alpes, Saint Martin D'Hères, France

D. Adzkiya Institut Teknologi Sepuluh Nopember, Surabaya, Indonesia

M. Al Khatib Laboratoire des Signaux et Systèmes (L2S), CNRS, CentraleSupélec, Université Paris-Sud, Université Paris-Saclay, Gif-sur-Yvette, France

V. Andrieu Univ. Lyon, Université Claude Bernard Lyon 1, CNRS, LAGEP UMR 5007, Villeurbanne, France

M. Baştuğ Centre de Recherche en Informatique, Signal, et Automatique de Lille (CRIStAL, CNRS UMR 9189), Centrale Lille, Villeneuve d'Ascq, France

J. Ben Rejeb Université de Lorraine, CNRS, CRAN, Nancy, France

D. P. Borgers Eindhoven University of Technology, Eindhoven, The Netherlands

D. Boskos ACCESS Linnaeus Center and Center for Autonomous Systems, KTH Royal Institute of Technology, Stockholm, Sweden

F. Ceragioli Politecnico di Torino, Torino, Italy

J. Daafouz Université de Lorraine, CNRS, CRAN, Nancy, France

T. Dang Verimag, Université Grenoble-Alpes-CNRS, Grenoble, France

C. De Persis ENTEG and Jan C. Willems Center for Systems and Control, University of Groningen, Groningen, The Netherlands

D. V. Dimarogonas ACCESS Linnaeus Center and Center for Autonomous Systems, KTH Royal Institute of Technology, Stockholm, Sweden

A. D'Innocenzo Department of Information Engineering, Computer Science and Mathematics, Center of Excellence DEWS, University of L'Aquila, L'Aquila, Italy

V. S. Dolk Eindhoven University of Technology, Eindhoven, The Netherlands

G. E. Dullerud University of Illinois, Urbana, IL, USA

L. Etienne Centre de Recherche en Informatique, Signal, et Automatique de Lille (CRIStAL, CNRS UMR 9189), Centrale Lille, Villeneuve d'Ascq, France

M. Fiacchini University Grenoble Alpes, CNRS, Gipsa-lab, Grenoble, France

C. Fiter Centre de Recherche en Informatique, Signal, et Automatique de Lille (CRIStAL, CNRS UMR 9189), Université de Lille 1, Villeneuve d'Ascq, France

P. Frasca Univ. Grenoble Alpes, CNRS, Inria, Grenoble INP, GIPSA-Lab, Grenoble, France

A. Girard Laboratoire des Signaux et Systèmes (L2S), CNRS, CentraleSupélec, Université Paris-Sud, Université Paris-Saclay, cedex, Gif-sur-Yvette, France

M. Guo ACCESS Linnaeus Center and Center for Autonomous Systems, KTH Royal Institute of Technology, Stockholm, Sweden

W. P. M. H. Heemels Eindhoven University of Technology, Eindhoven, The Netherlands

L. Hetel Centre de Recherche en Informatique, Signal, et Automatique de Lille (CRIStAL, CNRS UMR 9189), Centrale Lille, Villeneuve d'Ascq, France

C. Hop Delft University of Technology, Delft, The Netherlands

M. Jungers Université de Lorraine, CRAN, Lorraine, France; CNRS, CRAN, Lorraine, France

T.-E. Korabi Non-A team at INRIA Lille, Villeneuve d'Ascq, France

S. Lasaulce L2S, Gif-sur-Yvette, France

M. Mazo Jr. Delft University of Technology, Delft, The Netherlands

I.-C. Morărescu Université de Lorraine, CNRS, CRAN, Nancy, France

M. Nadri Univ. Lyon, Université Claude Bernard Lyon 1, CNRS, LAGEP UMR 5007, Villeurbanne, France

J. Peralez Univ. Lyon, Université Claude Bernard Lyon 1, CNRS, LAGEP UMR 5007, Villeurbanne, France

M. Petreczky Centre de Recherche en Informatique, Signal, et Automatique de Lille (CRIStAL, CNRS UMR 9189), Centrale Lille, Villeneuve d'Ascq, France

C. Prieur Grenoble INP, GIPSA-lab, Univ. Grenoble Alpes, CNRS, Grenoble, France

R. G. Sanfelice Department of Computer Engineering, University of California, Santa Cruz, CA, USA

D. Senejohnny ENTEG and Jan C. Willems Center for Systems and Control, University of Groningen, Groningen, The Netherlands

U. Serres Univ. Lyon, Université Claude Bernard Lyon 1, CNRS, LAGEP UMR 5007, Villeurbanne, France

A. Seuret LAAS-CNRS, Université de Toulouse, CNRS, Toulouse, France

A. Sharifi-Kolarijani Delft University of Technology, Delft, The Netherlands

A. Tanwani LAAS – CNRS, University of Toulouse, CNRS, Toulouse, France

S. Tarbouriech LAAS-CNRS, Université de Toulouse, Toulouse, France

A. R. Teel University of California, Santa Barbara, CA, USA

P. Tesi ENTEG and Jan C. Willems Center for Systems and Control, University of Groningen, Groningen, The Netherlands; DINFO, Università degli Studi di Firenze, Firenze, Italy

J. Tumova Robotics, Perception, and Learning Department, KTH Royal Institute of Technology, Stockholm, Sweden

L. Zaccarian LAAS-CNRS, Université de Toulouse, CNRS, Toulouse, France; Dipartimento di Ingegneria Industriale, University of Trento, Trento, Italy

Part I
Switched and Sampled-Data Systems

Chapter 1
Minimal- and Reduced-Order Models for Aperiodic Sampled-Data Systems

M. Baştuğ, L. Hetel and M. Petreczky

Abstract Networked and embedded control systems are ubiquitous nowadays in practical applications. The detailed models for such systems can be very complex due to the interactions of different subsystems in the network, the inherent complexity of the models of the subsystems, and due to the sampling phenomenon itself. In turn, simulations for control synthesis or performance specifications can easily become intractable. Hence, computing low-order accurate models for such systems can be of great importance. In this chapter, we precisely address this problem by studying two main topics for aperiodic sampled-data systems which are used as modeling abstractions for networked control systems: First, we formulate minimal order models capturing the whole behavior of the system with taking into account all possible sampling patterns. Second, we investigate a model reduction method to further simplify the models without losing much of the accuracy.

1.1 Introduction

Minimality of a state-space model of a system implies that there is no other model with a smaller state-space dimension (i.e., less number of state variables) which represents the input–output behavior of the system [23]. Hence, when only the input–output behavior of the system is of interest, converting a non-minimal state-space representation to a minimal one can be considered as the first, and most natural, step of model reduction. For models which are already minimal, such a simplification is of course not possible. In turn, for such models, model reduction techniques [2] can

M. Baştuğ (✉) · L. Hetel · M. Petreczky
Centre de Recherche en Informatique, Signal, et Automatique de Lille (CRIStAL, CNRS UMR 9189), Centrale Lille, 59650 Villeneuve d'Ascq, France
e-mail: mert.bastug@univ-lille1.fr

L. Hetel
e-mail: laurentiu.hetel@ec-lille.fr

M. Petreczky
e-mail: mihaly.petreczky@ec-lille.fr

S. Tarbouriech et al. (eds.), *Control Subject to Computational and Communication Constraints*, Lecture Notes in Control and Information Sciences 475, https://doi.org/10.1007/978-3-319-78449-6_1

be used to *approximate* the input–output behavior of a given system by reducing the number of states of an existing minimal state-space model.

The chapter considers these two problems, the formulation of minimality and model reduction of state-space models, for aperiodic sampled-data systems. Such systems are used as a modeling framework for networked and embedded control systems [9, 12, 18, 20, 22, 40]. Aperiodic sampled-data systems can be represented as impulsive systems [17, 28], time-delay systems [13], linear fractional representations (LFRs) [27, 29], and discrete-time linear parameter-varying (LPV) polytopic systems [10, 15, 21].

The aperiodic sampling patterns studied in this chapter cover two general cases. Namely, if the inter-sample time is time-varying and takes its values in between a fixed arbitrarily small minimum and maximum real value between two consecutive sampling instants, such systems can be represented by discrete-time LPV models through convex embedding. Whereas if the inter-sample time takes its value from a finite set of values in between any two consecutive sampling instants, linear time-invariant (LTI) systems sampled in such manner can be represented by discrete-time linear switched systems. Hence, first, the connection between the minimality of the sampled LTI system and the resulting model, discrete-time LPV or linear switched system depending on the sampling pattern, is investigated. Then, an approach of moment matching [2] based model reduction for the case of linear switched models of sampled-data systems is given. The given approach can be summarized as a moment matching based model reduction method on the linear switched model acquired from the original LTI plant. It is proven that for the given approach, as long as the original continuous-time LTI plant is stable, the resulting reduced-order linear switched model of the sampled-data system is quadratically stable. This model reduction approach is illustrated with two numerical examples containing a stable and an unstable plant model, respectively.

The results about minimality in this chapter can be related to the work in [25] where the controllability and observability of a class of sampled-data systems are inspected. However, in [25] the controllability and observability conditions are given for a very specific type of sampling pattern (namely, an asynchronous and periodical one). Some other papers dealing with the problem of reachability and/or observability of sampled-data systems are [1, 14, 37, 38]. Model reduction for sampled-data systems has been considered previously on [3, 34]. Both papers deal with the case of periodical sampling and are valid only for the case when the considered plant is stable. In contrast, in this chapter, the general aperiodic sampling case considered and the considered plant is allowed to be unstable.

The chapter represents an extended version of the Refs. [4, 7]. The main differences of the chapter with [4, 7] are as follows: The minimality argument in [7] is extended to the linear switched models of aperiodic sampled-data systems. The proofs of the lemmas and the main theorem related to the conservation of stability by the proposed model reduction method are presented in detail in this chapter, whereas [4] contains no proofs. In addition, the current chapter includes two numerical examples (one containing an unstable plant) which have not appeared previously in any of these references.

The chapter is structured as follows. In Sect. 1.2, we present a brief overview of modeling the input–output behavior of a sampled continuous-time LTI plant on sampling instants, for two general scenarios of aperiodic sampling. In Sect. 1.3, we provide a review of realization theory for LTI, LPV, and linear switched (LS) systems, as a preliminary for the subsequent discussion on minimality. In Sect. 1.4, we state the main result on minimality. In Sect. 1.5, we present a model reduction method based on moment matching, to approximate the high-order models of the sampled-data system. In Sect. 1.6, we prove that the given model reduction method preserves the stability of the original model. Finally, in Sect. 1.7, we illustrate the model reduction method with two numerical examples.

1.2 Modeling of Aperiodic Sampled-Data Systems

In this section, we review the process of modeling an aperiodically sampled continuous-time LTI system with a discrete-time model. In the following, we will use $\mathbb{Z}$, $\mathbb{N}$, and $\mathbb{R}_+$ to denote, respectively, the set of integers, the set of natural numbers including 0, and the set $[0, +\infty)$ of nonnegative real numbers.

An LTI SS representation Σ_{LTI} is a tuple $\Sigma_{\text{LTI}} = (A, B, C)$ with $A \in \mathbb{R}^{n\times n}$, $B \in \mathbb{R}^{n\times m}$, $C \in \mathbb{R}^{p\times n}$. The state $x(t) \in \mathbb{R}^n$ and the output $y(t) \in \mathbb{R}^p$ of the LTI system Σ_{LTI} at time $t \geqslant 0$ is defined by[1]

$$\Sigma_{\text{LTI}} \begin{cases} \dot{x}(t) = Ax(t) + Bu(t) \\ y(t) = Cx(t), \ \forall t \in \mathbb{R}_+. \end{cases} \tag{1.1}$$

In the following, $\dim(\Sigma_{\text{LTI}})$ will be used to denote the dimension n of the state space of Σ_{LTI}.

Let $\Sigma_{\text{LTI}} = (A, B, C)$ be a continuous-time LTI SS representation of the form (1.1). Let the state $x(t_k)$ and output $y(t_k)$ of Σ_{LTI} be sampled in arbitrary time instants t_k, $k \in \mathbb{N}$ such that $t_0 = 0$ and $t_{k+1} - t_k \in \mathscr{H} \subset \mathbb{R}$ for all $k \in \mathbb{N}$ to form the constant control signal $u(t) = u_k$ for all $t \in [t_k, t_{k+1})$, $k \in \mathbb{N}$. Note that the sequence t_k, $k \in \mathbb{N}$ is monotonically increasing. The resulting sampled-data system Σ_{SD} can be represented as follows:

$$\Sigma_{\text{SD}} \begin{cases} \dot{x}(t) = Ax(t) + Bu_k, \ t \in [t_k, t_{k+1}), \ k \in \mathbb{N} \\ y_k = Cx(t_k) \\ t_{k+1} = t_k + h_k, \ h_k \in \mathscr{H}. \end{cases} \tag{1.2}$$

[1]Unless stated otherwise, we take $x(0) = x_0 = 0$ for all classes of systems discussed in the chapter for notational simplicity. We remark that the given results can easily be extended to the case of nonzero initial states.

In (1.2), $x(t) \in \mathbb{R}^n$ is the state, $u_k \in \mathbb{R}^m$ is the constant input, and $y(t)$ is the output at time $t \in \mathbb{R}_+$; $A \in \mathbb{R}^{n\times n}$, $B \in \mathbb{R}^{n\times m}$, and $C \in \mathbb{R}^{p\times n}$ are the same as the system parameters (A, B, C) of Σ_{LTI}. We call the set $\mathscr{H}$ as the *sampling interval set* and the number h_k as the *kth sampling interval*. We will use the shorthand notation $\Sigma_{\mathrm{SD}} = (A, B, C)$ for the sampled-data system of the form (1.2). Note that different from the model (1.1) and (1.2) has also $h_k, k \in \mathbb{N}$ as the control parameter in addition to the input $u(t) = u_k,\ t \in [t_k, t_{k+1}),\ k \in \mathbb{N}$.

The state $x_k = x(t_k)$ and output y_k of the sampled-data system Σ_{SD} in (1.2) at sampling instants $t_k, k \in \mathbb{N}$ can be written by induction as

$$\begin{aligned} x_{k+1} &= x(t_{k+1}) = e^{Ah_k}x_k + \left(\int_0^{h_k} e^{As}ds\right)Bu_k,\ \forall k \in \mathbb{N}, \\ y_k &= Cx_k. \end{aligned} \tag{1.3}$$

Let

$$\Theta(h_k) = \int_0^{h_k} e^{As}ds. \tag{1.4}$$

It is easy to see that the following holds:

$$e^{Ah_k} = I_n + A\Theta(h_k). \tag{1.5}$$

Replacing (1.5) in (1.3) and defining the matrix functions $\Phi : \mathscr{H} \to \mathbb{R}^{n\times n}$ and $\Gamma : \mathscr{H} \to \mathbb{R}^{n\times m}$ as $\Phi(h_k) = e^{Ah_k} = I_n + A\Theta(h_k)$ and $\Gamma(h_k) = \Theta(h_k)B$, (1.3) can be rewritten as

$$\Sigma_{\mathrm{disc}} \begin{cases} x_{k+1} = \Phi(h_k)x_k + \Gamma(h_k)u_k, \\ y_k = Cx_k,\ \forall k \in \mathbb{N}. \end{cases} \tag{1.6}$$

With Eq. (1.6), the sampled LTI plant Σ_{SD} in (1.2) is modeled by a discrete-time, time-varying linear system Σ_{disc} whose read-out map (map represented by the matrix C) is time invariant. Here, the discrete-time instants $k \in \mathbb{N}$ of (1.6) correspond to the time instants $t_k \in \mathbb{R}_+, k \in \mathbb{N}$ for the original sampled-data system Σ_{SD}. In addition, the state x_k and the output y_k of (1.6) correspond to the state $x(t_k)$ and output $y(t_k)$ of Σ_{SD} at the sampling instants $t_k \in \mathbb{R}_+$ when $u(t) = u_k$ for $t \in [t_k, t_{k+1})$. Hence, we have built the relationship between the sampled-data system $\Sigma_{\mathrm{SD}} = (A, B, C)$ and the corresponding discrete-time, linear time-varying system representation Σ_{disc}.

In the following, we will consider two different sampling scenarios, described as follows:

Scenario 1: The sampling interval set $\mathscr{H}$ is infinite and defined as $\mathscr{H} = (0, \overline{h}]$ with $\overline{h} \in \mathbb{R}_+\backslash\{0\}$. In other words, the sampling interval between each sampling instants could take its value in between a fixed minimum and maximum value. One way of modeling the sampled plant with such a scenario is to create a discrete-time LPV SS representation through convex embedding of the value of the state transition matrix of the plant model, for different time intervals [10, 15, 21].

Scenario 2: The sampling interval set $\mathscr{H}$ is finite and defined as $\mathscr{H} = \{\hat{h}_1, \ldots, \hat{h}_M\}$ where $\hat{h}_1, \ldots, \hat{h}_M \in \mathbb{R}_+$ and $M \in \mathbb{N}\backslash\{0\}$. In other words, the sampling interval in between any two consecutive sampling instants takes its value from a finite set. Modeling the sampled plant for such a scenario could be achieved by creating a discrete-time LS SS model [11, 16, 39]. Intuitively, such a linear switched model would have as many discrete modes (linear subsystems) as the number of elements in the allowed finite sampling interval set.

In the following, we will use the terms *"the case when the sampling interval set $\mathscr{H}$ is infinite"* and *"the case when the sampling interval set $\mathscr{H}$ is finite"* to refer to **Scenario 1** and **Scenario 2**, respectively.

1.2.1 Modeling Aperiodic Sampled-Data Systems with LPV SS Representations

One important approach to design control for the model (1.6) for the case when $\mathscr{H} = (0, \bar{h}]$ is to embed the function $\Theta(h_k)$, $h_k \in \mathscr{H}$ (hence the functions $\Phi(h_k)$ and $\Gamma(h_k)$ also) inside a convex polytope. This approach results in a discrete-time LPV SS model with affine dependence of the state on the scheduling signal at each time instant. Below we summarize this procedure.

Notation 1 *Let $a, b \in \mathbb{N}$. In the following, we use $\mathbb{I}_a^b$ to denote the set $\mathbb{I}_a^b = \{c \in \mathbb{N} \mid a \leqslant c \leqslant b\}$.*

The procedure of LPV modeling of the sampled-data plant Σ_{SD} relies on the embedding of the function $\Theta(h_k)$ into a convex polytope [10, 15, 19]. The general idea is that the matrix function $\Theta(h_k) \in \mathbb{R}^{n\times n}$ can be written as a convex combination of N_v real matrix vertices $D_1, \ldots, D_{N_v} \in \mathbb{R}^{n\times n}$. Hence

$$\Theta(h_k) = \sum_{i=1}^{N_v} \alpha_i^k D_i, \ \forall h_k \in \mathscr{H}, k \in \mathbb{N}, \tag{1.7}$$

where $0 \leqslant \alpha_i^k \in \mathbb{R}$ for all $k \in \mathbb{N}$, $i \in \mathbb{I}_1^{N_v}$ and $\alpha_1^k + \cdots + \alpha_{N_v}^k = 1$ for all $k \in \mathbb{N}$. Note that the coefficients $\alpha_1^k, \ldots, \alpha_{N_v}^k$ are functions of h_k for each $k \in \mathbb{N}$. For notational simplicity, this dependency is not explicitly denoted. Using (1.7), (1.6) can be rewritten as the following SS representation:

$$\begin{aligned} x_{k+1} &= A^{\mathrm{LPV}}(p^k)x_k + B^{\mathrm{LPV}}(p^k)u_k \\ y_k &= Cx_k, \ \forall k \in \mathbb{N}, \ x_0 = 0, \end{aligned} \tag{1.8}$$

where $p^k = \left[\alpha_1^k \ \ldots \ \alpha_{N_v}^k\right]^{\mathrm{T}} \in \mathbb{R}^{N_v}$ is the scheduling signal at time $k \in \mathbb{N}$. The matrix functions $A^{\mathrm{LPV}}(p^k)$, $B^{\mathrm{LPV}}(p^k)$ are given by

$$A^{\mathrm{LPV}}(p^k) = A_0^{\mathrm{LPV}} + \sum_{i=1}^{N_v} A_i^{\mathrm{LPV}} \alpha_i^k, \quad B^{\mathrm{LPV}}(p^k) = B_0^{\mathrm{LPV}} + \sum_{i=1}^{N_v} B_i^{\mathrm{LPV}} \alpha_i^k \tag{1.9}$$

for all $k \in \mathbb{N}$, where $A_0^{\mathrm{LPV}} = I_n$, $B_0^{\mathrm{LPV}} = 0_{n\times m}$ and $A_i^{\mathrm{LPV}} = AD_i$, $B_i^{\mathrm{LPV}} = D_i B$ for all $i \in \mathbb{I}_1^{N_v}$. A state-space model of the form (1.8) belongs to the class of discrete-time LPV SS representations [32, 36]. Hence, from now on, we will refer to the system representations of the form (1.8) as LPV SS representations and formally define the tuple $\Sigma_{\mathrm{LPV}} = (\{(A_i^{\mathrm{LPV}}, B_i^{\mathrm{LPV}}, C)\}_{i=0}^{N_v})$ with $A_i^{\mathrm{LPV}} \in \mathbb{R}^{n\times n}$, $B_i^{\mathrm{LPV}} \in \mathbb{R}^{n\times m}$ for all $i \in \mathbb{I}_0^{N_v}$, $C \in \mathbb{R}^{p\times n}$ as an *LPV SS representation*. We remark that the discrete-time LPV SS representation described by (1.8) *completely* models the behavior of Σ_{SD} in sampling instants.

1.2.2 Modeling Aperiodic Sampled-Data Systems with LS SS Representations

In this subsection, we continue with the discussion of modeling aperiodic sampled-data systems for the case when the sampling interval set $\mathscr{H}$ is finite, i.e., $\mathscr{H} = \{\hat{h}_1, \ldots, \hat{h}_M\}$, $M \in \mathbb{N}\backslash\{0\}$. Since in this case, the sampling interval h_k between any two consecutive sampling instants can take its values only from the finite set $\mathscr{H} = \{\hat{h}_1, \ldots, \hat{h}_M\}$ one approach to design control for the model (1.6) is to create an LS SS model from (1.6). The idea is that since the set $\mathscr{H}$ has M elements, $\Theta(h_k)$ can only take M different values for all $k \in \mathbb{N}$. In turn, (1.6) can be used to create an LS SS representation with M discrete modes. Below we summarize this procedure.

Let the matrices $A_1^{\mathrm{LS}}, \ldots, A_M^{\mathrm{LS}} \in \mathbb{R}^{n\times n}$ and $B_1^{\mathrm{LS}}, \ldots, B_M^{\mathrm{LS}} \in \mathbb{R}^{n\times m}$ be defined by

$$\begin{aligned} A_i^{\mathrm{LS}} &= I_n + A\Theta(\hat{h}_i),\ \forall i \in \mathbb{I}_1^M, \\ B_i^{\mathrm{LS}} &= \Theta(\hat{h}_i)B,\ \forall i \in \mathbb{I}_1^M. \end{aligned} \tag{1.10}$$

Using (1.10), (1.6) can be rewritten as the following SS representation:

$$\Sigma_{\mathrm{LS}} \begin{cases} x_{k+1} = A_{q_k}^{\mathrm{LS}} x_k + B_{q_k}^{\mathrm{LS}} u_k \\ y_k = Cx_k,\ \forall k \in \mathbb{N}, \end{cases} \tag{1.11}$$

where $q_k \in \mathbb{I}_1^M$ is called the value of the *switching sequence* at time $k \in \mathbb{N}$.

Models of the form (1.11) are a subclass of discrete-time LS SS representations where the read-out map represented by the matrix C is constant and independent from the value of the switching signal q_k at each time instant $k \in \mathbb{N}$. Hence, from now on, we will refer to the system representations of the form (1.11) as LS SS representations and formally define the tuple $\Sigma_{\mathrm{LS}} = (\{(A_i^{\mathrm{LS}}, B_i^{\mathrm{LS}}, C)\}_{i=1}^{M})$ with $A_i^{\mathrm{LS}} \in \mathbb{R}^{n\times n}$, $B_i^{\mathrm{LS}} \in \mathbb{R}^{n\times m}$ for all $i \in \mathbb{I}_1^M$, $C \in \mathbb{R}^{p\times n}$ as an *LS SS representation*. We remark that when the switching sequence is chosen such that $q_k = i$ where $h_k = \hat{h}_i$, $i \in \mathbb{I}_1^M$, the discrete-time LS SS representation described by (1.11) *completely* models the behavior of Σ_{SD} in sampling instants for the case when $\mathscr{H}$ is finite. More clearly, note that each linear mode $(A_i^{\text{LS}}, B_i^{\text{LS}}, C)$, $i \in \mathbb{I}_1^M$ corresponds to the ith element of the sampling interval set $\mathscr{H} = \{\hat{h}_1, \ldots, \hat{h}_M\}$, i.e., if the kth sampling interval h_k, $k \in \mathbb{N}$ is chosen as $h_k = \hat{h}_i$, $i \in \mathbb{I}_1^M$, then the value of the switching signal at time instant k is $q_k = i$. In the following, analogous to the LTI case, $\dim(\Sigma_{\text{LS}})$ will be used to denote the dimension n of the state space of Σ_{LS} and the number n will be called the *order* of Σ_{LS}.

We provide the problem statement of the first part of the chapter as follows.

Problem 1.1 Given the minimal continuous-time LTI system Σ_{LTI} of the form (1.1), show that

(i) the discrete-time LPV SS representation (1.8) computed for sampling **Scenario 1** is also minimal.

(ii) the discrete-time LS SS representation (1.11) computed for sampling **Scenario 2** is also minimal.

In the next section, we will continue with a review of realization theory for discrete-time LTI, LPV, and LS SS representations to formalize the problem statement given above.

1.3 Review of Realization Theory for LTI, LPV, and LS SS Models

In this section, we present a brief summary of the concepts related to the realization theory of LTI, LPV, and LS SS representations. The provided review is expected to serve as a preliminary for the following discussion on minimality.

1.3.1 Review: Realization Theory of LTI SS Representations

First, we start with the review of some classical concepts related to the realization theory of LTI systems.

Notation 2 *Let* $r \in \mathbb{N}\backslash\{0\}$. *The set of* continuous *and* absolutely continuous *maps of the form* $\mathbb{R}_+ \to \mathbb{R}^r$ *is denoted by* $\boldsymbol{C}(\mathbb{R}_+, \mathbb{R}^r)$ *and* $\boldsymbol{AC}(\mathbb{R}_+, \mathbb{R}^r)$, *respectively; and the set of* Lebesgue measurable maps *of the form* $\mathbb{R}_+ \to \mathbb{R}^r$ *which are integrable on any compact interval is denoted by* $\boldsymbol{L_{loc}}(\mathbb{R}_+, \mathbb{R}^r)$.

We define the *input-to-state map* $X^{x_0}_{\Sigma_{\text{LTI}}}$ and *input-to-output map* $Y^{x_0}_{\Sigma_{\text{LTI}}}$ of a system Σ_{LTI} of the form (1.1) as the maps

$$X_{\Sigma_{\mathrm{LTI}}}^{x_0} : \mathbf{L_{loc}}(\mathbb{R}_+, \mathbb{R}^m) \to \mathbf{AC}(\mathbb{R}_+, \mathbb{R}^n);\ u \mapsto X_{\Sigma_{\mathrm{LTI}}}^{x_0}(u),$$
$$Y_{\Sigma_{\mathrm{LTI}}}^{x_0} : \mathbf{L_{loc}}(\mathbb{R}_+, \mathbb{R}^m) \to \mathbf{C}(\mathbb{R}_+, \mathbb{R}^p);\ u \mapsto Y_{\Sigma_{\mathrm{LTI}}}^{x_0}(u),$$

defined by letting $t \mapsto X_{\Sigma_{\mathrm{LTI}}}^{x_0}(u)(t)$ be the solution to the first equation of (1.1) with $x(0) = x_0$, and letting $Y_{\Sigma_{\mathrm{LTI}}}^{x_0}(u)(t) = CX_{\Sigma_{\mathrm{LTI}}}^{x_0}(u)(t)$ for all $t \in \mathbb{R}_+$ as in second equation of (1.1).

Let f be a map of the form

$$f : \mathbf{L_{loc}}(\mathbb{R}_+, \mathbb{R}^m) \to \mathbf{C}(\mathbb{R}_+, \mathbb{R}^p). \tag{1.12}$$

An LTI SS representation Σ_{LTI} of the form (1.1) is a realization of a map f of the form (1.12) if for all $u \in \mathbf{L_{loc}}(\mathbb{R}_+, \mathbb{R}^m)$; $f(u) = Y_{\Sigma_{\mathrm{LTI}}}^{0}(u)$.

Moreover, we say that Σ_{LTI} is a *minimal realization* of f, if for any other realization $\hat{\Sigma}_{\mathrm{LTI}}$ of f, $\dim(\Sigma_{\mathrm{LTI}}) \leqslant \dim(\hat{\Sigma}_{\mathrm{LTI}})$.

A state $x_f \in \mathbb{R}^n$ of an LTI SS realization Σ_{LTI} is called reachable from the zero initial state if there exists a time instant $T \in \mathbb{R}_+$ and an input $u \in \mathbf{L_{loc}}(\mathbb{R}_+, \mathbb{R}^m)$ such that $X_{\Sigma_{\mathrm{LTI}}}^{0}(u)(T) = x_f$. Let $\mathscr{R}_{\mathrm{LTI}}$ denote the set of all reachable states of an LTI SS realization Σ_{LTI}. The LTI SS realization Σ_{LTI} is called *reachable* if $\mathscr{R}_{\mathrm{LTI}} = \mathbb{R}^n$.

A state $x_1 \in \mathbb{R}^n$ of Σ_{LTI} is called unobservable if there exists another state $x_2 \in \mathbb{R}^n$ such that $Y_{\Sigma_{\mathrm{LTI}}}^{x_1} = Y_{\Sigma_{\mathrm{LTI}}}^{x_2}$. Let $\mathscr{O}_{\mathrm{LTI}}$ denote the set of all unobservable states of an LTI SS realization Σ_{LTI}. The LTI SS realization Σ_{LTI} is called *observable* if $\mathscr{O}_{\mathrm{LTI}} = \{0\}$.

We conclude with recalling that LTI SS realization Σ_{LTI} is minimal if and only if it is reachable and observable.

1.3.2 Review: Realization Theory of LPV SS Representations

This subsection is devoted to a brief overview of some concepts related to the realization theory of discrete-time LPV SS representations. A detailed discussion on the topic can be found in [33].

Notation 3 *Let $\Sigma_{\mathrm{LPV}} = (\{(A_i^{\mathrm{LPV}}, B_i^{\mathrm{LPV}}, C)\}_{i=0}^{N_v})$ be a discrete-time LPV SS representation of the form* (1.8) *of the plant $\Sigma_{\mathrm{LTI}} = (A, B, C)$ of the form* (1.1), *sampled with respect to* ***Scenario 1****, i.e., $\mathscr{H} = (0, \overline{h}]$. In the sequel, we use the following notation and terminology: If $s = s_0 \dots s_N$ is a sequence with $N + 1$ elements, $N \in \mathbb{N}$, we denote the number N as $|s| = N$ and call $|s|$ as the length of the sequence $|s|$. We use $\boldsymbol{U}$ to denote the set of finite sequences in $\mathbb{R}^m$, i.e., $\boldsymbol{U} = \{u = u_0 \dots u_N \mid u_0, \dots, u_N \in \mathbb{R}^m, N \in \mathbb{N}\}$; $\boldsymbol{P}$ to denote the set of finite sequences in $\mathbb{R}^{N_v}$, i.e., $\boldsymbol{P} = \{p = p_0 \dots p_N \mid p_0, \dots, p_N \in \mathbb{R}^{N_v}, N \in \mathbb{N}\}$; $\boldsymbol{X}$ to denote the set of finite sequences in $\mathbb{R}^n$, i.e., $\boldsymbol{X} = \{x = x_0 \dots x_N \mid x_0, \dots, x_N \in \mathbb{R}^n, N \in \mathbb{N}\}$; and $\boldsymbol{Y}$ to denote the set of finite sequences in $\mathbb{R}^p$, i.e., $\boldsymbol{Y} = \{y = y_0 \dots y_N \mid y_0, \dots, y_N \in \mathbb{R}^p, N \in \mathbb{N}\}$. In addition, we will write $\overline{\boldsymbol{U} \times \boldsymbol{P}} = \{(u, p) \in \boldsymbol{U} \times \boldsymbol{P} \mid |u| = |p|\}$.*

We define the *input-to-state map* $X^{x_0}_{\Sigma_{\mathrm{LPV}}}$ and *input-to-output map* $Y^{x_0}_{\Sigma_{\mathrm{LPV}}}$ of a system Σ_{LPV} of the form (1.8) as the maps

$$\overline{\mathbf{U} \times \mathbf{P}} \to \mathbf{X};\ (u, p) \mapsto X^{x_0}_{\Sigma_{\mathrm{LPV}}}(u, p) = x,$$
$$\overline{\mathbf{U} \times \mathbf{P}} \to \mathbf{Y};\ (u, p) \mapsto Y^{x_0}_{\Sigma_{\mathrm{LPV}}}(u, p) = y,$$

defined by letting $k \mapsto X^{x_0}_{\Sigma_{\mathrm{LPV}}}(u, p)_k$ be the solution to the first equation of (1.8) with $\hat{x}_0 = x_0$, and letting $Y^{x_0}_{\Sigma_{\mathrm{LPV}}}(u, p)_k = CX^{x_0}_{\Sigma_{\mathrm{LPV}}}(u, p)_k$ for all $k \in \mathbb{N}$ as in second equation of (1.8).

Let $\hat{f}$ be a map of the form

$$\hat{f} : \overline{\mathbf{U} \times \mathbf{P}} \to \mathbf{Y}. \tag{1.13}$$

An LPV SS representation Σ_{LPV} of the form (1.8) is a realization of a map $\hat{f}$ of the form (1.13) if for all $(u, p) \in \overline{\mathbf{U} \times \mathbf{P}}$; $f(u, p) = Y^0_{\Sigma_{\mathrm{LPV}}}(u, p)$.

Moreover, we say that Σ_{LPV} is a *minimal realization* of $\hat{f}$, if for any other realization $\bar{\Sigma}_{\mathrm{LPV}}$ of $\hat{f}$, $\dim(\Sigma_{\mathrm{LPV}}) \leqslant \dim(\bar{\Sigma}_{\mathrm{LPV}})$.

A state $x_f \in \mathbb{R}^n$ of an LPV SS realization Σ_{LPV} is called reachable from the zero initial state if there exists a time instant $T \in \mathbb{N}$, and a sequence $(u, p) \in \overline{\mathbf{U} \times \mathbf{P}}$ with $|u| = T$ such that $X^0_{\Sigma_{\mathrm{LPV}}}(u, p)_T = x_f$. Let $\mathscr{R}_{\mathrm{LPV}}$ denote the smallest subspace containing the set of all reachable states of an LPV SS realization Σ_{LPV}. The LPV SS realization Σ_{LPV} is called *reachable* if $\mathscr{R}_{\mathrm{LPV}} = \mathbb{R}^n$.

A state $x_1 \in \mathbb{R}^n$ is called unobservable if there exists another state $x_2 \in \mathbb{R}^n$ such that $Y^{x_1}_{\Sigma_{\mathrm{LPV}}}(u, p) = Y^{x_2}_{\Sigma_{\mathrm{LPV}}}(u, p)$ for all $(u, p) \in \overline{\mathbf{U} \times \mathbf{P}}$. Let $\mathscr{O}_{\mathrm{LPV}}$ denote the set of all unobservable states of an LPV SS realization Σ_{LPV}. The LPV SS realization Σ_{LPV} is called *observable* if $\mathscr{O}_{\mathrm{LPV}} = \{0\}$.

Finally, we recall that the LPV SS realization Σ_{LPV} is minimal if and only if it is reachable and observable [33].

1.3.3 Review: Realization Theory of LS SS Representations

In this subsection, we provide a brief review some concepts related to the realization theory of discrete-time LS SS representations. We follow the outline presented in [31].

Notation 4 *Let* $\Sigma_{\mathrm{LS}} = (\{(A^{\mathrm{LS}}_i, B^{\mathrm{LS}}_i, C)\}_{i=1}^M)$ *be a discrete-time LS SS representation of the form* (1.11) *of the plant* $\Sigma_{\mathrm{LTI}} = (A, B, C)$ *of the form* (1.1), *sampled with respect to* ***Scenario 2****, i.e.,* $\mathscr{H} = \{\hat{h}_1, \ldots, \hat{h}_M\}$, $M \in \mathbb{N}\backslash\{0\}$. *In addition to Notation 3, in the sequel, we use the following notation and terminology:* $\boldsymbol{Q}$ *to denote the set of finite sequences in* $\mathbb{I}_1^M = \{1, \ldots, M\}$*, i.e.,* $\boldsymbol{Q} = \{q = q_0 \ldots q_N \mid q_0, \ldots, q_N \in \mathbb{I}_1^M, N \in \mathbb{N}\}$. *In addition, we will write* $\overline{\boldsymbol{U} \times \boldsymbol{Q}} = \{(u, q) \in \boldsymbol{U} \times \boldsymbol{Q} \mid |u| = |q|\}$.

We define the *input-to-state map* $X^{x_0}_{\Sigma_{\mathrm{LS}}}$ and *input-to-output map* $Y^{x_0}_{\Sigma_{\mathrm{LS}}}$ of a system Σ_{LS} of the form (1.11) as the maps

$$\overline{\mathbf{U}\times\mathbf{Q}} \to \mathbf{X};\ (u,q) \mapsto X^{x_0}_{\Sigma_{\mathrm{LS}}}(u,q) = x,$$
$$\overline{\mathbf{U}\times\mathbf{Q}} \to \mathbf{Y};\ (u,q) \mapsto Y^{x_0}_{\Sigma_{\mathrm{LS}}}(u,q) = y,$$

defined by letting $k \mapsto X^{x_0}_{\Sigma_{\mathrm{LS}}}(u,q)_k$ be the solution to the first equation of (1.11) with $\hat{x}_0 = x_0$, and letting $Y^{x_0}_{\Sigma_{\mathrm{LS}}}(u,q)_k = CX^{x_0}_{\Sigma_{\mathrm{LS}}}(u,q)_k$ for all $k \in \mathbb{N}$ as in second equation of (1.11).

Let $\hat{f}$ be a map of the form

$$\hat{f} : \overline{\mathbf{U}\times\mathbf{Q}} \to \mathbf{Y}. \tag{1.14}$$

An LS SS representation Σ_{LS} of the form (1.11) is a realization of a map $\hat{f}$ of the form (1.14) if for all $(u,q) \in \overline{\mathbf{U}\times\mathbf{Q}}$; $f(u,q) = Y^{0}_{\Sigma_{\mathrm{LS}}}(u,q)$.

Moreover, we say that Σ_{LS} is a *minimal realization* of $\hat{f}$, if for any other realization $\bar{\Sigma}_{\mathrm{LS}}$ of $\hat{f}$, $\dim(\Sigma_{\mathrm{LS}}) \leqslant \dim(\bar{\Sigma}_{\mathrm{LS}})$.

A state $x_f \in \mathbb{R}^n$ of an LS SS realization Σ_{LS} is called reachable from the zero initial state if there exists a time instant $T \in \mathbb{N}$, and a sequence $(u,q) \in \overline{\mathbf{U}\times\mathbf{Q}}$ with $|u| = T$ such that $X^{0}_{\Sigma_{\mathrm{LS}}}(u,q)_T = x_f$. Let $\mathscr{R}_{\mathrm{LS}}$ denote the smallest subspace containing the set of all reachable states of an LS SS realization Σ_{LS}. The LS SS realization Σ_{LS} is called *reachable* if $\mathscr{R}_{\mathrm{LS}} = \mathbb{R}^n$.

A state $x_1 \in \mathbb{R}^n$ is called unobservable if there exists another state $x_2 \in \mathbb{R}^n$ such that $Y^{x_1}_{\Sigma_{\mathrm{LS}}}(u,q) = Y^{x_2}_{\Sigma_{\mathrm{LS}}}(u,q)$ for all $(u,q) \in \overline{\mathbf{U}\times\mathbf{Q}}$. Let $\mathscr{O}_{\mathrm{LS}}$ denote the set of all unobservable states of an LS SS realization Σ_{LS}. The LS SS realization Σ_{LS} is called *observable* if $\mathscr{O}_{\mathrm{LS}} = \{0\}$.

In [31], it is shown that the LS SS realization Σ_{LS} is minimal if and only if it is reachable and observable.

1.4 Preservation of Minimality for LPV and LS SS Models of Aperiodic Sampled-Data Systems

In this section, we state the main result related to minimality of sampled-data systems with aperiodic sampling. The following theorem states this result formally.

Theorem 1.1 *Let $\Sigma_{\mathrm{LTI}} = (A, B, C)$ be a continuous-time LTI SS representation of the form* (1.1) *and $\Sigma_{\mathrm{SD}} = (A, B, C)$ be the corresponding sampled-data system of the form* (1.2). *Let $\Sigma_{\mathrm{LPV}} = (\{(A^{\mathrm{LPV}}_i, B^{\mathrm{LPV}}_i, C)\}_{i=0}^{N_v})$ be the LPV SS representation of the form* (1.8) *modeling the sampled-data system Σ_{SD} for $\mathscr{H} = (0, \bar{h}]$ **(Scenario 1)**, and $\Sigma_{\mathrm{LS}} = (\{(A^{\mathrm{LS}}_i, B^{\mathrm{LS}}_i, C)\}_{i=1}^{M})$ be the LS SS representation of the form* (1.11) *modeling the sampled-data system Σ_{SD} for $\mathscr{H} = \{\hat{h}_1, \ldots, \hat{h}_M\}$ (**Scenario 2**). In addition, let the set $\mathscr{H}$ for **Scenario 2** be such that it contains at least one sampling interval $\hat{h}^*$ satisfying*

$$h^*(\lambda - \mu) \neq 2c\pi j, \; c = \pm 1, \pm 2, \ldots. \tag{1.15}$$

where λ and μ are any pair of eigenvalues of A.

(i) *If $\Sigma_{\text{LTI}} = (A, B, C)$ is minimal, then $\Sigma_{\text{LPV}} = (\{(A_i^{\text{LPV}}, B_i^{\text{LPV}}, C)\}_{i=0}^{N_v})$ is also minimal.*

(ii) *If $\Sigma_{\text{LTI}} = (A, B, C)$ is minimal, then $\Sigma_{\text{LS}} = (\{(A_i^{\text{LS}}, B_i^{\text{LS}}, C)\}_{i=1}^{M})$ is also minimal.*

The proof of this theorem is constructed on the classical result on reachability and observability under periodic sampling [24, 35], which is represented by the property (1.15). See [5] for the relation between the case of aperiodic sampling.

Proof (Theorem 1.1 *(i): Preservation of Minimality for LPV SS Models of Aperiodic Sampled-Data Systems)* The proof is based on showing if $\Sigma_{\text{LTI}} = (A, B, C)$ is reachable and observable, the resulting $\Sigma_{\text{LPV}} = (\{(A_i^{\text{LPV}}, B_i^{\text{LPV}}, C)\}_{i=0}^{N_v})$ will also be reachable and observable. See Sect. 1.3 for the formal definitions of reachability and observability of LTI and LPV SS representations. In addition, see [5] for the detailed proof of the theorem.

Proof (Theorem 1.1 *(ii): Preservation of Minimality for LS SS Models of Aperiodic Sampled-Data Systems)* Let $\mathscr{H} = \{\hat{h}_1, \ldots, \hat{h}_M\}$ be the sampling interval set. By the assumption of the theorem, there exists a $h = \hat{h}_i \in \mathscr{H}$, $i \in \mathbb{I}_1^M$ such that the pair $(\bar{A}, \bar{B})$ is reachable and the pair $(C, \bar{A})$ is observable, where $\bar{A} := e^{Ah}$ and $\bar{B} := (\int_0^h e^{As} ds)B$. Let the sampling interval sequence is chosen such that $h_k = t_{k+1} - t_k = h$ for all $k \in \mathbb{N}$. Then, the proof follows by similar arguments as in the proof of Theorem 1.1 *(i)* given in [5].

1.5 A Model Reduction Method for Aperiodic Sampled-Data Systems

The fact that the considered sampling interval set $\mathscr{H}$ is finite creates the possibility of developing a simple moment matching type model reduction method to compute a reduced-order approximation for the sampled-data system, from its minimal LS SS representation. Below we state the problem considered in this section.

Problem 1.2 Let Σ_{LTI} be a continuous-time LTI plant model of order n, which is to be sampled aperiodically with respect to the finite set $\mathscr{H} = \{\hat{h}_1, \ldots, \hat{h}_M\}$ to form the sampled-data system Σ_{SD}. Compute a discrete-time model $\bar{\Sigma}_{\text{LS}}$ of order $r < n$ which is an approximation of the input–output behavior of Σ_{SD} in sampling instants.

The proposed model reduction approach for the solution of this problem can be summarized as follows:

Let $\Sigma_{\text{SD}} = (A, B, C, \mathscr{H})$ be the sampled-data system with $\mathscr{H} = \{\hat{h}_1, \ldots, \hat{h}_M\}$ of the form (1.2) corresponding to the continuous-time LTI plant $\Sigma_{\text{LTI}} = (A, B, C)$ of the form (1.1). Let $\Sigma_{\text{LS}} = (\{(A_i^{\text{LS}}, B_i^{\text{LS}}, C)\}_{i=1}^{M})$ of the form (1.11) be the corresponding LS SS model for the sampled-data system Σ_{SD}. Compute from Σ_{LS} another LS SS model $\bar{\Sigma}_{\text{LS}} = (\{(\bar{A}_i^{\text{LS}}, \bar{B}_i^{\text{LS}}, \bar{C})\}_{i=1}^{M})$ of order $r < n$ who approximates the input–output behavior of Σ_{LS}.

First, we recall the concepts of Markov parameters and moment matching for LTI and LS SS representations, and the analogy between two cases. Then, we present the model reduction method in detail.

1.5.1 Review: Moment Matching for LTI SS Representations

Let $\Sigma_{\text{LTI}} = (A, B, C)$ be an LTI SS representation of the form (1.1) and its input–output map $Y_{\Sigma_{\text{LTI}}}^{x_0}$ be defined as in Sect. 1.3.1. A moment's reflection lets us see that the kth Taylor series coefficient M_k of $Y_{\Sigma_{\text{LTI}}}^{0}$ around $t = 0$ for the unit impulse input will be

$$M_k = CA^k B,\ k \in \mathbb{N} \tag{1.16}$$

where A^0 defined to be I_n. The coefficients $M_k, k \in \mathbb{N}$ are called the *Markov parameters* or the *moments* of the system Σ_{LTI}. Hence, it is possible to approximate the input–output behavior of Σ_{LTI} by another system $\bar{\Sigma}_{\text{LTI}}$ (possibly of reduced order), whose first some number of Markov parameters are equal to the corresponding ones of Σ_{LTI}. If this number is chosen to be $N \in \mathbb{N}$, we will call such approximations as *N-partial realizations* of Σ_{LTI}. More precisely, a continuous-time LTI SS representation $\bar{\Sigma}_{\text{LTI}} = (\bar{A}, \bar{B}, \bar{C})$ is an N-partial realization of another continuous-time LTI SS representation $\Sigma_{\text{LTI}} = (A, B, C)$ if

$$CA^k B = \bar{C}\bar{A}^k\bar{B}, k = 0, \ldots, N.$$

The problem of model reduction of LTI systems by moment matching can now be stated as follows. Consider an LTI system $\Sigma_{\text{LTI}} = (A, B, C)$ of the form (1.1) and fix $N \in \mathbb{N}$. Find another LTI system $\bar{\Sigma}_{\text{LTI}}$ of order r strictly less than n such that $\bar{\Sigma}_{\text{LTI}}$ is an N-partial realization of Σ_{LTI}.

For a detailed account on how to compute N-partial realizations for the LTI case, see [2]. For the purpose of this chapter, we will focus on the moment matching problem for the linear switched SS representations which model the sampled-data system with respect to the sampling interval set $\mathscr{H} = \{\hat{h}_1, \ldots, \hat{h}_M\}$.

1.5.2 Review: Moment Matching for LS SS Representations

Let $\Sigma_{\mathrm{LS}} = (\{(A_i^{\mathrm{LS}}, B_i^{\mathrm{LS}}, C)\}_{i=1}^{M})$ be an LS SS representation of the form (1.11) modeling the sampled-data system for the sampling interval set $\mathscr{H} = \{\hat{h}_1, \ldots, \hat{h}_M\}$ (**Scenario 2**) and its input–output map $Y_{\Sigma_{\mathrm{LS}}}^{x_0}$ be defined as in Sect. 1.3.3. Using (1.11), one can see that the coefficients appearing in the output of Σ_{LS} for any pair of input and switching sequences $(u, \sigma) \in \overline{\mathbf{U} \times \mathbf{Q}}$ are of the form

$$CB_j^{\mathrm{LS}}, \ j \in \mathbb{I}_1^D \tag{1.17}$$

and

$$CA_{k_1}^{\mathrm{LS}} \ldots A_{k_L}^{\mathrm{LS}} B_j^{\mathrm{LS}}; \ k_1, \ldots, k_L, j \in \mathbb{I}_1^D, \ L \in \mathbb{N}\backslash\{0\}. \tag{1.18}$$

In [30], it is shown that these coefficients uniquely define the map $Y_{\Sigma_{\mathrm{LS}}}^{x_0}$. Analogously to the linear case, we will call the coefficients of the form (1.17) and (1.18) as the *Markov parameters* of $\Sigma_{\mathrm{LS}} = (\{(A_i^{\mathrm{LS}}, B_i^{\mathrm{LS}}, C)\}_{i=1}^{M})$. Specifically, we will call the Markov parameters of the form (1.17) as the Markov parameters of length 0 and the Markov parameters of the form (1.18) as the Markov parameters of length L for any $L \in \mathbb{N}\backslash\{0\}$.

In [7, 8], it is shown that similarly to the LTI case, it is possible to approximate the input–output behavior of Σ_{LS} by another LS SS representation $\bar{\Sigma}_{\mathrm{LS}}$ (possibly of reduced order), whose Markov parameters up to a certain length $N \in \mathbb{N}$ are equal with the corresponding ones of Σ_{LS}.[2] Again, we will call such approximations as *N-partial realizations* of Σ_{LS}. More precisely, a discrete-time LS SS representation $\bar{\Sigma}_{\mathrm{LS}} = (\{(\bar{A}_i^{\mathrm{LS}}, \bar{B}_i^{\mathrm{LS}}, \bar{C})\}_{i=1}^{M})$ is an N-partial realization of another discrete-time LS SS representation $\Sigma_{\mathrm{LS}} = (\{(A_i^{\mathrm{LS}}, B_i^{\mathrm{LS}}, C)\}_{i=1}^{M})$ if

$$CB_j^{\mathrm{LS}} = \bar{C}\bar{B}_j^{\mathrm{LS}}, \ j \in \mathbb{I}_1^M$$

and

$$CA_{k_1}^{\mathrm{LS}} \ldots A_{k_L}^{\mathrm{LS}} B_j^{\mathrm{LS}} = \bar{C}\bar{A}_{k_1}^{\mathrm{LS}} \ldots \bar{A}_{k_L}^{\mathrm{LS}} \bar{B}_j^{\mathrm{LS}}$$

for all $k_1, \ldots, k_L, j \in \mathbb{I}_1^M$ and $L \in \mathbb{I}_1^N$. Note that an N-partial realization $\bar{\Sigma}_{\mathrm{LS}}$ of Σ_{LS} will have the *same* output with Σ_{LS} for all time instants up to N, i.e., $k \in \mathbb{I}_0^N$, for all input and switching sequences. The reason why the output of an N-partial realization is indeed an approximation for the output of the original system model for also the time instants $k > N$ can be found in [6].

The problem of model reduction of LS models of sampled-data systems by moment matching can now be stated as follows: Consider an LS SS model $\Sigma_{\mathrm{LS}} = (\{(A_i^{\mathrm{LS}}, B_i^{\mathrm{LS}}, C)\}_{i=1}^{M})$ of the form (1.11) of order n and fix $N \in \mathbb{N}$. Find another LS

---

[2]Even though the results in [7, 8] are stated in the continuous-time context, the analogous results on N-partial realizations of discrete-time LS SS representations are also valid. See [6] for an application of these results for the model reduction of affine LPV systems in the discrete-time context.

SS model $\bar{\Sigma}_{\mathrm{LS}}$ of order r strictly less than n such that $\bar{\Sigma}_{\mathrm{LS}} = (\{(\bar{A}_i^{\mathrm{LS}}, \bar{B}_i^{\mathrm{LS}}, \bar{C})\}_{i=1}^{M})$ is an N-partial realization of Σ_{LS}.

Next, we recall a theorem of model reduction with N-partial realizations for the LS case [7]. This theorem (Theorem 1.2) can be considered as the formulation of the solution of the moment matching problem in the LS case. It can be formulated in two dual ways [7], namely the reachability or the observability ways. However, here we present only the formulation based on partial reachability spaces for simplicity. For this purpose, we define the N-reachability space $\mathscr{R}_{\mathrm{LS}}^{N}$, $N \in \mathbb{N}$ of an LS SS realization $\Sigma_{\mathrm{LS}} = (\{(A_i^{\mathrm{LS}}, B_i^{\mathrm{LS}}, C)\}_{i=1}^{M})$ of the form (1.11) inductively as follows:

$$\begin{aligned} \mathscr{R}_{\mathrm{LS}}^{0} &= \operatorname{span} \bigcup_{i_0 \in \mathbb{I}_1^M} \operatorname{im}(B_{i_0}), \\ \mathscr{R}_{\mathrm{LS}}^{N} &= \mathscr{R}_{\mathrm{LS}}^{0} + \sum_{j \in \mathbb{I}_1^M} \operatorname{im}(A_j \mathscr{R}_{\mathrm{LS}}^{N-1}), \quad N \geqslant 1. \end{aligned} \tag{1.19}$$

Theorem 1.2 (Moment Matching for LS SS Representations, [7]) *Let* $\Sigma_{\mathrm{LS}} = (\{(A_i^{\mathrm{LS}}, B_i^{\mathrm{LS}}, C)\}_{i=1}^{M})$ *be a discrete-time LS SS representation of the form* (1.11) *which models the sampled-data system* Σ_{SD} *of the form* (1.2) *for the sampling interval set* $\mathscr{H} = \{\hat{h}_1, \ldots, \hat{h}_M\}$ *(**Scenario 2**). Let* $N \in \mathbb{N}$ *and* $V \in \mathbb{R}^{n \times r}$ *be a full column rank matrix such that*

$$\mathscr{R}_{\mathrm{LS}}^{N} = \operatorname{im}(V). \tag{1.20}$$

If $\bar{\Sigma}_{\mathrm{LS}} = (\{(\bar{A}_i^{\mathrm{LS}}, \bar{B}_i^{\mathrm{LS}}, \bar{C})\}_{i=1}^{M})$ *is an LS SS representation such that for each* $i \in \mathbb{I}_1^M$, *the matrices* $\bar{A}_i^{\mathrm{LS}}, \bar{B}_i^{\mathrm{LS}}, \bar{C}$ *are defined as*

$$\bar{A}_i^{\mathrm{LS}} = V^{-1} A_i^{\mathrm{LS}} V,\ \bar{B}_i^{\mathrm{LS}} = V^{-1} B_i^{\mathrm{LS}},\ \bar{C} = CV, \tag{1.21}$$

where V^{-1} *is a left inverse of* V, *then* $\bar{\Sigma}_{\mathrm{LS}}$ *is an* N*-partial realization of* Σ_{LS}.

Note that the key of model reduction lies in the number of columns of the full column rank projection matrix $V \in \mathbb{R}^{n \times r}$ such that $r < n$. Choosing the number N small enough such that the matrix V satisfies the condition (1.20) and it has $r < n$ columns results in the reduced-order N-partial realization $\bar{\Sigma}_{\mathrm{LS}}$ of order r. A simple algorithm with polynomial computational complexity to compute the matrix V in Theorem 1.2 is given in [7].

1.5.3 The Model Reduction Approach

Now the model reduction approach for aperiodic sampled-data systems can be stated in detail as follows:

Procedure 1 (Model Reduction of Sampled-Data Systems) *Let* $\Sigma_{\mathrm{SD}} = (A, B, C, \mathscr{H})$ *be the sampled-data system with* $\mathscr{H} = \{\hat{h}_1, \ldots, \hat{h}_M\}$ *of the form* (1.2) *corresponding to the continuous-time LTI plant* $\Sigma_{\mathrm{LTI}} = (A, B, C)$ *of the form* (1.1) *of order* n. *Let*

$\Sigma_{\mathrm{LS}} = (\{(A_i^{\mathrm{LS}}, B_i^{\mathrm{LS}}, C)\}_{i=1}^M)$ *of the form* (1.11) *be the corresponding LS SS model for the sampled-data system* Σ_{SD}. *Let* $N \in \mathbb{N}$ *and* $V \in \mathbb{R}^{n\times r}$ *be a full column rank matrix such that*

$$\mathscr{R}_{\mathrm{LS}}^N = \mathrm{im}(V). \tag{1.22}$$

Consider the matrices

$$\bar{\Phi}(\hat{h}_i) = V^{-1}\Phi(\hat{h}_i)V,\ \bar{\Gamma}(\hat{h}_i) = V^{-1}\Gamma(\hat{h}_i),\ \bar{C} = CV,\ \forall \hat{h}_i \in \mathscr{H},$$

where V^{-1} *is a left inverse of* V. *Then the time-varying model*

$$\bar{\Sigma}_{\mathrm{disc}} \begin{cases} \bar{x}_{k+1} = \bar{\Phi}(h_k)\bar{x}_k + \bar{\Gamma}(h_k)u_k, \\ \bar{y}_k = \bar{C}\bar{x}_k,\ \forall h_k \in \mathscr{H},\ \forall k \in \mathbb{N}, \end{cases} \tag{1.23}$$

is an N*-partial realization of the model* (1.11).

Note that the time-varying reduced-order model $\bar{\Sigma}_{\mathrm{disc}}$ on (1.23) can as well be written as the approximation LS SS model $\bar{\Sigma}_{\mathrm{LS}} = (\{(\bar{A}_i^{\mathrm{LS}}, \bar{B}_i^{\mathrm{LS}}, \bar{C})\}_{i=1}^M)$ of order $r < n$ for Σ_{LS} where

$$\bar{A}_i^{\mathrm{LS}} = \bar{\Phi}(\hat{h}_i),\ \bar{B}_i^{\mathrm{LS}} = \bar{\Gamma}(\hat{h}_i),\ \bar{C} = CV,\ i \in \mathbb{I}_1^M.$$

Hence, in the following, $\bar{\Sigma}_{\mathrm{LS}}$ is also used to refer to the reduced-order model of the sampled-data system $\bar{\Sigma}_{\mathrm{disc}}$ on (1.23), which is computed by Procedure 1.

Note also that the outputs y_k of Σ_{disc} in (1.6) and $\bar{y}_k$ of $\bar{\Sigma}_{\mathrm{disc}}$ in (1.23) for any input $u \in \mathbf{U}$ will be the same for all $k \in \mathbb{I}_0^N$. In other words,

$$C\Phi(h_k)\ldots\Phi(h_i)\Gamma(h_i) = \bar{C}\bar{\Phi}(h_k)\ldots\bar{\Phi}(h_i)\bar{\Gamma}(h_i),$$

for all $k \in \mathbb{I}_0^N$ and $i \in \mathbb{I}_0^k$, where $h_l \in \mathscr{H}$ for all $l \in \mathbb{I}_i^k$. The relationship between Σ_{disc} of (1.6) and $\bar{\Sigma}_{\mathrm{disc}}$ of (1.23) can be constructed by stating that

$$\bar{x}_k = V^{-1}x_k,\ \bar{\Phi}(h_k) = V^{-1}\Phi(h_k)V,\ \bar{\Gamma}(h_k) = V^{-1}\Gamma(h_k)$$

for all $h_k \in \mathscr{H}$ and $k \in \mathbb{N}$.

1.6 Conservation of Stability

In this section, we will show that as long as the original continuous-time LTI system Σ_{LTI} is stable, the reduced-order discrete-time LS SS representation $\bar{\Sigma}_{\mathrm{LS}}$ modeling the sampled-data system computed by Procedure 1 will be quadratically stable. As the final result of this section, we will extend this conservation of stability argument for the representations of the form (1.6) of aperiodic sampled-data systems. We will start with presenting two technical lemmas for the purpose of stating this result.

In the sequel, we denote the fact that a matrix G is positive definite (resp. positive semi-definite, negative definite, and negative semi-definite) with $G > 0$ (resp. $G \geqslant 0$, $G < 0$, and $G \leqslant 0$).

Definition 1.1 *(Quadratic stability)* Let $\Sigma_{\text{LS}} = (\{(A_i^{\text{LS}}, B_i^{\text{LS}}, C)\}_{i=1}^M)$ be an LS SS representation of the form (1.11). The LS SS representation Σ_{LS} is quadratically stable if and only if there exists a symmetric positive definite $P \in \mathbb{R}^{n\times n}$ such that

$$A_i^{\text{LS}^{\text{T}}} P A_i^{\text{LS}} - P < 0, \quad \forall i \in \mathbb{I}_1^M. \tag{1.24}$$

Lemma 1.1 *Let $\Sigma_{\text{LTI}} = (A, B, C)$ be an LTI SS representation. For any $\mathscr{H} = \{\hat{h}_1, \ldots, \hat{h}_M\}$ with $M \in \mathbb{N}\backslash\{0\}$, the LS SS model $\Sigma_{\text{LS}} = (\{(A_i^{\text{LS}}, B_i^{\text{LS}}, C)\}_{i=1}^M)$ of the sampled-data system $\Sigma_{\text{SD}} = (A, B, C, \mathscr{H})$ is quadratically stable if Σ_{LTI} is stable.*

Proof The stability of $\Sigma_{\text{LTI}} = (A, B, C)$ implies the stability of the autonomous system $\Sigma_{\text{LTI}}^{\text{aut}} = (A, 0, 0)$. Then there exists a $P > 0$, $P^{\text{T}} = P$ such that

$$x(t)^{\text{T}} P x(t) - x(0)^{\text{T}} P x(0) < 0, \tag{1.25}$$

for all $x(0) \in \mathbb{R}^n$, $x(0) \neq 0$ and $t \in \mathbb{R}_+\backslash\{0\}$. Replacing $x(t)$ with $e^{At}x(0)$ in (1.25) yields that for all $x(0) \in \mathbb{R}^n$, $x(0) \neq 0$ and for all $t \in \mathbb{R}_+\backslash\{0\}$

$$x(0)^{\text{T}} \left(e^{A^{\text{T}}t} P e^{At} - P\right) x(0) < 0. \tag{1.26}$$

In turn (1.26) implies

$$e^{A^{\text{T}}\hat{h}_i} P e^{A\hat{h}_i} - P < 0, \ \forall \hat{h}_i \in \mathscr{H}.$$

Using (1.6), we conclude that

$$\Phi^{\text{T}}(\hat{h}_i) P \Phi(\hat{h}_i) - P < 0, \ \forall \hat{h}_i \in \mathscr{H}.$$

The proof of the statement follows by noticing that $\Phi(\hat{h}_i) = A_i^{\text{LS}}$, for all $i \in \mathbb{I}_1^M$.

Lemma 1.1 establishes the connection between the stability of Σ_{LTI} and the quadratic stability of Σ_{LS}. The following lemma establishes the remaining part of the conservation of stability argument stated at the beginning of this section.

Lemma 1.2 *Let $\Sigma_{\text{LS}} = (\{(A_i^{\text{LS}}, B_i^{\text{LS}}, C)\}_{i=1}^M)$ be a quadratically stable LS SS representation of the form (1.11) and $P > 0$ be a solution of (1.24). If the left inverse V^{-1} of the matrix $V \in \mathbb{R}^{n\times r}$ in Theorem 1.2 is chosen as $V^{-1} = (V^{\text{T}} P V)^{-1} V^{\text{T}} P$, then $\bar{\Sigma}_{\text{LS}} = (\{(\bar{A}_i^{\text{LS}}, \bar{B}_i^{\text{LS}}, \bar{C})\}_{i=1}^D)$ in Theorem 1.2 is also quadratically stable.*

Proof Below, we will use the following simple claims:

(C1) If $S \in \mathbb{R}^{n \times n}$ is symmetric negative (respectively positive) definite, then $\hat{S} = V^{\mathrm{T}} S V$ is also symmetric negative (respectively positive) definite.

(C2) $V^{-1} = (V^{\mathrm{T}} P V)^{-1} V^{\mathrm{T}} P \implies V^{-1} P^{-1} (V^{-1})^{\mathrm{T}} = (V^{\mathrm{T}} P V)^{-1}$.

(CS) (Schur complement lemma for positive/negative definiteness). Let $S \in \mathbb{R}^{n \times n}$ be a symmetric positive definite matrix and $G \in \mathbb{R}^{n \times n}$. Then $G^{\mathrm{T}} S G - S < 0 \iff G S^{-1} G^{\mathrm{T}} - S^{-1} < 0$.

Note that, by the assumption of the theorem, (1.24) holds. Multiplying (1.24) by V^{T} from left and V from right for all $i \in \mathbb{I}_1^M$ and using *(C1)* yields

$$V^{\mathrm{T}} A_i^{\mathrm{LS}^{\mathrm{T}}} P A_i^{\mathrm{LS}} V - V^{\mathrm{T}} P V < 0, \quad \forall i \in \mathbb{I}_1^M.$$

By *(CS)* it follows that

$$A_i^{\mathrm{LS}} V (V^{\mathrm{T}} P V)^{-1} V^{\mathrm{T}} A_i^{\mathrm{LS}^{\mathrm{T}}} - P^{-1} < 0, \quad \forall i \in \mathbb{I}_1^M. \tag{1.27}$$

In turn, multiplying (1.27) by V^{-1} from left and $(V^{-1})^{\mathrm{T}}$ from right for all $i \in \mathbb{I}_1^M$ and using *(C1)* yields

$$V^{-1} A_i^{\mathrm{LS}} V (V^{\mathrm{T}} P V)^{-1} V^{\mathrm{T}} A_i^{\mathrm{LS}^{\mathrm{T}}} (V^{-1})^{\mathrm{T}} - V^{-1} P^{-1} (V^{-1})^{\mathrm{T}} < 0, \tag{1.28}$$

for all $i \in \mathbb{I}_1^M$. Using *(C2)* and choosing $\bar{P} = V^{\mathrm{T}} P V$, the inequality (1.28) can be rewritten as

$$\overline{A}_i^{\mathrm{LS}} \bar{P}^{-1} \overline{A}_i^{\mathrm{LS}^{\mathrm{T}}} - \bar{P}^{-1} < 0, \quad \forall i \in \mathbb{I}_1^M. \tag{1.29}$$

Finally, using *(CS)* one more time for (1.29) yields

$$\overline{A}_i^{\mathrm{LS}^{\mathrm{T}}} \bar{P} \overline{A}_i^{\mathrm{LS}} - \bar{P} < 0, \quad \forall i \in \mathbb{I}_1^M. \tag{1.30}$$

Since $\bar{P} = V^{\mathrm{T}} P V$ is symmetric and positive definite by *(C1)*, (1.30) proves the quadratic stability of $\bar{\Sigma}_{\mathrm{LS}} = (\{(\bar{A}_i^{\mathrm{LS}}, \bar{B}_i^{\mathrm{LS}}, \bar{C})\}_{i=1}^M)$.

Remark 1.1 Note that even though Lemma 1.2 is presented in this chapter as a step on proving the stability of the reduced-order models for the sampled-data systems computed with Procedure 1, on the condition of stability of the original plant; it can also be considered as an independent stability result for model reduction of discrete-time LS SS representations.

With Lemma 1.2 we have established the connection between the quadratic stability of Σ_{LS} and $\bar{\Sigma}_{\mathrm{LS}}$. As the main result of this section, with the following theorem, we relate the conservation of stability argument to the discrete-time, time-varying representations of the reduced-order aperiodic sampled-data systems of the form (1.23), respectively.

Theorem 1.3 *Let $\Sigma_{\mathrm{LTI}} = (A, B, C)$ be stable. Then, the model $\bar{\Sigma}_{\mathrm{disc}}$ of the form* (1.23) *corresponding to $\bar{\Sigma}_{\mathrm{LS}}$ computed with Procedure 1 is quadratically stable.*

Proof The proof follows directly from the subsequent application of Lemmas 1.1 and 1.2 to Σ_{LTI}.

1.7 Numerical Examples

In this section, two generic numerical examples are presented to illustrate the proposed model reduction method, Procedure 1.[3]

Example 1.1 For the first example, the procedure is applied to compute a reduced-order model for a single-input single-output (SISO), stable system $\Sigma_{\mathrm{LTI}} = (A, B, C)$ of order 50, sampled with respect to a finite $\mathscr{H}$ (**Scenario 2**) to form the sampled-data system $\Sigma_{\mathrm{SD}} = (A, B, C, \mathscr{H})$ with $\mathscr{H} = \{\hat{h}_1, \hat{h}_2, \hat{h}_3, \hat{h}_4\} = \{1, 1.5, 2, 3\}$. First, the original LTI SS representation Σ_{LTI} is used to construct to LS SS representation Σ_{LS} which models the behavior of the sampled-data system with respect to the sampling interval set $\mathscr{H}$. The model Σ_{LS} is then used to get the reduced-order LS SS representation $\bar{\Sigma}_{\mathrm{LS}}$ using Theorem 1.2. The reduced-order LS SS representation $\bar{\Sigma}_{\mathrm{LS}}$ in this case is a 2-partial realization of order 18. For simulation, the output sequence y_k of the original sampled-data system Σ_{SD} and $\bar{y}_k$ of the reduced-order LS model $\bar{\Sigma}_{\mathrm{LS}}$ are acquired for $k = \mathbb{I}_0^K$ where $K + 1$ is the number of sampling instants of the simulation; by applying the same white Gaussian noise input sequence $u = u_0 \dots u_K$, $u_k \in \mathscr{N}(0, 1)$ for all $k \in \mathbb{I}_0^K$ and sampling sequence $h = h_0 \dots h_K$, $h_k \in \mathscr{H}$ for all $k \in \mathbb{I}_0^K$. The total time horizon is chosen as $[0, 50]$. For each simulation, the distance of the values $\bar{y}_k$ to the values y_k, $k \in \mathbb{I}_0^K$ are compared with the best fit rate (BFR) [26] which is defined as

$$\mathrm{BFR} = 100\% \max\left(1 - \frac{\sqrt{\sum_{k=0}^{K} \| y_k - \bar{y}_k \|_2^2}}{\sqrt{\sum_{k=0}^{K} \| y_k - y_m \|_2^2}}, 0\right), \tag{1.31}$$

where y_m is the mean of the sequence $\{y_k\}_{k=0}^K$. Over 200 such simulations, the computed mean of the BFRs is 98.5037%, whereas the best is 99.5113% and the worst is 95.5948%. The output sequence $\{\bar{y}_k\}_{k=0}^K$ giving the closest value to the mean of the BFRs over this 200 simulations is illustrated in Fig. 1.1 together with the original output sequence $\{y_k\}_{k=0}^K$.

Example 1.2 In the second example, the procedure is applied to compute a reduced-order model for a SISO unstable system $\Sigma_{\mathrm{LTI}} = (A, B, C)$ of order 10 sampled to

[3]The implementation of the procedure on MATLAB is freely available (with the examples) for experimentation from https://sites.google.com/site/mertbastugpersonal/. The original system parameters used and reduced-order system parameters computed can be obtained from the same site.

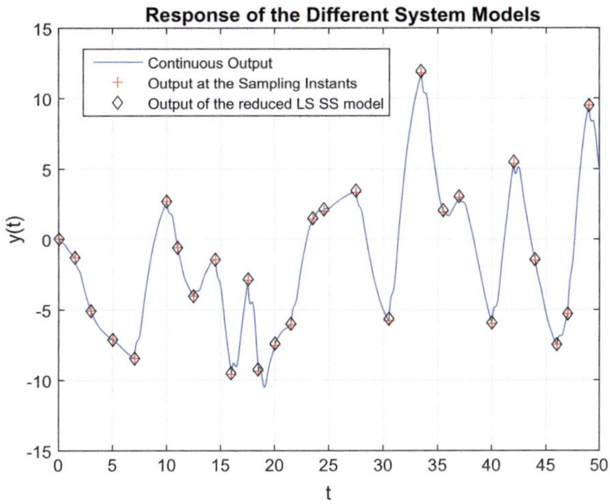

Fig. 1.1 (Example 1.1): The output of the reduced-order model $\bar{\Sigma}_{\text{LS}}$ compared with the output of the original model Σ_{LS}. For this simulation, the BFR is 98.5004%

form the sampled-data system $\Sigma_{\text{SD}} = (A, B, C, \mathscr{H})$ with $\mathscr{H} = \{\hat{h}_1, \hat{h}_2, \hat{h}_3, \hat{h}_4\} = \{0.1, 0.15, 0.2, 0.3\}$. First, the original LTI SS representation Σ_{LTI} is used to construct to LS SS representation Σ_{LS} which models the behavior of the sampled-data system with respect to the sampling interval set $\mathscr{H}$. The model Σ_{LS} is then used to get the reduced-order LS SS representation $\bar{\Sigma}_{\text{LS}}$ using Theorem 1.2. The reduced-order LS SS representation $\bar{\Sigma}_{\text{LS}}$ in this case is a 0-partial realization of order 4. The simulations are done with the input and sampling sequences with the specifications analogous to the ones given for **Example 1**. For this example, the total time horizon is chosen as [0, 5]. When such 200 simulations are done with this model, the mean of the BFRs is computed as 96.0565%, the best as 97.5042%, and the worst as 79.3883%. Again,

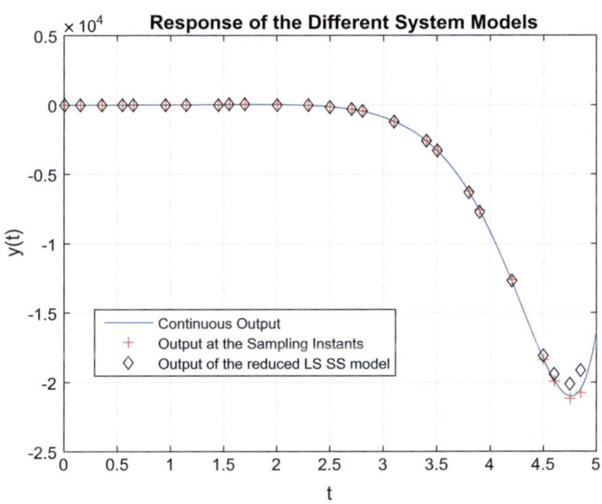

Fig. 1.2 (Example 1.2): The output of the reduced-order model $\bar{\Sigma}_{\text{LS}}$ compared with the output of the original model Σ_{LS}. For this simulation, the BFR is 96.0535%

the output sequence giving the closest value to the mean of the BFRs over this 200 simulations is illustrated in Fig. 1.2 together with the original output sequence.

1.8 Conclusions

The relation between the minimality property of a continuous-time LTI system and its discrete-time model, which models the behavior of the system in sampling instants, is investigated for two general scenarios of aperiodic sampling. For both of the considered sampling scenarios, it is shown that if the original continuous-time LTI SS model is minimal, i.e., reachable and observable, then the resulting discrete-time models are minimal as well. Then, a procedure for model reduction of sampled-data systems by moment matching is proposed for the latter sampling scenario. The procedure relies on computing a reduced-order discrete-time model from the LS SS model of the sampled-data system. The procedure is illustrated on some numerical examples. It is shown that the stability of the original continuous-time LTI plant guarantees the quadratic stability of the resulting reduced-order discrete-time LS SS model with respect to any finite allowed sampling interval set.

References

1. Ahmad, M.N., Mustafa, G., Khan, A.Q., Rehan, M.: On the controllability of a sampled-data system under nonuniform sampling. In: International Conference on Emerging Technologies (ICET), pp. 65–68. Islamabad (2014)
2. Antoulas, A.C.: Approximation of Large-Scale Dynamical Systems. SIAM, Philadelphia (2005)
3. Barb, F.D., Weiss, M.: Model reduction techniques for sampled-data systems. Numer. Algorithms **4**, 47–64 (1993)
4. Bastug, M., Hetel, L., Petreczky, M.: Model reduction for aperiodically sampled data systems. In: Proceedings of the 20th World Congress of the International Federation of Automatic Control (IFAC). Toulouse (2017). Accepted for publication
5. Bastug, M., Petreczky, M., Hetel, L.: Minimality of aperiodic sampled data systems. In: Proceedings of the American Control Conference (ACC), pp. 5750–5755. Seattle, WA (2017)
6. Bastug, M., Petreczky, M., Tóth, R., Wisniewski, R., Leth, J., Efimov, D.: Moment matching based model reduction for LPV state-space models. In: Proceedings of the 54th IEEE Conference on Decision and Control, pp. 5334–5339. Osaka (2015)
7. Bastug, M., Petreczky, M., Wisniewski, R., Leth, J.: Model reduction by moment matching for linear switched systems. In: Proceedings of the American Control Conference (ACC), pp. 3942–3947. Portland, OR (2014)
8. Bastug, M., Petreczky, M., Wisniewski, R., Leth, J.: Model reduction by nice selections for linear switched systems. IEEE Trans. Autom. Control **61**(11), 3422–3437 (2016)
9. Brockett, R.W.: Minimum attention control. In: Proceedings of the 36th IEEE Conference on Decision and Control (CDC), pp. 2628–2632. San Diego, CA (1997)
10. Cloosterman, M.B.G., Hetel, L., van de Wouw, N., Heemels, W.P.M.H., Daafouz, J., Nijmeijer, H.: Controller synthesis for networked control systems. Automatica **46**(10), 1584–1594 (2010)

11. Donkers, M.C.F., Hetel, L., Heemels, W.P.M.H., van de Wouw, N., Steinbuch, M.: In: Majumda, R., Tabuada, P. (eds.) Stability Analysis of Networked Control Systems Using a Switched Linear Systems Approach in Hybrid Systems: Computation and Control, pp. 150–164. Springer, Berlin (2009)
12. Donkers, M.C.F., Tabuada, P., Heemels, W.P.M.H.: On the minimum attention control problem for linear systems: a linear programming approach. In: Proceedings of the 50th IEEE Conference on Decision and Control and European Control Conference (CDC-ECC), pp. 4717–4722. Orlando, FL (2011)
13. Fridman, E.: Introduction to time-delay and sampled-data systems. In: Proceedings of the 2014 European Control Conference (ECC), pp. 1428–1433. Strasbourg (2014)
14. Fuster, A., Guillen, J.M.: Questions of controllability and observability for nonuniformly sampled discrete systems. IEE Proc. D - Control Theory Appl. **135**(4), 248–252 (1988)
15. Gielen, R.H., Olaru, S., Lazar, M., Heemels, W.P.M.H., van de Wouw, N., Niculescu, S.I.: On polytopic inclusions as a modeling framework for systems with time-varying delays. Automatica **46**(3), 615–619 (2010)
16. Gu, K., Kharitonov, V.L., Chen, J.: Stability of Time-Delay Systems. Springer, NY (2003)
17. Haddad, W.M., Chellaboina, V., Nersesov, S.G.: Impulsive and Hybrid Dynamical Systems: Stability, Dissipativity, and Control. Princeton University Press, NJ (2014)
18. Hespanha, J.P., Naghshtabrizi, P., Xu, Y.: A survey of recent results in networked control systems. IEEE Spec. Issue Technol. Netw. Control Syst. **95**(1), 138–162 (2007)
19. Hetel, L., Daafouz, J., Iung, C.: Stabilization of arbitrary switched linear systems with unknown time-varying delays. IEEE Trans. Autom. Control **51**(10), 1668–1674 (2006)
20. Hetel, L., Fiter, C., Omran, H., Seuret, A., Fridman, E., Richard, J.P., Niculescu, S.I.: Recent developments on the stability of systems with aperiodic sampling: an overview. Automatica **76**, 309–335 (2017)
21. Hetel, L., Kruszewski, A., Perruquetti, W., Richard, J.P.: Discrete and Intersample analysis of systems with aperiodic sampling. IEEE Trans. Autom. Control **56**(7), 1696–1701 (2011)
22. Hristu-Varsakelis, D., Levine, W. (eds.): Handbook of Networked and Embedded Control Systems. Birkhauser, Boston (2005)
23. Kalman, R.E.: Mathematical description of linear dynamical systems. J. Soc. Ind. Appl. Math. Ser. A Control **1**(2), 152–192 (1962)
24. Kalman, R., Ho, B.L., Narendra, K.: Controllability of linear dynamical systems. Contrib. Differ. Equ. **1**, 189–213 (1963)
25. Kreisselmeier, G.: On sampling without loss of observability/controllability. IEEE Trans. Autom. Control **44**(5), 1021–1025 (1999)
26. Ljung, L.: System Identification, Theory for the User. Prentice Hall, Englewood Cliffs (1999)
27. Mirkin, L.: Some remarks on the use of time-varying delay to model sample-and-hold circuits. IEEE Trans. Autom. Control **52**(6), 1109–1112 (2007)
28. Naghshtabrizi, P., Hespanha, J.P., Teel, A.R.: Exponential stability of impulsive systems with application to uncertain sampled-data systems. Syst. Control Lett. **57**(5), 378–385 (2008)
29. Oishia, Y., Fujioka, H.: Stability and stabilization of aperiodic sampled-data control systems using robust linear matrix inequalities. Automatica **46**(8), 1327–1333 (2010)
30. Petreczky, M.: Realization theory for linear and bilinear switched systems: formal power series approach - Part I: realization theory of linear switched systems. ESAIM Control Optim. Calc. Var. **17**, 410–445 (2011)
31. Petreczky, M., Bako, L., van Schuppen, J.H.: Realization theory of discrete-time linear switched systems. Automatica **49**(11), 3337–3344 (2013)
32. Petreczky, M., Mercère, G.: Affine LPV systems: Realization theory, input-output equations and relationship with linear switched systems. In: Proceedings of the IEEE 51st Conference on Decision and Control (CDC), pp. 4511–4516. Maui, HI (2012)
33. Petreczky, M., Tóth, R., Mercère, G.: Realization Theory for LPV State-Space Representations with Affine Dependence. IEEE Trans. Autom. Control (2016). https://doi.org/10.1109/TAC.2016.2629989. Accepted for publication

34. Shieh, L.S., Chang, Y.F.: Model simplification and digital design of multivariable sampled-data control systems via a dominant-data matching method. Appl. Math. Model. **8**(10), 355–364 (1984)
35. Sontag, E.D.: Mathematical Control Theory Deterministic Finite Dimensional Systems. Springer, New York (1990)
36. Tóth, R.: Modeling and Identification of Linear Parameter-Varying Systems. Springer, Germany (2010)
37. Wang, L.Y., Li, C., Yin, G.G., Guo, L., Xu, C.Z.: State observability and observers of linear-time-invariant systems under irregular sampling and sensor limitations. IEEE Trans. Autom. Control **56**(11), 2639–2654 (2011)
38. Yamé, J.J.: Some remarks on the controllability of sampled-data systems. SIAM J. Control Optim. **50**(4), 1775–1803 (2012)
39. Zhang, W.: Stability Analysis of Networked Control Systems. CASE Western Reserve University, Cleveland (2001)
40. Zhang, W., Branicky, M.S., Phillips, S.M.: Stability of networked control systems. IEEE Control Syst. Mag. **21**(1), 84–99 (2001)

Chapter 2
Stabilizability and Control Co-Design for Discrete-Time Switched Linear Systems

M. Fiacchini, M. Jungers, A. Girard and S. Tarbouriech

Abstract In this work we deal with the stabilizability property for discrete-time switched linear systems. First we provide a constructive necessary and sufficient condition for stabilizability based on set-theory and the characterization of a universal class of Lyapunov functions. Such a geometric condition is considered as the reference for comparing the computation-oriented sufficient conditions. The classical BMI conditions based on Lyapunov-Metzler inequalities are considered and extended. Novel LMI conditions for stabilizability, derived from the geometric ones, are presented that permit to combine generality with convexity. For the different conditions, the geometrical interpretations are provided and the induced stabilizing switching laws are given. The relations and the implications between the stabilizability conditions are analyzed to infer and compare their conservatism and their complexity. The results are finally extended to the problem of the co-design of a control policy, composed by both the state feedback and the switching control law, for discrete-time switched linear systems. Constructive conditions are given in form of LMI that are necessary and sufficient for the stabilizability of systems which are periodic stabilizable.

M. Fiacchini (✉)
University Grenoble Alpes, CNRS, Gipsa-lab, F-38000, Grenoble, France
e-mail: mirko.fiacchini@gipsa-lab.fr

M. Jungers
Université de Lorraine, CRAN, UMR 7039, 2 avenue de la forêt de Haye, Vandœuvre-lès-Nancy Cedex, 54516 Lorraine, France
e-mail: marc.jungers@univ-lorraine.fr

M. Jungers
CNRS, CRAN, UMR 7039, Lorraine, France

A. Girard
Laboratoire des signaux et systèmes (L2S), CNRS, CentraleSupélec, Université Paris-Sud, Université Paris-Saclay, 3, rue Joliot-Curie, 91192, Gif-sur-Yvette cedex, France
e-mail: Antoine.Girard@l2s.centralesupelec.fr

S. Tarbouriech
LAAS-CNRS, Université de Toulouse, Toulouse, France
e-mail: tarbour@laas.fr

S. Tarbouriech et al. (eds.), *Control Subject to Computational and Communication Constraints*, Lecture Notes in Control and Information Sciences 475, https://doi.org/10.1007/978-3-319-78449-6_2

2.1 Introduction

Switched systems are characterized by dynamics that may change along the time among a finite number of possible dynamical behaviors. Each behavior is determined by a mode and the active one is selected by means of a function of time, or state, or both, and referred to as switching law. The interest that such a kind of systems rose in the last decades lies in their capability of modeling complex real systems, as embedded or networked ones, and also for the theoretical issues involved, see [19, 20, 24].

Several conditions for stability have been proposed in the literature based on: switched Lyapunov functions [8]; the joint spectral radius [15]; path-dependent Lyapunov functions [18]; and the variational approach [21]. If the existence of polyhedral, hence convex, Lyapunov functions has been proved to be necessary and sufficient for stability [5, 22], convex functions result to be conservative for switched systems with switching law as control input, see [6, 24]. Thus, nonconvex functions must be considered for addressing stabilizability. Sufficient conditions for stabilizability have been provided in literature, mainly based on min-switching policies introduced in [26], developed in [17, 19] and leading to Lyapunov-Metzler inequalities [13, 14]. The fact that the existence of a min-switching control law is necessary and sufficient for exponential stabilizability has been claimed in [24]. In the same work, as well as in [6], it has been proved that the stabilizability of a switched system does not imply the existence of a convex Lyapunov function. Thus, for stabilizability, nonconvex Lyapunov functions might be considered, see for instance [14, 24].

We present here some recent results, mostly based on set-theory and convex analysis, on stabilizability and control co-design for switched linear systems, see [10–12]. We first propose a stabilizability approach based on set-theory and invariance, see [2, 3, 16]. A geometric necessary and sufficient condition for stabilizability and sufficient one for non-stabilizability of discrete-time linear switched systems are presented in [10]. A family of nonconvex, homogeneous functions is proved to be a universal class of Lyapunov functions for switched linear systems.

The geometric condition in [10] might, nonetheless, result to be often computationally unaffordable, although such a computational complexity appears to be inherent to the problem itself, hence unavoidable. In the literature, computation-oriented sufficient conditions for stabilizability have been provided that are based on min-switching policies and lead to nonconvex control Lyapunov functions in form of minimum of quadratics. Such functions are obtained as solutions to Lyapunov-Metzler BMI conditions, [1, 14], and through an LQR-based iterative procedure, [24]. New LMI conditions for stabilizability, which could conjugate computational affordability with generality, are proposed here, see [12]. The LMI conditions are proved to admit a solution if and only if the system is periodic stabilizable. Moreover, we provide geometrical and numerical insights on different stabilizability conditions to quantify their conservatism and the relations between them and with the necessary and sufficient ones.

The problem of co-designing both the switching law and the control input is even more involved than the problem of stabilizability of autonomous switched systems. This problem has been addressed in several works based on Lyapunov-Metzler BMI conditions, as in [9], or on techniques based on LQR control approximation in [1, 27, 28]. Constructive LMI conditions are given here that are necessary and sufficient for the stabilizability of systems which are periodic stabilizable, [11]. The conditions are constructive and provide the switching law and a family of state feedback gains stabilizing the system, although their complexity grows combinatorially with the maximal length of modes sequences considered.

Notation Given $n \in \mathbb{N}$, define $\mathbb{N}_n = \{j \in \mathbb{N} : 1 \leq j \leq n\}$. The Euclidean-norm in $\mathbb{R}^n$ is $\|x\|$. The ith element of a finite set of matrices is denoted as A_i. We use the shortcut $P > 0$ (resp. $P \geq 0$) to define a symmetric positive definite (resp. semi-definite) matrix, i.e., such that $P = P^T$ and its eigenvalues are positive (resp. non-negative). Given $P \in \mathbb{R}^{n \times n}$ with $P > 0$, define $\mathscr{E}(P) = \{x \in \mathbb{R}^n : x^T P x \leq 1\}$. Given $\theta \in \mathbb{R}$, $R(\theta) \in \mathbb{R}^2$ is the rotation matrix of angle θ. The set of q switching modes is $\mathscr{I} = \mathbb{N}_q$, all the possible sequences of modes of length N is $\mathscr{I}^N = \prod_{j=1}^{N} \mathscr{I}$, and $|\sigma| = N$ if $\sigma \in \mathscr{I}^N$. Given $N, M \in \mathbb{N}$ with $N \leq M$, denote $\mathscr{I}^{[N:M]} = \bigcup_{i=N}^{M} \mathscr{I}^i$ and then $N_{\mathscr{I}} = \sum_{k=1}^{N} q^k$ is the number of elements in $\mathscr{I}^{[1:N]}$. Given $\sigma \in \mathscr{I}^N$, define: $\mathbb{A}_\sigma = \prod_{j=1}^{N} A_{\sigma_j} = A_{\sigma_N} \ldots A_{\sigma_1}$, and define $\prod_{j=M}^{N} A_{\sigma_j} = I$ if $M > N$. Given $a \in \mathbb{R}$, the maximal integer smaller than or equal to a is $\lfloor a \rfloor$. The set of Metzler matrices of dimension N, i.e., matrices $\pi \in \mathbb{R}^{N \times N}$ whose elements are nonnegative and $\sum_{j=1}^{N} \pi_{ji} = 1$ for all $i \in \mathbb{N}_N$, is $\mathscr{M}_N$.

2.2 Stabilizability of Discrete-Time Linear Switched Systems

Consider the discrete-time switched system

$$x_{k+1} = A_{\sigma(k)} x_k, \tag{2.1}$$

where $x_k \in \mathbb{R}^n$ is the state at time $k \in \mathbb{N}$ and $\sigma : \mathbb{N} \to \mathbb{N}_q$ is the switching law that, at any instant, selects the transition matrix among the finite set $\{A_i\}_{i \in \mathbb{N}_q}$, with $A_i \in \mathbb{R}^{n \times n}$ for all $i \in \mathbb{N}_q$. Given the initial state x_0 and a switching law $\sigma(\cdot)$, we denote with $x_N^\sigma(x_0)$ the state of the system (2.1) at time N starting from x_0 by applying the switching law $\sigma(\cdot)$, that can be state-dependent, i.e., $\sigma(k) = \sigma(x(k))$ with slight abuse of notation.

Assumption 2.1 The matrices A_i, with $i \in \mathbb{N}_q$, are nonsingular.

Remark 2.1 Assumption 2.1 is not restrictive. In fact, the stable eigenvalues of the matrices A_i are beneficial from the stability point of view of the switched systems and poles in zero are related to the most contractive dynamics. Moreover, the results presented in the following can be extended to the general case with appropriate considerations.

A concept widely employed in the context of set-theory and invariance is the C-set, see [3, 5]. A C-set is a compact and convex set with $0 \in \text{int}(\Omega)$. We define an analogous concept useful for our purpose. For this, we first recall that a set Ω is a star-convex set if there exists $x_0 \in \Omega$ such that every convex combination of x and x_0 belongs to Ω for every $x \in \Omega$.

Definition 2.1 A set $\Omega \subseteq \mathbb{R}^n$ is a C*-set if it is the union of a finite number of C-sets. The gauge function of a C*-set $\Omega \subseteq \mathbb{R}^n$ is $\Psi_\Omega(x) = \min\limits_{\alpha \geq 0}\{\alpha \in \mathbb{R} : x \in \alpha\Omega\}$.

Notice that every C*-set is star-convex, i.e., there is $z \in \Omega$ such that every convex combination of x and z belongs to Ω for all $x \in \Omega$, but the converse is not true in general. Some basic properties of the C*-sets and their gauge functions are listed below, see also [23].

Property 2.1 *Any C-set is a C*-set. Given a C*-set $\Omega \subseteq \mathbb{R}^n$, we have that $\alpha\Omega \subseteq \Omega$ for all $\alpha \in [0, 1]$, and the gauge function $\Psi_\Omega(\cdot)$ is: homogeneous of degree one, i.e., $\Psi_\Omega(\alpha x) = \alpha\Psi_\Omega(x)$ for all $\alpha \geq 0$ and $x \in \mathbb{R}^n$; positive definite; defined on $\mathbb{R}^n$ and radially unbounded.*

The gauge functions induced by C-sets have been used in the literature as Lyapunov functions candidates, see [4]. In particular, it has been proved that they provide a universal class of Lyapunov functions for linear parametric uncertain systems, [5, 22], and switched systems with arbitrary switching, [20]. We prove that the gauge functions induced by C*-sets form a universal class of Lyapunov function for switched systems with switching control law. For this, we provide a definition of Lyapunov function for the particular context, analogous to the one given in [5].

Definition 2.2 A positive definite continuous function $V : \mathbb{R}^n \to \mathbb{R}$ is a global control Lyapunov function for (2.1) if there exist a positive $N \in \mathbb{N}$ and a switching law $\sigma(\cdot)$, defined on $\mathbb{R}^n$, such that V is non-increasing along the trajectories $x_k^\sigma(x)$ and decreasing after N steps, i.e., $V(x_1^\sigma(x)) \leq V(x)$ and $V(x_N^\sigma(x)) < V(x)$, for all $x \in \mathbb{R}^n$.

Definition 2.2 is a standard definition of global control Lyapunov function except for the N-steps decreasing requirement. On the other hand, such a function implies the convergence of every subsequence in $j \in \mathbb{N}$ of the trajectory, i.e., $x_{i+jN}^\sigma(x)$ for all $i < N$, then also the convergence of the trajectory itself. This, with the stability assured by $V(x_1^\sigma(x)) \leq V(x)$, ensures global asymptotic stabilizability.

Definition 2.3 The system (2.1) is globally exponentially stabilizable if there are $c \geq 0$ and $\lambda \in [0, 1)$ and, for all $x \in \mathbb{R}^n$, there exists a switching law $\sigma : \mathbb{N} \to \mathbb{N}_q$, such that

$$\|x_k^\sigma(x)\| \leq c\lambda^k\|x\|, \quad \forall k \in \mathbb{N}. \tag{2.2}$$

A periodic switching law is given by $\sigma(k) = i_{p(k)}$ and $p(k) = k - M\lfloor k/M \rfloor + 1$, with $M \in \mathbb{N}$ and $i \in \mathscr{I}^M$, which means that the sequence given by i repeats cyclically. We will consider conditions under which system (2.1) is stabilized by a periodic $\sigma(\cdot)$.

Definition 2.4 The system (2.1) is periodic stabilizable if there exist a periodic switching law $\sigma : \mathbb{N} \to \mathbb{N}_q$, $c \geq 0$ and $\lambda \in [0, 1)$ such that (2.2) holds for all $x \in \mathbb{R}^n$.

For stabilizability the switching function might be state-dependent whereas for periodic stabilizability it must be not dependent on the state.

Lemma 2.1 *The system (2.1) is periodic stabilizable if and only if there exist $M \in \mathbb{N}$ and $i \in \mathscr{I}^M$ such that $\mathbb{A}_i$ is Schur.*

2.2.1 Geometric Necessary and Sufficient Condition

It is proved in [22] that for an autonomous linear switched system, the origin is asymptotically stable if and only if there exists a polyhedral Lyapunov function, see also [5, 20]. Analogous results can be stated in the case where the switching sequence is a properly chosen selection, that is considering it as a control law. This contribution is based on the following algorithm.

Algorithm 1 Computation of a contractive C*-set for (2.1) satisfying Assumption 2.1.

- **Initialization** given the C*-set $\Omega \subseteq \mathbb{R}^n$, define $\Omega_0 = \Omega$ and $k = 0$;
- **Iteration** for $k \geq 0$: $\Omega_{k+1} = \bigcup_{i \in \mathbb{N}_q} \Omega_{k+1}^i$ with $\Omega_{k+1}^i = A_i^{-1}\Omega_k$ for all $i \in \mathbb{N}_q$;
- **Stop** if $\Omega \subseteq \text{int}\Big(\bigcup_{j \in \mathbb{N}_{k+1}} \Omega_j\Big)$; denote $\check{N} = k + 1$ and

$$\check{\Omega} = \bigcup_{j \in \mathbb{N}_{\check{N}}} \Omega_j. \tag{2.3}$$

From the geometrical point of view, Ω_{k+1}^i is the set of x mapped in Ω_k through A_i. Then Ω_{k+1} is the set of $x \in \mathbb{R}^n$ for which there exists a selection $i(x) \in \mathbb{N}_q$ such that $A_{i(x)}x \in \Omega_k$. Thus, Ω_k is the set of x that can be driven in Ω in at most k steps and hence $\check{\Omega}$ the set of x that can reach Ω in $\check{N}$ or less steps.

Proposition 2.1 *The sets Ω_k for all $k \geq 0$ are C*-sets.*

Algorithm 1 provides a C*-set $\check{\Omega}$ contractive in $\check{N}$ steps, for every initial C*-set $\Omega \in \mathbb{R}^n$, if and only if the switched system (2.1) is stabilizable, as stated below.

Theorem 2.2 [10] *There exists a Lyapunov function for the switched system (2.1) if and only if Algorithm 1 ends with finite* $\check{N}$.

Then finite termination of Algorithm 1 is a necessary and sufficient condition for the global asymptotic stabilizability of the switched system (2.1). An alternative formulation of such a necessary and sufficient condition is presented below.

Theorem 2.3 [10] *There exists a Lyapunov function for the switched system (2.1) if and only if there exists a C*-set whose gauge function is a Lyapunov function for the system.*

Theorem 2.3 states that the existence of a C*-set induced Lyapunov function is a necessary and sufficient condition for stabilizability of switched systems. Hence, such functions, nonconvex and homogeneous of order one, form a class of universal Lyapunov functions for the switched systems.

Remark 2.2 The Algorithm 1 terminates after a finite number of iterations only if the switched system is stabilizable, there is no guarantee of finite termination in general (which means it is a semi-algorithm, to be exact). An analogous, but just sufficient, constructive condition ensuring that there is not a switching law such that the system (2.1) converges to the origin is given in [10].

Besides a Lyapunov function, Algorithm 1 provides a stabilizing switching control law for system (2.1), if it terminates in finite time.

Proposition 2.2 [10] *If Algorithm 1 ends with finite* $\check{N}$ *then* $\Psi_{\check{\Omega}} : \mathbb{R}^n \to \mathbb{R}$ *is a Lyapunov function for the switched system (2.1) and given the set valued map*

$$\check{\Sigma}(x) = \arg\min_{(i,k)}\{\Psi_{\Omega_k^i}(x) : i \in \mathbb{N}_q,\ k \in \mathbb{N}_{\check{N}}\} \subseteq \mathbb{N}_q \times \mathbb{N}_{\check{N}}, \tag{2.4}$$

any switching law defined as $(\check{\sigma}(x), \check{k}(x)) \in \check{\Sigma}(x)$, *is a stabilizing switching law. Furthermore, one gets* $\Psi_{\check{\Omega}}(x^{\check{\sigma}}_{\check{k}(x)}(x)) \leq \check{\lambda}\Psi_{\check{\Omega}}(x)$ *and* $\Psi_{\check{\Omega}}(x^{\check{\sigma}}_j(x)) \leq \Psi_{\check{\Omega}}(x)$ *for all* $j \in \mathbb{N}_{\check{k}(x)}$, *with* $\check{\lambda} = \min_{\lambda}\{\lambda \geq 0 : \Omega \subseteq \lambda\check{\Omega}\} < 1$.

It could be reasonable, to speed up the convergence of the trajectory of the system to origin, to select among the elements of $\Sigma(x)$, those whose k is minimal.

Remark 2.3 If the system is stabilizable, then the algorithm ends with finite $\check{N}$ for all initial C*-set Ω. Clearly, the value of $\check{N}$ and the complexity of the set $\check{\Omega}$ depend on the choice of Ω. In particular, if Ω is the Euclidean norm ball (or the union of ellipsoids), the sets Ω_k^i and Ω_k, with $i \in \mathbb{N}_q$ and $k \in \mathbb{N}_{\check{N}}$, are unions of ellipsoids, and $\check{\Omega}$ also. Then, the switching law computation reduces to check the minimum among $x^T P_j x$ with $j \in \check{M}$, where $\{P_j\}_{j\in\check{M}}$ are the $\check{M}$ positive definite matrices defining $\check{\Omega}$, with $\check{M} = q + \cdots + q^{\check{N}} = (q^{\check{N}+1} - q)/(q-1)$, for $q > 1$ and $\check{M} = \check{N}$ for $q = 1$.

2.2.2 Duality Robustness-Control of Switched Systems

In this section, we recall some results from the literature on the stability of a switched linear system with arbitrary switching law $\sigma(\cdot)$ to highlight the analogies with the approaches proposed here for stabilizability.

Consider the linear switched system (2.1) and assume that the switching law is arbitrary. This would mean that the switching law might be regarded as a parametric uncertainty and the results in [4, 5, 22] on robust stability apply with minor adaptations, see also [20]. The following algorithm provides a polytopic contractive set, and then an induced polyhedral Lyapunov function, for this class of systems, see [3].

Algorithm 2 Computation of a λ-contractive C-set for (2.1) with arbitrary switching.

- **Initialization** given the C-set $\Gamma \subseteq \mathbb{R}^n$ and $\lambda \in [0, 1)$, define $\Gamma_0 = \Gamma$ and $k = 0$;
- **Iteration** for $k \geq 0$: $\Gamma_{k+1} = \Gamma \cap \bigcap_{i \in \mathbb{N}_q} \lambda A_i^{-1} \Gamma_k$;
- **Stop** if $\Gamma_k \subseteq \Gamma_{k+1}$; denote $\hat{N} = k$ and $\hat{\Gamma} = \Gamma_k$.

The set $\hat{\Gamma}$ is the maximal λ-contractive set in Γ for the switched system with arbitrary switching law. Provided that Algorithm 2 terminates with finite $\hat{N}$, it can be proved that the system is globally exponentially stable, see [5].

Remark 2.4 Notice the analogies between Algorithms 1 and 2: they share the same iterative structure and they both generate contractive sets which induce Lyapunov functions if they terminate in a finite time. The main substantial difference resides in the use of intersection/union operators and in the family of sets generated, C*-sets by Algorithm 1 and C-sets by Algorithm 2. Interestingly, the C-sets are closed under the intersection operation whereas C*-sets are closed under the union.

Finally, for linear parametric uncertain systems, the existence of a polyhedral Lyapunov function is a necessary and sufficient condition for asymptotic stability.

Theorem 2.4 [5, 22] *There is a Lyapunov function for a linear parametric uncertain system if and only if there is a polyhedral Lyapunov function for the system.*

The result in Theorem 2.4 holds for general parametric uncertainty and applies also for switched systems with arbitrary switching law, as remarked in [20].

Remark 2.5 As for the duality of Algorithms 1 and 2 highlighted in Remark 2.4, evident conceptual analogies hold between Theorems 2.3 and 2.4. Then the class of gauge functions induced by C*-sets is universal for linear switched systems with switching control law, in analogy with the class of polyhedral functions (i.e., induced by C-sets) for the case of arbitrary switching law, [4, 5].

2.3 Novel Conditions for Stabilizability and Comparisons

As seen above, system (2.1) is stabilizable if and only if there exists $N \in \mathbb{N}$ such that

$$\Omega \subseteq \operatorname{int}\Big(\bigcup_{i \in \mathscr{I}^{[1:N]}} \Omega_i \Big) \quad \text{with} \quad \Omega_i = \Omega_i(\Omega) = \{x \in \mathbb{R}^n : \mathbb{A}_i x \in \Omega\}. \tag{2.5}$$

Since the stabilizability property does not depend on the choice of the initial C*-set Ω, even if N does, focusing on the case $\Omega = \mathbf{B}$ and ellipsoidal pre-images entails no loss of generality, see [10]. Then condition (2.3) can be replaced by

$$\mathbf{B} \subseteq \operatorname{int}\Big(\bigcup_{i \in \mathscr{I}^{[1:N]}} \mathbf{B}_i \Big) \quad \text{with} \quad \mathbf{B}_i = \{x \in \mathbb{R}^n : x^T \mathbb{A}_i^T \mathbb{A}_i x \leq 1\}, \tag{2.6}$$

for what concerns stabilizability, although the value N might depend on the choice of Ω. The set inclusions (2.5) or (2.6) are the stopping conditions of the algorithm and then must be numerically checked at every step. The main computational issue consists in determining if a C*-set Ω is included into the interior of the union of some C*-sets. This problem is very complex in general, also in the case of ellipsoidal sets where it relates to quantifier elimination over real closed fields [7]. On the other hand, the condition given by Theorem 2.2 provides an exact characterization of the complexity inherent to the problem of stabilizing a switched linear system.

The objective here is to consider alternative conditions for stabilizability to provide geometrical and numerical insights and analyze their conservatism by comparison with the necessary and sufficient one given in Theorem 2.2.

2.3.1 *Lyapunov-Metzler BMI Conditions*

The condition we are considering first is related to the Lyapunov-Metzler inequalities that is sufficient and given by a set of BMI inequalities involving Metzler matrices.

Theorem 2.5 [14] *If there exist $P_i > 0$, with $i \in \mathscr{I}$, and $\pi \in \mathscr{M}_q$ such that*

$$A_i^T \left(\sum_{j=1}^{q} \pi_{ji} P_j \right) A_i - P_i < 0, \qquad \forall i \in \mathscr{I}, \tag{2.7}$$

holds, then the switched system (2.1) is stabilizable.

As proved in [14], the satisfaction of (2.7) implies that the homogeneous function induced by the set $\bigcup_{i \in \mathscr{I}} \mathscr{E}(P_i)$ is a control Lyapunov function. A first relation between the Lyapunov-Metzler condition (2.7) and the geometric one (2.5) is provided below. We prove that the satisfaction of (2.7) implies that the condition given by Theorem 2.2 holds for the particular case of $\Omega = \bigcup_{i\in\mathcal{I}} A_i\mathcal{E}(P_i)$ and $N = 1$.

Proposition 2.3 [12] *If the Lyapunov-Metzler condition (2.7) holds then (2.5) holds with $N = 1$ and $\Omega = \bigcup_{i\in\mathcal{I}} A_i\mathcal{E}(P_i)$.*

Proposition 2.3 provides a geometrical meaning of the Lyapunov-Metzler condition and a first relation with the necessary and sufficient condition given in Theorem 2.2. In fact, for the general case of $q \in \mathbb{N}$ the Lyapunov-Metzler condition is just sufficient for $\bigcup_{i\in\mathcal{I}} A_i\mathcal{E}(P_i) \subseteq \operatorname{int}\left(\bigcup_{i\in\mathcal{I}} \mathcal{E}(P_i)\right)$ to hold. Moreover, it is proved in [12] that the condition is also necessary for $q = 2$.

2.3.2 Generalized Lyapunov-Metzler Conditions

Two generalizations of the Lyapunov-Metzler condition can be given, by relaxing the intuitive but unnecessary constraint stating that the number of ellipsoids and the number of modes are equal.

Proposition 2.4 [12] *If there exist $M \in \mathbb{N}$ and $P_i > 0$, with $i \in \mathcal{I}^{[1:M]}$, and $\pi \in \mathcal{M}_{M_\mathcal{I}}$ such that*

$$\mathbb{A}_i^T\left(\sum_{j\in\mathcal{I}^{[1:M]}} \pi_{ji}P_j\right)\mathbb{A}_i - P_i < 0, \qquad \forall i \in \mathcal{I}^{[1:M]},$$

holds, then the switched system (2.1) is stabilizable.

Proposition 2.4 extends the Lyapunov-Metzler condition (2.7), which is recovered for $M = 1$. Another possible extension is obtained by maintaining the sequence length in 1 but increasing the number of ellipsoids involved.

Proposition 2.5 [12] *If for every $i \in \mathcal{I}$ there exist a set of indices $\mathcal{K}_i = \mathbb{N}_{h_i}$, with $h_i \in \mathbb{N}$; a set of matrices $P_k^{(i)} > 0$, with $k \in \mathcal{K}_i$, and there are $\pi_{m,k}^{(p,i)} \in [0, 1]$, satisfying $\sum_{p\in\mathcal{I}} \sum_{m\in\mathcal{K}_p} \pi_{m,k}^{(p,i)} = 1$ for all $k \in \mathcal{K}_i$, such that*

$$A_i^T\left(\sum_{p\in\mathcal{I}} \sum_{m\in\mathcal{K}_p} \pi_{m,k}^{(p,i)} P_m^{(p)}\right) A_i - P_k^{(i)} < 0, \quad \forall i \in \mathcal{I}, \forall k \in \mathcal{K}_i,$$

holds, then the switched system (2.1) is stabilizable.

Geometrically, Proposition 2.5 provides a condition under which there exists a C*-set composed by a finite number of ellipsoids that is contractive.

2.3.3 LMI Sufficient Condition

The main drawback of the necessary and sufficient condition for stabilizability is its inherent complexity. The Lyapunov-Metzler-based approach leads to a more practical BMI sufficient condition. Nevertheless, the complexity could be still computationally prohibitive, see [25]. Our next aim is to formulate an alternative condition that could be checked by convex optimization algorithms.

Theorem 2.6 [12] *The switched system (2.1) is stabilizable if there exist $N \in \mathbb{N}$ and $\eta \in \mathbb{R}^{N_{\mathscr{I}}}$ such that $\eta \geq 0$, $\sum_{i \in \mathscr{I}^{[1:N]}} \eta_i = 1$ and*

$$\sum_{i \in \mathscr{I}^{[1:N]}} \eta_i \mathbb{A}_i^T \mathbb{A}_i < I. \tag{2.8}$$

We wonder now if the sufficient condition given in Theorem 2.6 is also necessary. The answer is negative, in general, as proved by the following counter-example.

Example 2.1 The aim of this illustrative example is to show a case for which the inclusion condition (2.6) is satisfied with $N = 1$, but there is not a finite value of $\hat{N} \in \mathbb{N}$ for which condition (2.8) holds. Consider the three modes given by the matrices

$$A_1 = \begin{bmatrix} a & 0 \\ 0 & a^{-1} \end{bmatrix}, \quad A_2 = \begin{bmatrix} a & 0 \\ 0 & a^{-1} \end{bmatrix} R\left(\frac{2\pi}{3}\right), \quad A_3 = \begin{bmatrix} a & 0 \\ 0 & a^{-1} \end{bmatrix} R\left(\frac{-2\pi}{3}\right),$$

with $a = 0.6$. Set $\Omega = \mathbf{B}$. By geometric inspection, condition (2.6) holds at the first step, i.e., for $N = 1$, see [12]. On the other hand, A_i are such that $\det(A_i^T A_i) = a^2 a^{-2} = 1$ and $\mathrm{tr}(A_i^T A_i) = a^2 + a^{-2} = 3.1378$ while the determinant and trace of the matrix defining $\mathbf{B}$ are 1 and 2, respectively. Notice that $a^2 + a^{-2} > 2$ for every a different from 1 or -1 and $a^2 + a^{-2} = 2$ otherwise.

For every N and every $\mathbf{B}_i$ with $i \in \mathscr{I}^{[1:N]}$, the related $\mathbb{A}_i$ is such that $\det(\mathbb{A}_i^T \mathbb{A}_i) = 1$ and $\mathrm{tr}(\mathbb{A}_i^T \mathbb{A}_i) \geq 2$. Notice that, for all the matrices $Q > 0$ in $\mathbb{R}^{2 \times 2}$ such that $\det(Q) = 1$, then $\mathrm{tr}(Q) \geq 2$ and $\mathrm{tr}(Q) = 2$ if and only if $Q = I$, since the determinant is the product of the eigenvalues and the trace its sum. Thus, for every subset of the ellipsoids $\mathbf{B}_i$, determined by a subset of indices $K \subseteq \mathscr{I}^{[1:N]}$, we have that $\sum_{i \in K} \eta_i \mathbb{A}_i^T \mathbb{A}_i < I$, cannot hold, since either $\mathrm{tr}(\mathbb{A}_i^T \mathbb{A}_i) > 2$ or $\mathbb{A}_i^T \mathbb{A}_i = I$. Thus the LMI condition (2.8) is sufficient but not necessary.

Another interesting implication that follows from Example 2.1 concerns the stabilizability through periodic switching sequences.

Proposition 2.6 [12] *The existence of a stabilizing periodic switching law is sufficient but not necessary for the stabilizability of the system (2.1).*

In the proof of Proposition 2.6 we used the fact that the existence of a stabilizing periodic switching law implies the satisfaction of the LMI condition, see [12]. One might wonder if there exists an equivalence relation between periodic stabilizability and condition (2.8). The answer is provided below.

Theorem 2.7 [12] *A stabilizing periodic switching law for the system (2.1) exists if and only if condition (2.8) holds.*

Note that, although periodic stabilizability and condition (2.8) are equivalent from the stabilizability point of view, the computational aspects and the resulting controls are different. Indeed, checking periodic stabilizability consists of an eigenvalue test for a number of matrices exponential in M, see Lemma 2.1, while condition (2.8) is an LMI that grows exponentially with N. On the other hand, M is always greater or equal than N, much greater in general. Finally, notice that the periodic law is in open loop whereas (2.8) leads to a state-dependent switching law.

The LMI condition (2.8) can be used to derive the controller synthesis techniques. If (2.8) holds, then there is $\mu \in [0, 1)$ such that

$$\sum_{i \in \mathscr{I}^{[1:N]}} \eta_i \mathbb{A}_i^T \mathbb{A}_i \leq \mu^2 I. \tag{2.9}$$

A stabilizing controller does not necessarily select at each time step $k \in \mathbb{N}$ the input to be applied. This can be done only at $\{k_p\}_{p \in \mathbb{N}}$ with $k_0 = 0$, and $k_p < k_{p+1} \leq k_p + N$, for all $p \in \mathbb{N}$. At time k_p, the controller selects the sequence of inputs to be applied up to step $k_{p+1} - 1$. The instant k_{p+1} is also determined by the controller at time k_p. More precisely, the controller acts as follows for all $p \in \mathbb{N}$, let

$$i_p = \arg \min_{i \in \mathscr{I}^{[1:N]}} (x_{k_p}^T \mathbb{A}_i^T \mathbb{A}_i x_{k_p}). \tag{2.10}$$

Then, the next instant k_{p+1} is given by

$$k_{p+1} = k_p + l(i_p), \tag{2.11}$$

with $l(i_p)$ length of i_p, and the controller applies the sequence of inputs

$$\sigma_{k_p+j-1} = i_{p,j}, \quad \forall j \in \{1, \ldots, l(i_p)\}. \tag{2.12}$$

Theorem 2.8 [12] *Assume that (2.8) holds, and consider the control given by (2.10)–(2.12). For all $x_0 \in \mathbb{R}^n$ and $k \in \mathbb{N}$, we have $\|x_k\| \leq \mu^{k/N-1} L^{N-1} \|x_0\|$, where $L \geq \|A_i\|$, for all $i \in \mathscr{I}$ and $L \geq 1$, and the controlled system is globally exponentially stable.*

From Theorem 2.8, the LMI condition (2.8) implies that the switched system with the switching rule given by (2.10)–(2.12) is globally exponentially stable. Nevertheless, neither the Euclidean norm of x nor the function $\min_{i \in \mathscr{I}^{[1:N]}} (x^T \mathbb{A}_i^T \mathbb{A}_i x)$ are monotonically decreasing along the trajectories. On the other hand a positive definite homogeneous nonconvex function decreasing at every step can be inferred for a different switching rule.

Proposition 2.7 [12] *Given the switched system (2.1), suppose there exist* $N \in \mathbb{N}$ *and* $\eta \in \mathbb{R}^{N_{\mathcal{I}}}$ *such that* $\eta \geq 0$, $\sum_{i \in \mathcal{I}^{[1:N]}} \eta_i = 1$ *and (2.8) hold. Then there is* $\lambda \in [0, 1)$ *such that the function*

$$V(x) = \min_{i \in \mathcal{I}^{[1:N]}} (x^T \lambda^{-n_i} \mathbb{A}_i^T \mathbb{A}_i x), \tag{2.13}$$

where n_i *is the length of* $i \in \mathcal{I}^{[1:N]}$, *satisfies* $V(A_{\sigma(x)}x) \leq \lambda V(x)$ *for all* $x \in \mathbb{R}^n$, *with*

$$i^*(x) = \arg \min_{i \in \mathcal{I}^{[1:N]}} (x^T \lambda^{-n_i} \mathbb{A}_i^T \mathbb{A}_i x), \tag{2.14}$$

and $\sigma(x) = i_1^*(x)$.

Remark 2.6 If the LMI (2.8) has a solution, then there exists a scalar $\mu \in [0, 1)$, such that (2.9) is verified. The value of μ induces the rate of convergence λ for the Lyapunov function (2.13). Thus one might solve the optimization problem $\min_{\mu^2, \eta} \mu^2$ subject to (2.9), to get higher convergence rate.

2.3.4 Stabilizability Conditions Relations

The implications between the stabilizability conditions, whose proofs can be found in [12], are summarized in Fig. 2.1. Remark that, compared to the Lyapunov-Metzler inequalities (2.7), the LMI condition (2.8) concerns a convex problem and it is less conservative. On the other hand, the dimension of the LMI problem might be consistently higher than the BMI one. The direct extension to the case of output-based switching design is not straightforward and requires further research. Nevertheless, since the LMI condition and the periodic stabilizability are equivalent, if (2.8) has a solution then an open-loop stabilizing switching sequence can be designed, and no output is necessary to stabilize the system.

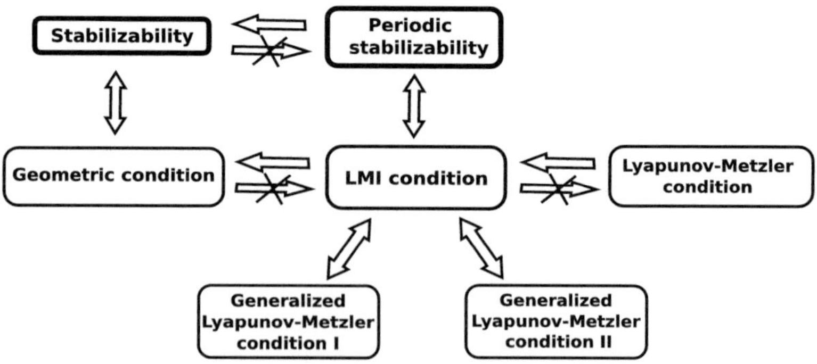

Fig. 2.1 Implications diagram of stabilizability conditions

2.4 Control Co-design for Discrete-Time Switched Linear Systems

Consider now the discrete-time controlled switched linear system

$$x_{k+1} = A_{\sigma_k} x_k + B_{\sigma_k} u_k, \tag{2.15}$$

where $x_k \in \mathbb{R}^n$ and $u_k \in \mathbb{R}^m$ are the state and the control input at time $k \in \mathbb{N}$, respectively; $\sigma : \mathbb{N} \to \mathscr{I}$ is the switching law and $\{A_i\}_{i\in\mathscr{I}}$ and $\{B_i\}_{i\in\mathscr{I}}$, with $A_i \in \mathbb{R}^{n\times n}$ and $B_i \in \mathbb{R}^{n\times m}$ for all $i \in \mathscr{I}$. A time-varying control policy $\nu : \mathbb{R}^n \times \mathbb{N} \to \mathscr{I} \times \mathbb{R}^{m\times n}$, is such that $\nu(x, k) = \big(\sigma(x, k),\ K(x, k)\big) \in \mathscr{I} \times \mathbb{R}^{m\times n}$, where $K(x, k)$ is the state feedback gain that may change at every instant, i.e., such that $u_k(x_k) = K(x_k, k)x_k$.

Remark 2.7 As proved in [27], see Theorems 5 and 7 in particular, the attention can be restricted without loss of generality to static control policies of the form

$$\nu(x) = \big(\sigma(x),\ K(x)\big) \in \mathscr{I} \times \mathbb{R}^{m\times n}, \tag{2.16}$$

such that $\nu(ax) = \nu(x)$ for all $x \in \mathbb{R}^n$ and $a \in \mathbb{R}$, and to piecewise quadratic Lyapunov functions. Moreover $K(x)$ belongs to a finite set, i.e., $K(x) \in \mathscr{K} = \{\kappa_i\}_{i\in\mathbb{N}_M}$.

The switched system in closed loop with (2.16) reads

$$x_{k+1} = \big(A_{\sigma(x_k)} + B_{\sigma(x_k)} K(x_k)\big)\, x_k, \tag{2.17}$$

where $\sigma(x_k) = \sigma_k$. We denote with $x_k^\nu(x_0) \in \mathbb{R}^n$ the state of the system (2.15) at time k starting from $x(0) = x_0$ by applying the control policy ν. Given $\sigma \in \mathscr{I}^D$ we denote with $x_k^\sigma(x_0)$ the state of (2.17) at time $k \leq D$ starting at x_0 under the switching sequence σ. The dependence of x_k^ν and x_k^σ on the initial conditions will be dropped.

Definition 2.5 The system (2.15) is globally exponentially stabilizable if there are a control policy $\nu(x)$ as in (2.16), $c \geq 0$ and $\lambda \in [0, 1)$ such that $\|x_k^\nu(x_0)\| \leq c\lambda^k \|x_0\|$, for all $x_0 \in \mathbb{R}^n$, with x_k state of (2.17).

In Sects. 2.3.3 and 2.3.4 we proved that, for autonomous systems as (2.1), periodic stabilizability is more conservative than generic stabilizability. On the other hand, the equivalent condition is much more computationally tractable. Indeed, the condition in case of periodic stabilizability is an LMI in the parameter N that might by much smaller than the periodic cycle length. Hereafter we focus on a condition analogous to the LMI one (2.8) for the controlled switched system (2.15) to determine a stabilizing control policy (2.16) for periodic stabilizable systems.

From Remark 2.7, the problem of co-design is equivalent to determine a stabilizing static control policy as in (2.16), with finite number of feedback gains, and a piecewise quadratic Lyapunov function for the system (2.17). Applying Theorem 2.7, the objective is to search for sequences of modes and feedback gains, fulfilling the LMI condition (2.8) in the context of co-design. That is, given a sequence $\vartheta \in \mathscr{I}$, of length J, and a time instant $j \in \mathbb{N}_J$, a gain among the finite set $\mathscr{K}$ can be applied, denoted as K_j^ϑ and whose value has to be designed. Then, with a slight abuse of notation, given $J \in \mathbb{N}$ and a sequence $\vartheta \in \mathscr{I}^J$, we denote

$$\mathbb{F}_\vartheta = \prod_{j=1}^{J} F_{\vartheta_j} = F_{\vartheta_J} \dots F_{\vartheta_1} = (A_{\vartheta_J} + B_{\vartheta_J} K_J^\vartheta) \dots (A_{\vartheta_1} + B_{\vartheta_1} K_1^\vartheta). \tag{2.18}$$

Thus a set of $N_\mathscr{I} = \sum_{k=1}^{N} q^k$ matrices $\mathbb{F}_\vartheta$, one for every $\vartheta \in \mathscr{I}^{[1:N]}$, can be defined as in (2.18) that are parameterized in the gains $\{K_j^\vartheta\}_{j\in\mathbb{N}_{|\vartheta|}}$. We focus on the control policy for (2.15) of the form (2.16) where $K(x)$ belongs to one of the elements of a sequence associated to a mode in $\mathscr{I}^{[1:N]}$. Then, $K(x)$ is a gain among the $\sum_{k=1}^{N} kq^k$ possible, i.e., $K(x) \in \mathscr{K}$ where

$$\mathscr{K} = \{\kappa_i\}_{i\in\mathbb{N}_M} = \{K_j^\vartheta \in \mathbb{R}^{m\times n} : \vartheta \in \mathscr{I}^{[1:N]},\ j \in \mathbb{N}_{|\vartheta|}\}, \tag{2.19}$$

with $M = \sum_{k=1}^{N} kq^k$. Given a switching law $\vartheta : \mathbb{N} \to \mathscr{I}$ and a sequence of feedback gains $K^\vartheta : \mathbb{N} \to \mathbb{R}^{m\times n}$, we denote with $x_k^\vartheta(x)$ the state at time k starting at x if the control $\nu_k = (\vartheta_k, K_k^\vartheta)$ is applied at k for all $k \in \mathbb{N}$. As for the case without control input, the concept of periodic ϑ-stabilizability can be given for the system (2.15).

Definition 2.6 The system (2.15) is periodic ϑ-stabilizable if there exist: a periodic switching law $\vartheta : \mathbb{N} \to \mathscr{I}$ and a periodic sequence $K^\vartheta : \mathbb{N} \to \mathbb{R}^{m\times n}$, both of cycle length $D \in \mathbb{N}$; $c \geq 0$ and $\lambda \in [0, 1)$ such that $\|x_k^\vartheta(x)\| \leq c\lambda^k \|x\|$ holds for all $x \in \mathbb{R}^n$ and $k \in \mathbb{N}$.

Clearly periodic ϑ-stabilizability is sufficient for exponential stabilizability of (2.15) as in Definition 2.5. From Definition 2.6 and Theorem 2.7, the conditions

$$\sum_{i\in\mathscr{I}^{[1:N]}} \eta_i = 1, \tag{2.20}$$

$$\sum_{j\in\mathscr{I}^{[1:N]}} \eta_j \mathbb{F}_j^T \mathbb{F}_j < I, \tag{2.21}$$

are necessary and sufficient for periodic ϑ-stabilizability of system (2.15). Thus, condition (2.21) is an LMI that provides the exact characterization of ϑ-stabilizability, together with (2.20). Below we give a convex condition equivalent to (2.21).

Proposition 2.8 [11] *Given $N \in \mathbb{N}$, $\eta \in \mathbb{R}^{N_\mathscr{I}}$ with $\eta > 0$, and the set of gains (2.19), condition (2.21) holds if and only if for every $j \in \mathscr{I}^{[1:N]}$ there exist $|j| - 1$ nonsingular matrices $G_{j,k} \in \mathbb{R}^{n\times n}$ with $k \in \mathbb{N}_{|j|-1}$ and $R_j \in \mathbb{R}^{n\times n}$ such that $R_j = R_j^T > 0$ and*

$$\begin{bmatrix} \eta_j I & X_{j,|j|} & 0 & \dots & 0 & 0 & 0 \\ X_{j,|j|}^T & Y_{j,|j|-1} & X_{j,|j|-1} & \dots & 0 & 0 & 0 \\ 0 & X_{j,|j|-1}^T & Y_{j,|j|-1} & \dots & 0 & 0 & 0 \\ \dots & \dots & \dots & \dots & \dots & \dots \\ 0 & 0 & 0 & \dots & Y_{j,2} & X_{j,2} & 0 \\ 0 & 0 & 0 & \dots & X_{j,2}^T & Y_{j,1} & X_{j,1} \\ 0 & 0 & 0 & \dots & 0 & X_{j,1}^T & R_j \end{bmatrix} > 0 \tag{2.22}$$

for every $j \in \mathscr{I}^{[1:N]}$ *with* $X_{j,1} = \eta_j \mathbb{F}_{j_1}$ *and* $X_{j,k+1} = \mathbb{F}_{j_{k+1}} G_{j,k}$ *and* $Y_{j,k} = G_{j,k} + G_{j,k}^T$ *for all* $k \in \mathbb{N}_{|j|-1}$ *and*

$$\sum_{j \in \mathscr{I}^{[1:N]}} R_j < I. \tag{2.23}$$

The following theorems, based on Proposition 2.8, provide a necessary and sufficient LMI condition for periodic ϑ-stabilizability of the controlled system (2.15), see their proofs in [11]. Moreover, the explicit form of the control law (2.16) is given.

Theorem 2.9 [11] *The system (2.15) is periodically* ϑ*-stabilizable if and only if there exist* $N \in \mathbb{N}$; $\eta \in \mathbb{R}^{N_\mathscr{I}}$ *such that* $\eta > 0$ *and (2.20) holds; and for every* $j \in \mathscr{I}^{[1:N]}$ *there are:*

- $|j| - 1$ *nonsingular matrices* $G_{j,k} \in \mathbb{R}^{n \times n}$*, with* $k \in \mathbb{N}_{|j|-1}$*;*
- $|j|$ *matrices* $Z_{j,k} \in \mathbb{R}^{m \times n}$ *with* $k \in \mathbb{N}_{|j|}$*;*
- *a symmetric positive definite matrix* $R_j \in \mathbb{R}^{n \times n}$*;*

such that (2.22) and (2.23) hold with

$$\begin{aligned} &X_{j,1} = \eta_j A_{j_1} + B_{j_1} Z_{j,1}, \\ &X_{j,k+1} = A_{j_{k+1}} G_{j,k} + B_{j_{k+1}} Z_{j,k+1}, && \forall k \in \mathbb{N}_{|j|-1}, \\ &Y_{j,k} = G_{j,k} + G_{j,k}^T, && \forall k \in \mathbb{N}_{|j|-1}, \end{aligned} \tag{2.24}$$

and gains

$$\begin{aligned} &K_1^j = \eta_j^{-1} Z_{j,1}, \\ &K_{k+1}^j = Z_{j,k+1} G_{j,k}^{-1}, && \forall k \in \mathbb{N}_{|j|-1}, \end{aligned} \tag{2.25}$$

for all $j \in \mathscr{I}^{[1:N]}$.

The following theorem provides a ϑ-stabilizability condition, a control policy and a bound on the decreasing of the Euclidean norm every N steps at most.

Theorem 2.10 [11] *Suppose there exist* $\alpha > 1$ *and* $N \in \mathbb{N}$; $\eta \in \mathbb{R}^{N_\mathscr{I}}$ *such that* $\eta > 0$; *matrices* $G_{j,k} \in \mathbb{R}^{n \times n}$ *with* $k \in \mathbb{N}_{|j|-1}$, $Z_{j,k} \in \mathbb{R}^{m \times n}$ *with* $k \in \mathbb{N}_{|j|}$ *and* $R_j \in \mathbb{R}^{n \times n}$ *as defined in Theorem 2.9 such that (2.22)–(2.24) hold and*

$$\sum_{i \in \mathscr{I}^{[1:N]}} \eta_i = \alpha. \tag{2.26}$$

Then system (2.15) is periodically ϑ-stabilizable and $\|\mathbb{F}_{\vartheta(x)}x\|_2 < \lambda\|x\|_2$ holds for all $x \in \mathbb{R}^n$, with

$$\vartheta = \vartheta(x) = \arg \min_{j \in \mathcal{I}[1:N]} (x^T \mathbb{F}_j^T \mathbb{F}_j x),$$

and $\lambda = \alpha^{-1/2}$. Given $x(t) = x$, the stabilizing control policy is defined from (2.25) within an horizon of length $|\vartheta|$ as

$$\nu(x,k) = (\sigma(x,k),\ K(x,k)) = \left(\vartheta_k,\ K_k^{\vartheta}\right) \tag{2.27}$$

to be applied at time $t + k - 1$, for all $k \in \mathbb{N}_{|\vartheta|}$.

From Theorem 2.10, the value α is related to λ and then could serve for obtaining the fastest decreasing rate, for a given N, by solving the following LMI problem

$$\begin{aligned} \alpha = & \sup_{\alpha,\eta,G_{j,k},Z_{j,k},R_j} \sum_{j \in \mathcal{I}[1:N]} \eta_j \\ & \text{s.t. } (2.22) - (2.23) - (2.24), \end{aligned} \tag{2.28}$$

with η, $G_{j,k}$, $Z_{j,k}$, R_j as defined in Theorem 2.9.

Remark 2.8 A nonconvex control Lyapunov function $V(x)$, decreasing at every step, analogous to (2.13), and a state-dependent control policy $\nu(x)$ as in (2.16) can be defined by solving on-line an LMI problem, see [11].

The interested reader is referred to [11] for a detailed comparison analysis, in terms of conservatism and complexity, of this approach with respect to methods from the literature, such as those presented in [9, 14, 27, 28].

2.5 Numerical Examples

Some illustrative examples, taken from [10–12], follow.

Example 2.2 Consider the system (2.1) with $q = 4$, $n = 2$ and

$$A_1 = \begin{bmatrix} 1.5 & 0 \\ 0 & -0.8 \end{bmatrix}, \quad A_2 = 1.1\, R(\frac{2\pi}{5}) \quad A_3 = 1.05\, R(\frac{2\pi}{5} - 1), \quad A_4 = \begin{bmatrix} -1.2 & 0 \\ 1 & 1.3 \end{bmatrix}.$$

The matrices A_i, with $i \in \mathbb{N}_4$, are not Schur, which implies that the system (2.1) is not stabilizable by any constant switching law. We apply Algorithm 1 with $\Omega = \mathbb{B}^2$. The algorithm stops at the fifth iteration. Figure 2.2 left, emphasizes that $\mathbb{B}^2$ is included in $\bigcup_{k \in \mathbb{N}_5} \Omega_k$. A stabilizing switching law and the related Lyapunov function are given in Fig. 2.2 right, for the initial condition $x_0 = [-3\ \ 3]^T$.

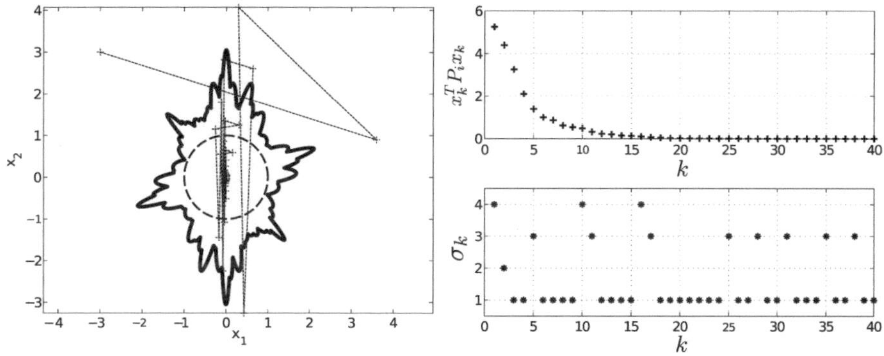

Fig. 2.2 Left: ball $\mathbb{B}^2$ in dashed and $\bigcup_{k\in\mathbb{N}_5}\Omega_k$ in solid line. Trajectory starting from $x_0 = (-3, 3)^T$ in dotted line. Right: Lyapunov function and switching control law in time

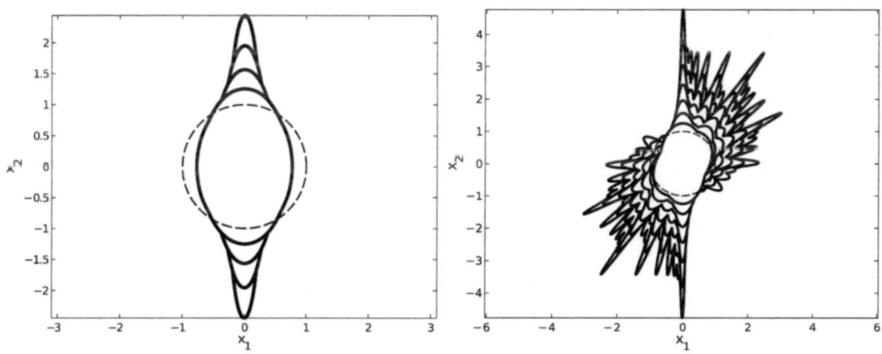

Fig. 2.3 Ball $\mathbb{B}^2$ in dashed and $\cup_{k\in\mathbb{N}_i}\Omega_k$, for $i \in \mathbb{N}_i$ in solid line for Example 2.3, with $\theta = 0$ (left) and $\theta = \frac{\pi}{5}$ (right)

Example 2.3 Consider the system (2.1) with $q = 2$, $n = 2$ and

$$A_1 = \begin{bmatrix} 1.3 & 0 \\ 0 & 0.9 \end{bmatrix} R(\theta), \quad A_2 = \begin{bmatrix} 1.4 & 0 \\ 0 & 0.8 \end{bmatrix},$$

non-Schur. From Fig. 2.3, left, one can infer that the system is not stabilizable if $\theta = 0$. Nevertheless, taking $\theta = \frac{\pi}{5}$, Algorithm 1 stops after seven steps implying the stabilizability of the system, see Fig. 2.3, right.

Example 2.4 Consider (2.1) with $q = 2$, $n = 2$, $x_0 = [-3\ \ 3]^T$ and the non-Schur matrices

$$A_1 = 1.01R\left(\tfrac{\pi}{5}\right), \quad A_2 = \begin{bmatrix} -0.6 & -2 \\ 0 & -1.2 \end{bmatrix}.$$

Four switching laws are designed and compared: the geometric condition given in Theorem 2.2, proving the stabilizability of the system; the min-switching strategy

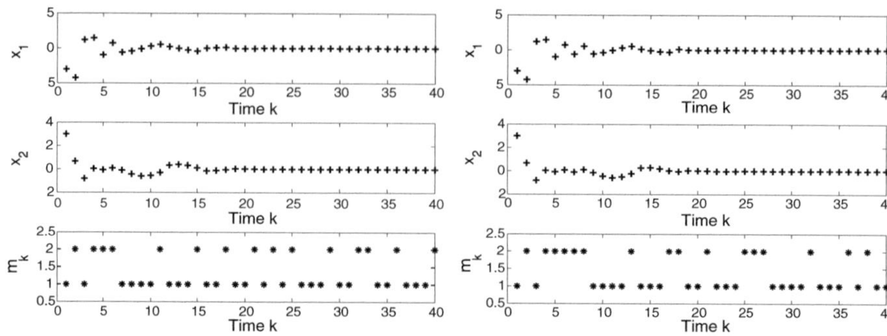

Fig. 2.4 State evolution and switching control induced by the geometric condition (2.6) (left), and min-switching control (2.10)–(2.12) (right)

(2.10)–(2.12) related to the LMI condition (2.8); the switching control law given in Proposition 2.7 and the periodic switching law, that exists from Theorem 2.7.

As noticed in [10, 14], for systems with $q = 2$ the Lyapunov-Metzler inequalities become two linear matrix inequalities once two parameters, both contained in [0, 1], are fixed. Such LMIs have been checked for this example to be infeasible on a grid of these two parameters, with step of 0.01. It is then reasonable to conclude that the Lyapunov-Metzler inequalities are infeasible for this numerical example.

Then, an iterative procedure is applied to determine $N \in \mathbb{N}$ such that (2.6) is satisfied. The result is that (2.6) holds with $N = 5$ and then the homogeneous function induced by the obtained set is a control Lyapunov function and the related min-switching rule is a stabilizing law. The state evolution and the switching law are depicted in Fig. 2.4, left.

The LMI condition (2.8) is solved with $N = 7$ and the min-switching law (2.10)–(2.12) is applied to the system at first. The control results in the concatenation of elements of $\mathscr{I}^{[1,7]}$, respectively of lengths $\{7, 6, 5, 7, 7, \ldots\}$. The time-varying length of the switching subsequences is a consequence of the state dependence of the min-switching strategy. The resulting behavior is depicted in Fig. 2.4, right. Then, the control law defined in Proposition 2.7, namely (2.14) with $\lambda = 0.9661$, is applied and the result is shown in Fig. 2.5, left. The value of λ is obtained by solving the optimization problem described in Remark 2.6.

The periodic switching law of length $M = 4$ is then obtained, by searching the shorter sequence of switching modes which yields a Schur matrix $\mathbb{A}_i$. The resulting evolution is represented in Fig. 2.5, right.

Finally a comparison between the switching laws is provided in Fig. 2.6, where the time-evolution of the Euclidean distance of the state from the origin is depicted.

Example 2.5 Consider Example 2 in [27], that is a 4-dimensional system with 4 modes whose matrices are

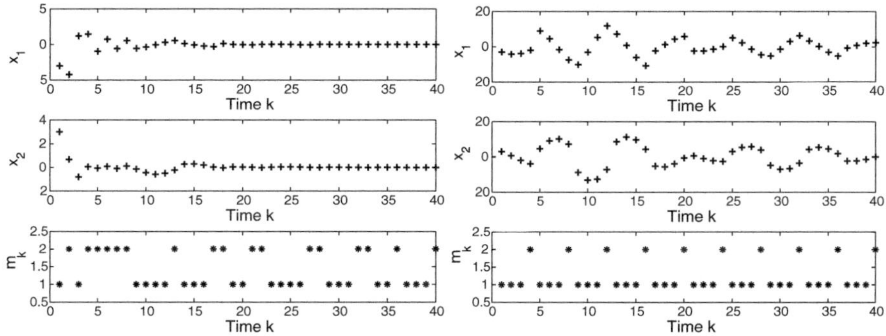

Fig. 2.5 State evolution and min-switching control (2.14) (left) and periodic switching control with $M = 4$ (right)

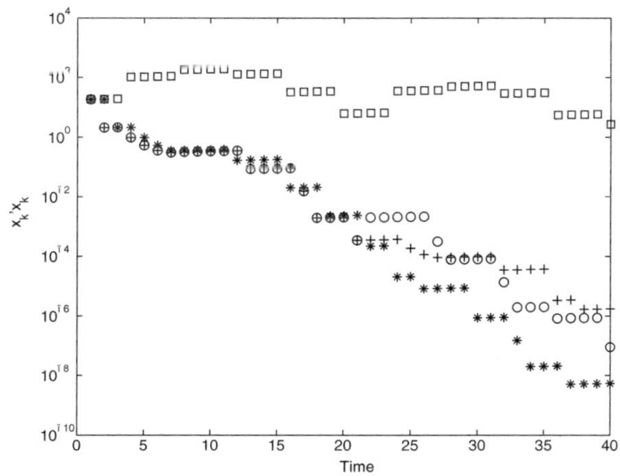

Fig. 2.6 Comparison between the evolution of the Euclidean norm of the state for the different switching laws: induced by geometric condition (2.6) (star); min-switching law (2.10)–(2.12) (cross); min-switching control (2.14) (circle) and periodic rule (square)

$$A_1 = \begin{bmatrix} 0.5 & -1 & 2 & 3 \\ 0 & -0.5 & 2 & 4 \\ 0 & -1 & 2.5 & 2 \\ 0 & 0 & 0 & 1.5 \end{bmatrix}, \; A_2 = \begin{bmatrix} -0.5 & -1 & 2 & 1 \\ 0 & 1.5 & -2 & 0 \\ 0 & 0 & 0.5 & 0 \\ -2 & -1 & 2 & 2.5 \end{bmatrix}, \; A_3 = \begin{bmatrix} 1.5 & 0 & 0 & 0 \\ 1 & 1 & 0.5 & -0.5 \\ 0 & 0.5 & 1 & -0.5 \\ 1 & 0 & 0 & 0.5 \end{bmatrix}$$
$$A_4 = \begin{bmatrix} 0.5 & 1 & 0 & 0 \\ 0 & 0.5 & 0 & 0 \\ 0 & 0 & 0.5 & 0 \\ 0 & 2 & -2 & 0.5 \end{bmatrix}, \quad B_1 = \begin{bmatrix} 1 \\ 2 \\ 3 \\ 4 \end{bmatrix}, \quad B_2 = \begin{bmatrix} 4 \\ 3 \\ 2 \\ 1 \end{bmatrix}, \quad B_3 = \begin{bmatrix} 4 \\ 3 \\ 2 \\ 1 \end{bmatrix}, \quad B_4 = \begin{bmatrix} 1 \\ 2 \\ 3 \\ 4 \end{bmatrix}. \tag{2.29}$$

The conditions of Theorem 2.10 are satisfied with horizon $N = 3$. Besides the inherent computational benefit of having a stabilization condition in form of LMI with respect to the algorithmic method presented in [27], also the control obtained is substantially simpler and more efficient. Actually, in [27] stabilizability is proved by means of an algorithm which inspects control horizons of length 7 resulting in a piecewise quadratic function determined by 13 matrices. Moreover, a much faster

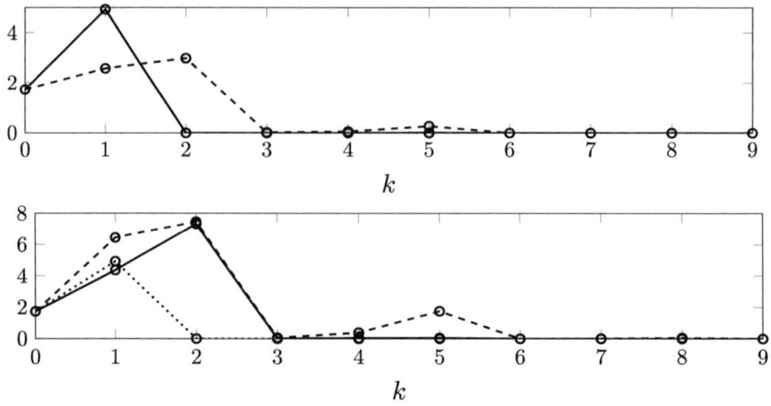

Fig. 2.7 Evolutions of $\|x\|_2$ with control (2.27), in solid, with min-switching of Remark 2.8 in dashed and with the periodic control in dotted with (2.29) (top) and with A_4 multiplied by 2.5, (bottom). In the top figure the solid and dotted lines are overlapped

convergence rate is obtained by solving the LMI problem (2.28), if compared with the results in [27], see Fig. 2.7 where $x_0 = [1\ 1\ 0\ -1]^T$ and Fig. 4 in [27], top.

Finally, A_4 being Schur, with 4 eigenvalues equal to 0.5, a trivial stabilizing solution exists. Then we define a new A_4 by multiplying the one in (2.29) by 2.5. All the eigenvalues of the new A_4 are then equal to 1.25. The evolutions of the Euclidean norm of the state, for $x_0 = [1\ 1\ 0\ -1]^T$, under the obtained controls are depicted in Fig. 2.7, bottom.

2.6 Conclusions

We considered the problems of stabilizability and control co-design for switched linear systems. Via a set-theory approach, a geometric necessary and sufficient condition for the stabilizability have been provided, proving that the family of nonconvex, homogeneous functions induced by a C*-set is a universal class of Lyapunov functions. Then, a novel LMI condition has been presented that overcomes the computational issues related to the geometric condition. Such a condition, together with others from the literature, have been analyzed and compared in terms of conservatism and computational complexity. Finally, an LMI condition is given for control co-design that is proved to be necessary and sufficient for the stabilizability of switched systems that admit periodic control policies.

References

1. Antunes, D., Heemels, W.P.M.H.: Linear quadratic regulation of switched systems using informed policies. IEEE Trans. Autom. Control **62**, 2675–2688 (2017)
2. Bertsekas, D.P.: Infinite-time reachability of state-space regions by using feedback control. IEEE Trans. Autom. Control **17**, 604–613 (1972)
3. Blanchini, F., Miani, S.: Set-Theoretic Methods in Control. Birkhäuser (2008)
4. Blanchini, F.: Ultimate boundedness control for discrete-time uncertain systems via set-induced Lyapunov functions. IEEE Trans. Autom. Control **39**, 428–433 (1994)
5. Blanchini, F.: Nonquadratic Lyapunov functions for robust control. Automatica **31**, 451–461 (1995)
6. Blanchini, F., Savorgnan, C.: Stabilizability of switched linear systems does not imply the existence of convex Lyapunov functions. Automatica **44**, 1166–1170 (2008)
7. Bochnak, J., Coste, M., Roy, M.-F.: Real Algebraic Geometry. Springer, Berlin (1998)
8. Daafouz, J., Riedinger, P., Iung, C.: Stability analysis and control synthesis for switched systems : a switched Lyapunov function approach. IEEE Trans. Autom. Control **47**, 1883–1887 (2002)
9. Deaecto, G.S., Geromel, J.C., Daafouz, J.: Dynamic output feedback H_∞ control of switched linear systems. Automatica **47**(8), 1713–1720 (2011)
10. Fiacchini, M., Jungers, M.: Necessary and sufficient condition for stabilizability of discrete-time linear switched systems: a set-theory approach. Automatica **50**(1), 75–83 (2014)
11. Fiacchini, M., Tarbouriech, M.: Control co-design for discrete-time switched linear systems. Automatica **82**, 181–186 (2017)
12. Fiacchini, M., Girard, A., Jungers, M.: On the stabilizability of discrete-time switched linear systems: novel conditions and comparisons. IEEE Trans. Autom. Control **61**(5), 1181–1193 (2016)
13. Geromel, J.C., Colaneri, P.: Stability and stabilization of continuous-time switched linear systems. SIAM J. Control Optim. **45**(5), 1915–1930 (2006)
14. Geromel, J.C., Colaneri, P.: Stability and stabilization of discrete-time switched systems. Int. J. Control **79**(7), 719–728 (2006)
15. Jungers, R.M.: The Joint Spectral Radius: Theory and Applications. Springer, Heidelberg (2009)
16. Kolmanovsky, I., Gilbert, E.G.: Theory and computation of disturbance invariant sets for discrete-time linear systems. Math. Probl. Eng. **4**, 317–367 (1998)
17. Kruszewski, A., Bourdais, R., Perruquetti, W.: Converging algorithm for a class of BMI applied on state-dependent stabilization of switched systems. Nonlinear Anal. Hybrid Syst. **5**, 647–654 (2011)
18. Lee, J.W., Dullerud, G.E.: Uniformly stabilizing sets of switching sequences for switched linear systems. IEEE Trans. Autom. Control **52**, 868–874 (2007)
19. Liberzon, D.: Switching in Systems and Control. Birkhäuser, Boston (2003)
20. Lin, H., Antsaklis, P.J.: Stability and stabilizability of switched linear systems: a survey of recent results. IEEE Trans. Autom. Control **54**(2), 308–322 (2009)
21. Margaliot, M.: Stability analysis of switched systems using variational principles: an introduction. Automatica **42**, 2059–2077 (2006)
22. Molchanov, A.P., Pyatnitskiy, Y.S.: Criteria of asymptotic stability of differential and difference inclusions encounterd in control theory. Syst. Control Lett. **13**, 59–64 (1989)
23. Rubinov, A.M., Yagubov, A.A.: The space of star-shaped sets and its applications in nonsmooth optimization. In: Demyanov, V.F., Dixon, L.C.W. (eds.) Quasidifferential Calculus, pp. 176–202. Springer, Heidelberg (1986)
24. Sun, Z., Ge, S.S.: Stability Theory of Switched Dynamical Systems. Springer, Berlin (2011)
25. VanAntwerp, J.G., Braatz, R.D.: A tutorial on linear and bilinear matrix inequalities. J. Process Control **10**(4), 363–385 (2000)
26. Wicks, M.A., Peleties, P., De Carlo, R.A.: Construction of piecewise Lyapunov functions for stabilizing switched systems. Proceedings of the 33rd IEEE Conference on Decision and Control, pp. 3492–3497 (1994)

27. Zhang, W., Abate, A., Hu, J., Vitus, M.P.: Exponential stabilization of discrete-time switched linear systems. Automatica **45**(11), 2526–2536 (2009)
28. Zhang, W., Hu, J., Abate, A.: Infinite-horizon switched lqr problems in discrete time: a suboptimal algorithm with performance analysis. IEEE Trans. Autom. Control **57**(7), 1815–1821 (2012)

Chapter 3
Stability Analysis of Singularly Perturbed Switched Linear Systems

J. Ben Rejeb, I.-C. Morărescu, A. Girard and J. Daafouz

Abstract This chapter proposes a methodology for stability analysis of singularly perturbed switched linear systems. We emphasize that, besides the stability of each subsystem, we need a dwell-time condition to guarantee the overall system stability. The main results of this study provide a characterization of an upper bound on the dwell time ensuring the overall system's stability. Remarkably, this bound is the sum of two terms. The first one is an upper bound on the dwell time ensuring stability of the reduced-order linear switched system, which is zero if all the reduced modes share a common Lyapunov function. The magnitude of the second term is of order of the parameter defining the ratio between the two timescales of the singularly perturbed system.

3.1 Introduction

Switched systems are a class of hybrid systems consisting of several dynamical subsystems and a switching signal that specifies the active subsystem at each instant of time. They are encountered in many engineering applications, such as motor engine control [1] or networked control systems [2]. Significant results related to stability analysis of switched systems can be found in [3–5]. A remarkable fact is that even

J. Ben Rejeb · I.-C. Morărescu (✉) · J. Daafouz
Université de Lorraine, CNRS, CRAN, F-54000 Nancy, France
e-mail: Constantin.Morarescu@univ-lorraine.fr

J. Ben Rejeb
e-mail: Jihene.Ben-Rejeb@univ-lorraine.fr

J. Daafouz
e-mail: Jamal.Daafouz@univ-lorraine.fr

A. Girard
Laboratoire des Signaux et Systèmes (L2S), CNRS, CentraleSupélec,
Université Paris-Sud, Université Paris-Saclay, 3, rue Joliot-Curie,
91192 cedex, Gif-sur-Yvette, France
e-mail: Antoine.Girard@l2s.centralesupelec.fr

S. Tarbouriech et al. (eds.), *Control Subject to Computational and Communication Constraints*, Lecture Notes in Control and Information Sciences 475, https://doi.org/10.1007/978-3-319-78449-6_3

when all the subsystems are asymptotically stable, the overall switched system may diverge under certain switching signals [6]. Nevertheless, when all the switching modes share a common Lyapunov function, the overall system is asymptotically stable independently of the switching signal.

Another feature that characterizes many real systems is the presence of processes that evolve on different timescales [7–10]. In this case, the standard stability analysis becomes more difficult and singular perturbation theory [11, 12] has to be used. This theory is based on Tikhonov approach that proposes to approximate the dynamics by decoupling the slow dynamical processes from the faster ones. The stability analysis is done separately for each timescale and under appropriate assumptions one can conclude on the stability of the overall system. Significant results related to stability analysis and approximation of solutions of singularly perturbed systems can be found in [13–15].

Stability analysis and stabilization of singularly perturbed linear switched systems are considered in [16, 17]. Interestingly, it is shown in [17] that even though the switched dynamics on each timescale are stable, the overall system may be destabilized by fast switching signals. Clearly, this is in contrast with classical results on continuous singularly perturbed linear systems [12] and is a motivation for developing dedicated techniques for stability analysis of singularly perturbed hybrid systems. Moreover, the switching signal can destabilize the overall dynamics even if all the reduced-order slow systems share a common Lyapunov function [17]. This contradicts the intuition that stabilizing the switched reduced-order systems we obtain stability of original system.

In this paper, we present a different stability methodology that results in a better characterization of the upper bound on the dwell time ensuring the overall system's stability. This bound is given as the sum of two terms. The first one corresponds to an upper bound on the minimum dwell time ensuring the stability of the reduced-order linear hybrid system describing the slow dynamics. The order of magnitude of the second term is determined by that of the parameter ε defining the ratio between the two timescales of the singularly perturbed system. In particular, it follows that when the reduced-order system has a common quadratic Lyapunov function, the first term is zero and the minimum dwell time ensuring the stability of the overall system goes to zero as fast as ε when the timescale parameter ε goes to zero.

Notation

Throughout this paper, $\mathbb{R}_+$, $\mathbb{R}^n$, and $\mathbb{R}^{n\times m}$ denote, respectively, the set of nonnegative real numbers, the n-dimensional Euclidean space, and the set of all $n \times m$ real matrices. The identity matrix of dimension n is denoted by $\mathbf{I}_n$. We also denote by $\mathbf{0}_{n,m} \in \mathbb{R}^{n\times m}$ the matrix whose components are all 0. For a matrix $A \in \mathbb{R}^{n\times n}$, $\|A\|$ denotes the spectral norm i.e., induced 2-norm. $A \geq \mathbf{0}$ ($A \leq \mathbf{0}$) means that A is positive semidefinite (negative semidefinite). We write $A^\top$ and A^{-1} to, respectively, denote the transpose and the inverse of A. For a symmetric matrix $A \geq \mathbf{0}$, $A^{\frac{1}{2}}$ is the unique symmetric matrix $B \geq 0$ such that $B^2 = A$. The matrix A is said to be Hurwitz if all its eigenvalues have negative real parts. A is said to be Schur if all its eigenvalues have modulus smaller than one. The matrix A is said to be positive if all its coefficients are positive. We also use $x(t^-) = \lim_{\delta \to 0,\ \delta > 0} x(t - \delta)$. Given a function $\eta : (0, \varepsilon^*) \to \mathbb{R}$, we say that $\eta(\varepsilon) = \mathcal{O}(\varepsilon)$ if and only if there exists $\varepsilon_0 \in (0, \varepsilon^*)$ and $c > 0$, such that for all $\varepsilon \in (0, \varepsilon_0)$, $|\eta(\varepsilon)| \leq c\varepsilon$.

3.2 Problem Formulation

In this paper, we consider a switched system of the form:

$$\begin{pmatrix} \dot{x}(t) \\ \varepsilon \dot{z}(t) \end{pmatrix} = A^{\sigma_k} \begin{pmatrix} x(t) \\ z(t) \end{pmatrix}, \quad \forall t \in (t_k, t_{k+1}],\ k \in \mathbb{N} \tag{3.1}$$

where $x(t) \in \mathbb{R}^{n_x}$, $z(t) \in \mathbb{R}^{n_z}$, $0 = t_0 < t_1 < \ldots$ are the instants of switches, $\sigma_k \in \mathcal{I}$ with $\mathcal{I}$ finite set of indices, A^i are the matrices of appropriate dimensions for all $i \in \mathcal{I}$, and $\varepsilon > 0$ is a small parameter characterizing the timescale separation between the slow dynamics of x and the fast dynamics of z.

For $i, i' \in \mathcal{I}$ let

$$A^i = \begin{pmatrix} A^i_{11} & A^i_{12} \\ A^i_{21} & A^i_{22} \end{pmatrix},$$

where $A^i_{11} \in \mathbb{R}^{n_x \times n_x}$, and A^i_{22}, A^i_{12}, A^i_{21} are of appropriate dimensions.

Let us impose the following standard assumption [12] in the singular perturbation theory framework:

Assumption 3.1 A^i_{22} is non-singular for all $i \in \mathcal{I}$.

Then, we perform the following time-dependent change of variable:

$$\begin{pmatrix} x(t) \\ y(t) \end{pmatrix} = P_{\sigma_k} \begin{pmatrix} x(t) \\ z(t) \end{pmatrix}, \quad \forall t \in [t_k, t_{k+1}),\ k \in \mathbb{N} \tag{3.2}$$

where, for all $i \in \mathcal{I}$

$$P_i = \begin{pmatrix} \mathbf{I}_{n_x} & \mathbf{0}_{n_x,n_z} \\ (A^i_{22})^{-1} A^i_{21} & \mathbf{I}_{n_z} \end{pmatrix}.$$

It is worth noting that the matrix P_i is invertible and for all $i \in \mathcal{I}$

$$P_i^{-1} = \begin{pmatrix} \mathbf{I}_{n_x} & \mathbf{0}_{n_x,n_z} \\ -(A^i_{22})^{-1} A^i_{21} & \mathbf{I}_{n_z} \end{pmatrix}.$$

The switching dynamics (3.1) in the variables x, y becomes

$$\begin{pmatrix} \dot{x}(t) \\ \varepsilon \dot{y}(t) \end{pmatrix} = \begin{pmatrix} A_0^{\sigma_k} & B_1^{\sigma_k} \\ \varepsilon B_2^{\sigma_k} & A_{22}^{\sigma_k} + \varepsilon B_3^{\sigma_k} \end{pmatrix} \begin{pmatrix} x(t) \\ y(t) \end{pmatrix}, \quad \forall t \in [t_k, t_{k+1}), \ k \in \mathbb{N} \tag{3.3}$$

where for all $i \in \mathscr{I}$ one has

$$\begin{aligned} A_0^i &= A_{11}^i - A_{12}^i (A_{22}^i)^{-1} A_{21}^i, & B_1^i &= A_{12}^i, \\ B_2^i &= (A_{22}^i)^{-1} A_{21}^i A_0^i, & B_3^i &= (A_{22}^i)^{-1} A_{21}^i A_{12}^i. \end{aligned}$$

It is noteworthy that, when (3.1) is expressed in the variables x, y, at all the switching instants t_k, $k \in \mathbb{N}$, an instantaneous jump will occur:

$$\begin{pmatrix} x(t_k) \\ y(t_k) \end{pmatrix} = R^{\sigma_{k-1} \to \sigma_k} \begin{pmatrix} x(t_k^-) \\ y(t_k^-) \end{pmatrix}, \quad \forall k \geq 1 \tag{3.4}$$

where

$$R^{i \to j} = \begin{pmatrix} \mathbf{I}_{n_x} & \mathbf{0}_{n_x, n_z} \\ (A_{22}^j)^{-1} A_{21}^j - (A_{22}^i)^{-1} A_{21}^i & \mathbf{I}_{n_z} \end{pmatrix}.$$

Therefore, the switching linear system (3.1) is equivalent with the switching linear impulsive system (3.3)–(3.4). One can formally introduce the reduced-order switched system with single timescale:

$$\dot{x}(t) = A_0^{\sigma_k} x(t), \ \forall t \in [t_k, t_{k+1}), \ k \in \mathbb{N}. \tag{3.5}$$

The goal of this chapter is to investigate the stability of the singularly perturbed switched linear system (3.3)–(3.4) for small values of the parameter ε, and its relation to the stability of the reduced-order model (3.5). In particular, we aim at characterizing an upper bound on the minimum dwell time ensuring stability.

3.3 Preliminaries

In this section, we provide some results on the Lyapunov stability of singularly perturbed linear systems, which will be used in the next sections to prove the main results of the paper concerning the stability of (3.3)–(3.4).

Let us consider the singularly perturbed linear system:

$$\begin{cases} \dot{x}(t) = A_{11} x(t) + A_{12} z(t) \\ \varepsilon \dot{z}(t) = A_{21} x(t) + A_{22} z(t), \end{cases} \tag{3.6}$$

where $x(t) \in \mathbb{R}^{n_x}$, $z(t) \in \mathbb{R}^{n_z}$ and $\varepsilon > 0$ is a small parameter. Let us assume that A_{22} is non-singular and proceed with the change of variable

$$\begin{pmatrix} x(t) \\ y(t) \end{pmatrix} = \begin{pmatrix} \mathbf{I}_{n_x} & \mathbf{0}_{n_x,n_z} \\ A_{22}^{-1}A_{21} & \mathbf{I}_{n_z} \end{pmatrix} \begin{pmatrix} x(t) \\ z(t) \end{pmatrix}. \tag{3.7}$$

In the variables x, y the system becomes

$$\begin{cases} \dot{x}(t) = A_0 x(t) + B_1 y(t) \\ \varepsilon \dot{y}(t) = A_{22} y(t) + \varepsilon (B_2 x(t) + B_3 y(t)), \end{cases} \tag{3.8}$$

where

$$\begin{aligned} A_0 &= A_{11} - A_{12} A_{22}^{-1} A_{21}, \ B_1 = A_{12}, \\ B_2 &= A_{22}^{-1} A_{21} A_0, \ B_3 = A_{22}^{-1} A_{21} A_{12}. \end{aligned}$$

Let us make the following assumption:

Assumption 3.2 A_0 and A_{22} are Hurwitz.

Under the previous assumption, there exist symmetric positive definite matrices $Q_s \geq \mathbf{I}_{n_x}$, $Q_f \geq \mathbf{I}_{n_z}$ and positive numbers λ_s and λ_f such that

$$\begin{aligned} A_0^\top Q_s + Q_s A_0^\top &\leq -2\lambda_s Q_s \\ A_{22}^\top Q_f + Q_f A_{22}^\top &\leq -2\lambda_f Q_f \end{aligned}$$

Then, let us define $b_1 = \| Q_s^{\frac{1}{2}} B_1 Q_f^{-\frac{1}{2}} \|$, $b_2 = \| Q_f^{\frac{1}{2}} B_2 Q_s^{-\frac{1}{2}} \|$ and $b_3 = \| Q_f^{\frac{1}{2}} Q_f B_3 Q_f^{-\frac{1}{2}} \|$.

Proposition 3.1 *Under Assumption 3.2,*

$$V(x, y) = x^\top Q_s x + y^\top Q_f y$$

is a Lyapunov function for system (3.8) *for all* $\varepsilon \in (0, \varepsilon_1]$ *where*

$$\varepsilon_1 = \frac{\lambda_f}{\frac{(b_1+b_2)^2}{4\lambda_s} + b_3}. \tag{3.9}$$

Proof By computing the time derivative of V along the trajectories of (3.8), one has

$$\begin{aligned} \dot{V} &= 2x^\top Q_s \dot{x} + 2y^\top Q_f \dot{y} \\ &= 2x^\top Q_s A_0 x + \frac{2}{\varepsilon} y^\top Q_f A_{22} y + 2x^\top Q_s B_1 y + 2y^\top Q_f B_2 x + 2y^\top Q_f B_3 y \\ &\leq -2\lambda_s x^\top Q_s x - \frac{2\lambda_f}{\varepsilon} y^\top Q_f y + 2(b_1 + b_2)\|x\| \|y\| + 2b_3 \|y\|^2 \\ &\leq -2\lambda_s x^\top Q_s x - \frac{2\lambda_f}{\varepsilon} y^\top Q_f y + 2(b_1 + b_2)\sqrt{x^\top Q_s x}\sqrt{y^\top Q_f y} + 2b_3 y^\top Q_f y, \end{aligned}$$

where $b_1 = \|Q_s B_1\|$, $b_2 = \|Q_f B_2\|$ and $b_3 = \|Q_f B_3\|$. Then, it follows that

$$\dot{V} \leq -(\frac{2\lambda_f}{\varepsilon} - 2b_3 - \frac{(b_1+b_2)^2}{2\lambda_s})y^\top Q_f y.$$

Let $\varepsilon_1 > 0$ be given by

$$\varepsilon_1 = \frac{\lambda_f}{\frac{(b_1+b_2)^2}{4\theta\lambda_s} + b_3} \tag{3.10}$$

Then, for all $\varepsilon \in (0, \varepsilon_1]$,

$$\dot{V} \leq 0 \tag{3.11}$$

■

In the following, let us denote $W_s(t) = \sqrt{x(t)^\top Q_s x(t)}$ and $W_f(t) = \sqrt{y(t)^\top Q_f y(t)}$.

Proposition 3.2 *Under Assumption 3.2, let ε_1 be given by (3.9), then for all $\varepsilon \in (0, \varepsilon_1]$ and $t \geq 0$*

$$W_f(t) \leq W_f(0)e^{-\frac{\lambda_f}{\varepsilon}t} + \varepsilon\beta_1\sqrt{V(0)},$$

where $\beta_1 = \frac{\sqrt{b_2^2+b_3^2}}{\lambda_f}$.

Proof Computing the time derivative of W_f gives

$$\begin{aligned}
\dot{W}_f &= \frac{2y^\top Q_f \dot{y}}{2\sqrt{y^\top Q_f y}} \leq \frac{-\frac{\lambda_f}{\varepsilon}y^\top Q_f y + y^\top Q_f(B_2 x + B_3 y)}{\sqrt{y^\top Q_f y}} \\
&\leq -\frac{\lambda_f}{\varepsilon}W_f + \frac{\|y\|(b_2\|x\| + b_3\|y\|)}{\sqrt{y^\top Q_f y}} \leq -\frac{\lambda_f}{\varepsilon}W_f + (b_2\|x\| + b_3\|y\|) \\
&\leq -\frac{\lambda_f}{\varepsilon}W_f + \sqrt{b_2^2+b_3^2}\sqrt{\|x\|^2+\|y\|^2} \leq -\frac{\lambda_f}{\varepsilon}W_f + \sqrt{b_2^2+b_3^2}\sqrt{V}
\end{aligned}$$

Using (3.11) it follows that

$$\dot{W}_f(t) \leq -\frac{\lambda_f}{\varepsilon}W_f(t) + \sqrt{b_2^2+b_3^2}\sqrt{V(0)}.$$

Integrating this we get the desired inequality. ■

Proposition 3.3 *Under Assumption 3.2, let ε_1 be given by (3.9), and let $\varepsilon_2 \in (0, \varepsilon_1] \cap (0, \frac{\lambda_f}{\lambda_s})$ then for all $\varepsilon \in (0, \varepsilon_2]$ and $t \geq 0$*

$$W_s(t) \leq W_s(0)e^{-\lambda_s t} + \varepsilon\beta_2 W_f(0) + \varepsilon\beta_3\sqrt{V(0)},$$

where $\beta_2 = \frac{b_1}{\lambda_f - \varepsilon_2\lambda_s}$ and $\beta_3 = \frac{b_1\beta_1}{\lambda_s}$.

Proof Computing the time derivative of W_s gives

$$\begin{aligned}\dot{W}_s &= \frac{2x^\top Q_s \dot{x}}{2\sqrt{x^\top Q_s x}} \leq \frac{-\lambda_s x^\top Q_s x + x^\top Q_s B_1 y}{\sqrt{x^\top Q_s x}} \\ &\leq -\lambda_s W_s + \frac{b_1 \|x\| \|y\|}{\sqrt{x^\top Q_s x}} \leq -\lambda_s W_s + b_1 W_f.\end{aligned}$$

Using Proposition 3.2, one gets

$$\dot{W}_s(t) \leq -\lambda_s W_s(t) + b_1 W_f(0) \mathrm{e}^{\frac{-\lambda_f}{\varepsilon}t} + \varepsilon b_1 \beta_1 \sqrt{V(0)}.$$

Then, we have

$$\begin{aligned}W_s(t) \leq\, & \mathrm{e}^{-\lambda_s t} W_s(0) + b_1 W_f(0) \int_0^t \mathrm{e}^{-\frac{\lambda_f}{\varepsilon}s} \mathrm{e}^{-\lambda_s(t-s)} ds \\ & + \varepsilon b_1 \beta_1 \sqrt{V(0)} \int_0^t \mathrm{e}^{-\lambda_s(t-s)} ds \\ \leq\, & \mathrm{e}^{-\lambda_s t} W_s(0) + \frac{b_1}{\frac{\lambda_f}{\varepsilon} - \lambda_s} W_f(0) \left(\mathrm{e}^{-\lambda_s t} - \mathrm{e}^{-\frac{\lambda_f}{\varepsilon}t}\right) \\ & + \frac{\varepsilon b_1 \beta_1}{\lambda_s} \sqrt{V(0)} \left(1 - \mathrm{e}^{-\lambda_s t}\right).\end{aligned}$$

Then, $\varepsilon \leq \varepsilon_2 < \frac{\lambda_f}{\lambda_s}$ gives

$$W_s(t) \leq \mathrm{e}^{-\lambda_s t} W_s(0) + \frac{b_1 \varepsilon}{\lambda_f - \varepsilon_2 \lambda_s} W_f(0) + \frac{\varepsilon b_1 \beta_1}{\lambda_s} \sqrt{V(0)}.$$

■

3.4 Stability Analysis

We now study the stability of system (3.3)–(3.4). In the rest of the chapter, we impose the following additional assumption on the singularly perturbed system at hand, related to the stability of the slow and fast dynamics of each mode.

Assumption 3.3 A_0^i and A_{22}^i are Hurwitz for all $i \in \mathscr{I}$.

From the previous assumption, we can deduce that there exist symmetric positive definite matrices $Q_s^i \geq \mathbf{I}_{n_x}$, $Q_f^i \geq \mathbf{I}_{n_z}$, $i \in \mathscr{I}$, and positive numbers λ_s^i and λ_f^i such that for all $i \in \mathscr{I}$:

$$\begin{aligned}A_0^{i\top} Q_s^i + Q_s^i A_0^i &\leq -2\lambda_s^i Q_s^i \\ A_{22}^{i\top} Q_f^i + Q_f^i A_{22}^i &\leq -2\lambda_f^i Q_f^i.\end{aligned}$$

We denote $\lambda_s = \min_{i\in\mathscr{I}} \lambda_s^i$ and $\lambda_f = \min_{i\in\mathscr{I}} \lambda_f^i$. For each $i \in \mathscr{I}$, let $b_1^i = \|(Q_s^i)^{\frac{1}{2}} B_1 (Q_f^i)^{-\frac{1}{2}}\|$, $b_2^i = \|(Q_f^i)^{\frac{1}{2}} B_2 (Q_s^i)^{-\frac{1}{2}}\|$, $b_3^i = \|(Q_f^i)^{\frac{1}{2}} Q_f B_3 (Q_f^i)^{-\frac{1}{2}}\|$ and $b_j = \max_{i\in\mathscr{I}} b_j^i$, $j = 1, \ldots, 3$.

Let ε_1 be given by (3.9), then it follows from Proposition 3.1 that the linear dynamics of (3.3) are all Lyapunov stable, for $\varepsilon \in (0, \varepsilon_1]$. Let $\varepsilon_2 \in (0, \varepsilon_1] \cap (0, \frac{\lambda_f}{\lambda_s})$ and $\beta_1, \beta_2, \beta_3$ be defined as in Propositions 3.2 and 3.3. Let us also introduce the following functions:

$$\begin{cases} W_s(t) = \sqrt{x(t)^\top Q_s^{\sigma_k} x(t)} \\ W_f(t) = \sqrt{y(t)^\top Q_f^{\sigma_k} y(t)} \end{cases}, \quad \forall t \in [t_k, t_{k+1}), \; k \in \mathbb{N}.$$

Lemma 3.1 *Under Assumption 3.3, let $\varepsilon \in (0, \varepsilon_2]$, and let $\tau_k = t_{k+1} - t_k$ for a sequence $(t_k)_{k\geq 0}$ of event times. Then for all $k \in \mathbb{N}$,*

$$\begin{aligned} W_s(t_{k+1}^-) &\leq W_s(t_k)(e^{-\lambda_s \tau_k} + \varepsilon\beta_3) + W_f(t_k)\varepsilon(\beta_2 + \beta_3) \\ W_f(t_{k+1}^-) &\leq W_s(t_k)\varepsilon\beta_1 + W_f(t_k)\big(e^{-\frac{\lambda_f}{\varepsilon}\tau_k} + \varepsilon\beta_1\big). \end{aligned}$$

Proof This is straightforward from Propositions 3.2 and 3.3 by remarking that $\sqrt{V} \leq W_s + W_f$.

In the following, we complete the characterization of the variation of W_s and W_f by analyzing their behavior when a switch occurs. Let $\gamma_{11}, \gamma_{21}, \gamma_{22}$ be defined as

$$\begin{aligned} \gamma_{11} &= \max_{i,i'\in\mathscr{I}} \|(Q_s^{i'})^{\frac{1}{2}} (Q_s^i)^{-\frac{1}{2}}\|, \\ \gamma_{21} &= \max_{i,i'\in\mathscr{I}} \|(Q_f^{i'})^{\frac{1}{2}} \big((A_{22}^j)^{-1} A_{21}^j - (A_{22}^i)^{-1} A_{21}^i\big)(Q_s^i)^{-\frac{1}{2}}\|, \\ \gamma_{22} &= \max_{i,i'\in\mathscr{I}} \|(Q_f^{i'})^{\frac{1}{2}} (Q_f^i)^{-\frac{1}{2}}\|. \end{aligned} \tag{3.12}$$

Note that $\gamma_{11} = 1$ and $\gamma_{22} = 1$ if and only if $Q_s^i = Q_s^{i'}$ and $Q_f^i = Q_f^{i'}$ for all $i, i' \in \mathscr{I}$. Then, we have the following result.

Lemma 3.2 *Let a sequence $(t_k)_{k\geq 0}$ of event times, then for all $k \geq 1$,*

$$\begin{aligned} W_s(t_k) &\leq \gamma_{11} W_s(t_k^-) \\ W_f(t_k) &\leq \gamma_{21} W_s(t_k^-) + \gamma_{22} W_f(t_k^-). \end{aligned}$$

Proof The first inequality can be proven as follows:

$$\begin{aligned} W_s(t_k) &= \sqrt{x(t_k)^\top Q_s^{\sigma_k} x(t_k)} = \|(Q_s^{\sigma_k})^{\frac{1}{2}} x(t_k)\| \\ &= \|(Q_s^{\sigma_k})^{\frac{1}{2}} x(t_k^-)\| = \|(Q_s^{\sigma_k})^{\frac{1}{2}} (Q_s^{\sigma_{k-1}})^{-\frac{1}{2}} (Q_s^{\sigma_{k-1}})^{\frac{1}{2}} x(t_k^-)\| \\ &\leq \gamma_{11} \|(Q_s^{\sigma_{k-1}})^{\frac{1}{2}} x(t_k)\| = \gamma_{11} W_s(t_k^-). \end{aligned}$$

The second inequality is obtained similarly using the jump map given in (3.4). ■

In order to keep the notation simple, we introduce the positive matrix parameterized by $\tau > 0$:

$$M_\tau = \begin{pmatrix} e^{-\lambda_s \tau} + \varepsilon\beta_3 & \varepsilon(\beta_2 + \beta_3) \\ \varepsilon\beta_1 & e^{-\frac{\lambda_f}{\varepsilon}\tau} + \varepsilon\beta_1 \end{pmatrix}.$$

Let us also consider the positive matrix

$$\Gamma = \begin{pmatrix} \gamma_{11} & 0 \\ \gamma_{21} & \gamma_{22} \end{pmatrix}.$$

Lemma 3.3 *Under Assumption 3.3, let $\varepsilon \in (0, \varepsilon_2]$, and let $\tau_k = t_{k+1} - t_k$ for a sequence $(t_k)_{k\geq 0}$ of event times. Then for all $k \in \mathbb{N}$,*

$$\begin{pmatrix} W_s(t_{k+1}) \\ W_f(t_{k+1}) \end{pmatrix} \leq \Gamma M_{\tau_k} \begin{pmatrix} W_s(t_k) \\ W_f(t_k) \end{pmatrix}.$$

Proof The result is obtained by simply combining Lemmas 3.2 and 3.3. ■

Lemma 3.4 *Under Assumption 3.3, let $\varepsilon \in (0, \varepsilon_2]$ and let $\tau^* \geq 0$ such that the positive matrix ΓM_{τ^*} is Schur. Then, for all sequences $(t_k)_{k\geq 0}$ of event times satisfying the dwell-time property $\tau_k \geq \tau^*$, for all $k \in \mathbb{N}$, the system* (3.3)–(3.4) *is globally asymptotically stable.*

Proof Let τ^* be such that $\tau_k \geq \tau^*$, for all $k \in \mathbb{N}$. From Lemma 3.3, it follows that for all $k \in \mathbb{N}$,

$$\begin{pmatrix} W_s(t_k) \\ W_f(t_k) \end{pmatrix} \leq \Gamma M_{\tau_{k-1}} \dots \Gamma M_{\tau_0} \begin{pmatrix} W_s(t_0) \\ W_f(t_0) \end{pmatrix}.$$

Remarking that the coefficient of the positive matrix M_τ is decreasing with respect to τ, it follows that

$$\begin{pmatrix} W_s(t_k) \\ W_f(t_k) \end{pmatrix} \leq \left(\Gamma M_{\tau^*}\right)^k \begin{pmatrix} W_s(t_0) \\ W_f(t_0) \end{pmatrix}.$$

Hence, if the positive matrix ΓM_{τ^*} is Schur, then both sequences $(W_s(t_k))_{k\geq 0}$ and $(W_f(t_k))_{k\geq 0}$ go to 0, and the system (3.1) is globally asymptotically stable. ■

Hence, the stability of system (3.3)–(3.4) can be investigated by studying the spectral properties of the positive matrix ΓM_{τ^*}. Let us remark that values τ^* such that ΓM_{τ^*} is Schur provide upper bounds on the minimal dwell time between two events that ensure the stability of the singularly perturbed linear hybrid system. In the following, we establish sufficient conditions for deriving such values τ^*.

Theorem 3.4 *Under Assumption 3.3, let $\gamma_{11} > 1$. Then, there exists $\varepsilon^* > 0$ and a function $\eta : (0, \varepsilon^*) \to \mathbb{R}^+$ with $\eta(\varepsilon) = \mathcal{O}(\varepsilon)$, such that for all $\varepsilon \in (0, \varepsilon^*)$, for all* *sequences* $(t_k)_{k\geq 0}$ *of event times satisfying a dwell-time property* $\tau_k \geq \tau^*$, *for all* $k \in \mathbb{N}$, *where*

$$\tau^* > \frac{\ln(\gamma_{11})}{\lambda_s} + \eta(\varepsilon),$$

the system (3.1) *is globally asymptotically stable.*

Proof Let us remark that

$$\Gamma M_{\tau^*} = \begin{pmatrix} \gamma_{11} e^{-\lambda_s \tau^*} + \varepsilon\delta_1 & \varepsilon\delta_2 \\ \gamma_{21} e^{-\lambda_s \tau^*} + \varepsilon\delta_3 & \gamma_{22} e^{-\frac{\lambda_f}{\varepsilon}\tau^*} + \varepsilon\delta_4 \end{pmatrix},$$

where

$$\begin{aligned} \delta_1 &= \gamma_{11}\beta_3, & \delta_2 &= \gamma_{11}(\beta_2 + \beta_3), \\ \delta_3 &= \gamma_{21}\beta_3 + \gamma_{22}\beta_1, & \delta_4 &= \gamma_{21}(\beta_2 + \beta_3) + \gamma_{22}\beta_1. \end{aligned}$$

Moreover, the positive matrix ΓM_{τ^*} is Schur if and only if there exists $p \in \mathbb{R}^2_+$, such that $\left(\Gamma M_{\tau^*}\right)^\top p < p$ (see e.g., [18]). Let us look for p under the form $(1, a\varepsilon)^\top$ with $a > \delta_2$. Then, $\left(\Gamma M_{\tau^*}\right)^\top p < p$ is equivalent to

$$\begin{cases} \gamma_{11} e^{-\lambda_s \tau^*} + \varepsilon\delta_1 + a\varepsilon\gamma_{21} e^{-\lambda_s \tau^*} + a\varepsilon^2\delta_3 < 1 \\ \varepsilon\delta_2 + a\varepsilon\gamma_{22} e^{-\frac{\lambda_f}{\varepsilon}\tau^*} + a\varepsilon^2\delta_4 < a\varepsilon. \end{cases} \tag{3.13}$$

The first inequality of (3.13) is equivalent to

$$\tau^* > \frac{-1}{\lambda_s} \ln\left(\frac{1 - \varepsilon\delta_1 - a\varepsilon^2\delta_3}{\gamma_{11} + a\varepsilon\gamma_{21}}\right) = \frac{\ln(\gamma_{11})}{\lambda_s} + \eta(\varepsilon),$$

where

$$\eta(\varepsilon) = \frac{1}{\lambda_s}\Big(\ln(1 + \frac{a\varepsilon\gamma_{21}}{\gamma_{11}}) - \ln(1 - \varepsilon\delta_1 - a\varepsilon^2\delta_3)\Big). \tag{3.14}$$

It is clear that $\eta(\varepsilon) = \mathcal{O}(\varepsilon)$. Moreover, let us remark that $\eta(\varepsilon)$ is only defined if $1 - \varepsilon\delta_1 - a\varepsilon^2\delta_3 > 0$, that is, if $\varepsilon < \varepsilon_3$ where

$$\varepsilon_3 = \frac{-\delta_1 + \sqrt{\delta_1^2 + 4a\delta_3}}{2a\delta_3}.$$

The second inequality of (3.13) is equivalent to

$$\begin{aligned} &\tau^* > \frac{-\varepsilon}{\lambda_f} \ln\left(\frac{a - \delta_2 - a\varepsilon\delta_4}{a\gamma_{22}}\right) \\ \Longleftrightarrow\ &\tau^* > \frac{\varepsilon}{\lambda_f} \ln\left(\frac{a\gamma_{22}}{a - \delta_2 - a\varepsilon\delta_4}\right). \end{aligned}$$

As $\tau^* > \frac{\ln(\gamma_{11})}{\lambda_s} + \eta(\varepsilon) \geq \frac{\ln(\gamma_{11})}{\lambda_s}$, then the previous inequality holds if

$$\frac{\ln(\gamma_{11})}{\lambda_s} > \frac{\varepsilon}{\lambda_f} \ln\left(\frac{a\gamma_{22}}{a - \delta_2 - a\varepsilon\delta_4}\right). \tag{3.15}$$

By remarking that the right-hand side of the inequality goes to 0 when ε goes to 0, one concludes that that exists $\varepsilon_4 > 0$ such that for all $\varepsilon \in (0, \varepsilon_4)$, (3.15) holds. Then, the theorem is proved by setting $\varepsilon^* = \min(\varepsilon_2, \varepsilon_3, \varepsilon_4)$. ■

Remark 3.1 Theorem 3.4 shows that the dwell time ensuring stability of the singularly perturbed switched system (3.1) can be written as the sum of a constant part and of a function of ε, which goes to 0 when ε goes to 0. Interestingly, the constant part only depends on λ_s and γ_{11}, which can be determined from the reduced-order model (3.5). Moreover, it is easy to show that the system (3.5) is globally asymptotically stable for all switching signals with dwell time $\tau^* > \frac{\ln(\gamma_{11})}{\lambda_s}$.

Theorem 3.4 deals with the case when $\gamma_{11} > 1$. When $\gamma_{11} = 1$, the following result can be proved using similar reasonings as in Theorem 3.4.

Theorem 3.5 *Under Assumption 3.3, let $\gamma_{11} = 1$. Then, there exists $\varepsilon^* > 0$ and a function $\eta : (0, \varepsilon^*) \to \mathbb{R}^+$ with $\eta(\varepsilon) = \mathcal{O}(\varepsilon)$, such that for all $\varepsilon \in (0, \varepsilon^*)$, for all sequences $(t_k)_{k\geq 0}$ of event times satisfying a dwell-time property $\tau_k \geq \tau^*$, for all $k \in \mathbb{N}$, where*

$$\tau^* > \eta(\varepsilon),$$

the system (3.1) *is globally asymptotically stable.*

Remark 3.2 The previous theorems show that when $\gamma_{11} = 1$, the dwell time ensuring stability of the singularly perturbed switched system (3.1) goes to 0 when ε goes to 0. It is interesting to remark that in that case, the reduced-order system (3.5) is globally asymptotically stable for all switching signals without any dwell-time condition.

3.5 Numerical Examples

In this section, we provide numerical illustrations of the previous results. First, we reconsider the numerical example in [9] to show that an arbitrary switching law may destabilize the overall dynamics even if all the reduced-order slow systems share a common Lyapunov function. Moreover, by applying the previous analysis, we give a dwell time that explicitly depends on the timescale parameter. Second, we consider the case when the reduced-order systems (3.5) do not share a common Lyapunov function.

Consider the following singularly perturbed switched system [9] :

$$\begin{pmatrix} \dot{x}(t) \\ \varepsilon\dot{z}(t) \end{pmatrix} = A^{\sigma_k} \begin{pmatrix} x(t) \\ z(t) \end{pmatrix}, \quad \forall t \in [t_k, t_{k+1}), \ k \in \mathbb{N} \tag{3.16}$$

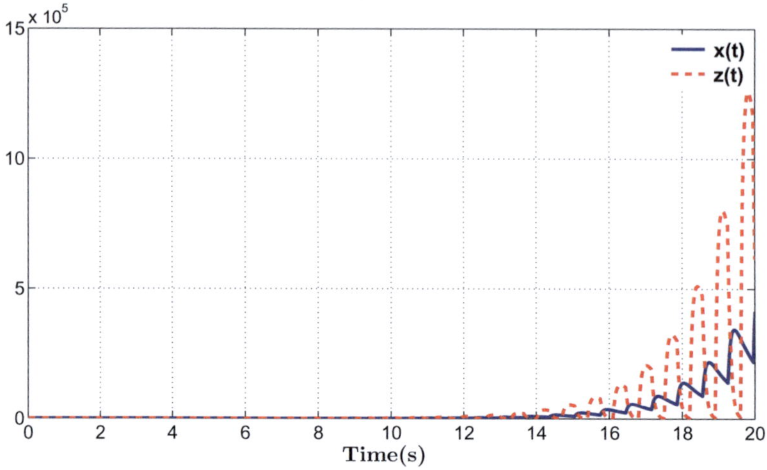

Fig. 3.1 State's trajectories for (3.16) with A^1, A^2 defined by (3.17) and $t_{k+1} - t_k = 0.35\,\text{s}$

where $\sigma_k \in \mathcal{I} = \{1, 2\}$ and the state matrices take the following numerical values :

$$A^1 = \begin{pmatrix} -1 & 0 \\ 5 & -1 \end{pmatrix} \; ; \; A^2 = \begin{pmatrix} -1 & 5 \\ 0 & -1 \end{pmatrix}. \tag{3.17}$$

Since $A_0^1 = A_0^2 = -1$ and $A_{22}^1 = A_{22}^2 = -1$, Assumption 3.3 holds.

Then, let us consider that the two switching modes share a common Lyapunov function. So that

$$Q_s^1 = Q_s^2 = 1, \quad \text{and} \quad Q_f^1 = Q_f^2 = 1.$$

Our aim is to show numerically that asymptotic stability of each subsystem $i \in \mathcal{I}$ is not a sufficient condition for the stability of a switched system. Indeed, a switching law with a sufficiently high switching frequency may destabilize the singularly perturbed switched system. This result is illustrated in Fig. 3.1 with $\varepsilon = 0.076$ and the initial condition $[x(0), z(0)]^\top = [1, 1]^\top$.

Simulating the system with $t_{k+1} - t_k = 0.35\,\text{s}$, we can see that the switched dynamics diverge.

Using the change of variable (3.2) with

$$P_1 = \begin{pmatrix} 1 & 0 \\ -5 & 1 \end{pmatrix}, \quad P_2 = \begin{pmatrix} 1 & 0 \\ 0 & 1 \end{pmatrix},$$

it is easy to check that

$$R^{1\to 2} = \begin{pmatrix} 1 & 0 \\ 5 & 1 \end{pmatrix}, \quad R^{2\to 1} = \begin{pmatrix} 1 & 0 \\ -5 & 1 \end{pmatrix}.$$

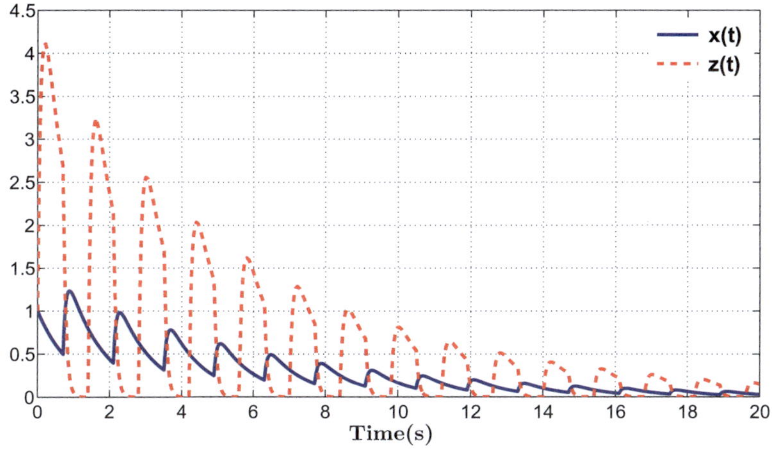

Fig. 3.2 State's trajectories for (3.16) with A^1, A^2 defined by (3.17) and $t_{k+1} - t_k = 0.7001$ s

Then, $\gamma_{11} = 1$ and following Theorem 3.5 an upper bound on the minimum stabilizing dwell time is $0.588 = \mathcal{O}(\varepsilon)$ which is better than the dwell time given in [9]. Simulating the system with $t_{k+1} - t_k = 0.7001$ s, Fig. 3.2 shows that the expected stability is obtained.

Let us now consider another choice for the state matrices A^1, A^2 in (3.16):

$$A_{11}^1 = \begin{pmatrix} -0.92 & -2.8 \\ -8.25 & -6.5 \end{pmatrix}, A_{12}^1 = \begin{pmatrix} 1 & -2 \\ 1 & 3 \end{pmatrix}, A_{21}^1 = \begin{pmatrix} 0.5 & 7 \\ 4 & 0 \end{pmatrix}, A_{22}^1 = \begin{pmatrix} -2 & 0 \\ 0 & -1.5 \end{pmatrix};$$
$$A_{11}^2 = \begin{pmatrix} 2.7 & -1 \\ 0.34 & -6 \end{pmatrix}, A_{12}^2 = \begin{pmatrix} 0.5 & 4 \\ 1 & 1 \end{pmatrix}, A_{21}^2 = \begin{pmatrix} 1 & 3 \\ -2 & 0 \end{pmatrix}, A_{22}^2 = \begin{pmatrix} -1.5 & 0 \\ 0 & -2 \end{pmatrix}. \tag{3.18}$$

Again, one can easily observe that Assumption 3.3 holds:

$$A_0^1 = \begin{pmatrix} -6 & 0 \\ 0 & -2 \end{pmatrix}, A_0^2 = \begin{pmatrix} -1 & 0 \\ 0 & -4 \end{pmatrix}, A_{22}^1 = \begin{pmatrix} -2 & 0 \\ 0 & -1.5 \end{pmatrix}, A_{22}^2 = \begin{pmatrix} -1.5 & 0 \\ 0 & -2 \end{pmatrix}$$

Letting $\lambda_s = 8$ and $\lambda_f = 2.5$, we obtain $\gamma_{11} = 1.554 > 1$.
According to Theorem 3.4, an upper bound on the minimum stabilizing dwell time is given by $\frac{\ln(\gamma_{11})}{\lambda_s} + \mathcal{O}(\varepsilon)$ where $\frac{\ln(\gamma_{11})}{\lambda_s} = 0.0441$ s.

For $\varepsilon = 10^{-3}$ and the initial condition $[x(0), z(0)]^\top = [2, -3, -4, 1]^\top$, Figs. 3.3 and 3.4 show the convergence to zero of the singularly perturbed switched system with $t_{k+1} - t_k = 0.0447$ s,

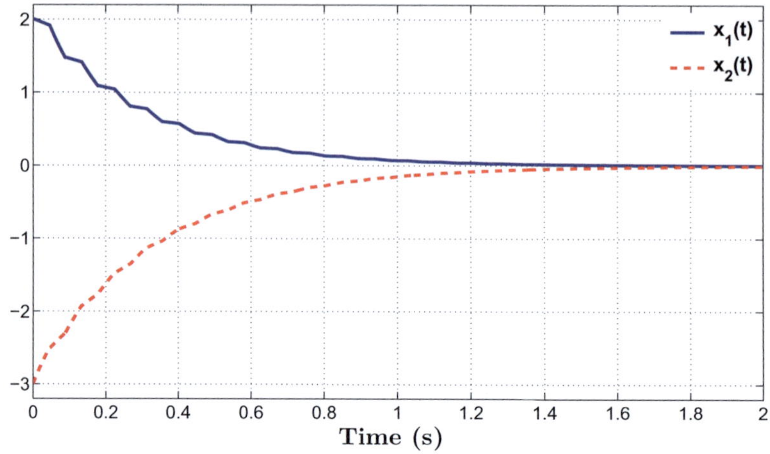

Fig. 3.3 Slow state's trajectories for (3.16) with A^1, A^2 defined by (3.18) and $t_{k+1} - t_k = 0.0447$ s

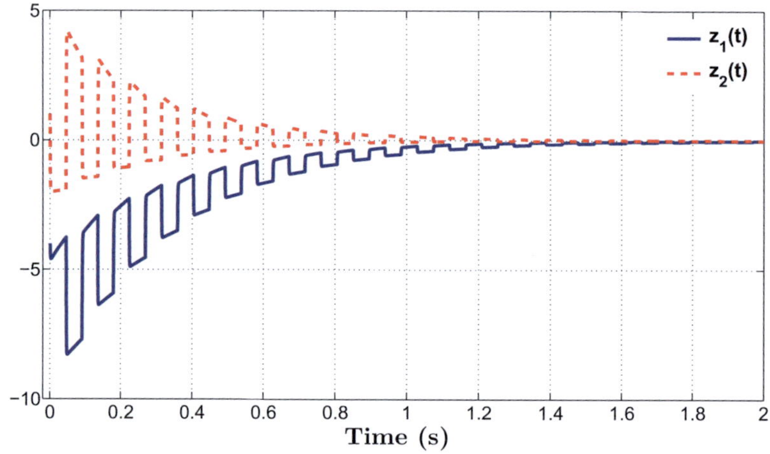

Fig. 3.4 Fast state's trajectory for (3.16) with A^1, A^2 defined by (3.18) and $t_{k+1} - t_k = 0.0447$ s

3.6 Conclusion

In this paper, we analyzed a class of singularly perturbed switched linear systems. Starting from the known fact that switches can destabilize the overall system although each mode is stable, we investigated the dwell-time condition preserving stability. We showed that the dwell time required to ensure stability of the overall system is the sum of two terms. The first one essentially consists of a dwell time ensuring the reduced-order system which is zero when reduced-order systems share a common Lyapunov function. The second term depends on the scale parameter defining the ratio between the two timescales and goes to zero when the parameter goes to zero.

Acknowledgements This work was funded by the ANR project COMPACS—"Computation Aware Control Systems", ANR-13-BS03-004.

References

1. Balluchi, A., .Di Benedetto, M., Pinello, C., Rossi, C., Sangiovanni-Vincentelli, A.: Cut-off in engine control: a hybrid system approach. In: Proceedings of the 36th IEEE Conference on Decision and Control, 1997, vol. 5, pp. 4720–4725 (1997)
2. Donkers, M.C.F., Heemels, W.P.M.H., van de Wouw, N., Hetel, L.: Stability analysis of networked control systems using a switched linear systems approach. IEEE Trans. Autom. Control **56**(9), 2101–2115 (2011). Sept
3. Branicky, M.S.: Multiple lyapunov functions and other analysis tools for switched and hybrid systems. IEEE Trans. Autom. Control **43**(4), 475–482 (1998). Apr
4. Dayawansa, W.P., Martin, C.F.: A converse lyapunov theorem for a class of dynamical systems which undergo switching. IEEE Trans. Autom. Control **44**(4), 751–760 (1999). April
5. Liberzon, D.: Switching in Systems and Control. Springer, Boston (2003)
6. Malloci, I., Daafouz, J., Iung, C., Bonidal, R., Szczepanski, P.: Switched system modeling and robust steering control of the tail end phase in a hot strip mill. Nonlinear Anal. Hybrid Syst. **3**(3), 239–250 (2009)
7. Chen, L., Aihara, K.: A model of periodic oscillation for genetic regulatory systems. IEEE Trans. Circuits Syst. **49**(10), 1429–1436 (2002)
8. Hodgkin, A.L., Huxley, A.F.: A quantitative description of membrane current and its application to conduction and excitation in nerve. J. Physiol. **117**, 500–544 (1952)
9. Malloci, J.: Two time scale switched systems: application to steering control in hot strip mills. PhD thesis, University of Lorraine - CRAN UMR 7039 (2009)
10. Sanfelice, R.G., Teel, A.R.: On singular perturbations due to fast actuators in hybrid control systems. Automatica **47**, 692–701 (2011)
11. H.K. Khalil. *Nonlinear Systems (Third Edition)*. Prentice Hall, 2001
12. Kokotović, P., Khalil, H.K., O'Reilly, J.: Singular perturbation methods in control: analysis and design. SIAM Series in Classics and Applied Mathematics. SIAM, Philadelphia (1999)
13. Balachandra, M., Sethna, P.R.: A generalization of the method of averaging for systems with two time scales. Arch. Ration. Mech. Anal. **58**(3), 261–283 (1975)
14. Nesic, D., Teel, A.R.: Input-to-state stability for nonlinear time-varying systems via averaging. Math. Control Signals Syst. **14**, 257–280 (2001)
15. Teel, A.R., Moreau, L., Nesic, D.: A unified framework for input-to-state stability in systems with two time scales. IEEE Trans. Autom. Control **48**(9), 1526–1544 (2003)
16. Alwan, M., Liu, X., Ingalls, B.: Exponential stability of singularly perturbed switched systems with time delay. Nonlinear Anal. Hybrid Syst. **2**(3), 913–921 (2008)
17. Malloci, J., Daafouz, J., Iung, C.: Stabilization of continuous-time singularly perturbed switched systems. In: Proceedings of the 48th IEEE Conference on Decision and Control (2009)
18. Rantzer, A.: Distributed control of positive systems. In: 50th IEEE Conference on Decision and Control and European Control Conference (CDC-ECC), pp. 6608–6611. IEEE (2011)

Chapter 4
Stability of LTI Systems with Distributed Sensors and Aperiodic Sampling

C. Fiter, T.-E. Korabi, L. Etienne and L. Hetel

Abstract This chapter is dedicated to the stability analysis of sampled-data linear time-invariant systems with asynchronous sensors and aperiodic sampling. The study is performed using an input/output interconnection modeling, and tools from the robust control theory. Two approaches are presented. One is based on the small gain theorem, while the other is based on the dissipativity theory. Tractable stability criteria that allow an estimation of the maximal admissible sampling period are obtained for both approaches. Finally, experimental results performed on an inverted pendulum benchmark are presented. They confirm the applicability of both approaches and allow for some comparisons between both results.

4.1 Introduction

Networked control systems (NCSs) are often required to share limited communication and computation resources, which may lead to fluctuations of the sampling interval [2, 19]. These variations bring up new challenges from the control theory point of view, since they may be sources of unstability. In order to cope with this phenomenon, several researches have recently been conducted to analyze the stability of systems with arbitrarily time-varying sampling intervals (see the survey [10]).

C. Fiter (✉)
Centre de Recherche en Informatique, Signal, et Automatique de Lille (CRIStAL, CNRS UMR 9189), Université de Lille 1, 59650 Villeneuve d'Ascq, France
e-mail: christophe.fiter@univ-lille1.fr

T.-E. Korabi · L. Hetel
Centre de Recherche en Informatique, Signal, et Automatique de Lille (CRIStAL, CNRS UMR 9189), Centrale Lille, 59650 Villeneuve d'Ascq, France
e-mail: taki-inhiste@hotmail.fr

L. Hetel
e-mail: laurentiu.hetel@ec-lille.fr

L. Etienne
Non-A team at INRIA Lille, 59650 Villeneuve d'Ascq, France
e-mail: lucien.etienne@inria.fr

S. Tarbouriech et al. (eds.), *Control Subject to Computational and Communication Constraints*, Lecture Notes in Control and Information Sciences 475, https://doi.org/10.1007/978-3-319-78449-6_4

Time-delay approaches [5, 6, 17], hybrid systems approaches [12, 15], discrete-time approaches [3, 4, 9], and input/output stability approaches [7, 11, 14, 16] have been used to address this problem.

Most of the studies have concerned the stability analysis of systems with one sensor measuring the whole system's state. A problem that must be addressed in practical applications is how to guarantee the stability of a system with multiple sensors. In the case of periodic sampling, this problem has been solved a while ago in the context of multirate systems [1]. However, measurements are usually aperiodic and asynchronous. To the best of our knowledge, this problem has not yet been studied in the literature.

In this work, we propose to tackle this issue in the case of linear time-invariant (LTI) systems, using two approaches from the robust control theory. Both approaches are based on a modeling of the system as an interconnection between a continuous-time (sampling-free) closed-loop system and an operator representing the sampling effects. This new model can be seen as an extension of the one proposed in [7] in the single sensor case. The first result, based on the small gain theorem, provides simple numerical tools to design an overall upper bound on the sampling intervals that guarantees the system's stability. The second result, which is based on the dissipativity theory, allows through the use of Linear Matrix Inequalities (LMIs) to compute an upper bound on the sampling interval for each and every sensor that guarantees the system's stability.

The chapter is constructed as follows. First, in Sect. 4.2, we present the system under study and formulate the problem. In Sect. 4.3, we propose the new modeling of the system that will be used for the whole analysis. Then, in Sects. 4.4 and 4.5, we present the two stability analysis approaches, based, respectively, on the small gain theorem and on the dissipativity theory. Finally, in Sect. 4.6, we illustrate our results with experiments on an inverted pendulum benchmark, before concluding in Sect. 4.7.

Notation: I_M denotes the set $\{1, \ldots, M\}$. The Euclidean norm of a vector x is denoted by $|x|$, and its $\mathscr{L}_2$ norm is denoted by $\|x\|_{L_2}$. The $\mathscr{L}_2$ induced norm of an operator G is denoted by $\|G\|_{L_2-L_2}$. A function $\beta : \mathbb{R}_{\geq 0} \to \mathbb{R}_{\geq 0}$ is said to be of class $\mathscr{K}$ if it is continuous, zero at zero and strictly increasing. It is said to be of class $\mathscr{K}_\infty$ if it is of class $\mathscr{K}$, and it is unbounded. A function $\beta : \mathbb{R}_{\geq 0} \times \mathbb{R}_{\geq 0} \to \mathbb{R}_{\geq 0}$ is said to be of class $\mathscr{KL}$ if $\beta(., t)$ is of class $\mathscr{K}$ for each $t \geq 0$, and $\beta(s, .)$ is nonincreasing and satisfies $\lim_{t\to\infty} \beta(s, t) = 0$ for each $s \geq 0$. We use $diag(N_1, \ldots, N_M)$ to denote a block diagonal matrix which contains the blocks N_i $(i \in I_M)$ on its diagonal, and zeros everywhere else.

4.2 Problem Formulation

We consider the linear time-invariant (LTI) system with distributed sensors presented in Fig. 4.1.

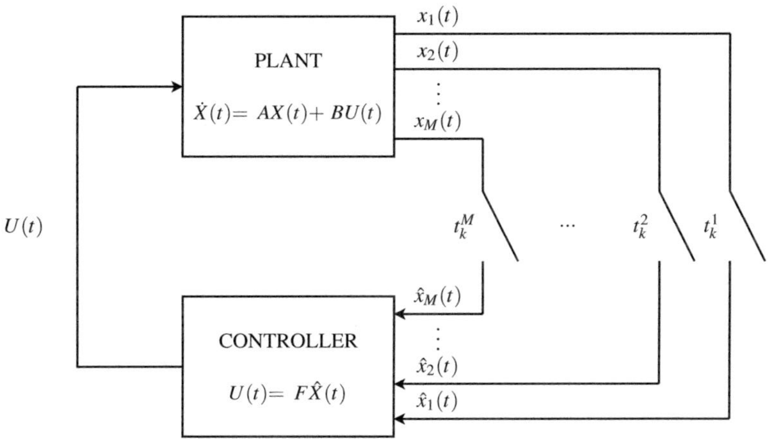

Fig. 4.1 System description

This system is mathematically defined by

$$\begin{aligned} \dot{X}(t) &= AX(t) + BU(t), \forall t \geq 0, \\ X(0) &= X_0 \in \mathbb{R}^n, \end{aligned} \tag{4.1}$$

with $X \in \mathbb{R}^n$ the system state, $X_0 \in \mathbb{R}^n$ the initial state, $U \in \mathbb{R}^m$ the input, and A and B matrices of appropriate dimensions.

We consider that the state X is divided into M subparts x_i, $i \in I_M$, each linked to one sensor:

$$X(t) = \left[x_1^T(t) \ldots x_M^T(t)\right]^T,$$

with $x_i \in \mathbb{R}^{n_i}$ and $\sum_{i=1}^{M} n_i = n$.

These subparts of the state are transmitted to the controller in an asynchronous manner. For each of them, we consider a monotonously increasing sampling sequence with bounded sampling intervals

$$\sigma^i = \left\{t_k^i\right\}_{k\in\mathbb{N}} \text{ with } t_{k+1}^i - t_k^i \in (0, h_i]. \tag{4.2}$$

We assume that $t_0^i = 0$ for all $i \in I_M$.

The sampled state is denoted $\hat{X}(t) = \left[\hat{x}_1^T(t) \ldots \hat{x}_M^T(t)\right]^T$, with

$$\hat{x}_i(t) = x_i(t_k^i), \forall t \in \left[t_k^i, t_{k+1}^i\right), k \in \mathbb{N}, i \in I_M. \tag{4.3}$$

The control is considered as a linear state-feedback using the known, sampled version of the state:

$$U(t) = F\hat{X}(t), \tag{4.4}$$

where $F \in \mathbb{R}^{m\times n}$ is a given feedback gain.

The aim of this work is to design methods for checking the stability of system (4.1)–(4.4).

4.3 System Modeling

In this section, adapting the approach proposed in [11] in the single sensor case, and further used in [7, 14, 16], we remodel the system as an interconnection between a continuous-time closed-loop system and an operator representing the sampling effect on all the subparts of the system's state.

We define the sampling system error

$$E(t) = \hat{X}(t) - X(t) = \left[e_1^T(t) \ \ldots \ e_M^T(t)\right]^T, \tag{4.5}$$

where, for all $i \in I_M$,

$$\begin{aligned} e_i(t) &= \hat{x}_i(t) - x_i(t), \ \forall t \geq 0, \\ &= x_i(t_k^i) - x_i(t), \ \forall t \in \left[t_k^i, t_{k+1}^i\right), \ k \in \mathbb{N}. \end{aligned} \tag{4.6}$$

Let us remark that the sampling error can be described as a reset integrator

$$e_i(t) = -\int_{t_k^i}^{t} \dot{x}_i(\theta)d\theta, \ \forall t \in \left[t_k^i, t_{k+1}^i\right), \ k \in \mathbb{N}.$$

For each $i \in I_M$, we define the operator

$$\begin{aligned} \Delta_i : \mathscr{L}_{2e}^{n_i}[0,\infty) &\mapsto \mathscr{L}_{2e}^{n_i}[0,\infty) \\ z &\mapsto \Delta_i z \end{aligned}$$

$$(\Delta_i z)(t) = -\int_{t_k^i}^{t} z(\theta)d\theta, \forall i \in I_M, \forall t \in \left[t_k^i, t_{k+1}^i\right). \tag{4.7}$$

Finally, the system (4.1)–(4.4) can be represented by the interconnection of $G : E \mapsto Y$, the LTI system described by

$$G : \begin{cases} \dot{X}(t) = A_{\text{cl}}X(t) + B_{\text{cl}}E(t), \\ Y(t) = C_{\text{cl}}X(t) + D_{\text{cl}}E(t), \end{cases} \tag{4.8}$$

with $A_{\text{cl}} = C_{\text{cl}} = A + BF$, $B_{\text{cl}} = D_{\text{cl}} = BF$, $Y(t) = \left[y_1^T(t) \ \ldots \ y_M^T(t)\right]^T$, and the operator $\Delta : Y \mapsto E$ defined by

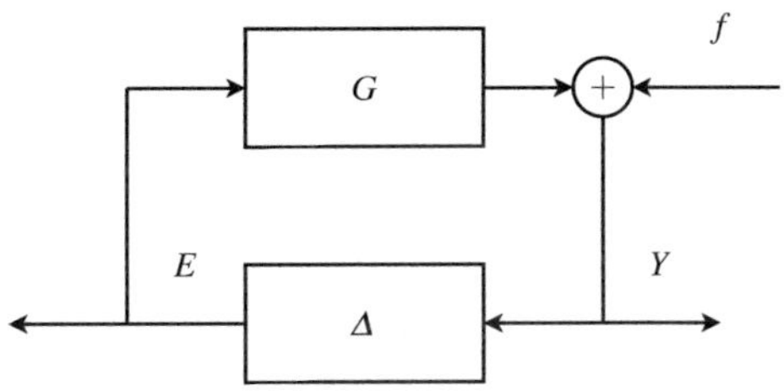

Fig. 4.2 System's representation as an interconnected system

$$\Delta : E = \Delta Y, \tag{4.9}$$

with $\Delta = diag(\Delta_1, \ldots, \Delta_M)$.

Here, A_{cl} represents the nominal (closed-loop) state matrix, and E the sampling-induced error described by operator Δ in (4.7).

This representation of the system as an interconnection (4.8) and (4.9) will be used throughout the rest of the chapter for the stability analysis.

Note that this interconnection can also be rewritten so as to explicitly see the influence of the initial state x_0. Indeed, let G : $\mathscr{L}_2^\infty[0, \infty) \mapsto \mathscr{L}_2^\infty[0, \infty)$ be the linear operator described by the transfer function

$$\hat{G}(s) = C_{\text{cl}}(sI - A_{\text{cl}})^{-1}B_{\text{cl}} + D_{\text{cl}} = s(sI - A_{\text{cl}})^{-1}B_{\text{cl}}.$$

If we consider the free response of system (4.8)

$$f(t) = A_{\text{cl}}e^{A_{\text{cl}}t}x_0, \forall t \geq 0,$$

we can rewrite the system as the interconnection

$$\begin{aligned} Y &= GE + f, \\ E &= \Delta Y. \end{aligned} \tag{4.10}$$

This interconnection is presented in Fig. 4.2.

4.4 Small Gain Approach

The first stability approach we consider is based on the small gain theorem [13]. Before presenting the main result of this approach, we need to provide a technical lemma.

Lemma 4.1 *The $\mathscr{L}_2$ induced norm of the operator Δ is upper bounded by h_{max}, where $h_{max} = \max_{i \in I_M} \{h_i\}$.*

Proof Using (4.6) for all $i \in I_M$ and $t \in [t_k^i, t_{k+1}^i)$, $k \in \mathbb{N}$, we consider

$$e_i(t) = -\int_{t_k^i}^{t} y_i(\theta)d\theta,\ \forall t \in [t_k^i, t_{k+1}^i),\ k \in \mathbb{N}.$$

Then, using Jensen's inequality [8], we get

$$\begin{aligned}
e_i^T(t)e_i(t) &= \left(\int_{t_k^i}^{t} y_i(\theta)d\theta\right)^T \left(\int_{t_k^i}^{t} y_i(\theta)d\theta\right), \\
&\leq (t - t_k^i)\left(\int_{t_k^i}^{t} y_i^T(\theta)y_i(\theta)d\theta\right), \\
&\leq h_i\left(\int_{t_k^i}^{t} y_i^T(\theta)y_i(\theta)d\theta\right).
\end{aligned}$$

Applying further inequalities, we can compute an upper bound on $\|e_i\|_{L_2}$:

$$\begin{aligned}
\|e_i\|_{L_2}^2 &= \int_0^{\infty} e_i^T(t)e_i(t)dt, \\
&= \sum_{k=0}^{\infty}\int_{t_k^i}^{t_{k+1}^i} e_i^T(t)e_i(t)dt, \\
&\leq \sum_{k=0}^{\infty}\int_{t_k^i}^{t_{k+1}^i} h_i \int_{t_k^i}^{t} y_i^T(\theta)y_i(\theta)d\theta dt, \\
&\leq h_i \sum_{k=0}^{\infty}\int_{t_k^i}^{t_{k+1}^i}\int_{t_k^i}^{t_{k+1}^i} y_i^T(\theta)y_i(\theta)d\theta dt, \\
&= h_i \sum_{k=0}^{\infty}(t_{k+1}^i - t_k^i)\int_{t_k^i}^{t_{k+1}^i} y_i^T(\theta)y_i(\theta)d\theta, \\
&\leq h_i^2 \sum_{k=0}^{\infty}\int_{t_k^i}^{t_{k+1}^i} y_i^T(\theta)y_i(\theta)d\theta, \\
&= h_i^2 \int_0^{\infty} y_i^T(\theta)y_i(\theta)d\theta, \\
&= h_i^2 \|y_i\|_{L_2}^2.
\end{aligned}$$

We have found that

$$\|e_i\|_{L_2}^2 \leq h_i^2 \|y_i\|_{L_2}^2, \forall i \in I_M. \tag{4.11}$$

From this equation, considering the sampling error E as described in (4.5), and using some majorations, we may also compute an upper bound on $\|E\|_{L_2}$:

$$
\begin{aligned}
\|E\|_{L_2}^2 &= \int_0^\infty E^T(t)E(t)dt,\\
&= \int_0^\infty \sum_{i=1}^{M} e_i^T(t)e_i(t)dt,\\
&= \sum_{i=1}^{M} \int_0^\infty e_i^T(t)e_i(t)dt,\\
&\leq \sum_{i=1}^{M} h_i^2 \int_0^\infty y_i^T(\theta)y_i(\theta)d\theta,\\
&\leq h_{max}^2 \sum_{i=1}^{M} \int_0^\infty y_i^T(\theta)y_i(\theta)d\theta,\\
&= h_{max}^2 \int_0^\infty \sum_{i=1}^{M} y_i^T(\theta)y_i(\theta)d\theta,\\
&= h_{max}^2 \int_0^\infty Y^T(\theta)Y(\theta)d\theta,\\
&= h_{max}^2 \|Y\|_{L_2}^2 .
\end{aligned}
$$

Therefore, we have obtained that

$$
\|E\|_{L_2}^2 \leq h_{max}^2 \|Y\|_{L_2}^2 ,
$$

which ends the proof. □

Before giving the main result of this section, we recall the following definition [13].

Definition 4.1 ($\mathscr{L}_2$*-stability*) A linear system $\mathbf{F}$ is said to be finite-gain $\mathscr{L}_2$-stable from w to $\mathbf{F}w$ with an induced gain less than γ if $\mathbf{F}$ is a linear operator from $\mathscr{L}_2$ to $\mathscr{L}_2$ and if there exist positive real constants γ and ξ such that for all $w \in \mathscr{L}_2$,

$$
\|\mathbf{F}w\|_{L_2} \leq \gamma \|w\|_{L_2} + \xi.
$$

Theorem 4.1 *The interconnection* (4.10) *is* $\mathscr{L}_2$*-stable if* $\|G\|_\infty < \frac{1}{h_{max}}$.

Proof We use the small gain theorem to provide the stability condition. A direct consequence of this theorem is the fact that if

$$
\|G\|_{L_2-L_2} \|\Delta\|_{L_2-L_2} < 1,
$$

the interconnection (4.10) is $\mathscr{L}_2$-stable.

Since G is linear, its $\mathscr{L}_2$-induced norm can be calculated [20] using the H_∞ norm of its transfer function:

$$\|G\|_{L_2-L_2} = \|G\|_\infty := \sup_{w\in\mathbb{R}} \bar{\sigma}(\hat{G}(jw)).$$

Since $\|\Delta\|_{L_2} \leq h_{\max}$, according to Lemma 4.1, the interconnection (4.10) is $\mathscr{L}_2$-stable if

$$\|G\|_{L_2-L_2} = \|G\|_\infty < \frac{1}{h_{\max}}.$$

□

Remark 4.1 The condition proposed in the previous theorem can be easily implemented in MATLAB. It depends solely on the system, and on the overall upper bound on the sampling intervals h_{max}. However, this approach does not make it possible to take into account the difference in the upper bounds of each sensor ($\{h_1, \ldots, h_M\}$) in the analysis. In the next section, we propose an approach that solves this problem.

4.5 Dissipativity-Based Approach

In this section, we propose another stability approach inspired by the dissipativity theory [18] for the interconnected system (4.8) and (4.9). We first start by providing a few tools and definitions.

Let us define a sequence γ that includes all the sampling sequences $\sigma^i = \left\{t_k^i\right\}_{k\in\mathbb{N}}$, $i \in I_M$, in chronological order:

$$\gamma = \{t_s\}_{s\in\mathbb{N}}, \text{ with } t_0 = 0 \text{ and } t_{s+1} = \min_{i\in I_M,\, k\in\mathbb{N}} \left\{t_k^i : t_k^i > t_s\right\}, \ \forall s \in \mathbb{N}. \quad (4.12)$$

For any $i \in I_M$ and $s \in \mathbb{N}$, let us also define the coefficient

$$k_{i,s} = \max\left\{k \in \mathbb{N} : t_k^i \leq t_s\right\}, \quad (4.13)$$

which represents the index of the last sample sent to the controller before the time t_s, by sensor i. Intuitively, the time $t^i_{k_{i,s}}$, for $i \in I_M$ and $s \in \mathbb{N}$ represents the last time before t_s that sensor i sent a measure to the controller.

It is important to remark that for all $s \in \mathbb{N}$, $t_{s+1} - t_s \in (0, h_{\min}]$, where $h_{\min} = \min_{i\in I_M} \{h_i\}$.

In order to give a better understanding of the notation that were just introduced, we show in Fig. 4.3 an example of construction of a sequence γ in the case of a system with two sensors. The values of the scalars $k_{i,s}$ obtained from these sequences are shown in Table 4.1.

Before giving the main results, we recall the following stability definition:

Definition 4.2 (*GUAS*) The equilibrium point $X = 0$ of system (4.1)–(4.4) is globally uniformly asymptotically stable (GUAS) if there exists a class $\mathscr{K}\mathscr{L}$ function $\beta(.,.)$, such that

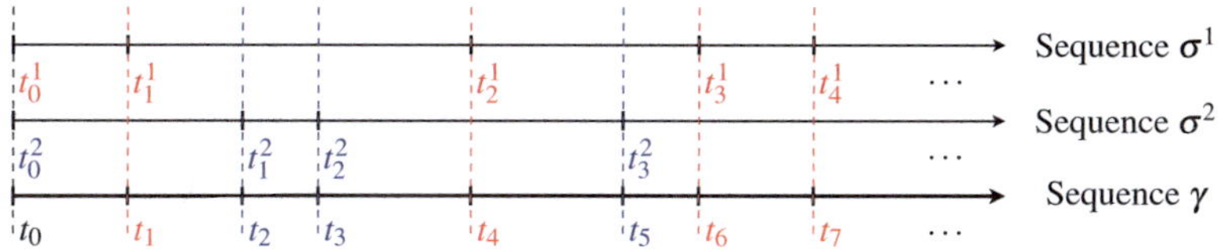

Fig. 4.3 An example of sampling sequences with 2 sensors

Table 4.1 Value of the coefficients $k_{i,s}$ in the example

	$s=0$	$s=1$	$s=2$	$s=3$	$s=4$	$s=5$	$s=6$	$s=7$	...
$i=1$ (sensor 1)	0	1	1	1	2	2	3	4	...
$i=2$ (sensor 2)	0	0	1	2	2	3	3	3	...

$$|X(t)| \leq \beta(|X(t_0)|, t-t_0), \ \forall t \geq t_0, \ \forall X(t_0) \in \mathbb{R}^n. \tag{4.14}$$

Theorem 4.2 *Consider the sampled-data system* (4.1)–(4.4) *and the equivalent representation* (4.8) *and* (4.9). *Consider the definition of $e_i(t)$ in* (4.6). *Assume that*

1. *There exist M continuous functions $\mathscr{S}_i(y_i, e_i)$ which satisfy the integral property*

$$\int_{t_k^i}^{t} \mathscr{S}_i(y_i(\theta), e_i(\theta))d\theta \leq 0, \forall t \in \left[t_k^i, t_{k+1}^i\right), k \in \mathbb{N}, \ i \in I_M. \tag{4.15}$$

2. *There exist a differentiable positive define function $V : \mathbb{R}^n \to \mathbb{R}$ and class $\mathscr{K}$ functions β_1 and β_2 verifying*

$$\beta_1(|X|) \leq V(X) \leq \beta_2(|X|), \forall X \in \mathbb{R}^n. \tag{4.16}$$

3. *There exists a scalar $\alpha > 0$ such that*

$$\dot{V}(X(t)) + \alpha V(X(t)) \leq \sum_{i=1}^{M} e^{-\alpha\tau_i(t)} \mathscr{S}_i(y_i(t), e_i(t)),$$
$$\forall t \in \left[t_s, t_{s+1}\right), \ \forall s \in \mathbb{N}, \ \textit{with } \tau_i(t) = t - t_{k_{i,s}}^i, \textit{ where } k_{i,s} \textit{ is defined in (4.13).} \tag{4.17}$$

Then the equilibrium point $X = 0$ of the system (4.1)–(4.4) *is GUAS.*

Proof Consider $s \in \mathbb{N}$ and $t \in [t_s, t_{s+1})$. From (4.17), we have

$$e^{\alpha t}\left(\dot{V}(X(t)) + \alpha V(X(t)\right) \leq \sum_{i=1}^{M} e^{\alpha t_{k_{i,s}}^i} \mathscr{S}_i(y_i(t), e_i(t)).$$

By integrating this equation on the interval $[t_s, t]$, we get

$$e^{\alpha t} V(X(t)) - e^{\alpha t_s} V(X(t_s)) \leq \sum_{i=1}^{M} e^{\alpha t^i_{k_{i,s}}} \int_{t_s}^{t} \mathscr{S}_i(y_i(\theta), e_i(\theta)) d\theta,$$

which leads to

$$V(X(t)) \leq e^{-\alpha(t-t_s)} V(X(t_s)) + e^{-\alpha t} \sum_{i=1}^{M} e^{\alpha t^i_{k_{i,s}}} \int_{t_s}^{t} \mathscr{S}_i(y_i(\theta), e_i(\theta)) d\theta. \quad (4.18)$$

Therefore, we can see that for any $s \in \mathbb{N}$, we have

$$V(X(t_{s+1})) \leq e^{-\alpha(t_{s+1}-t_s)} V(X(t_s)) + e^{-\alpha t_{s+1}} \sum_{i=1}^{M} e^{\alpha t^i_{k_{i,s}}} \int_{t_s}^{t_{s+1}} \mathscr{S}_i(y_i(\theta), e_i(\theta)) d\theta. \quad (4.19)$$

Replacing $V(X(t_s))$ in (4.18) by its expression (4.19) leads to

$$\begin{aligned} V(X(t)) \leq e^{-\alpha(t-t_s)} \Big[& e^{-\alpha(t_s - t_{s-1})} V(X(t_{s-1})) \\ & + e^{-\alpha t_s} \sum_{i=1}^{M} e^{\alpha t^i_{k_{i,s-1}}} \int_{t_{s-1}}^{t_s} \mathscr{S}_i(y_i(\theta), e_i(\theta)) d\theta \Big] \\ & + e^{-\alpha t} \sum_{i=1}^{M} e^{\alpha t^i_{k_{i,s}}} \int_{t_s}^{t} \mathscr{S}_i(y_i(\theta), e_i(\theta)) d\theta, \end{aligned}$$

and thus

$$\begin{aligned} V(X(t)) \leq e^{-\alpha(t-t_{s-1})} V(X(t_{s-1})) + e^{-\alpha t} \sum_{i=1}^{M} \Big[e^{\alpha t^i_{k_{i,s-1}}} \int_{t_{s-1}}^{t_s} \mathscr{S}_i(y_i(\theta), e_i(\theta)) d\theta \\ + e^{\alpha t^i_{k_{i,s}}} \int_{t_s}^{t} \mathscr{S}_i(y_i(\theta), e_i(\theta)) d\theta \Big]. \end{aligned} \quad (4.20)$$

Then, by recursivity, replacing $V(X(t_{s-1}))$ in (4.20) by its expression (4.19) and so on, we can show that

$$\begin{aligned} V(X(t)) \leq e^{-\alpha(t-t_0)} V(X(t_0)) + e^{-\alpha t} \sum_{i=1}^{M} \Bigg[\left(\sum_{j=1}^{s} e^{\alpha t^i_{k_{i,j-1}}} \int_{t_{j-1}}^{t_j} \mathscr{S}_i(y_i(\theta), e_i(\theta)) d\theta \right) \\ + e^{\alpha t^i_{k_{i,s}}} \int_{t_s}^{t} \mathscr{S}_i(y_i(\theta), e_i(\theta)) d\theta \Bigg]. \end{aligned} \quad (4.21)$$

Let us now consider some $k \in \mathbb{N}$ and $i \in I_M$. By construction, there exist positive integers s_1 and s_2, with $s_1 < s_2$, such that $t^i_k = t_{s_1}$ and $t^i_{k+1} = t_{s_2}$. From the definition of elements t_s and $k_{i,s}$ in (4.12) and (4.13), respectively (see also Fig. 4.3 and Table 4.1 for an example), one can see that for any $\bar{s} \in \{s_1, s_1 + 1, \ldots, s_2 - 1\}$, $k_{i,\bar{s}} = k$, and thus $t^i_{k_{i,\bar{s}}} = t^i_k$. Therefore, using this property, we can rewrite (4.21) as

$$V(X(t)) \le \mathrm{e}^{-\alpha(t-t_0)}V(X(t_0)) + \mathrm{e}^{-\alpha t}\sum_{i=1}^{M}\left[\left(\sum_{k=0}^{k_{i,s}-1}\mathrm{e}^{\alpha t_k^i}\int_{t_k^i}^{t_{k+1}^i}\mathscr{S}_i(y_i(\theta),e_i(\theta))d\theta\right)\right.$$
$$\left. +\mathrm{e}^{\alpha t_{k_{i,s}}^i}\int_{t_{k_{i,s}}^i}^{t}\mathscr{S}_i(y_i(\theta),e_i(\theta))d\theta\right]. \tag{4.22}$$

Then, using assumption (4.15), one can see that all the integral terms are negative, and thus one gets

$$V(X(t)) \le \mathrm{e}^{-\alpha(t-t_0)}V(X(t_0)). \tag{4.23}$$

Finally, assumption (4.16) leads to

$$\begin{aligned} |X(t)| &\le \beta_1^{-1}(V(X(t_0))e^{-\alpha(t-t_0)}), \\ &\le \beta_1^{-1}(\beta_2(|X(t_0)|)e^{-\alpha(t-t_0)}), \\ &:= \beta(|X(t_0)|, t-t_0), \ \forall t > t_0, \end{aligned} \tag{4.24}$$

which concludes the proof. □

Theorem 4.2 is based on the existence of M functions $\mathscr{S}_i$, $i \in I_M$, satisfying assumption (4.15). In the following, we show that such functions exist.

First, we start by providing a technical lemma.

Lemma 4.2 *Consider the set of operators Δ_i defined in* (4.7) *for any $z_i \in \mathscr{L}_{2e}^{n_i}$ and $t \in [t_k^i, t_{k+1}^i)$, $k \in \mathbb{N}$. We have the following inequality:*

$$\int_{t_k^i}^{t}(\Delta_i z_i)(\rho)^T(\Delta_i z_i)(\rho) - h_i^2 z_i^T(\rho)z_i(\rho)d\rho \le 0. \tag{4.25}$$

It follows that the M functions $\mathscr{S}_i$ such that

$$\mathscr{S}_i(y_i(\theta),e_i(\theta)) = e_i^T(\theta)e_i(\theta) - h_i^2 y_i^T(\theta)y_i(\theta),$$

with $e_i = \Delta_i y_i$, as defined in (4.7), *satisfy the assumption* (4.15).

Proof Using the same steps as in the proof of Lemma 4.1, we can show that

$$(\Delta_i z_i)^T(\rho)(\Delta_i z_i)(\rho) \le h_i\left(\int_{t_k^i}^{\rho} z_i^T(\theta)z_i(\theta)d\theta\right), \ \forall \rho \in [t_k^i, t_{k+1}^i], k \in \mathbb{N}.$$

Therefore we have, for all $t \in [t_k^i, t_{k+1}^i]$ and $k \in \mathbb{N}$,

$$\int_{t_k^i}^t (\Delta_i z_i)^T(\rho)(\Delta_i z_i)(\rho)d\rho \leq \int_{t_k^i}^t h_i \int_{t_k^i}^\rho z_i^T(\theta)z_i(\theta)d\theta d\rho,$$
$$\leq h_i \int_{t_k^i}^t \int_{t_k^i}^t z_i^T(\theta)z_i(\theta)d\theta d\rho,$$
$$= h_i(t - t_k^i) \int_{t_k^i}^t z_i^T(\theta)z_i(\theta)d\theta,$$
$$\leq h_i^2 \int_{t_k^i}^t z_i^T(\theta)z_i(\theta)d\theta,$$

which ends the proof. □

Lemma 4.2 proves the existence of functions $\mathscr{S}_i$ satisfying condition (4.15) in Theorem 4.2. In the following, we show the existence of a more general, parameter-dependent, class of functions $\mathscr{S}_i$ satisfying that condition. The idea is to use a scaling matrix R_i that gives an additional degree of liberty in the choice of the $\mathscr{S}_i$ functions. This will be useful later to reduce the conservatism of the stability conditions.

Lemma 4.3 *Consider the set of operators Δ_i in* (4.7) *and* (4.9)*. Then for any $z_i \in \mathscr{L}_{2e}^{n_i}$, $0 < R_i^T = R_i \in \mathbb{R}^{m\times m}$ we have the following inequality:*

$$\mathscr{N}(t) = \int_{t_k^i}^t \left[(\Delta_i z_i)^T(\rho)R_i(\Delta_i z_i)(\rho) - h_i^2 z_i^T(\rho)R_i z_i(\rho)\right]d\rho \leq 0, \forall t \in [t_k^i, t_{k+1}^i), k \in \mathbb{N}. \tag{4.26}$$

It follows that the M functions $\mathscr{S}_i$ such that

$$\mathscr{S}_i(y_i(\theta), e_i(\theta)) = e_i^T(\theta)R_i e_i(\theta) - h_i^2 y_i^T(\theta)R_i y_i(\theta),$$

with $e_i = \Delta_i y_i$, as defined in (4.7)*, satisfy the assumption* (4.15).

Proof First of all, we note that since $R_i^T = R_i > 0$, then there exists $U_i \in \mathbb{R}^{n\times n}$ such that $R_i = U_i^T U_i$. Therefore, for any $t \in [t_k^i, t_{k+1}^i)$, $k \in \mathbb{N}$ we have

$$\mathscr{N}(t) = \int_{t_k^i}^t \left[(U_i(\Delta_i z_i)(\rho))^T(U_i(\Delta_i z_i)(\rho)) - h_i^2(U_i y_i(\rho))^T(U_i y_i(\rho))\right]d\rho.$$

From (4.7) we can see that $U_i(\Delta_i z_i) = \Delta_i(U_i z_i)$. Then

$$\mathscr{N}(t) = \int_{t_k^i}^t \left[(\Delta_i(U_i z_i)(\rho))^T(\Delta_i(U_i z_i)(\rho)) - h_i^2(U_i y_i(\rho))^T(U_i y_i(\rho))\right]d\rho.$$

Therefore, we can write

$$\mathscr{N}(t) = \int_{t_k^i}^t \left[(\Delta_i \psi_i(\rho))^T(\Delta_i \psi_i(\rho)) - h_i^2 \psi_i^T(\rho)\psi_i(\rho)\right]d\rho,$$

with $\psi_i(\rho) = U_i z_i(\rho)$, which can be seen to be negative from Lemma 4.2. □

We have now shown that the $\mathscr{S}_i$ functions satisfying the assumption (4.15) exist. The next step consists on showing that we can obtain a function V which satisfies conditions (4.16) and (4.17), based on the $\mathscr{S}_i$ functions in Lemma 4.3.

Remark 4.2 The $\mathscr{S}_i$ functions presented in Lemmas 4.2 and 4.3 were inspired by the ones proposed in [16], where it was studied the stability of bilinear systems with aperiodic sampling and a single sensor measuring the whole state.

4.5.1 Numerical Criteria

The goal of this section is to provide tractable LMI conditions that guarantee the stability conditions of Theorem 4.2, using Lemma 4.3. The main result is as follows.

Theorem 4.3 *If there exist symmetric positive definite matrices $R_i \in \mathbb{R}^{n_i \times n_i}$ ($i \in I_M$), $P \in \mathbb{R}^{n \times n}$, and a scalar $\alpha > 0$ such that the LMIs*

$$\begin{bmatrix} A_{cl}^T P + P A_{cl} + \alpha P + A_{cl}^T \bar{L}(\lambda) A_{cl} & P B_{cl} + A_{cl}^T \bar{L}(\lambda) B_{cl} \\ * & -\bar{R}(\lambda) + B_{cl}^T \bar{L}(\lambda) B_{cl} \end{bmatrix} \leq 0 \tag{4.27}$$

are satisfied for all $\lambda = \left[\lambda_1, \ldots, \lambda_M\right]^T \in \{0, 1\}^M$, with

$$\bar{R}(\lambda) = diag(e^{-\alpha h_1 \lambda_1} R_1, \ldots, e^{-\alpha h_M \lambda_M} R_M),$$

and

$$\bar{L}(\lambda) = diag(e^{-\alpha h_1 \lambda_1} h_1^2 R_1, \ldots, e^{-\alpha h_M \lambda_M} h_M^2 R_M),$$

then the equilibrium $X = 0$ of system (4.1)–(4.4) *is GUAS.*

Proof The goal is to provide sufficient conditions for the existence of functions V and $\mathscr{S}_i$ that satisfy the conditions of Theorem 4.2. Consider a quadratic function $V(X(t)) = X^T(t) P X(t)$ and the representation of the system in (4.8) where $Y(t) = A_{\text{cl}} X(t) + B_{\text{cl}} E(t)$. Multiplying the LMIs (4.27) by $\left(X^T(t), E^T(t)\right)$ on the left and by its transpose on the right implies that

$$\begin{aligned} &2X^T(t) P (A_{\text{cl}} X(t) + B_{\text{cl}} E(t)) + \alpha X^T(t) P X(t) \\ &\leq E^T(t) \bar{R}(\lambda) E(t) - Y^T(t) \bar{L}(\lambda) Y(t), \ \forall \lambda \in \{0, 1\}^M. \end{aligned} \tag{4.28}$$

Let us now define

$$\tilde{R}(t) = \text{diag}(\text{e}^{-\alpha \tau_1(t)} R_1, \ldots, \text{e}^{-\alpha \tau_M(t)} R_M),$$

and

$$\tilde{L}(t) = \mathrm{diag}(\mathrm{e}^{-\alpha\tau_1(t)}h_1^2 R_1, \ldots, \mathrm{e}^{-\alpha\tau_M(t)}h_M^2 R_M).$$

Taking into account the fact that for any $t \in [t_s, t_{s+1})$, $s \in \mathbb{N}$, one has $\tau_i(t) = t - t^i_{k_{i,s}} \in [0, h_i)$ for all $i \in I_M$, it is clear that $\tilde{L}(t) \in \mathrm{conv}\{\bar{L}(\lambda) : \lambda \in \{0,1\}^M\}$ and $\tilde{R}(t) \in \mathrm{conv}\{\bar{R}(\lambda) : \lambda \in \{0,1\}^M\}$. Then, since (4.28) is satisfied, by convexity we obtain

$$\begin{aligned} &2X^T(t)P(A_{\mathrm{cl}}X(t) + B_{\mathrm{cl}}E(t)) + \alpha X^T(t)PX(t) \\ &\leq E^T(t)\tilde{R}(t)E(t) - Y^T(t)\tilde{L}(t)Y(t), \ \forall t \in [t_s, t_{s+1}), \ s \in \mathbb{N}. \end{aligned} \tag{4.29}$$

This leads to

$$\begin{aligned} &2X^T(t)P(A_{\mathrm{cl}}X(t) + B_{\mathrm{cl}}E(t)) + \alpha X^T(t)PX(t) \\ &\leq \sum_{i=1}^{M} \mathrm{e}^{-\alpha\tau_i(t)}(e_i^T(t)R_i e_i(t) - h_i^2 y_i^T(t)R_i y_i(t)), \end{aligned} \tag{4.30}$$

which is equivalent to

$$\dot{V}(X(t)) + \alpha V(X(t)) \leq \sum_{i=1}^{M} \mathrm{e}^{-\alpha\tau_i(t)}\mathscr{S}_i(y_i(t), e_i(t)), \ \forall t \in \left[t_s, t_{s+1}\right), \ s \in \mathbb{N}, \tag{4.31}$$

with the functions $\mathscr{S}_i$ defined in Lemma 4.3 that satisfy (4.15). Therefore, assumption (4.17) in Theorem 4.2 is satisfied.

Furthermore, we can show that using Lemma 4.3 in [13] for a quadratic function $V(X(t)) = X^T(t)PX(t)$, assumption (4.16) is satisfied with functions β_1 and β_2 defined such that $\beta_1(|X|) = \lambda_{min}(P)|X|^2$ and $\beta_2(|X|) = \lambda_{max}(P)|X|^2$.

Therefore, we can see that the quadratic function V and the $\mathscr{S}_i$ functions in Lemma 4.3 satisfy all the conditions of Theorem 4.2, if the LMIs in (4.27) are feasible, which ends the proof. □

Remark 4.3 Theorem 4.3 provides tractable LMI conditions for LTI systems with distributed, asynchronous sensors. Unlike the result obtained in the previous section, in Theorem 4.1, which depends on the overall upper bound $h_{max} = \max_{i \in I_M}\{h_i\}$, the conditions of Theorem depend explicitly on each sensor's sampling upper bound h_i, $i \in I_M$.

4.6 Validation of the Results on an Inverted Pendulum Benchmark

In this section, we show some experimentations performed on an inverted pendulum benchmark, using the results from the previous sections. The system is obviously nonlinear, so we will use for analysis a linearization of the model around its unstable equilibrium point (upper position).

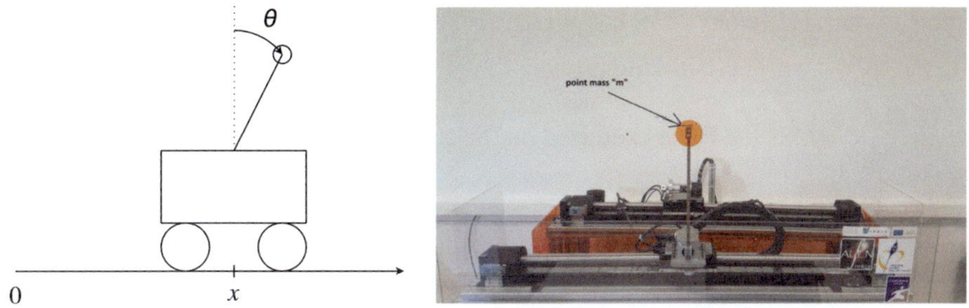

Fig. 4.4 Inverted pendulum at CRIStAL

4.6.1 Benchmark Description

The inverted pendulum (see Fig. 4.4) consists of a cart which is driven by a linear motor and a pendulum. The pendulum is fixed and left free on the cart. The system also consists of two sensors: the first one measures the linear position of the cart where the second one measures the angular position of the pendulum. An estimation of the linear and angular velocity is calculated using a filtered derivative. The communication between the system and the calculator (computer) is assured by a Dspace card. The control task is performed using SIMULINK and the ControlDesk software, which can allow us to see in real time the informations coming from the sensors and also to send commands to the linear motor.

4.6.2 Dynamic Model

The Lagrangian of the system is represented by the following equation:

$$L = \frac{1}{2}Mv_1^2 + \frac{1}{2}mv_2^2 - mgl\cos\theta, \tag{4.32}$$

where θ stands for the angular displacement measured from the equilibrium position. M and m represent the mass of the cart and the point mass, respectively. v_1 and v_2 are, respectively, the velocity of the cart and the velocity of the point mass m. l is the length of the rod, and g is the gravitational constant. After replacing v_1 and v_2 by their expression, and simplifying the result, the Lagrangian is given by

$$L = \frac{1}{2}(M+m)\dot{x}^2 - ml\dot{x}\dot{\theta}\cos\theta + \frac{1}{2}ml^2\dot{\theta}^2 - mgl\cos\theta, \tag{4.33}$$

where x denotes the cart's position. Using the equation of Euler–Lagrange, one can easily calculate the equations of motion of the system

$$
\begin{aligned}
&(M+m)\ddot{x} - ml\sin\theta\dot{\theta}^2 + ml\cos\theta\ddot{\theta} = N,\\
&ml^2\ddot{\theta} + ml\cos\theta\ddot{x} - mgl\sin\theta = 0,
\end{aligned}
\tag{4.34}
$$

where $N = \alpha U$ is the force exercised on the cart. The matrix representation of these equations is

$$
\begin{bmatrix}\dot{x}\\ \ddot{x}\\ \dot{\theta}\\ \ddot{\theta}\end{bmatrix} = \begin{bmatrix}\dot{x}\\ \dfrac{(-mgl\cos\theta + ml^2\dot{\theta}^2)\sin\theta}{l(M+m\sin^2\theta)}\\ \dot{\theta}\\ \dfrac{((M+m)g - ml\cos\theta\dot{\theta}^2)\sin\theta}{l(M+m\sin^2\theta)}\end{bmatrix} + \begin{bmatrix}0\\ \dfrac{l}{l(M+m\sin^2\theta)}\\ 0\\ \dfrac{-\cos\theta}{l(M+m\sin^2\theta)}\end{bmatrix} N. \tag{4.35}
$$

We will consider the linearization of (4.35) at the upper position:

$$
\begin{bmatrix}\dot{x}\\ \ddot{x}\\ \dot{\theta}\\ \ddot{\theta}\end{bmatrix} = \begin{bmatrix}0 & 1 & 0 & 0\\ 0 & 0 & \dfrac{-mg}{M} & 0\\ 0 & 0 & 0 & 1\\ 0 & 0 & \dfrac{(m+M)g}{Ml} & 0\end{bmatrix}\begin{bmatrix}x\\ \dot{x}\\ \theta\\ \dot{\theta}\end{bmatrix} + \begin{bmatrix}0\\ \dfrac{\alpha}{M}\\ 0\\ \dfrac{-\alpha}{Ml}\end{bmatrix} U. \tag{4.36}
$$

The parameters describing the linear model (4.36) are given in Table 4.2.

4.6.3 Theoretical Results

In the following, we consider a feedback control gain (4.4):

$$
F = \left[5.825\ 5.883\ 24.94\ 5.140\right],
$$

which was obtained using a placement of the poles $\{-100, -2+2i, -2-2i, -2\}$.

Table 4.2 Inverted pendulum benchmark parameters

Parameters	Values	Characteristics
M	3.9249 Kg	Mass of the cart
m	0.2047 Kg	Mass of the m point
l	0.2302 m	Length of the rod
g	9.81 N/kg	Gravitational constant
α	25.3 N/V	Motor's gain

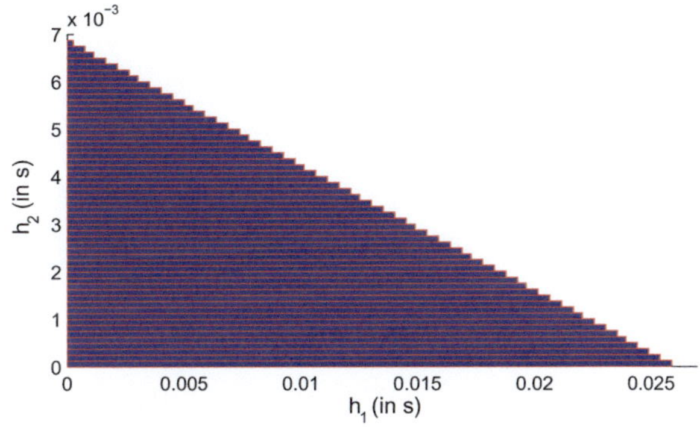

Fig. 4.5 Stability domain obtained with the dissipativity-based approach

We consider the case of $M = 2$ sensors, one measuring $x_1 = \left[x\ \dot{x}\right]^T$, and one measuring $x_2 = \left[\theta\ \dot{\theta}\right]^T$.

Using the result from the small gain approach (Theorem 4.1), we can show that the interconnected system (4.10) is $\mathscr{L}_2$-stable if $h_{max} = \max_{i \in I_M} \{h_i\} < \frac{1}{\|G\|_\infty} = 1.3\,\text{ms}$.

Note that although this approach allows to design an admissible upper bound h_{max} for the sampling intervals, it does not make it possible to take into account the difference in the upper bounds of the two sensors (h_1 and h_2) in the analysis.

Using the LMI result from the dissipativity-based approach (Theorem 4.3), it is possible to design admissible upper bounds on the sampling intervals for each sensor. The obtained stability domain is shown in Fig. 4.5.

On top of allowing the design of each sensor's admissible sampling interval upper bound separately, this approach seems to be much less conservative than the previous one.

4.6.4 Experimental Results

In this section, we present the experimentations performed using our benchmark. The results have been obtained using randomly varying sampling intervals with upper bounds $h_1 = 7\,\text{ms}$ and $h_2 = 5\,\text{ms}$ (see Fig. 4.6 to see the profile of the sampling intervals). Note that these upper bounds are within the stability domain obtained with the dissipativity approach, which is presented in Fig. 4.5.

The evolution of the position x and angular position θ is presented in Fig. 4.7, where large perturbations have been introduced at $t = 3.1\,\text{s}$, $t = 16.4\,\text{s}$, and $t = 25.9\,\text{s}$, by acting on the pendulum manually. One can see that the pendulum stabilizes even in the presence of sampling intervals variations and asynchronicity of both sensors. The absence of oscillating patterns can be explained by the presence of dry friction.

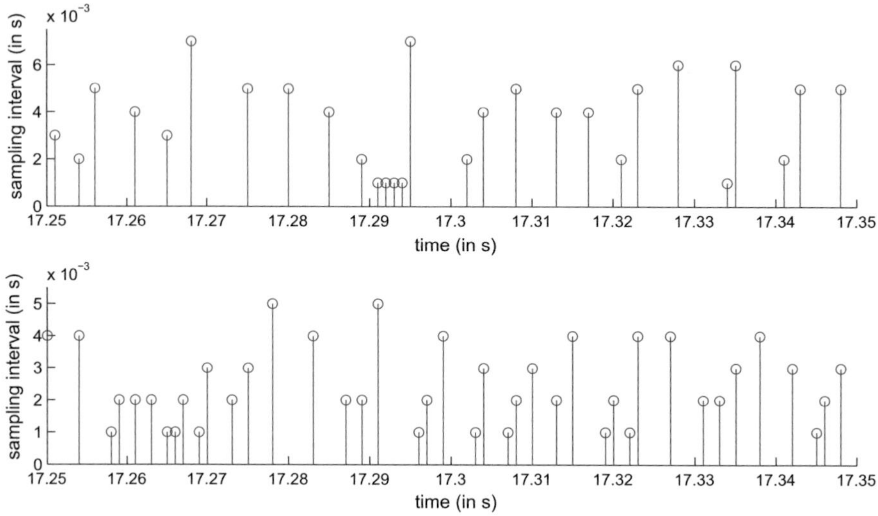

Fig. 4.6 Profile of the sampling of x and $\dot{x}$ (on top), and of θ and $\dot{\theta}$ (at the bottom)

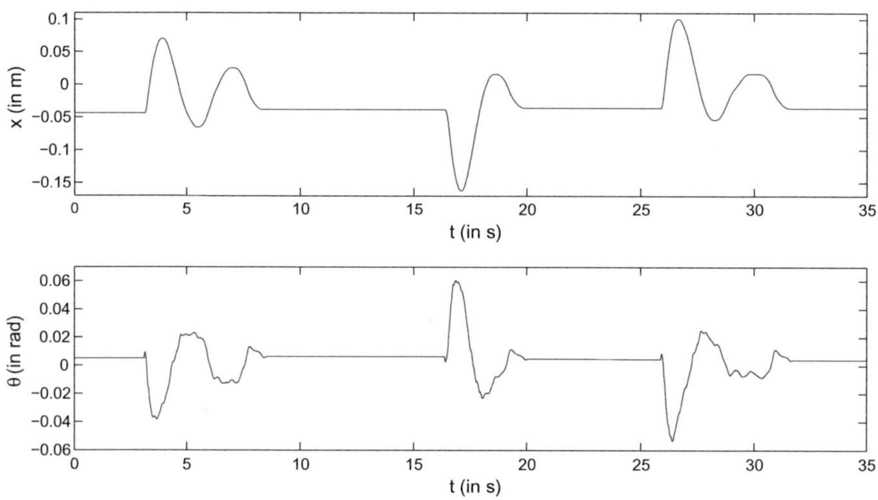

Fig. 4.7 Evolution of the position x (top) and the angular position θ (bottom) of the inverted pendulum subject to perturbations for $h_1 = 7\,\text{ms}$ and $h_2 = 5\,\text{ms}$

4.7 Conclusion

This chapter was dedicated to the stability analysis of LTI sampled-data systems with asynchronous sensors and aperiodic sampling. The study is based on tools from the robust control theory, and on a modeling of the system as an input/output interconnection between an operator, representing the sampling error, and a continuous-time (i.e., sampling-free) closed-loop system. Two approaches have been presented. The first one is based on the small gain theorem and provides a tractable estimation of the overall (with respect to all the sensors) maximum allowable sampling period. The second approach is based on the dissipativity theory. It provides simple numerical stability criterion using LMIs. An advantage of that approach compared to the previous one is that it allows for computing an estimation of the maximum allowable sampling for each and every sensor of the system. Experimentations have been performed on an inverted pendulum benchmark and confirm the applicability of both approaches. The results seem to indicate that the dissipativity-based approach is less conservative than small gain-based one. In further studies, it would be interesting to consider additional phenomena in one study, such as delays or quantization.

References

1. Araki, M., Yamamoto, K.: Multivariable multirate sampled-data systems: state-space description, transfer characteristics and Nyquist criterion. IEEE Trans. Autom. Control **31**(2), 145–154 (1986)
2. Chen, T., Francis, B.A.: Optimal Sampled-Data Control Systems. Springer, Berlin (1993)
3. Cloosterman, M.-B.-G., Hetel, L., van de Wouw, N., Heemels, W.-P.-M.-H., Daafouz, J., Nijmeijer, H.: Controller synthesis for networked control systems. Automatica **46**(10), 1584–1594 (2010)
4. Fiter, C., Hetel, L., Perruquetti, W., Richard, J.-P.: A state dependent sampling for linear state feedback. Automatica **48**(8), 1860–1867 (2012)
5. Fridman, E.: A refined input delay approach to sampled-data control. Automatica **46**(2), 421–427 (2010)
6. Fridman, E., Seuret, A., Richard, J.P.: Robust sampled-data stabilization of linear systems: an input delay approach. Automatica **40**(8), 1441–1446 (2004)
7. Fujioka, H.: Stability analysis of systems with aperiodic sample-and-hold devices. Automatica **45**(3), 771–775 (2009)
8. Gu, K., Kharitonov, V., Chen, J.: Stability of Time-Delay Systems. Birkhauser, Boston (2003)
9. Hetel, L., Daafouz, J., Iung, C.: Stabilization of arbitrary switched linear systems with unknown time-varying delays. IEEE Trans. Autom. Control **51**(10), 1668–1674 (2006)
10. Hetel, L., Fiter, C., Omran, H., Seuret, A., Fridman, E., Richard, J.-P., Niculescu, S.-I.: Recent developments on the stability of systems with aperiodic sampling: an overview. Automatica **76**, 309–335 (2017)
11. Kao, C.Y., Lincoln, B.: Simple stability criteria for systems with time-varying delays. Automatica **40**(8), 1429–1434 (2004)
12. Karafyllis, I., Kravaris, C.: Global stability results for systems under sampled-data control. Int. J. Robust Nonlinear Control **19**(10), 1105–1128 (2009)
13. Khalil, H.: Nonlinear Systems, 3rd edn, p. 9. Prentice Hall, New Jewsey (2002)
14. Mirkin, L.: Some remarks on the use of time-varying delay to model sample-and-hold circuits. IEEE Trans. Autom. Control **52**(6), 1109–1112 (2007)

15. Nesic, D., Teel, A., Carnevale, D.: Explicit computation of the sampling period in emulation of controllers for nonlinear sampled-data systems. IEEE Trans. Autom. Control **54**(3), 619–624 (2009)
16. Omran, H., Hetel, L., Richard, J.P., Lamnabhi-Lagarrigue, F.: Stability analysis of bilinear systems under aperiodic sampled-data control. Automatica **50**(4), 1288–1295 (2014)
17. Seuret, A.: A novel stability analysis of linear systems under asynchronous samplings. Automatica **48**(1), 177–182 (2012)
18. Willems, J.C.: Dissipative dynamical systems. part I: general theory. Arch. Ration. Mech. Anal. **45**(5), 321–351 (1972)
19. Zhang, W., Branicky, M.-S., Phillips, S.-M.: Stability of networked control systems. IEEE Control Syst. Mag. **21**(1), 84–99 (2001)
20. Zhou, K., Doyle, J.C., Glover, K.: Robust and Optimal Control, vol. 40. Prentice hall, New Jersey (1996)

Chapter 5
Template Complex Zonotope Based Stability Verification

A. Adimoolam and T. Dang

Abstract In this paper, we consider the problem of verifying stability of computer control systems whose behavior can be modeled by nearly periodic linear impulsive systems. In these systems, the eigenstructure and stability of the dynamics are closely related. A recently introduced set representation called complex zonotopes could utilize the possibly complex eigenstructure of the dynamics to define contractive sets for stability verification, which demonstrated good accuracy on some benchmark examples. However, complex zonotopes had the drawback that it is not easy to refine them in the stability verification procedure, while also we had to guess the contractive complex zonotope instead of systematically synthesizing it. Overcoming this drawback, in this paper we introduce a more general set representation called template complex zonotope, which has the advantage that it is easy to refine and also the contractive set can be systematically synthesized. We corroborate the efficiency of our approach by experimenting on some benchmark examples.

5.1 Introduction

Embedded control systems combine computer software with the physical world, and their global behavior can be modeled using hybrid systems. To assert correctness of such systems, their global behavior under all possible nondeterminism resulting from the interaction between continuous and discrete dynamics should be accurately analyzed. Thus, one of the key ingredients for safe design and verification of embedded control systems is a set representation which, on one hand, is expressive enough to describe the evolution of sets of hybrid trajectories and, on the other hand, can be manipulated by time-efficient algorithms. For most hybrid systems with nontrivial

---

A. Adimoolam (✉)
VERIMAG, Université Grenoble-Alpes, 700 Avenue Centrale,
38400 Saint Martin D'Hères, France
e-mail: santosh.adimoolam@univ-grenoble-alpes.fr

T. Dang
CNRS, VERIMAG, Université Grenoble Alpes, 700 Avenue Centrale,
38400 Saint Martin D'Hères, France
e-mail: thao.dang@univ-grenoble-alpes.fr

S. Tarbouriech et al. (eds.), *Control Subject to Computational and Communication Constraints*, Lecture Notes in Control and Information Sciences 475, https://doi.org/10.1007/978-3-319-78449-6_5

continuous dynamics, exact computation of hybrid trajectories is impossible, so the focus is put on approximate computation of reachable states. In the area of abstract interpretation within program verification, there is a similar need for data structures for set manipulation, called abstract domains, which should be fine-grained enough to be accurate, yet computationally tractable to deal with complex programs.

Two classical abstract domains are intervals [12] and convex polyhedra [13], and their variants have been developed to achieve a good compomise between computational speed and precision, such as zones [28], octagons [29], linear templates [35], zonotopes [19], and tropical polyhedra [4]. For hybrid model checking, convex polyhedra and their special classes such as parallelotopes and zonotopes are also among popular set representations. Beyond polyhedral set representations, ellipsoids can be used for reachable set computations [27]. In abstract interpretation, polynomial inequalities are used for invariant computation via their reduction to linear inequalities in [7] and polynomial equalities via Gröner basis methods [34]. Quadratic templates are also proposed, where semi-definite relaxations are used for deriving nonlinear invariants (for instance quadratic invariants inspired by Lyapunov functions) [2, 15]. Recently, complex zonotopes [1] extended usual zonotopes to the complex domain, which geometrically speaking are Minkowski sum of line segments and some ellipsoids. Other extensions of zonotopes, such as quadratic [3] and more general polynomial zonotopes [6] have been proposed. Complex zonotopes are however different from polynomial zonotopes because while a polynomial zonotope is a set-valued polynomial function of *intervals*, a complex zonotope is a set-valued function of unit *circles* in the complex plane.

Complex zonotopes can utilize the possibly complex eigenstructure of the dynamics of linear impulsive systems to define contractive sets for stability verification. This is an advantage over polytopes or usual zonotopes that can only utilize the real eigenstructure but not the complex eigenstructure. For stability verification of nearly periodic linear impulsive systems, complex zonotopes demonstrated good accuracy on some benchmark examples. However, a drawback of complex zonotopes is that adding more generators to it can violate the property of contraction with respect to the dynamics. Therefore, it is not easy to refine complex zonotopes for verifying stability for larger intervals of sampling times. Moreover, we had to heuristically guess a suitable complex zonotope for stability verification instead of systematically synthesizing it.

Overcoming the aforementioned drawback, in this paper, we introduce a more general set representation called *template complex zonotopes*. In a template complex zonotope, the bounds on the complex combining coefficients, called *scaling factors*, are treated as variables, while the directions for the generators are fixed a priori by a *template* of complex vectors. This allows us to systematically synthesize a suitable template complex zonotope for stability verification, instead of guessing it like in the case of complex zonotopes. Furthermore, template complex zonotopes can be refined easily by adding any arbitrary set of vectors to the existing template, because the scaling factors can be adjusted accordingly. We present experiments on some benchmark examples where template complex zonotopes contract faster than complex zonotopes, resulting in faster verification.

Basic notation. We represent integers by $\mathbb{Z}$, real numbers by $\mathbb{R}$ and complex numbers by $\mathbb{C}$. For integers p and q, the set of $p \times q$ matrices with entries drawn from a set Ψ is denoted $\mathbb{M}_{p\times q}(\Psi)$. If z is a vector of complex numbers, then $real(z)$ denotes the real part of z and the imaginary part is denoted $img(z)$. For a positive integer i, the ith component of z is denoted by z_i. For a matrix X and positive integers i, j, X_{ij} is the ith row and jth column entry of X. The diagonal square matrix containing entries of z along the diagonal is denoted by $\mathscr{D}(z)$. If c is a scalar complex number, its absolute value is $|c| = (|real(c)|^2 + |img(c)|^2)^{1/2}$. The infinity norm of a possibly complex n-dimensional vector z is $\|z\|_\infty = \max_{i=1}^{n} |z_i|$. The infinity norm of a possibly complex $n \times m$ matrix X is $\|X\|_\infty = \max_{i=1}^{n} \sum |X_{ij}|$.

The rest of the paper is organized as follows. We introduce template complex zonotopes in Sect. 5.2 and discuss important operations on them like linear transformation, Minkowski sum and inclusion checking. In Sect. 5.3, we first define a nearly periodic linear impulsive system and the problem of verifying global exponential stability. Using Proposition 5.1, we relate the contraction of a template complex zonotope to the eigenstructure of the dynamics, which is a motivation for using template complex zonotopes for stability verification. Later in that section, we discuss how to find suitable template complex zonotopes and verify their contraction to establish exponential stability of the system. In Sect. 5.4, we describe experiments on two benchmark examples and their results which corroborate the efficiency of our approach.

5.2 Template Complex Zonotopes

A template complex zonotope is a set representation in which each point is described as a linear combination of a set of complex-valued vectors, called as a *template*, such that the complex combining coefficients are bounded in absolute values by a set of positive bounds called *scaling factors*.

Definition 5.1 (*Template complex zonotope*) For $n, m \in \mathbb{Z}_{>0}$, let $V \in \mathbb{M}_{n\times m}(\mathbb{C})$ be a template, $c \in \mathbb{C}^n$ be a center point and $s \in \mathbb{R}^m_{\geq 0}$ be scaling factors. Then we define a template complex zonotope as

$$\mathscr{C}(V, c, s) = \left\{ V\zeta + c : \zeta \in \mathbb{C}^m \wedge \forall i \in \{1, \dots, m\}, |\zeta_i| \leq s_i \right\}.$$

Complex zonotopes, which were introduced in [1], are a special case of template complex zonotopes where $s_i = 1$ for all $i \in \{1, \dots, m\}$. The real projection of a template complex zonotope can represent, in addition to polytopic zonotopes, non-polyhedral convex sets. Therefore, they are also more expressive than usual (real-valued) zonotopes. To illustrate, Fig. 5.1 represents the non-polyhedral real projection of the template complex zonotope $\mathscr{C}(V, 0, s)$ where $V = \begin{pmatrix} (1+2i) & 1 & (2+i) \\ (1-2i) & 1 & (2-i) \end{pmatrix}$ and $s = [1\ 1\ 1]^T$.

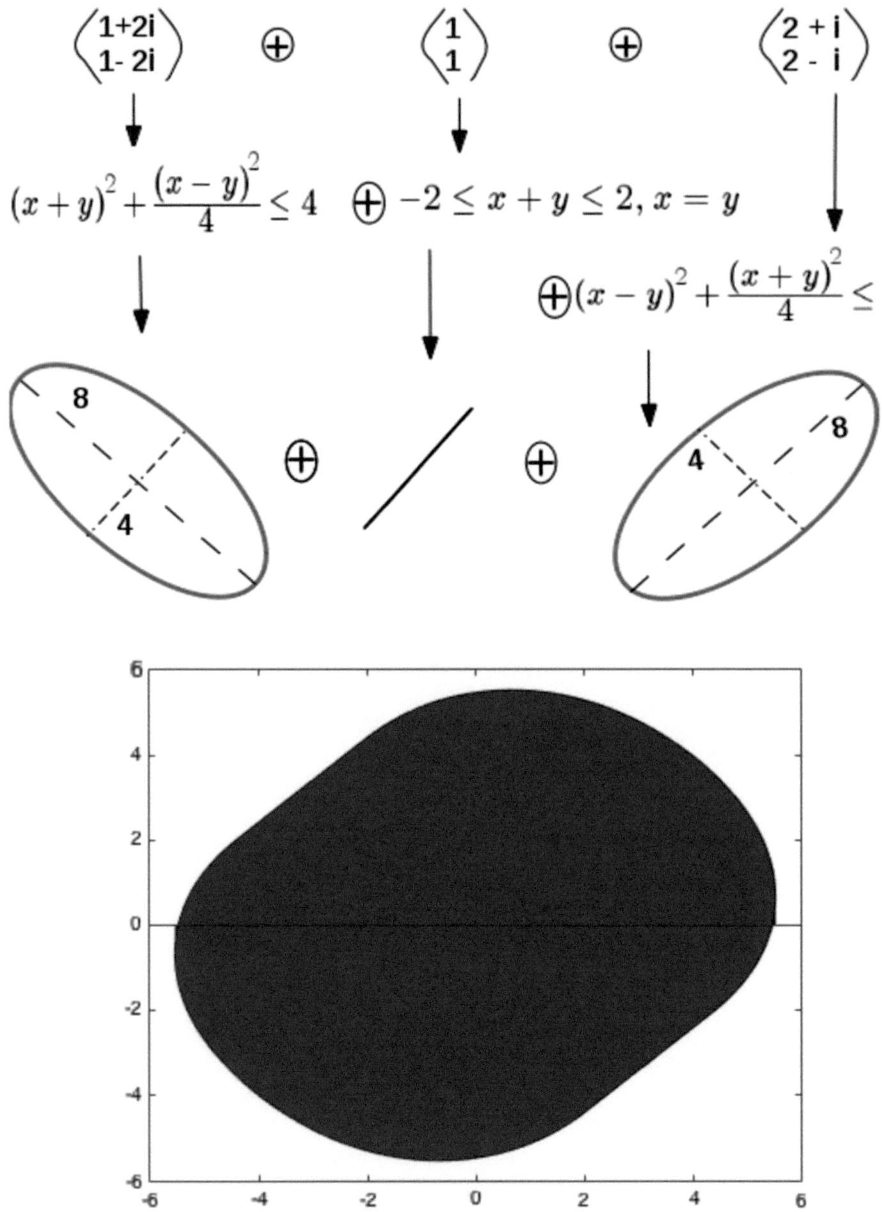

Fig. 5.1 Real projection of a template complex zonotope

In the rest of the paper, unless otherwise stated, we assume that V is an $n \times m$ complex matrix, s is an $m \times 1$ column vector and c is a $n \times 1$ column vector for $m, n \in \mathbb{Z}_{>0}$.

Operations on template complex zonotopes. Concerning linear transformation and the Minkowski sum, computations on template complex zonotopes are similar to those on usual (real-valued) zonotopes as stated in the following results, which can be proved by the same techniques as that of a usual zonotope.

Lemma 5.1 (Linear transformation and Minkowski sum) *Let* $A \in \mathbb{M}_{n\times n}(\mathbb{C})$, $V \in \mathbb{M}_{n\times m}(\mathbb{C})$, $G \in \mathbb{M}_{n\times r}(\mathbb{C})$, $c, d \in \mathbb{C}^n$, $s \in \mathbb{R}^m_{\geq 0}$, $h \in \mathbb{R}^r_{\geq 0}$ *for some* $n, m, r \in \mathbb{Z}_{>0}$.

1. Linear transformation: $A\mathscr{C}(V, c, s) = \mathscr{C}(AV, Ac, s)$.
2. Minkowski sum: $\mathscr{C}(V, c, s) \oplus \mathscr{C}(G, d, h) = \mathscr{C}\left([V \;\; G], (c+d), \begin{pmatrix} s \\ h \end{pmatrix}\right)$.

Checking inclusion is a fundamental problem in reachability computation. For stability verification, we are interested in efficiently checking the inclusion between two template complex zonotopes centered at the origin. Although this problem is non-convex in general, we derive an easily verifiable convex condition which is sufficient (but not necessary), as follows. The inclusion checking method we propose in the following can be extended to zonotopes centered anywhere.

Let us consider that we want to check the inclusion of a template complex zonotope $\mathscr{C}(V^a, 0, s^a)$ inside a template complex zonotope $\mathscr{C}(V^b, 0, s^b)$, where $V^a \in \mathbb{M}_{n\times r}(\mathbb{C})$, $V^b \in \mathbb{M}_{n\times m}(\mathbb{C})$, $s^a \in \mathbb{R}^r_{\geq 0}$ and $s^b \in \mathbb{R}^m_{\geq 0}$ for some $n, m, r \in \mathbb{Z}_{>0}$. We relate any point in V^a to a point in V^b as follows. Consider $X \in \mathbb{M}_{r\times m}(\mathbb{C})$ as a matrix solving $V^a\mathscr{D}(s^a) = V^bX$, which we call a *transfer matrix* from $\mathscr{C}(V^a, 0, s^a)$ to $\mathscr{C}(V^b, 0, s^b)$. Recall that $\mathscr{D}(s^a)$ is the diagonal square matrix containing entries of s^a along its diagonal. Let any point z in $\mathscr{C}(V^a, 0, s)$ be written as $z = V^a\zeta$ where ζ is a combining coefficient whose absolute values are bounded by s^a. By normalizing with the scaling factors, we can write $\zeta = \mathscr{D}(s)\varepsilon$ for some $\varepsilon : \|\varepsilon\|_\infty \leq 1$. So, we get $z = V^a\mathscr{D}(s)\varepsilon$. Then using the relation for the transfer matrix, we rewrite $z = V^bX\varepsilon$. If the absolute value of every component of $X\varepsilon$ is less than the corresponding value of s^b, then $X\varepsilon$ can be treated as a combining coefficient of $\mathscr{C}(V^b, 0, s^b)$ for the point z and then z also belongs to $\mathscr{C}(V^b, 0, s^b)$. This is true if $\sum_{j=1}^m |X_{ij}| \leq s^b_i \; \forall i \in \{1, \ldots, m\}$. This gives us a sufficient condition, stated in the following theorem for inclusion of $\mathscr{C}(V^a, 0, s^a)$ in $\mathscr{C}(V^b, 0, s^b)$.

Theorem 5.1 (Inclusion) *Let* $V^a \in \mathbb{M}_{n\times r}(\mathbb{C})$, $V^b \in \mathbb{M}_{n\times m}(\mathbb{C})$, $s^a \in \mathbb{R}^r_{\geq 0}$ *and* $s^b \in \mathbb{R}^m_{\geq 0}$ *for some* $n, m, r \in \mathbb{Z}_{>0}$. *Then* $\mathscr{C}(V^a, 0, s^a) \subseteq \mathscr{C}(V^b, 0, s^b)$ *if all the following statements are true.*

$$\exists X \in \mathbb{M}_{m\times r}(\mathbb{C}) :$$

$$V^bX = V^a\mathscr{D}(s^a) \;\; \textit{and} \;\; \forall i \in \{1, \ldots, m\} \;\; \left(\sum_{j=1}^r |X_{ij}|\right) \leq s^b_i. \tag{5.1}$$

For fixed V_a and V_b, the constraints in Theorem 5.1 are second-order conic constraints in the variables s^a, s^b, and X. Many convex optimization solvers can solve such constraints efficiently upto a high numerical precision.

5.3 Nearly Periodic Linear Impulsive System

A nearly periodic linear impulsive system is a hybrid system whose state evolves continuously by a linear differential equation for some bounded time period, after which there is an instantaneous linear impulse. Formally, a linear impulsive system is specified by a tuple $\mathscr{L} = \langle A_c, A_r, \Delta \rangle$ where A_c and A_r are $n \times n$ real matrices called the linear field matrix and impulse matrix, respectively. The positive integer n is the dimension of the state space and $\Delta = [t_{min}, t_{max}]$ is an interval of nonnegative reals, called the sampling period interval. The dynamics of the nearly periodic linear impulsive system is described as follows. A function $\mathbf{x} : \mathbb{R}_{\geq 0} \rightarrow \mathbb{R}^n$ is called a trajectory of the system if there exists a sequence of sampling timesdefinition $(t_i)_{i=1}^{\infty}$ satisfying all the following:

$$\begin{aligned} &(t_{i+1} - t_i) \in \Delta \;\; \forall i \in \mathbb{Z}_{>0} \quad \text{(uncertainty in sampling period)} \\ &\dot{\mathbf{x}}(t) = A_c \mathbf{x}(t) \;\; \forall t \in \mathbb{R}_{\geq 0} : t \neq t_i \; \forall i \in \mathbb{Z}_{>0} \;\; \text{(continuous)} \\ &\mathbf{x}(t_i^+) = A_r \mathbf{x}(t_i^-) \;\; \forall i \in \mathbb{Z}_{>0} \quad \text{(linear impulse)} \end{aligned} \tag{5.2}$$

We say that the linear impulsive system is globally exponentially stable (GES) if all the trajectories of the system beginning at any point in the state space eventually reach arbitrarily close to the origin at an exponential rate, as follows.

Definition 5.2 (*Global exponential stability*) The system (5.2) is globally exponentially stable (GES) if there exist positive scalars $c > 0$ and $\lambda \in [0, 1)$ such that for all $t \in \mathbb{R}_{\geq 0}$, $\mathbf{x}(t) \leq c\lambda^t \|\mathbf{x}(0)\|$.

We state the stability verification problem as follows.

Problem 5.1 (*Stability verification problem*) proposition Given A_c, A_r, t_{min}, find the largest upper bound t_{max} on the sampling time to guarantee exponential stability.

Related work on stability verification of nearly periodic linear impulsive systems

A common approach to this problem is extending Lyapunov techniques, which results in Lyapunov Krakovskii functionals [37, 39] (using the framework of time-delay systems), and discrete-time Lyapunov functions [36]. Stability with respect to time-varying input delay can also be handled by input/output approach [24]. Stability verification problem for time-varying impulsive systems can also be formulated in a hybrid systems framework [8, 20, 32], for which various Lyapunov-based methods including discontinuous time-independent [31] or time-dependent Lyapunov functions [17] were developed. Another approach involves using convex embedding [18, 21, 22]. In this approach, stability conditions can be checked using parametric Linear Matrix Inequalities (LMIs) [21], or as set contractiveness (such as, polytopic set contractiveness) [10, 16, 25]. Inspired by these results on set contractiveness, conditions [1] provides a stability condition, expressed in terms of complex zonotopes, which is more conservative but can be efficiently verified. The novelty of this work is in the extension of complex zonotopes to template complex zonotopes which allows a systematic way to synthesize contractive zonotopic sets to verify stability.

The state reached after a linear impulse followed by a continuous evolution for time t is $e^{A_c t} A_r x$. So, let us denote $H_t = e^{A_c t} A_r$ for any positive real t, which we call as the reachability operator if t lies in the sampling period interval Δ. Given a set $\Psi \subset \mathbb{R}^n$, the set of all reachable points of Ψ, when acted upon by an impulse followed by continuous evolution for sampling time period $t \in \Delta$ until before the next impulse, is $\bigcup_{t \in \Delta} H_t \Psi$. It was shown previously in [16, 25] that a necessary and sufficient condition for exponential stability is the existence of a convex, compact, and closed set containing the origin in its interior, called a C-set, that contracts between subsequent impulses. In other words, we can establish exponential stability by finding a C-set Ψ such that $H_t \Psi \subseteq \lambda \Psi$ for some $\lambda \in [0, 1)$ and for all $t \in \Delta$. In this paper, we want to find contractive C-sets represented as template complex zonotopes. We define *contraction* of a template complex due to a linear operator as follows, which when less than one, implies that the zonotope is contractive.

Definition 5.3 For a template complex zonotope $\mathscr{C}(V, 0, s)$, the amount of contraction by a square matrix $J \in \mathbb{M}_{n \times n}(\mathbb{R})$, denoted by $\chi(V, s, J)$ is

$$\chi(V, s, J) = \min \left\{ \lambda \in \mathbb{R}_{\geq 0} : J\mathscr{C}(V, 0, s) \subseteq \lambda \mathscr{C}(V, 0, s) \right\}.$$

Our motivation for considering template complex zonotopes can be inferred from the following proposition.

Proposition 5.1 *Let V contain only the eigenvectors of H_t as its column vectors and μ be the vector of eigenvalues corresponding the columns of V. Then $H_t \mathscr{C}(V, 0, s) = \mathscr{C}(V, 0, \mathscr{D}(|\mu|)s)$.*

For a fixed sampling time period, i.e., when $t_{min} = t_{max} = t$, we can infer from the above proposition that the contraction of the template complex zonotope formed by the eigenvector template is bounded by the largest absolute value of the eigenvalues. Therefore, when the sampling period is fixed, we can find contractive template complex zonotopes for exponentially stable systems by choosing the template as the collection of eigenvectors. However, we are interested in the case where the sampling time period varies in the interval Δ, i.e., there are uncountably many reachability operators parametrized over the time interval Δ. Motivated by the above analysis, when the sampling period is uncertain, we choose the template as the collection of eigenvectors of a few reachability operators and try to synthesize suitable scaling factors for which the template complex zonotope contracts with respect to the chosen finite set of reachability operators. However, later we also verify that the synthesized template complex zonotope actually contracts with respect to all the (uncountably many) reachability operators. First, we describe the procedure to synthesize the template complex zonotope.

Synthesizing a candidate template complex zonotope. This step of systematically synthesizing a suitable template complex zonotope that is likely to be contractive constitutes the main improvement over the procedure proposed in [1]. Our criterion for synthesizing the template complex zonotope is that it has to be contractive with respect to a few reachability operators (called reference operators), which is a necessary condition for contraction with respect to the overall system dynamics. Since the eigenstructure of the reachability operators is related to the stability of the system, we include the eigenvectors of a few reachability operators in the template. For a fixed number $k \in \mathbb{Z}_{>0}$ of reference operators, they can be chosen incrementally as follows. Define k-sampled time points as $\Lambda_k = \left\{t_{min} + i\frac{(t_{max}-t_{min})}{k} : i \in \{0, \ldots, k-1\}\right\}$. Then, we define the set of k-sampled reachability operators as $\Gamma_k = \{H_t : t \in \Lambda_k\}$. Let us denote the template of collection of eigenvectors of all operators in Γ_k as E_k. For this template, we synthesize suitable scaling factors based on the following theorem. The derivation of this theorem uses the inclusion checking condition from Theorem 5.1.

Theorem 5.2 *If E_k have rank n. For a vector of scaling factors $s \in \mathbb{R}^m_{\geq 0}$, the template complex zonotope $\mathscr{Z} = \mathscr{C}(E_k, 0, s)$ would represent a C-set and also $H_t\mathscr{Z} \subseteq \lambda\mathscr{Z}$ for $\lambda \in (0,1)$ if following are all satisfied.*

$$
\begin{aligned}
& s \in \mathbb{R}^n_{\geq 1} \quad \textit{(sufficient condition for representing C-set)} \\
& \exists X_t \in \mathbb{M}_{m\times m}(\mathbb{C}) \;\; \forall t \in \Lambda_k \;\; \textit{s.t.} \\
& E_k X_t = H_t E_k \mathscr{D}(s) \quad \textit{(transfer matrix condition)} \\
& \sum_{j=1}^{m} \left|(X_t)_{ij}\right| \leq \lambda s_i \;\; \forall i \in \{1, \ldots, m\} \;\; \textit{(bounding contraction)}
\end{aligned}
\tag{5.3}
$$

Therefore, we can synthesize a template complex zonotope that is contractive with respect to a finite chosen number $k > 0$ of reference operators by solving for the scaling factors satisfying the second-order conic constraints in (5.3).

Verifying contraction. To verify that this synthesized template complex zonotope actually contracts with respect to all the reachability operators H_t, where t is parametrized over the whole sampling time interval Δ, we divide the sampling interval into small enough subintervals and verify contraction in each interval. To bound the amount of contraction in small intervals, we use some useful properties of contraction, which were earlier derived in [1].

Let $H_{t+\rho}$ be an operator where ρ lies in the interval $(0, \varepsilon)$. For an order of Taylor expansion $r \in \mathbb{Z}_{>0}$ and some $\delta \in [0, \varepsilon]$, define

$$P_r^t(\rho) = \sum_{i=0}^{r} \frac{A_c^i \rho^i}{i!} H_t \quad \text{and} \quad E_r^t(\delta) = \frac{A_c^{r+1}\delta^{r+1}}{(r+1)!} H_t.$$

Then based on Taylor expansion, we get that

$$H_{t+\rho} = P_r^t(\rho) + theorem E_r^t(\delta).$$

Furthermore, a bound on contraction as sum of contractions depending on ε is derived in [1] as

$$\chi\left(V, s, H_{t+\rho}\right) \leq \left(\max_{i=0}^{r} \chi\left(V, s, P_r^t(\varepsilon)\right)\right) + \frac{\varepsilon^{r+1}}{(r+1)!}\chi\left(V, s, A_c^{r+1} H_t\right) \tag{5.4}$$

The right-hand side of (5.4) can be bounded if we know a bound on the contraction under a linear operation, which is derived as follows.

Lemma 5.2 *Let $J \in \mathbb{M}_{n\times n}(\mathbb{R})$. Define*

$$\beta(V, s, J) = \min\{\|X\|_\infty : X \in \mathbb{M}_{m\times m}(\mathbb{C}) \wedge V\mathscr{D}(s)X = JV\mathscr{D}(s)\}$$

Then we have $\chi(V, s, J) \leq \beta(V, s, J)$.

Proof If $JV\mathscr{D}(s) = V\mathscr{D}(s)X$, we want to prove that $J\mathscr{C}(V, 0, s) = \mathscr{C}(JV, 0, s) \subseteq \|X\|_\infty \mathscr{C}(V, 0, s)$. We deduce $\mathscr{C}(JV, 0, s) = \{JV\zeta : \forall i \in \{1, \ldots, m\} |\zeta_i| \leq s_i\} = \{JV\mathscr{D}(s)\zeta' : \|\zeta'\|_\infty \leq 1\} = \{V\mathscr{D}(s)X\zeta' : \|\zeta'\|_\infty \leq 1\} \subseteq \|X\|_\infty\{V\mathscr{D}(s)\zeta' : \|\zeta'\|_\infty \leq 1\} = \|X\|_\infty\{V\zeta : \forall i \in \{1, \ldots, m\} |\zeta_i| \leq s_i\} = \|X\|_\infty \mathscr{C}(V, 0, s)$.

Then the contraction for sampling time interval $(t, t+\delta)$ can be bounded as follows.

Theorem 5.3 *Let $\rho \in (t, t+\varepsilon)$ and for $r \in \mathbb{Z}_{>0}$, $P_r^t(\rho) = \sum_{i=0}^{r} \frac{A_c^t \rho^i}{i!} H_t$. Define $\eta_r(V, s, t, \varepsilon) = \left(\max_{i=0}^{r} \beta\left(V, s, P_r^t(\varepsilon)\right)\right) + \frac{\varepsilon^{r+1}}{(r+1)!}\beta\left(V, s, A_c^{r+1} H_t\right)$, where the bound $\beta(.)$ is defined in Lemma 5.2. Then $\chi\left(V, s, H_{t+\rho}\right) \leq \eta_r(V, s, t, \varepsilon)$*

Verification algorithm. We begin with $k = 3$ reference operators that correspond to the two end points of the sampling interval and the middle point. The algorithm first finds suitable scaling factors s that gives a template complex zonotope which contracts with respect to these reference operators. Next, it checks whether the contraction of the template complex zonotope $\mathscr{C}(E_k, 0, s)$ with respect to all the reachability operators is less than one. For checking contraction, we use the algorithm earlier proposed in [1]. If successful, then exponential stability of the system is verified. Otherwise, k is increased, the algorithm synthesizes the template complex zonotope for the increased k and then checks contraction. If unsuccessful after a maximum value of k, the algorithm stops and the result is inconclusive. This is described in Algorithm 1. This algorithm has been implemented and we defer experimental results to Sect. 5.4.

5.4 Experiments

We evaluated our algorithm on two benchmark examples of linear impulsive systems below and compared it with other state-of-the-art approaches. For convex optimization, we use CVX version 2.1 with MATLAB 8.5.0.197613 (R2015a). The reported experimental results were obtained on Intel(R) Core(TM) i5-3470 CPU @ 3.20GHz.

Algorithm 1 Exponential stability verification of system (5.2)

Initialize $k = 3$.
Choose $M \in \mathbb{Z}_{>0}$ as the largest value of k and tol as discretization parameter.
while $k \leq M$ **do**
  Initialize $t = t_{min}$, $h = tol$, r as order of Taylor expansion (typically ≤ 2).
  while $h \geq tol$ and $t < t_{max}$ **do**
    if $\eta_r(E_k, s, t, h) < 1$ **then**
      $t \leftarrow t + h;\ h \leftarrow h + tol$
    else
      $h \leftarrow h - tol$
    end if
  end while
  if $t \geq t_{max}$ **then**
    —BreakLoop—
  else
    $k \leftarrow k + 1.$
  end if
end while
if $k \leq M$ **then**
  System is exponentially stable
else
  Inconclusive
end if

Example 1 We consider a networked control system with uncertain but bounded transmission period. A networked control system is composed of a plant and a controller that interact with each other by transmission of feedback input from controller to the plant. If the system dynamics is linear with linear feedback, then for uncertain but bounded transmission period, we can equivalently represent it as a linear impulsive system where $A_c = \begin{pmatrix} A_p & 0 & B_p \\ 0 & 0 & 0 \\ 0 & 0 & 0 \end{pmatrix}$, $A_r = \begin{pmatrix} \mathbb{I} & 0 & 0 \\ B_o C_p & A_o & 0 \\ D_o C_p & C_o & 0 \end{pmatrix}$ for some parameter matrices A_p, B_p, B_o, C_p, A_o, C_o, and D_o. The sampling interval Δ of the linear impulsive system specifies bounds on the transmission interval. Our example of a networked control system is taken from Björn et al. [38]. The system is originally described by discrete-time transfer functions, which has an equivalent state-space representation with parameter matrices $A_p = \begin{pmatrix} -1 & 0 \\ 1 & 0 \end{pmatrix}$, $B_p = \begin{pmatrix} 1 \\ 0 \end{pmatrix}$, $C_p = \begin{pmatrix} 0 & 1 \end{pmatrix}$, $A_o = 0.4286$, $B_o = -0.8163$, $C_o = -1$ and $D_o = -3.4286$. Given the lower bound on the transmission period as $t_{min} = 0.8$, we want to find as high a value of t_{max} as possible for which the system is GES.

Example 2 We consider the following linear impulsive system from Hetel et. al. [22] that describes an LMI-based approach to verify stability. The specification is given by $A_c = \begin{pmatrix} 0 & -3 & 1 \\ 1.4 & -2.6 & 0.6 \\ 8.4 & -18.6 & 4.6 \end{pmatrix}$ and $A_r = \begin{pmatrix} 1 & 0 & 0 \\ 0 & 1 & 0 \\ 0 & 0 & 0 \end{pmatrix}$.

Table 5.1 Example 1

Reference	t_{min}	t_{max}
Value recommended in [38]	0.08	0.22
NCS toolbox [8]	0.08	0.4
Complex zonotope [1]	0.08	0.5
Template complex zonotope	0.08	0.58

Table 5.2 Example 2

Reference	t_{min}	t_{max}
Lyapunov, parametric LMI [22]	0.1	0.3
Polytopic set contractiveness [16]	0.1	0.475
Khatib et al. [25]	0.1	0.514
Complex zonotope [1]	0.1	0.49
Template complex zonotope	0.1	0.496

Table 5.3 Template complex zonotopes (TCZ) vs Complex zonotopes (CZ) [1]

	CZ	TCZ
Finding suitable zonotope	Requires guessing	Systematically synthesized
No. of impulses for contraction	2 (both examples)	1 (both examples)
Computation time (Example 1)	27.41 s	14.9443 s
Computation time (Example 2)	74.04 s	10.6097 s

Setting and Results. While implementing the algorithm for stability verification, we used order Taylor expansion, a tolerance of $tol = 0.01$ for Example 1 and $tol = 0.006$ for Example 2. We required $k = 3$ number of reachability operators for both examples, for synthesizing a suitable template complex zonotope used in checking contraction. We could verify exponential stability in a sampling interval $[0.08, 0.58]$ for Example 1 and $[0.1, 0.496]$ for Example 2. The comparison of our approach with the state-of-the-art NCS toolbox [8] and also other approaches is presented in Tables 5.1 and 5.2. For the first example, our method outperforms other approaches, while it is competitive with other approaches on the second example. Furthermore, Table 5.3 shows that the template complex zonotope based approach is also faster than the complex zonotope approach of [1].

5.5 Conclusion

We extended complex zonotopes to template complex zonotopes in order to improve the efficiency of the computation of contractive sets and positive invariants. Template complex zonotopes retain a useful feature of complex zonotopes, which is the scope to incorporate the eigenvectors of linear dynamics among the generators because the eigenstructure is related to existence of positive invariants. In addition, compared to complex zonotopes, the advantage template complex zonotopes have is the ability to regulate the contribution of each generator to the set by using the scaling factors. Accordingly, we proposed a systematic and more efficient procedure for verification of stability of nearly periodic impulsive systems. The advantage of this new set representation is attested by the experimental results that are better or competitive, compared to the state-of-the-art methods and tools on benchmark examples. This work also contributes a method for exploiting the eigenstructure of linear dynamics to algorithmically determine template directions, required by most verification approaches using template-based set representations. A number of directions for future research can be identified. First, we intend to extend these techniques to analysis to switched systems under constrained switching laws. Also computationally speaking, our approach is close in spirit to abstract interpretation. Indeed, the operations used to find contractive sets can be extended to invariant computation for general hybrid systems with state-dependent discrete transitions.

Acknowledgements This work is partially supported by the ANR MALTHY project (grant ANR-12-INSE-003).

References

1. Adimoolam, A., Dang, T.: Using complex zonotopes for stability verification. In: American Control Conference. (2016). https://sites.google.com/site/cztopepubs/
2. Adje, A., Gaubert, S., Goubault, E.: Coupling policy iteration with semi-definite relaxation to compute accurate numerical invariants in static analysis. ESOP **6012**, 23–42 (2010)
3. Adjé, A., Garoche, P.-L., Werey, A.: Quadratic zonotopes - an extension of zonotopes to quadratic arithmetics. APLAS **2015**, 127–145 (2015)
4. Allamigeon, X., Gaubert, S., Goubault, E.: Inferring min and max invariants using max-plus polyhedra. In: SAS. LNCS, vol. 5079, pp. 189–204. Springer, Berlin (2008)
5. Allamigeon, X., Gaubert, S., Goubault, E., Putot, S., Stott, N.: A scalable algebraic method to infer quadratic invariants of switched systems. In: Embedded Software, pp. 75–84. IEEE Press (2015)
6. Althoff, M.: Reachability analysis of nonlinear systems using conservative polynomialization and non-convex sets. In: HSCC 2013, pp. 173–182. ACM, New York (2013)
7. Bagnara, R., Rodríguez-Carbonell, E., Zaffanella, E.: Generation of basic semi-algebraic invariants using convex polyhedra. In: SAS 2005. LNCS, vol. 3672, pp. 19–34. Springer, Berlin (2005)
8. Bauer, N.W., van Loon, S.J.L.M., Donkers, M.C.F., van de Wouw, N., Heemels, W.P.M.H.: Networked control systems toolbox: Robust stability analysis made easy. In: IFAC Workshop on Distributed Estimation and Control in Networked Systems (NECSYS), pp. 55–60 (2012)

9. Blanchini, F., Miani, S.: Switching and switched systems. In: Set-Theoretic Methods in Control, pp. 405–466. Springer, Berlin (2015)
10. Briat, C.: Convex conditions for robust stability analysis and stabilization of linear aperiodic impulsive and sampled-data systems under dwell-time constraints. Automatica **49**(11), 3449–3457 (2013)
11. Cai, C., Goebel, R., Teel, A.R.: Smooth lyapunov functions for hybrid systems part ii: (pre)asymptotically stable compact sets. IEEE Trans. Autom. Control. **53**(3), 734–748 (2008)
12. Cousot, P., Cousot. R.: Static determination of dynamic properties of programs. In: 2nd International Symposium Program, pp. 106–130. Dunod (1976)
13. Cousot, P., Halbwachs, N.: Automatic discovery of linear restraints among variables of a program. In: POPL, pp. 84–96 (1978)
14. Dang, T., Gawlitza, T.M.: Template-based unbounded time verification of affine hybrid automata. In: APLAS 2011. LNCS, vol. 7078, pp. 34–49. Springer, Berlin (2011)
15. Feron, E.: From control systems to control software: integrating lyapunov-theoretic proofs within code. IEEE Control. Syst. Mag. **1**, 50–71 (2010)
16. Fiacchini, M., Morarescu, I.-C.: Set theory conditions for stability of linear impulsive systems. In: CDC 2014. IEEE (2014)
17. Fridman, E.: A refined input delay approach to sampled-data control. Automatica **46**(2), 421–427 (2010)
18. Fujioka, H.: A discrete-time approach to stability analysis of systems with aperiodic sample-and-hold devices. IEEE Trans. Autom. Control. **54**(10), 2440–2445 (2009)
19. Girard, A.: Reachability of uncertain linear systems using zonotopes. In: HSCC, pp. 291–305 (2005)
20. Goebel, R., Sanfelice, R.G., Teel, A.: Hybrid dynamical systems. IEEE Control. Syst. **29**(2), 28–93 (2009)
21. Hetel, L., Daafouz, J., Iung, C.: Stabilization of arbitrary switched linear systems with unknown time-varying delays. IEEE Trans. Autom. Control. **51**(10), 1668–1674 (2006)
22. Hetel, L., Daafouz, J., Tarbouriech, S., Prieur, C.: Stabilization of linear impulsive systems through a nearly-periodic reset. Nonlinear Anal. Hybrid Syst. **7**(1), 4–15 (2013)
23. Hu, L., Lam, J., Cao, Y., Shao, H.: A LMI approach to robust h2 sampled- data control for linear uncertain systems. IEEE Trans. Syst. Man Cybern. **33**(1), 149–155 (2003)
24. Kao, C.-Y., Wu, D.-R.: On robust stability of aperiodic sampled-data systems - an integral quadratic constraint approach. ACC **2014**, 4871–4876 (2014)
25. Al Khatib, M., Girard, A., Dang, T.: Stability verification of nearly periodic impulsive linear systems using reachability analysis. In: ADHS 2015. ACM, New York (2015)
26. Kouramas, K.I., Raković, S.V., Kerrigan, E.C., Allwright, J.C., Mayne, D.Q.: On the minimal robust positively invariant set for linear difference inclusions. In: CDC-ECC'05, pp. 2296–2301. IEEE (2005)
27. Kurzhanski, A., Varaiya, P.: Ellipsoidal techniques for reachability analysis. In: HSCC. LNCS, vol. 1790, pp. 202–214. Springer, Berlin (2000)
28. Miné, A.: A new numerical abstract domain based on difference-bound matrices. In: PADO, pp. 155–172 (2001)
29. Miné, A.: The octagon abstract domain. High. Order Symb. Comput. **19**(1), 31–100 (2006)
30. Naghshtabrizi, P.: Delay Impulsive Systems: A Framework for Modeling Networked Control Systems. University of California, Santa Barbara (2007)
31. Naghshtabrizi, P., Hespanha, J.P., Teel, A.R.: Exponential stability of impulsive systems with application to uncertain sampled-data systems. Syst. Control. Lett. **57**(5), 378–385 (2008)
32. Nešić, D., Teel, A.R.: A framework for stabilization of nonlinear sampled-data systems based on their approximate discrete-time models. IEEE Trans. Autom. Control. **49**(7), 1103–1122 (2004)
33. Polyak, B.T., Nazin, A.V., Topunov, M.V., Nazin, S.: Rejection of bounded disturbances via invariant ellipsoids technique. In: CDC 2006, pp. 1429–1434. IEEE (2006)
34. Rodríguez-Carbonell, E., Kapur, D.: Automatic generation of polynomial invariants of bounded degree using abstract interpretation. Sci. Comput. Program **64**(1), 54–75 (2007)

35. Sankaranarayanan, S., Sipma, H.B., Manna, Z.: Scalable analysis of linear systems using mathematical programming. In: VMCAI, pp. 25–41 (2005)
36. Seuret, A.: A novel stability analysis of linear systems under asynchronous samplings. Automatica **48**(1), 177–182 (2012)
37. Teel, A., Nesic, D., Kokotovic, P.V.: A note on input-to-state stability of sampled-data nonlinear systems. In: CDC 1998, pp. 2473–2478. IEEE (1998)
38. Wittenmark, B., Astrom, K.J., Arzen, K.-E.: Computer control: an overview. IFAC Professional Brief, 1 (2002)
39. Mikheev, Y.V., Sobolev, V.A., Fridman, E.M.: Asymptotic analysis of digital control systems. Autom. Remote. Control. **49**(9), 1175–1180 (1988)
40. Grant, M., Boyd, S., Ye, Y.: CVX: Matlab software for disciplined convex programming (2008)

Chapter 6
Timing Contracts for Multi-Core Embedded Control Systems

M. Al Khatib, A. Girard and T. Dang

Abstract In physical dynamical systems equipped with embedded controllers, timing contracts specify the time instants at which certain operations are performed such as sampling, computation, and actuation. In the first part of this chapter, we present a class of timing contracts specifying bounds on the sampling-to-actuation delay and on the sampling period. We then review existing techniques that can handle the problem of stability verification: given models of the physical plant and of the controller and a timing contract, we verify that the resulting dynamical system is stable. In the second part of the chapter, we consider the scheduling problem of embedded controllers on a multiple core computational platform: given a set of controllers, each of which is subject to a timing contract, we synthesize a dynamic scheduling policy, which guarantees that each timing contract is satisfied and that each of the shared computational resources is allocated to at most one embedded controller at any time. The approach is based on a timed game formulation whose solution provides a suitable schedule.

6.1 Introduction

Physical systems equipped with embedded controllers have a long history (aircrafts, cars, robots, etc.) and are becoming ever more complex and pervasive (smart buildings, autonomous vehicles, etc.). Efficient usage of the computational resources in embedded control systems while providing formal guarantees of stability requires a

M. Al Khatib · A. Girard (✉)
Laboratoire des Signaux et Systèmes (L2S), CNRS, CentraleSupélec, Université Paris-Sud, Université Paris-Saclay, 91192 Gif-sur-Yvette, France
e-mail: antoine.girard@l2s.centralesupelec.fr

M. Al Khatib
e-mail: mohammad.alkhatib@l2s.centralesupelec.fr

T. Dang
Verimag, Université Grenoble-Alpes-CNRS, 38000 Grenoble, France
e-mail: thao.dang@imag.fr

S. Tarbouriech et al. (eds.), *Control Subject to Computational and Communication Constraints*, Lecture Notes in Control and Information Sciences 475, https://doi.org/10.1007/978-3-319-78449-6_6

profound understanding of the interaction between their computational and physical components. Models faithfully describing such cyber-physical integration combine continuous as well as discrete dynamics whereby the former originates from the behavior of the physical systems whereas the latter results from the behavior of components like sensors, actuators, and other computation and communication resources. One direction in modeling timing of events (sampling, computation, and actuation) in this orchestration is given by timing contracts [11]. Under such contracts, the control engineers are responsible for designing a control law that is robust to all possible timing variations specified in the contract while the software engineers can focus on implementing the proposed control law so as to satisfy the timing contract. Consequently, we propose techniques that are useful within this framework.

In the first part of this chapter, we present a class of parameterized timing contracts specifying bounds on the sampling-to-actuation delay and on the sampling period. We then review existing techniques [6, 9, 10] that can handle the problem of stability verification: given models of the physical plant and of the controller and a timing contract, verify that the resulting dynamical system is stable. In this context, we briefly present our approach [2, 3], based on the notion of reachable set, and which is built on efficient over-approximation algorithms developed over the past decade (see e.g., [19]).

In the second part, we consider the scheduling problem of embedded controllers on a multiple core computational platform. Given a set of controllers, each of which is subject to a timing contract, we synthesize a dynamic scheduling policy, which guarantees that each timing contract is satisfied and that each of the shared computational resources is allocated to at most one embedded controller at any time. The approach is based on a timed game formulation [5] whose solution provides a suitable schedule. Results on this second problem partially appear, in the case of single core computational platforms in [4].

Notation

Let $\mathbb{R}$, $\mathbb{R}_0^+$, $\mathbb{R}^+$, $\mathbb{N}$, $\mathbb{N}^+$ denote the sets of reals, nonnegative reals, positive reals, nonnegative integers, and positive integers, respectively. For $I \subseteq \mathbb{R}_0^+$, let $\mathbb{N}_I = \mathbb{N} \cap I$. Finally, for a set S, we denote the set of all subsets of S by 2^S.

6.2 Problem Formulation

The model considered in the chapter is represented by the block diagram given by Fig. 6.1. Typically, the plant's state z flows under continuous dynamics. Then, at each sampling instant t_k^s, $k \in \mathbb{N}$, the plant's state is sampled by the sampler and is passed through a network to the controller. The latter computes the control input u based on $z(t_k^s)$ and updates the plant's input at instant t_k^a, $k \in \mathbb{N}$. The plant's input is then held constant by a zero-order hold until the next update arrives via the network.

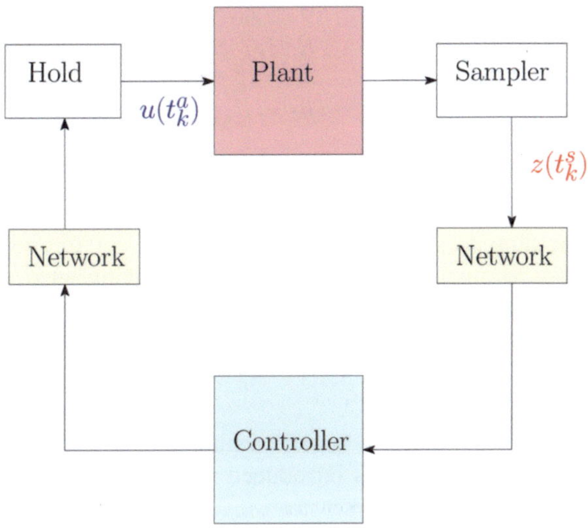

Fig. 6.1 Block diagram of a sampled-data system

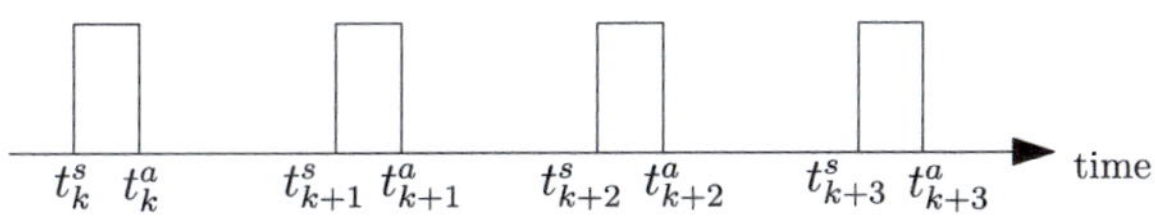

Fig. 6.2 Periodic sampled-data systems

The following model allows us to capture the continuous dynamics of the plant as well as the discrete dynamics, introduced by the sampler and the zero-order hold.

$$\dot{z}(t) = Az(t) + Bu(t), \quad \forall t \in \mathbb{R}_0^+ \tag{6.1a}$$

$$u(t) = Kz(t_k^s), \qquad t_k^a < t \leq t_{k+1}^a \tag{6.1b}$$

where $z(t) \in \mathbb{R}^p$ is the state of the system, $u(t) \in \mathbb{R}^m$ is the control input, the matrices $A \in \mathbb{R}^{p\times p}$, $B \in \mathbb{R}^{p\times m}$, $K \in \mathbb{R}^{m\times p}$, and $k \in \mathbb{N}$. In addition, it is assumed that K is designed such that the matrix $A + BK$ is Hurwitz and that for all $t \in [0, t_0^a]$, $u(t) = 0$.

Traditionally, controllers assume that sampling is performed periodically and that actuation is performed with as little latency as possible. This scenario is shown in Fig. 6.2 where the sampling instants are given by $t_k^s = kh$ for all $k \in \mathbb{N}$ and h being the sampling period. However outside this ideal case, variations in the timing sampling and actuation instants can be captured by *timing contracts*, which make it possible to take into account the temporal nondeterminism of the sequences of sampling and actuation instants $(t_k^s)_{k\in\mathbb{N}}$ and $(t_k^a)_{k\in\mathbb{N}}$.

We assume that the sequences of sampling and actuation instants $(t_k^s)_{k\in\mathbb{N}}$ and $(t_k^a)_{k\in\mathbb{N}}$ satisfy a *timing contract* $\theta(\underline{\tau}, \overline{\tau}, \underline{h}, \overline{h})$ given by

$$
\begin{aligned}
0 &\leq t_0^s, \\
t_k^s &\leq t_k^a \leq t_{k+1}^s, && \forall k \in \mathbb{N} \\
\tau_k &= t_k^a - t_k^s \in [\underline{\tau}, \overline{\tau}], && \forall k \in \mathbb{N} \\
h_k &= t_{k+1}^s - t_k^s \in [\underline{h}, \overline{h}], && \forall k \in \mathbb{N}
\end{aligned}
\tag{6.2}
$$

where $\underline{\tau} \in \mathbb{R}_0^+$, $\overline{\tau} \in \mathbb{R}_0^+$, $\underline{h} \in \mathbb{R}^+$, and $\overline{h} \in \mathbb{R}^+$ provide bounds on the sampling-to-actuation delays (which include time for computation of the control law) and sampling periods. Note that we impose $\underline{h} \neq 0$ to prevent Zeno behavior. Moreover, these parameters must belong to the following set $\mathscr{C}$ so that the time intervals given in (6.2) are always non-empty and it is always possible to choose $t_{k+1}^s \geq t_k^a$:

$$
\mathscr{C} = \left\{(\underline{\tau}, \overline{\tau}, \underline{h}, \overline{h}) \in \mathbb{R}_0^+ \times \mathbb{R}_0^+ \times \mathbb{R}^+ \times \mathbb{R}^+ : \underline{\tau} \leq \overline{\tau} \leq \overline{h}, \underline{h} \leq \overline{h}\right\}.
$$

Contract (6.2) is a general timing contract which includes or over-approximates the different contracts introduced in [11]. Their relation to the timing contract (6.2) is described as follows:

1. **ZET Contract**: The Zero Execution Time contract is given by (6.2) with $\underline{\tau} = \overline{\tau} = 0$ and $\underline{h} = \overline{h} = h \in \mathbb{R}^+$. In other words, the contract states that the sampling and actuation instants are periodic and simultaneous such that $t_k^s = t_k^a = kh$ for $k \in \mathbb{N}$. As mentioned in [11], this contract is hardly achievable in practice since computation always takes time in between the sampling and actuation instants.
2. **LET Contract**: The Logical Execution Time contract is given by (6.2) with $\underline{\tau} = \overline{\tau} = \underline{h} = \overline{h} = h \in \mathbb{R}^+$. The contract states that the sampling and actuation instants are periodic such that $t_0^s = 0$ and $t_k^s = t_{k-1}^a = kh$ for $k \in \mathbb{N}^+$.
3. **DET Contract**: The Deadline Execution Time contract is given by (6.2) with $\underline{\tau} = 0$ and $\underline{h} = \overline{h} = h \in \mathbb{R}^+$. The contract states that the sampling instants are periodic, or $t_k^s = kh$ for $k \in \mathbb{N}$, and actuation instants are at some point t_k^a in the interval $[t_k^s, t_k^s + \overline{\tau}]$, with $\overline{\tau} \leq h$.
4. **TOL Contract**: The Timing Tolerance contract is defined by a nominal sampling period $h \in \mathbb{R}^+$, nominal sampling-to-actuation delay $\tau \in \mathbb{R}_0^+$, and two jitters $J^h, J^\delta \in \mathbb{R}_0^+$ with $J^\tau \leq \tau$ and $J^h + J^\tau + \tau \leq h$, such that $t_k^s \in [kh, kh + J^h]$ and $t_k^a \in [t_k^s + \tau - J^\tau, t_k^s + \tau + J^\tau]$, for $k \in \mathbb{N}$ (refer to Fig. 6.3). We cannot exactly model this contract using (6.2). However, we can over-approximate it using (6.2) with $\underline{\tau} = \tau - J^\tau$, $\overline{\tau} = \tau + J^\tau$, $\underline{h} = h - J^h$, and $\overline{h} = h + J^h$.

In the following, we formulate the two problems discussed in this chapter.

6.2.1 *Stability Verification Problem*

In our problem formulation, we consider the following notion of stability for system (6.1)–(6.2), that guarantees the exponential convergence of the state to the origin, i.e., $z = 0$, with a predefined rate $\beta \in \mathbb{R}^+$:

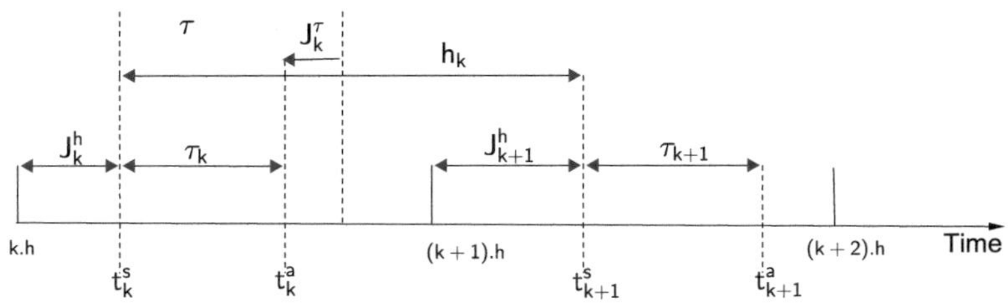

Fig. 6.3 Time variables included in a *TOL* contract. $J_k^h \in [0, J^h]$ and $J_k^\tau \in [-J^\tau, J^\tau]$

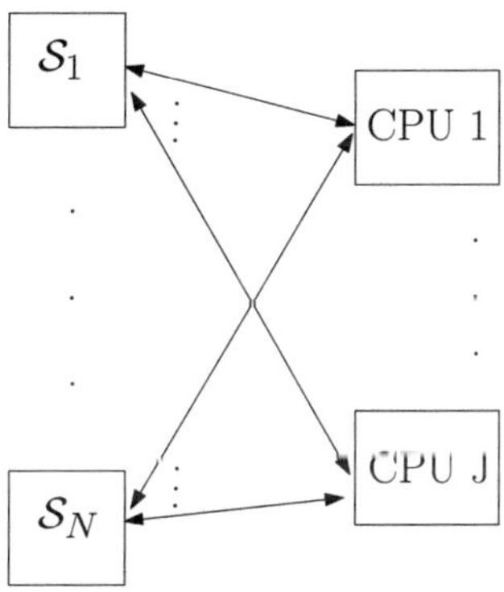

Fig. 6.4 Block diagram of N sampled-data systems sharing J CPUs

Definition 6.1 (*β-Stability*) Let $\beta \in \mathbb{R}^+$, system (6.1)–(6.2) is β-stable if there exist $C \in \mathbb{R}^+$ and $\varepsilon' \in \mathbb{R}^+$ such that

$$|z(t)| \leq Ce^{-(\beta+\varepsilon')(t-t_0^s)}|z(t_0^s)|, \ \forall t \in \mathbb{R}^+. \tag{6.3}$$

Consequently, in this work, we consider the following problem:

Problem 6.1 (*Stability verification*) Given $\beta \in \mathbb{R}^+$, $A \in \mathbb{R}^{p\times p}$, $B \in \mathbb{R}^{p\times m}$, $K \in \mathbb{R}^{m\times p}$, $(\underline{\tau}, \overline{\tau}, \underline{h}, \overline{h}) \in \mathscr{C}$, verify that (6.1)–(6.2) is β-stable.

The reader is referred to Sect. 6.3 where we provide an overview of existing techniques that can solve Problem 6.1, and we present our own approach.

6.2.2 Scheduling Problem on Multiple CPUs

Consider a collection of $N \in \mathbb{N}^+$ sampled-data systems $\{\mathscr{S}_1, \ldots, \mathscr{S}_N\}$ of the form (6.1) where each system $\mathscr{S}_i = (A_i, B_i, K_i)$ is subject to a timing contract $\theta(\underline{\tau}^i, \overline{\tau}^i, \underline{h}^i, \overline{h}^i)$ of the form (6.2), with parameters $(\underline{\tau}^i, \overline{\tau}^i, \underline{h}^i, \overline{h}^i) \in \mathscr{C}$, $i \in \mathbb{N}_{[1,N]}$.

In addition, we assume that these systems share J CPUs, as shown in Fig. 6.4, to compute the value of their control inputs given by (6.1b). Note that no communication exists in between the CPUs or between the systems, but there exists communication only between the systems and all J CPUs. Furthermore, the time required by CPU j to compute inputs of system S_i is assumed to belong to some known interval $[\underline{c}^i_j, \overline{c}^i_j]$ with $0 \le \underline{c}^i_j \le \overline{c}^i_j$, $i \in \mathbb{N}_{[1,N]}$, and $j \in \mathbb{N}_{[1,J]}$, where $\underline{c}^i_j$ and $\overline{c}^i_j$ denote the best and worst case execution time, respectively.

The timing of events in the kth control cycle of system $\mathscr{S}_i$ starts at instant $t^{s_i}_k$ when sampling occurs. Then, system $\mathscr{S}_i$ gains access to the CPU j at instant $t^{b_i}_k$, at which computation of the control input value begins. The CPU is released at instant $t^{e_i}_k$, at which computation of the control input value ends. After that, actuation occurs at instant $t^{a_i}_k$. We denote by $\mathbb{N}(i, j)$ the set gathering indexes of the control cycles, at which system $\mathscr{S}_i$ accesses the CPU j, where $\bigcup_{j\in\mathbb{N}_{[1,J]}} \mathbb{N}(i, j) = \mathbb{N}$ for all $i \in \mathbb{N}_{[1,N]}$.

Then, the sequences $(t^{s_i}_k)_{k\in\mathbb{N}}$, $(t^{b_i}_k)_{k\in\mathbb{N}}$, $(t^{e_i}_k)_{k\in\mathbb{N}}$, and $(t^{a_i}_k)_{k\in\mathbb{N}}$ satisfy the following constraints for all $i \in \mathbb{N}_{[1,N]}$:

$$
\begin{aligned}
0 &\le t^{s_i}_0 \\
t^{s_i}_k &\le t^{b_i}_k \le t^{e_i}_k \le t^{a_i}_k \le t^{s_i}_{k+1}, \ \forall k \in \mathbb{N} \\
c^i_k &= t^{e_i}_k - t^{b_i}_k \in [\underline{c}^i_j, \overline{c}^i_j], \quad \forall k \in \mathbb{N}(i, j), \forall j \in \mathbb{N}_{[1,J]} \\
\tau^i_k &= t^{a_i}_k - t^{s_i}_k \in [\underline{\tau}^i, \overline{\tau}^i], \quad \forall k \in \mathbb{N} \\
h^i_k &= t^{s_i}_{k+1} - t^{s_i}_k \in [\underline{h}^i, \overline{h}^i], \quad \forall k \in \mathbb{N}.
\end{aligned}
\tag{6.4}
$$

In addition, a conflict arises if several systems request access to one of the J CPUs at the same time. Let us define the following time sets, for $i \in \mathbb{N}_{[1,N]}$ and $j \in \mathbb{N}_{[1,J]}$:

$$\mathsf{Com}(\mathscr{S}_i, j) = \bigcup_{k\in\mathbb{N}(i,j)} [t^{b_i}_k, t^{e_i}_k).$$

$\mathsf{Com}(\mathscr{S}_i, j)$ is the union of time intervals when CPU j is used by system $\mathscr{S}_i$. Then, in order to prevent conflicting accesses to the CPU the following property must hold:

$$
\begin{aligned}
&\forall (m, n, j) \in \mathbb{N}^2_{[1,N]} \times \mathbb{N}_{[1,J]} \text{ with } m \ne n, \\
&\mathsf{Com}(\mathscr{S}_m, j) \cap \mathsf{Com}(\mathscr{S}_n, j) = \emptyset.
\end{aligned}
\tag{6.5}
$$

Remark 6.1 It is straightforward to verify that for any sequences $(t^{s_i}_k)_{k\in\mathbb{N}}$, $(t^{b_i}_k)_{k\in\mathbb{N}}$, $(t^{e_i}_k)_{k\in\mathbb{N}}$, and $(t^{a_i}_k)_{k\in\mathbb{N}}$ satisfying (6.4)–(6.5), the sequences $(t^{s_i}_k)_{k\in\mathbb{N}}$ and $(t^{a_i}_k)_{k\in\mathbb{N}}$ satisfy the timing contract $\theta(\underline{\tau}^i, \overline{\tau}^i, \underline{h}^i, \overline{h}^i)$.

We aim at synthesizing a dynamic scheduling policy, generating sequences of timing events satisfying (6.4)–(6.5). The scheduler has control over the sampling and actuation instants $(t^{s_i}_k)_{k\in\mathbb{N}}$, $(t^{a_i}_k)_{k\in\mathbb{N}}$ and over the instants $(t^{b_i}_k)_{k\in\mathbb{N}}$ when computation begins. Also, the scheduler assigns a CPU to compute the control input for each system $\mathscr{S}_i$ at each control cycle $k \in \mathbb{N}$. However, the execution times $(c^i_k)_{k\in\mathbb{N}}$, and thus the instants when computation ends $(t^{e_i}_k)_{k\in\mathbb{N}}$, are determined by the environment and are therefore uncontrollable from the point of view of the scheduler. Next, given

Table 6.1 Methods that can solve instances of Problem 6.1 with description of the modeling and computational approaches, list of restrictions, and possible extensions

References	Models	Algorithm	Restrictions	Extensions
[9]	Difference inclusions	LMI	–	$\tau_k > h_k$; controller synthesis
[10]		LMI	–	Scheduling
[15]		LMI	$\underline{\tau} = \overline{\tau} = 0$	Controller synthesis
[16]		LMI	$\underline{\tau} = \overline{\tau} = 0$	–
[22]		SOS	$\underline{\tau} = \overline{\tau} = 0$	–
[12]		Invariance	$\underline{\tau} = \overline{\tau} = 0$	–
[18]	Time-delay systems	LMI	$\underline{h} = 0$	$\tau_k > h_k$; scheduling
[14]		LMI	$\underline{h} = \overline{h}, \underline{\tau} = 0$	Controller synthesis; quantization
[20]		LMI	$\underline{\tau} = \overline{\tau} = 0$	–
[13]	Interconnected systems	LMI	$\underline{h} = \underline{\tau} = \overline{\tau} = 0$	–
[6]	Hybrid systems	SOS	–	Nonlinear dynamics; scheduling
[17]		LMI	$\underline{\tau} = 0, \underline{h} = 0$	Scheduling

that a task T_i, a task-set $\mathscr{T}$, and timing contracts Θ are characterized as

$$T_i = \left((\underline{c}_1^i, \overline{c}_1^i), \ldots, (\underline{c}_J^i, \overline{c}_J^i)\right), i \in \mathbb{N}_{[1,N]} \tag{6.6a}$$

$$\mathscr{T} = \{T_1, \ldots, T_N\}, \tag{6.6b}$$

$$\Theta = \{\theta(\underline{\tau}^1, \overline{\tau}^1, \underline{h}^1, \overline{h}^1), \ldots, \theta(\underline{\tau}^N, \overline{\tau}^N, \underline{h}^N, \overline{h}^N)\}, \tag{6.6c}$$

we define the scheduling problem informally, at this point of the chapter, as

Problem 6.2 (*Schedulability verification*) Given a set of control tasks $\mathscr{T}$ and timing contracts $\Theta = \{\theta(\underline{\tau}^1, \overline{\tau}^1, \underline{h}^1, \overline{h}^1), \ldots, \theta(\underline{\tau}^N, \overline{\tau}^N, \underline{h}^N, \overline{h}^N)\}$ as in (6.6), verify whether or not there exists a scheduling policy with sequences of timing events satisfying (6.4)–(6.5).

A precise formulation of the schedulability of the task-set $\mathscr{T}$ is provided in Sect. 6.4 along with a solution to the schedulability verification problem based on safety games over timed game automata.

6.3 Stability Verification

Several approaches are presented in the literature to solve instances of Problem 6.1. A non-exhaustive list is given in Table 6.1. From the modeling perspective, the problem can be tackled using difference inclusions, time-delay systems, or hybrid systems. On the computational side, the approaches are based on semi-definite programming (Linear Matrix Inequalities (LMI) or Sum Of Squares (SOS) formulations), invariant sets, or reachability analysis. Let us remark that approaches [6, 9, 10] appear to be able to address all instances of Problem 6.1.

Regarding our approach to Problem 6.1, we solve the same problem for a more general class of dynamic systems, given by a difference inclusion, and conclude on the stability of system (6.1)–(6.2). Meanwhile, one essential ingredient that is used in our study is the approximation scheme developed for over-approximating the reachable set of (6.1)–(6.2) from a given initial set. Such an over-approximation is provided in a previous work [2]. Let us first start by rewriting the system in terms of impulsive systems, in order to interpret the reachable set we use in the sequel.

6.3.1 Reformulation Using Impulsive Systems

In our analysis, it is more practical to transform (6.1) into an impulsive system with two types of resets each referring to a sampling or actuation instant. Such a reformulation is convenient to develop stability conditions based on reachability analysis. The system is thus given by

$$\begin{aligned} \dot{x}(t) &= A_c x(t), \quad t \neq t_k^s, t \neq t_k^a \\ x(t_k^{s+}) &= A_s x(t_k^s), \quad k \in \mathbb{N} \\ x(t_k^{a+}) &= A_a x(t_k^a), \quad k \in \mathbb{N} \end{aligned} \tag{6.7}$$

where $x(t) \in \mathbb{R}^n$ is the state of the system with $n = p + 2m$, $(t_k^s)_{k\in\mathbb{N}}$ and $(t_k^a)_{k\in\mathbb{N}}$ are given by (6.2), $x(t^+) = \lim\limits_{\tau\to 0, \tau>0} x(t+\tau)$, and

$$\begin{aligned} A_c &= \begin{pmatrix} A & 0 & B \\ 0 & 0 & 0 \\ 0 & 0 & 0 \end{pmatrix}, \quad A_s = \begin{pmatrix} I_p & 0 & 0 \\ K & 0 & 0 \\ 0 & 0 & I_m \end{pmatrix}, \\ A_a &= \begin{pmatrix} I_p & 0 & 0 \\ 0 & I_m & 0 \\ 0 & I_m & 0 \end{pmatrix}, \quad x(t) = \begin{pmatrix} z(t) \\ Kz(\theta^s(t)) \\ u(t) \end{pmatrix}, \end{aligned} \tag{6.8}$$

with $\theta^s(t) = t_k^s$ for $t \in (t_k^s, t_{k+1}^s]$. We consider in the following, system (6.7) under timing contract (6.2).

A notion for stability of the impulsive system guaranteeing the exponential convergence of the state to the origin with a predefined rate $\beta \in \mathbb{R}^+$ is given by

Definition 6.2 (*β-Stability*) Let $\beta \in \mathbb{R}^+$, system (6.2)–(6.7) is β-stable if there exist $C \in \mathbb{R}^+$ and $\varepsilon^* \in \mathbb{R}^+$ such that

$$|x(t)| \leq Ce^{-(\beta+\varepsilon^*)(t-t_0^s)}|x(t_0^s)|, \ \forall t \in \mathbb{R}^+. \tag{6.9}$$

Note that β-stability of system (6.2)–(6.7) is equivalent to the β-stability of (6.2)–(6.1). We are now interested in verifying stability of embedded control systems in the form given by (6.7) under one of the general timing contracts defined previously in Sect. 6.2. Indeed, we can easily show that system (6.7) under the ZET and LET contracts is stable if and only if the eigenvalues of the matrix $e^{hA_c}A_aA_s$ and $A_ae^{hA_c}A_s$ are inside the unit circle, respectively. As for the DET or TOL contracts, we have that stability of system (6.2)–(6.1) is guaranteed by the stability of system (6.2)–(6.7) with an adequate choice of the timing contract parameters. It is noteworthy that in the case of the TOL contract, stability of system (6.2)–(6.7) is only sufficient when the parameters of the over-approximating timing contract are chosen as explained in Sect. 6.2. Consequently, in this work, we consider an equivalent to Problem 6.1:

Problem 6.3 (*Stability verification*) Given $\beta \in \mathbb{R}^+$, $A_c, A_s, A_a \in \mathbb{R}^{n\times n}$, $(\underline{\tau}, \overline{\tau}, \underline{h}, \overline{h}) \in \mathscr{C}$, verify that (6.2)–(6.7) is β-stable.

6.3.2 A Stability Verification Approach Based on Difference Inclusions

Our stability verification approach to solve Problem 6.3 is based on a reformulation of the linear impulsive systems (6.2)–(6.7) in the general framework of difference inclusions. Then, for a fairly large class of difference inclusions, we recall necessary and sufficient conditions for stability, established in [3]. These conditions are based on the successive images of a set under the dynamics of the difference inclusion. For linear impulsive systems (6.2)–(6.7), these conditions allow us to design a stability verification algorithm using reachability analysis techniques developed in [2].

Let us introduce first a general formulation based on difference inclusions and later show how linear impulsive systems in the form of (6.2)–(6.7) can be embedded in this framework. We consider discrete-time dynamical systems modeled by the following difference inclusion:

$$\xi_{k+1} \in \Phi(\{\xi_k\}), \ k \in \mathbb{N} \tag{6.10}$$

where $\xi_k \in \mathbb{R}^n$ is the state of the system, and $\Phi : 2^{\mathbb{R}^n} \to 2^{\mathbb{R}^n}$ is a set-valued map. Stability for systems of the form (6.10) is considered in the following sense:

Definition 6.3 (*GES*) System (6.10) is globally exponentially stable (GES) if there exists $(C, \varepsilon) \in \mathbb{R}^+ \times (0, 1)$ such that for all trajectories $(\xi_k)_{k\in\mathbb{N}}$ of (6.10), we have

$$|\xi_k| \leq C\varepsilon^k |\xi_0|, \ \forall k \in \mathbb{N}. \tag{6.11}$$

Next we verify the stability of a difference inclusion of the form (6.10). We make first the following assumptions on the map Φ.

Assumption 6.1 For all $\mathscr{S} \subseteq \mathbb{R}^n$, $\lambda \in \mathbb{R}_0^+$, the following assertions hold:

(i) $\Phi(\mathscr{S}) = \bigcup_{z\in\mathscr{S}} \Phi(\{z\})$;
(ii) $\Phi(\lambda\mathscr{S}) \subseteq \lambda\Phi(\mathscr{S})$;
(iii) if $\mathscr{S}$ is bounded, then $\Phi(\mathscr{S})$ is bounded.

Under item (i) of Assumption 6.1, for all $\mathscr{S}, \mathscr{S}' \subseteq \mathbb{R}^n$, it follows that $\Phi(\mathscr{S} \cup \mathscr{S}') = \Phi(\mathscr{S}) \cup \Phi(\mathscr{S}')$. Also, if $\mathscr{S} \subseteq \mathscr{S}'$, then $\Phi(\mathscr{S}) \subseteq \Phi(\mathscr{S}')$. We define the iterates of Φ as $\Phi^0(\mathscr{S}) = \mathscr{S}$ for all $\mathscr{S} \subseteq \mathbb{R}^n$, and $\Phi^{k+1} = \Phi \circ \Phi^k$ for all $k \in \mathbb{N}$. Let $(\xi_k)_{k\in\mathbb{N}}$ be a trajectory of (6.10) such that $\xi_0 \in \mathscr{S}$, then under item (i) of Assumption 6.1, for all $k \in \mathbb{N}$, $\Phi^k(\mathscr{S})$ is the set of all possible values of ξ_k.

Then, the stability verification problem, for systems of the form (6.10), can be formulated as follows:

Problem 6.4 (*Stability verification*) Under Assumption 6.1, verify that system (6.10) is GES.

Let $\beta \in \mathbb{R}^+$ and suppose that $\mathscr{S} \subseteq \mathbb{R}^n$ represents all the states of the system at sampling instant t_k^s. Then, a scaled reachable set of system (6.2)–(6.7) at instant t_{k+1}^s is given by the map $\Phi : 2^{\mathbb{R}^n} \to 2^{\mathbb{R}^n}$ such that

$$\Phi(\mathscr{S}) = \bigcup_{\tau\in[\underline{\tau},\overline{\tau}]} \bigcup_{w\in[\max(0,\underline{h}-\tau),\overline{h}-\tau]} e^{(w+\tau)\beta} e^{wA_c} A_a e^{\tau A_c} A_s \mathscr{S}. \tag{6.12}$$

The following proposition establishes the equivalence between stability of systems (6.2)–(6.7) and (6.10).

Proposition 6.1 *Given $\beta \in \mathbb{R}^+$. System (6.2)–(6.7) is β-stable if and only if system (6.10) is GES with Φ given by (6.12).*

The next proposition shows that the map Φ in (6.12) satisfies the previous assumptions.

Proposition 6.2 *Let Φ be given by (6.12), then Φ satisfies Assumption 6.1.*

It follows from Propositions 6.1 and 6.2 that Problem 6.3 can be reduced to Problem 6.4. Therefore, in the next sections, we present necessary and sufficient theoretical conditions for stability verification and an algorithm to solve Problem 6.4.

6.3.2.1 Stability Verification: Theoretical Results

This section presents necessary and sufficient conditions, taken from [3], for stability of system (6.10). The following result characterizes the stability of system (6.10) in terms of the map Φ.

Theorem 6.2 *Let $\mathscr{S} \subseteq \mathbb{R}^n$ be bounded with 0 in its interior, under Assumption 6.1, the following statements are equivalent:*

(a) System (6.10) is GES;
(b) There exists $(k, j, \rho) \in \mathbb{N}^+ \times \mathbb{N}_{[0,k-1]} \times (0, 1)$ such that $\Phi^k(\mathscr{S}) \subseteq \rho\Phi^j(\mathscr{S})$;
(c) There exists $(k, \rho) \in \mathbb{N}^+ \times (0, 1)$ such that $\Phi^k(\mathscr{S}) \subseteq \rho \bigcup_{j=0}^{k-1} \Phi^j(\mathscr{S})$.

6.3.2.2 An Algorithm for Stability Verification

In this section, we present an algorithm for verifying the stability of system (6.10). Indeed, the maps Φ involved in Theorem 6.2 can be impractical to compute exactly. This is the case of linear impulsive system (6.2)–(6.7), which requires the computation of the reachable set given by (6.12). In that case, we may use an over-approximation $\overline{\Phi} : 2^{\mathbb{R}^n} \to 2^{\mathbb{R}^n}$, which is easier to compute and satisfies the following assumption:

Assumption 6.3 For all $\mathscr{S} \subseteq \mathbb{R}^n$, the following assertions hold:

(i) $\Phi(\mathscr{S}) \subseteq \overline{\Phi}(\mathscr{S})$;
(ii) if $\mathscr{S}$ is bounded then $\overline{\Phi}(\mathscr{S})$ is bounded.

The iterates of $\overline{\Phi}$ are defined similarly to those of Φ. We now derive sufficient conditions for stability of system (6.10) based on $\overline{\Phi}$.

Corollary 6.1 *Under Assumptions 6.1 and 6.3, if $\mathscr{S} \subseteq \mathbb{R}^n$ bounded with 0 in its interior, and $(k, i, \rho) \in \mathbb{N}^+ \times \mathbb{N}_{[0,k-1]} \times (0, 1)$ such that $\overline{\Phi}^k(\mathscr{S}) \subseteq \rho\overline{\Phi}^i(\mathscr{S})$, then system (6.10) is GES.*

Now, we propose a stability verification algorithm to solve Problem 6.4 based on the sufficient condition given in Corollary 6.1. The algorithm consists of an initialization step and a main loop. In the initialization step, we compute an initial set $\mathscr{S}$, which is then propagated in the main loop using the map $\overline{\Phi}$ to check the stability condition given by Corollary 6.1. The choice of the initial set is important in order to try to minimize the value of the integer k such that the stability condition given by Corollary 6.1 holds. Detailed approaches to compute the initial set $\mathscr{S}$ and the over-approximation $\overline{\Phi}$ can be found in [2].

6.4 Scheduling of Embedded Controllers Under Timing Contracts

Our aim in this section is to solve Problem 6.2 on schedulability verification.

6.4.1 Timed Game Automata and Safety Games

This section is intended to briefly introduce timed automata [1], timed game automata [21], and safety games.

6.4.1.1 Timed and Timed Game Automata

Let C be a finite set of real-valued variables called clocks. We denote by $\mathscr{B}(C)$ the set of conjunctions of clock constraints of the form $c \sim \alpha$ where $\alpha \in \mathbb{R}_0^+$, $c \in C$ and $\sim \in \{<, \leq, =, >, \geq\}$. We define a timed automaton (TA) and a timed game automaton (TGA) as in [8]:

Definition 6.4 A *timed automaton* is a sextuple (L, l_0, Act, C, E, I) where

- L is a finite set of locations;
- $l_0 \in L$ is the initial location;
- Act is a set of actions;
- C is a finite set of real-valued clocks;
- $E \subseteq L \times \mathscr{B}(C) \times Act \times 2^C \times L$ is the set of edges;
- $I : L \to \mathscr{B}(C)$ is a function that assigns invariants to locations.

Definition 6.5 A *timed game automaton* is a septuple $(L, l_0, Act_c, Act_u, C, E, I)$ such that $(L, l_0, Act_c \cup Act_u, C, E, I)$ is a timed automaton and $Act_c \cap Act_u = \emptyset$, where Act_c defines a set of controllable actions and Act_u defines a set of uncontrollable actions.

Formal semantics of TA and TGA are stated in [8]. Informally, semantics of a TA is described by a transition system whose state consists of the current location and value of the clocks. Then, the execution of a TA can be described by two types of transitions defined as follows:

- time progress: the current location $l \in L$ is maintained and the value of the clocks grow at unitary rate; these transitions are enabled as long as the value of the clocks satisfies $Inv(l)$.
- discrete transition: an instantaneous transition from the current location $l \in L$ to a new location $l' \in L$ labeled by an action $a \in Act$ is triggered; these transitions are enabled if there is an edge $(l, G, a, C', l') \in E$, such that the value of the clocks satisfies G; in that case, the value of the clocks belonging to C' is reset to zero.

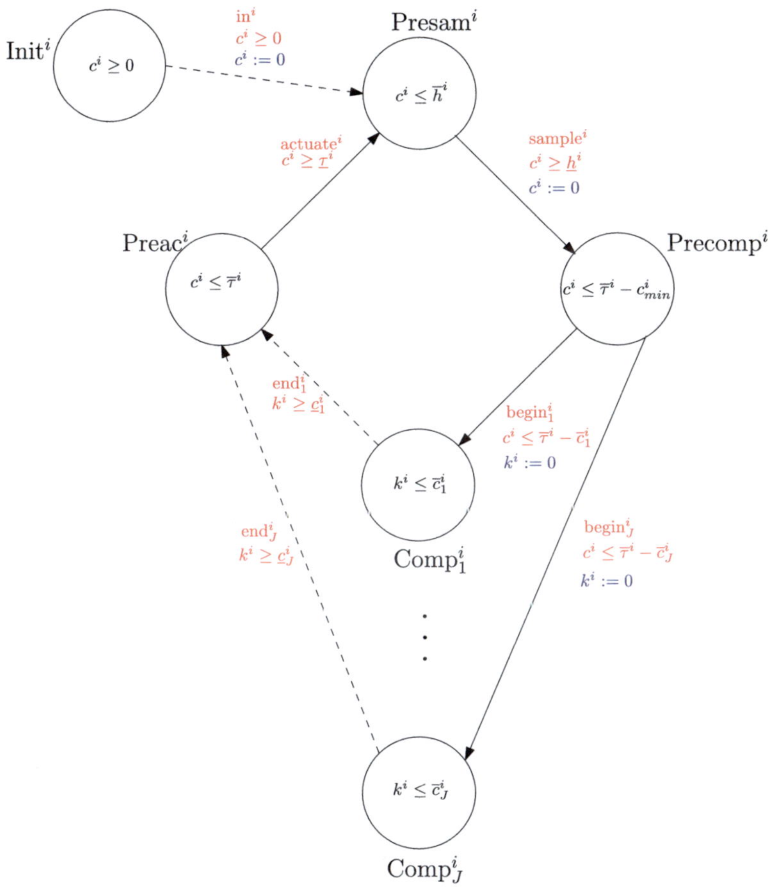

Fig. 6.5 TGA_i, where plain and dashed edges correspond to controllable and uncontrollable actions, respectively

The semantics of TGA is similar to that of TA with the specificity that discrete transitions labeled by a controllable actions (i.e., $a \in Act_c$) are triggered by a controller, while discrete transitions labeled by uncontrollable actions (i.e., $a \in Act_u$) are triggered by the environment/opponent.

6.4.1.2 Safety Games

Safety games (see, e.g., [8]) are defined by a timed game automaton and a set of unsafe locations $L_u \subseteq L$. A solution to the safety game is given by a winning strategy for the controller such that under any behavior of the environment/opponent, the set of unsafe locations is avoided by all controlled executions of the TGA.

6.4.2 Reformulation into TGA

We propose a reformulation of the schedulability verification problem using timed game automata and safety games.

We first associate to each control task and timing contract a timed game automaton, as shown in Fig. 6.5 and formally defined as follows:

Definition 6.6 Let $i \in \mathbb{N}_{[1,N]}$, the timed game automaton generated by control task $T_i = \big((\underline{c}_1^i, \overline{c}_1^i), \ldots, (\underline{c}_J^i, \overline{c}_J^i)\big)$ and timing contract $\theta(\underline{\tau}^i, \overline{\tau}^i, \underline{h}^i, \overline{h}^i)$ is

$$\mathsf{TGA}_i = (L^i, l_0^i, Act_c^i, Act_u^i, C^i, E^i, Inv^i),$$

where

- $L^i = \{Init^i, Presam^i, Precomp^i, Preac^i, Comp_1^i, \ldots, Comp_J^i\}$;
- $l_0^i = Init^i$;
- $Act_c^i = \{sample^i, begin_1^i, \ldots, begin_J^i, actuate^i\}$;
- $Act_u^i = \{end_1^i, \ldots, end_J^i, in^i\}$;
- $C^i = \{c^i, k^i\}$;
- $E^i = \{(Init^i, c^i \geq 0, in^i, \{c^i\}, Presam^i),$
  $(Presam^i, c^i \geq \underline{h}^i, sample^i, \{c^i\}, Precomp^i),$
  $(Precomp^i, c^i \leq \overline{\tau}^i - \overline{c}_1^i, begin_1^i, \{k^i\}, Comp_1^i), \ldots,$
  $(Precomp^i, c^i \leq \overline{\tau}^i - \overline{c}_J^i, begin_J^i, \{k^i\}, Comp_J^i),$
  $(Comp_1^i, k^i \geq \underline{c}_1^i, end_1^i, \emptyset, Preac^i), \ldots, (Comp_J^i, k^i \geq \underline{c}_J^i, end_J^i, \emptyset, Preac^i),$
  $(Preac^i, c^i \geq \underline{\tau}^i, actuate^i, \emptyset, Presam^i)\}$;
- $Inv^i(Init^i) = \{c^i \geq 0\},$
  $Inv^i(Presam^i) = \{c^i \leq \overline{h}^i\},$
  $Inv^i(Precomp^i) = \{c^i \leq \overline{\tau}^i - c_{min}^i\}, with\ c_{min}^i = \min_{j \in \mathbb{N}_{[1,J]}}(\overline{c}_j^i),$
  $Inv^i(Comp_1^i) = \{k^i \leq \overline{c}_1^i\}, \ldots, Inv^i(Comp_J^i) = \{k^i \leq \overline{c}_J^i\},$
  $Inv^i(Preac^i) = \{c^i \leq \overline{\tau}^i\}.$

Intuitively, the set of locations L^i denotes all the possible situations that a control task T_i may be in and E^i denotes all the possible transitions between locations. If we assume that the control loop has not started yet then this is modeled by the location $Init^i$. After that the control loop starts at a certain time that is determined by the environment and thus an uncontrollable transition $(Init^i, c^i \geq 0, in^i, \{c^i\}, Presam^i)$ takes place, where the task has to wait until sampling could occur. The latter is realized by the location $Presam^i$. Then whenever possible, a controller (which is the scheduler) has to decide when sampling must occur. When sampling takes place, the control task will be waiting until a CPU is assigned to compute its control input. This waiting situation is realized by the $Precomp^i$ location. The mission of assigning a CPU for task T_i is that of the scheduler, thus a possible controllable transition occurs when the assignment of CPU_j takes place declaring that the task is in a new situation realized in TGA_i by the location $Comp_J^i$. The task rests in this situation until its execution on the CPU finishes which means that this duration is decided by the environment (which is the CPU and not the scheduler) and thus an uncontrollable transition from $Comp^i_j$ to a new location $Preac^i$ means that the execution has terminated and the control task is in the situation where actuation is to happen next. The latter decision is taken by the scheduler, and thus is controllable, where the control input is fed to the plant and the control task is back again in the pre-sampling situation realized as before by the $Presam^i$ location. In such a case, the control loop is closed and the behavior of the control task is repeated infinitely. Note that all the executions of TGA_i explained informally above must respect the semantics of the timed game automata introduced in Sect. 6.4.1.

Now let the sequences $(t_k^{s_i})$, $(t_k^{a_i})$, $(t_k^{b_i})$ and $(t_k^{e_i})$ be given by the instants of the discrete transitions labeled by actions $sample^i$, $actuate^i$, $begin^i$ and end^i, respectively. It is easy to see that these sequences satisfy the constraints given by (6.4). Conversely, one can check that all sequences satisfying (6.4) can be generated by executions of TGA_i. Moreover, let us restate that the controllable actions are $sample^i$, $actuate^i$, $begin^i$, which means that the scheduler determines the instants when sampling and actuation occur and when computation begins. However, end^i is uncontrollable, which means that the execution time, and thus the instant at which computation ends is determined by the environment.

Finally, CPU j is used by system $\mathcal{S}_i$ if the current location of TGA_i is $Comp^i_j$, with $j \in \mathbb{N}_{[1,J]}$. To take into account the constraint given by (6.5), stating that two systems cannot access any of the J CPUs at the same time, we need to define the composition of the timed game automata defined above:

Definition 6.7 The timed game automaton generated by the set of control tasks $\mathcal{T} = \{T_1, \ldots, T_N\}$, with $T_i = \big((\underline{c}^i_1, \overline{c}^i_1), \ldots, (\underline{c}^i_J, \overline{c}^i_J)\big)$ for all $i \in \mathbb{N}_{[1,N]}$, and timing contracts $\Theta = \{\theta(\underline{\tau}^1, \overline{\tau}^1, \underline{h}^1, \overline{h}^1), \ldots, \theta(\underline{\tau}^N, \overline{\tau}^N, \underline{h}^N, \overline{h}^N)\}$ is given by $\mathsf{TGA} = (\overline{L}, \overline{l}_0, \overline{Act_c}, \overline{Act_u}, \overline{C}, \overline{E}, \overline{Inv})$ where

- $\overline{L} = L^1 \times \cdots \times L^N$, thus $l = (l^1, \ldots, l^N) \in L$ denotes the location of TGA;
- $\overline{l}_0 = (Init^1, \ldots, Init^N)$;
- $\overline{Act_c} = \bigcup_{i=1}^N Act_c^i$;
- $\overline{Act_u} = \bigcup_{i=1}^N Act_u^i$;
- $\overline{C} = \bigcup_{i=1}^N C^i$;
- $\overline{E} = \{(l_m, \lambda, act, C', l_n) \in \overline{L} \times \mathcal{B}(\overline{C}) \times (\overline{Act_c} \cup \overline{Act_u}) \times \overline{L} : \exists i \in \mathbb{N}_{[1,N]},$ $l^j_m = l^j_n \ \ \forall j \neq i$ and $(l^i_m, \lambda, act, C', l^i_n) \in E^i\}$;
- $Inv(l) = \bigwedge_{i=1}^N Inv^i(l^i), i \in \mathbb{N}_{[1,N]}$.

TGA describes the parallel evolution of the $\mathsf{TGA}_1, \ldots, \mathsf{TGA}_N$ and thus models the concurrent execution of the control tasks $T_1, \ldots, T_N$.

6.4.3 Scheduling as a Safety Game

In our setting, we denote the safety game by $(\mathsf{TGA}, \overline{L}_u)$, where the set of locations corresponding to conflicting accesses to the CPUs $\overline{L}_u \subseteq \overline{L}$ is defined by

$$\overline{L}_u = \{l \in \overline{L} : \exists(m, n, j) \in \mathbb{N}^2_{[1,N]} \times \mathbb{N}_{[1,J]}, m \neq n, \\ (l^m = Comp^m_j) \wedge (l^n = Comp^n_j)\}. \tag{6.13}$$

From the previous discussions, we define the following property:

Definition 6.8 (*Schedulability*) $\mathscr{T}$ is *schedulable* under timing contracts Θ if and only if there is a winning strategy to (TGA, $\overline{L}_u$).

From the practical point of view, the safety game, and thus Problem 6.2, can be solved using the tool UPPAAL-TIGA [5]. The latter synthesizes also a winning strategy when it exists, which provides us with a dynamic scheduling policy for generating the sequences $(t_k^{s_i})_{k\in\mathbb{N}}$, $(t_k^{b_i})_{k\in\mathbb{N}}$, $(t_k^{e_i})_{k\in\mathbb{N}}$, and $(t_k^{a_i})_{k\in\mathbb{N}}$ satisfying (6.4)–(6.5), for all $i \in \mathbb{N}_{[1,N]}$.

6.5 Illustrative Example

In this section, we are interested in synthesizing schedules for a given number N of sampled-data systems, which are subject to timing contracts and whose control input is computed by J shared CPUs, with $J < N$. Indeed, the schedule should guarantee the stability of each system. We implemented the scheduling approach presented in Sect. 6.4 using UPPAAL-TIGA [5], and used the stability verification algorithm from [2] to verify stability.

6.5.1 *One Processor*

Example 6.1 We take $N = 2$ where the two systems $\mathscr{S}_1 = (A_1, B_1, K_1)$ and $\mathscr{S}_2 = (A_2, B_2, K_2)$ are taken from [7] and are given by the following matrices:

$$A_1 = \begin{pmatrix} 0 & 1 \\ 0 & -0.1 \end{pmatrix}, \quad B_1 = \begin{pmatrix} 0 \\ 0.1 \end{pmatrix}, \quad K_1 = \begin{pmatrix} -3.75 & -11.5 \end{pmatrix}. \tag{6.14}$$

$$A_2 = \begin{pmatrix} 0 & 1 \\ -2 & 0.1 \end{pmatrix}, \quad B_2 = \begin{pmatrix} 0 \\ 1 \end{pmatrix}, \quad K_2 = \begin{pmatrix} 1 & 0 \end{pmatrix}. \tag{6.15}$$

6.5.1.1 Stability Verification

After setting $\beta = 0$, we use the stability verification algorithm in [2] to verify that systems $\mathscr{S}_1$ and $\mathscr{S}_2$ are β-stable under timing contracts $\theta(0.1, 0.35, 0.3, 0.85)$ and $\theta(0.2, 0.6, 0.8, 1.15)$, respectively. This means obviously that each of the two systems with any synthesized scheduling policy on a shared CPU, respecting the above timing contracts, is guaranteed to be stable. The computation times required for stability verification are 1.96 and 1.5 s, respectively.

6.5.1.2 Scheduling

Now, we consider the set of control tasks $\mathscr{T} = \{T_1, T_2\}$ running on a single processor, or $J = 1$. After setting the best and worst case execution times for each task as $\underline{c}_1^1 = 0.12$, $\overline{c}_1^1 = 0.35$, $\underline{c}^2 = 0.04$, and $\overline{c}^2 = 0.12$ we define task $T_1 = ((\underline{c}_1^1, \overline{c}_1^1))$, task $T_2 = ((\underline{c}_1^2, \overline{c}_1^2))$, and the same set of timing contracts as in the previous section $\Theta = \{\theta(0.1, 0.35, 0.3, 0.85), \theta(0.2, 0.6, 0.8, 1.15)\}$.

In order to solve the scheduling problem, we associate to $\mathscr{T}$ the timed game automaton TGA as given in Definition 6.7. Following the approach in Sect. 6.4, we solve the safety game on TGA to find a strategy (if it exists) for the triggering of controllable actions that occur at $(t_k^{s_i})_{k\in\mathbb{N}}$, $(t_k^{b_i})_{k\in\mathbb{N}}$, and $(t_k^{a_i})_{k\in\mathbb{N}}$, with $i \in \mathbb{N}_{[1,2]}$, guaranteeing that the set of bad states L_u of the system, given by (6.13), is never reached regardless of when uncontrollable actions occurring at $(t_k^{e_i})_{k\in\mathbb{N}}$, $i \in \mathbb{N}_{[1,2]}$, are exactly taken.

Using UPPAAL-TIGA, we successfully proved that $\mathscr{T}$ is schedulable under timing contracts Θ, and thus a scheduling policy was found. The computation time required to solve the game was 1.37 s.

Figure 6.6 shows the timing of events resulting from this scheduling policy. The first and second plots show that the timing contracts $\theta(0.1, 0.35, 0.3, 0.85)$ and $\theta(0.2, 0.6, 0.8, 1.15)$ are respected for both systems $\mathscr{S}_1$ and $\mathscr{S}_2$, respectively. The third plot shows that only one of the two systems gains access to the shared processor at a time since it appears clearly that

$$\forall (m, n) \in \mathbb{N}_{[1,2]}^2 \text{ with } m \neq n,$$
$$\mathsf{Com}(\mathscr{S}_m, 1) \cap \mathsf{Com}(\mathscr{S}_n, 1) = \emptyset.$$

One can notice that in the first three control cycles of $\mathscr{S}_2$, the beginning of the computation has to be delayed until the CPU is released by $\mathscr{S}_1$.

Using this scheduling policy, Fig. 6.7 shows results of simulating $\mathscr{S}_1$ and $\mathscr{S}_2$, when they share a single processor to compute the value of their control inputs, for the initial states $z_0^1 = \binom{2}{3}$ and $z_0^2 = \binom{2}{3}$ with $t_0^{s_1} = 0.4$ and $t_0^{s_2} = 0.9$. As shown, trajectories of both systems converge to zero and therefore the scheduling policy in this case guarantees the exponential stability of each system.

6.5.2 *Two Processors*

Example 6.2 We take $N = 3$, where we have two identical systems $\mathscr{S}_1$ and $\mathscr{S}_2$ whose matrices are given by (6.14) and another system $\mathscr{S}_3$ with matrices given by (6.15).

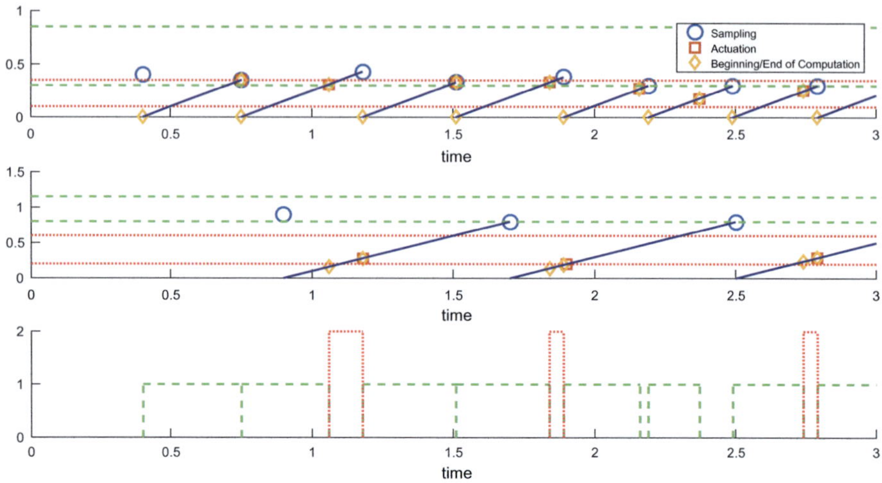

Fig. 6.6 Timing of events (sampling, beginning/end of computation, and actuation) for systems $\mathscr{S}_1$ (first plot) and $\mathscr{S}_2$ (second plot) during the first 3 s; dotted lines represent constraints on actuation instants, while dashed lines represent constraints on sampling instants. In the third plot, the dotted line represents $\mathsf{Com}(\mathscr{S}_2, t)$ (less frequent) and the dashed line represents $\mathsf{Com}(\mathscr{S}_1, t)$ (more frequent)

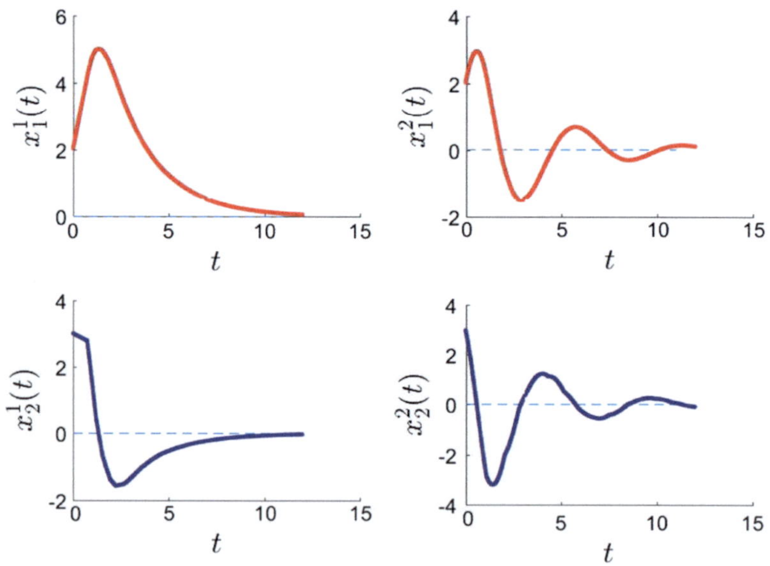

Fig. 6.7 Trajectories for systems $\mathscr{S}_1$ (left) and $\mathscr{S}_2$ (right) using the synthesized scheduling policy

First, we consider a single processor to compute the control input of the three systems (i.e., $J = 1$) where control tasks T_1, T_2, and T_3 are given by $T_1 = T_2 = (0.12, 0.25)$ and $T_3 = (0.04, 0.1)$. We consider the set of contracts $\Theta_a = \{\theta(0.1, 0.35, 0.1, 0.35), \theta(0.1, 0.35, 0.1, 0.35)\}$, $\Theta_b = \{\theta(0.1, 0.35, 0.1, 0.35),\ \theta(0.1, 0.2, 0.1, 0.2)\}$, and $\Theta_c = \{\theta(0.1, 0.35, 0.1, 0.35), \theta(0.1, 0.35, 0.1, 0.35), \theta(0.1, 0.2, 0.1, 0.2)\}$. Following the approach in Sect. 6.4 we can prove that each of the task-set $\{T_1, T_2\}$, the task-set $\{T_2, T_3\}$, and obviously the task-set $\{T_1, T_2, T_3\}$ is not schedulable under timing contracts Θ_a, Θ_b, and Θ_c respectively. On the other hand, this does not mean that systems $\mathscr{S}_1$, $\mathscr{S}_2$, and $\mathscr{S}_3$ cannot share two processors to compute their control input.

Now, we consider two CPUs, or $J = 2$, and define the task-set $\mathscr{T} = \{T_1, T_2, T_3\}$ with $T_1 = T_2 = ((0.12, 0.25), (0.12, 0.25))$ and $T_3 = ((0.04, 0.1), (0.04, 0.1))$. Then we associate to $\mathscr{T}$ the TGA as given in Definition 6.7 and solve the safety game on TGA to find a strategy (if it exists) for the triggering of controllable actions that occur at $(t_k^{s_i})_{k\in\mathbb{N}}$, $(t_k^{b_i})_{k\in\mathbb{N}}$, and $(t_k^{a_i})_{k\in\mathbb{N}}$, with $i \in \mathbb{N}_{[1,3]}$, guaranteeing that the set of bad states $\overline{L}_u$ of the system is never reached regardless of when uncontrollable actions occurring at $(t_k^{e_i})_{k\in\mathbb{N}}$, $i \in \mathbb{N}_{[1,3]}$, are exactly taken.

Using UPPAAL-TIGA, we successfully proved that $\mathscr{T}$ is schedulable under timing contracts Θ_c, and thus a scheduling policy was found. The computation time required to solve the game and output the scheduling policy was 10 s.

Figure 6.8 shows the timing of events resulting from this scheduling policy. The first three plots show that the contracts $\theta(0.1, 0.35, 0.1, 0.35)$, $\theta(0.1, 0.35, 0.1, 0.35)$, and $\theta(0.1, 0.2, 0.1, 0.2)$ are respected for systems $\mathscr{S}_1$, $\mathscr{S}_2$, and $\mathscr{S}_3$, respectively. The fourth and fifth plots show that only one of the three systems gains access to each of the two shared processors at a time since it appears clearly that

$$\forall (m, n, j) \in \mathbb{N}^2_{[1,3]} \times \mathbb{N}_{[1,2]} \text{with } m \neq n,$$
$$\mathsf{Com}(\mathscr{S}_m, j) \cap \mathsf{Com}(\mathscr{S}_n, j) = \emptyset.$$

At this point, we should mention that we followed the stability verification approach in [2] and proved that for $\beta = 0$, β-stability is guaranteed for systems $\mathscr{S}_1$, $\mathscr{S}_2$, and $\mathscr{S}_3$ under timing contracts $\theta(0.1, 0.35, 0.1, 0.35)$, $\theta(0.1, 0.35, 0.1, 0.35)$, and $\theta(0.1, 0.2, 0.1, 0.2)$ respectively. Using this scheduling policy, Fig. 6.9 shows results of simulating $\mathscr{S}_1$, $\mathscr{S}_2$, and $\mathscr{S}_3$ when they share two processors to compute the value of their control inputs, for the initial states $z_0^1 = z_0^2 = z_0^3 = \binom{2}{3}$ with $t_0^{s_1} = t_0^{s_2} = t_0^{s_3} = 0.01s$. As shown, trajectories of the three systems converge to zero and therefore the scheduling policy in this case guarantees the exponential stability of each system.

6.6 Conclusion

In this chapter, we proposed an approach for verifying stability and scheduling embedded control systems under timing contracts on a multi-core platform using

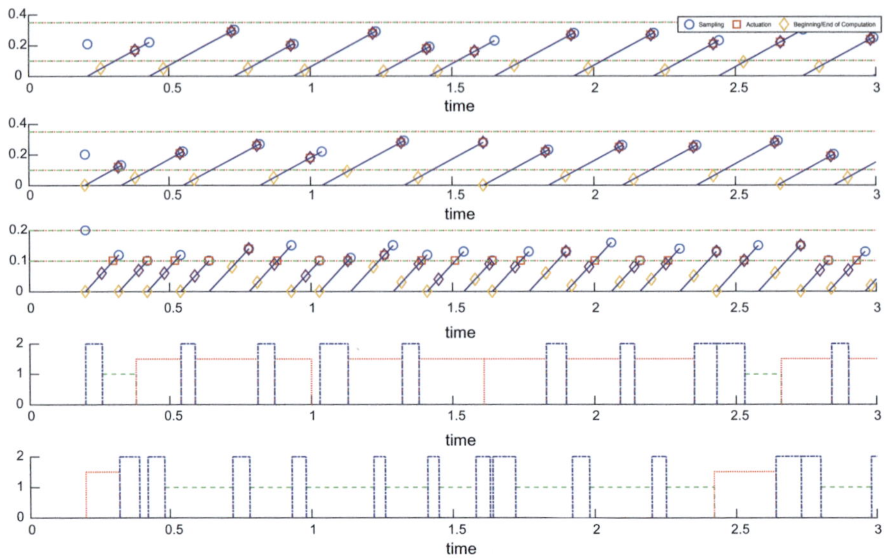

Fig. 6.8 Timing of events (sampling, beginning/end of computation, and actuation) for systems $\mathscr{S}_1$ (first plot), $\mathscr{S}_2$ (second plot), $\mathscr{S}_3$ (third plot) during the first 3 s; dotted lines represent constraints on actuation instants, while dashed lines represent constraints on sampling instants. In the fourth and fifth plot, the dashed line (magnitude 1) represents $\mathsf{Com}(\mathscr{S}_1, j)$, dotted line (magnitude 1.5) represents $\mathsf{Com}(\mathscr{S}_1, j)$, and the dotted-dashed line (magnitude 2) represents $\mathsf{Com}(\mathscr{S}_3, j)$ for $j = 1$ (fourth plot) and $j = 2$ (fifth plot)

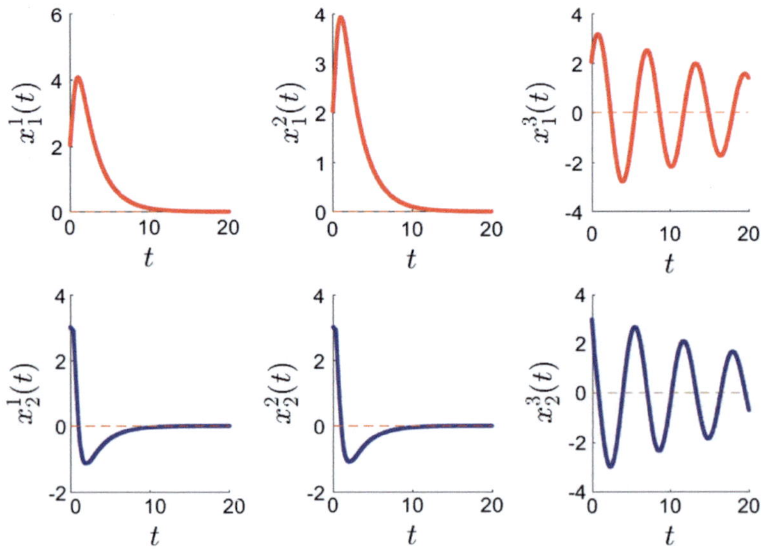

Fig. 6.9 Trajectories for systems $\mathscr{S}_1$ (left), $\mathscr{S}_2$ (middle), and $\mathscr{S}_3$ (right) using the synthesized scheduling policy

reachability analysis and safety timed games, respectively. As a future work, it would be interesting to consider preemptive scheduling since it is not trivial to extend the present work to scheduling with preemption. Another direction for improving our approach is to find optimal schedules in the sense that the control loop is to be closed as soon as possible for each task to have the best possible performance.

Acknowledgements This work was supported by the Agence Nationale de la Recherche (COMPACS project ANR-13-BS03-0004) and by the Labex DigiCosme, Université Paris-Saclay (CODECSYS project).

References

1. Alur, R., Dill, D.L.: A theory of timed automata. Theor. Comput. Sci. **126**(2), 183–235 (1994)
2. Al Khatib, M., Girard, A., Dang. T.: Verification and synthesis of timing contracts for embedded controllers. Hybrid Systems: Computation and Control, pp. 115–124. ACM (2016)
3. Al Khatib, M., Girard, A., Dang, T., Dang, T.: Stability verification and timing contract synthesis for linear impulsive systems using reachability analysis. Nonlinear Anal. Hybrid Syst, **25**, 211–226 (2017)
4. Al Khatib, M., Girard, A., Dang, T.: Scheduling of embedded controllers under timing contracts. Hybrid Systems: Computation and Control, pp. 131–140. ACM (2017)
5. Behrmann, G., Cougnard, A., David, A., Fleury, E., Larsen, K.G., Lime, D.: UPPAAL-TIGA: time for playing games! Computer Aided Verification, pp. 121–125. Springer (2007)
6. Bauer, N.W., Maas, P.J.H., Heemels, W.P.M.H.: Stability analysis of networked control systems: a sum of squares approach. Automatica **48**(8), 1514–1524 (2012)
7. Briat, C.: Convex conditions for robust stability analysis and stabilization of linear aperiodic impulsive and sampled-data systems under dwell-time constraints. Automatica **49**(11), 3449–3457 (2013)
8. Cassez, F., David, A., Fleury, E., Larsen, K.G., Lime, D.: Efficient on-the-fly algorithms for the analysis of timed games. Concurrency Theory, pp. 66–80. Springer, Berlin (2005)
9. Cloosterman, M.B.G., Hetel, L., Van De Wouw, N., Heemels, W.P.M.H., Daafouz, J., Nijmeijer, H.: Controller synthesis for networked control systems. Automatica **46**(10), 1584–1594 (2010)
10. Donkers, M.C.F., Heemels, W.P.M.H., Van De Wouw, N., Hetel, L.: Stability analysis of networked control systems using a switched linear systems approach. IEEE Trans. Autom. Control **56**(9), 2101–2115 (2011)
11. Derler, P., Lee, E.A., Tripakis, S., Törngren, M.: Cyber-physical system design contracts. In: International Conference on Cyber-Physical Systems, pages 109–118 (2013)
12. Fiacchini, M., Morărescu, I.-C.: Constructive necessary and sufficient condition for the stability of quasi-periodic linear impulsive systems. IEEE Trans. Autom. Control **61**(9), 2512–2517 (2016)
13. Fujioka, H.: Stability analysis of systems with aperiodic sample-and-hold devices. Automatica **45**(3), 771–775 (2009)
14. Gao, H., Meng, X., Chen, T., Lam, J.: Stabilization of networked control systems via dynamic output-feedback controllers. SIAM J. Control Optim. **48**(5), 3643–3658 (2010)
15. Hetel, L., Daafouz, J., Tarbouriech, S., Prieur, C.: Stabilization of linear impulsive systems through a nearly-periodic reset. Nonlinear Anal. Hybrid Syst. **7**(1), 4–15 (2013)
16. Hetel, L., Kruszewski, A., Perruquetti, W., Richard, J.-P.: Discrete and intersample analysis of systems with aperiodic sampling. IEEE Trans. Autom. Control **56**(7), 1696–1701 (2011)
17. Heemels, W.P.M.H., Teel, A.R., Van de Wouw, N., Nešić, D.: Networked control systems with communication constraints: tradeoffs between transmission intervals, delays and performance. IEEE Trans. Autom. Control **55**(8), 1781–1796 (2010)

18. Liu, K., Fridman, E., Hetel, L.: Networked control systems in the presence of scheduling protocols and communication delays. SIAM J. Control Optim. **53**(4), 1768–1788 (2015)
19. Le Guernic, C., Girard, A.: Reachability analysis of linear systems using support functions. Nonlinear Anal. Hybrid Syst. **4**(2), 250–262 (2010)
20. Liu, K., Suplin, V., Fridman, E.: Stability of linear systems with general sawtooth delay. IMA J. Math. Control Inf. **27**(4), 419–436 (2010)
21. Maler, O., Pnueli, A., Sifakis, J.: On the synthesis of discrete controllers for timed systems. In: Annual Symposium on Theoretical Aspects of Computer Science, pp. 229–242. Springer (1995)
22. Seuret, A., Peet, M.: Stability analysis of sampled-data systems using sum of squares. IEEE Trans. Autom. Control **58**(6), 1620–1625 (2013)

Part II
Event-Triggered Architectures

Chapter 7
Time-Regularized and Periodic Event-Triggered Control for Linear Systems

D. P. Borgers, V. S. Dolk, G. E. Dullerud, A. R. Teel and W. P. M. H. Heemels

Abstract In this chapter, we provide an overview of our recent results for the analysis and design of Event-Triggered controllers that are tailored to *linear systems* as provided in Heemels et al., IEEE Trans Autom Control 58(4):847–861, 2013, Heemels et al., IEEE Trans Autom Control 61(10):2766–2781, 2016, Borgers et al., IEEE Trans Autom Control, 2018. In particular, we discuss two different frameworks for the stability and contractivity analysis and design of (static) periodic Event-Triggered control (PETC) and time-regularized continuous Event-Triggered control (CETC) systems: the lifting-based framework of Heemels et al., IEEE Trans Autom Control 61(10):2766–2781, 2016, which applies to PETC systems, and the Riccati-based framework of Heemels et al., IEEE Trans Autom Control 58(4):847–861, 2013, Borgers et al., IEEE Trans Autom Control (2018), which applies to both PETC systems and time-regularized CETC systems. Moreover, we identify the connections and differences between the two frameworks. Finally, for PETC and time-regularized CETC systems, we show how the Riccati-based analysis leads to new designs for *dynamic* Event-Triggered controllers, which (for identical stability and contractivity guarantees) lead to a significantly reduced consumption of communication and energy resources compared to their static counterparts.

D. P. Borgers (✉) · V. S. Dolk · W. P. M. H. Heemels
Eindhoven University of Technology, Eindhoven, The Netherlands
e-mail: d.p.borgers@tue.nl

V. S. Dolk
e-mail: v.s.dolk@tue.nl

W. P. M. H. Heemels
e-mail: m.heemels@tue.nl

G. E. Dullerud
University of Illinois, Urbana, IL, USA
e-mail: dullerud@illinois.edu

A. R. Teel
University of California, Santa Barbara, CA, USA
e-mail: teel@ece.ucsb.edu

S. Tarbouriech et al. (eds.), *Control Subject to Computational and Communication Constraints*, Lecture Notes in Control and Information Sciences 475, https://doi.org/10.1007/978-3-319-78449-6_7

7.1 Introduction

In most digital control systems, the measured output of the plant is periodically transmitted to the controller, regardless of the state the system is in. This possibly leads to a waste of (e.g., computation, communication, and energy) resources, as many of the transmissions are actually not needed to achieve the desired control performance guarantees. In recent years, many *Event-Triggered control* (ETC) strategies have been proposed, which generate the transmission (event) times based on the current state or output of the system and the most recently transmitted measurement data, thereby bringing feedback into the process of deciding when control tasks are executed and corresponding measurement and control data is transmitted. In contrast, in periodic time-triggered control, the control execution process could be considered as an open-loop mechanism. By using feedback in the control execution process, measurement data is only transmitted to the controller when this is really necessary in order to be able to guarantee the required stability and performance properties of the system. Clearly, in the interconnected world we live in with many networked control applications including cooperative robotics, vehicle platooning, Internet-of-things, and so on, it is important to use the available (computation, communication, and energy) resources of the system carefully in order to avoid congesting the computational devices or communication networks, or draining batteries. The use of ETC can play an important role in achieving this.

A major challenge in the design of ETC strategies is meeting certain *control performance specifications* (quality-of-control), such as global asymptotic stability, bounds on convergence rates, or $\mathscr{L}_p$-gain requirements, while simultaneously satisfying constraints on the resource utilization (required quality-of-service), including a guaranteed positive lower bound on the inter-event times and thus the absence of Zeno behaviour (an infinite number of events in finite time). In [5, 15], it was shown that this combination of quality-of-control and (required) quality-of-service specifications is hard to achieve, especially for *continuous* Event-Triggered control (CETC) schemes, in which the event condition is continuously monitored (which also requires continuous measuring of the state or output of the plant), as proposed in, e.g., [12, 20, 21, 24, 32, 33, 41, 51].

In the recent years, two main solutions were proposed to tackle this problem:

- CETC schemes that adopt a minimal waiting time between two event times ("time-regularization"), see, e.g., [1, 2, 13, 14, 18, 24, 29, 39, 42–44] and the references therein;
- Periodic Event-Triggered control (PETC) schemes that check the event conditions only at periodic sampling times that are equidistantly distributed along the time axis, see, e.g., [24, 25, 28, 29, 36] and the references therein.

In this chapter, we provide an overview of our recent results for the analysis and design of PETC and time-regularized CETC schemes that are tailored to *linear systems* as provided in [8, 25, 26]. In particular, we discuss two different analysis and design frameworks: the framework as developed in [26], which uses ideas from lifting [4, 10, 16, 45, 46, 53], and the framework as developed in [6–8, 25], which exploits matrix Riccati differential equations.

The lifting-based framework of [26] applies to PETC systems, and leads to the important result that the stability and contractivity in $\mathscr{L}_2$-sense (meaning that the $\mathscr{L}_2$-gain is smaller than 1) of PETC closed-loop systems (which are hybrid systems) is equivalent to the stability and contractivity in ℓ_2-sense (meaning that the ℓ_2-gain is smaller than 1) of an appropriate discrete-time piecewise linear system [26]. These new insights are obtained by adopting a lifting-based perspective on this analysis problem, which leads to computable ℓ_2-gain (and thus $\mathscr{L}_2$-gain) conditions, despite the fact that the linearity assumption, which is usually needed in the lifting literature, is not satisfied.

The Riccati-based framework of [6–8, 25] applies both to PETC systems and to time-regularized CETC systems, and exploits matrix Riccati differential equations for the construction of appropriate Lyapunov/storage functions in the stability and performance analysis. For the PETC case, we identify the connections and differences between the Riccati-based and lifting-based approaches. Moreover, for PETC and time-regularized CETC systems, we show how the Riccati-based analysis leads to new designs for *dynamic* Event-Triggered controllers. Interestingly, the inclusion of a dynamic variable in the event-generator can lead to a significantly reduced consumption of communication and energy resources while leading to identical guarantees on stability and performance as their static counterparts, see also [13, 14, 21, 37, 38] in which designs of dynamic ETC schemes for general nonlinear systems were proposed for the first time.

Both frameworks lead to computationally friendly semi-definite programming conditions, and can also be used for applications in many other domains, including reset control, networked control systems, and switching sampled-data controllers [4, 9–11, 19, 45, 46, 53].

The chapter is organized as follows. In Sect. 7.2, we introduce the considered Event-Triggered control setups. For PETC systems, we introduce the lifting-based framework in Sect. 7.3, and the Riccati-based framework in Sect. 7.4. In Sect. 7.5, we show how the Riccati-based framework of Sect. 7.4 can be modified in order to analyze stability and contractivity of time-regularized CETC systems. We illustrate the results by a numerical example in Sect. 7.6, which also shows that our new frameworks tailored to linear systems are much less conservative than our previous results for nonlinear systems in [13, 14], in the sense that tighter performance bounds can be obtained. Finally, we discuss several directions of extensions of the two frameworks in Sect. 7.7, and summarize the chapter in Sect. 7.8.

7.1.1 Notation

By $\mathbb{N}$ we denote the set of natural numbers including zero, i.e., $\mathbb{N} := \{0, 1, 2, \dots\}$. For vectors $x_i \in \mathbb{R}^{n_i}$, $i \in \{1, 2, \dots, N\}$, we denote by $(x_1, x_2, \cdots, x_N)$ the vector $[x_1^\top x_2^\top \cdots x_N^\top]^\top \in \mathbb{R}^n$ with $n = \sum_{i=1}^N n_i$. For a matrix $P \in \mathbb{R}^{n\times n}$, we write $P \succ 0$

($P \succeq 0$) if P is symmetric and positive (semi-)definite, and $P \prec 0$ ($P \preceq 0$) if P is symmetric and negative (semi-)definite. By I and O we denote the identity and zero matrix of appropriate dimensions, respectively. For brevity, we sometimes write symmetric matrices of the form $\left[\begin{smallmatrix} A & B \\ B^\top & C \end{smallmatrix}\right]$ as $\left[\begin{smallmatrix} A & B \\ \star & C \end{smallmatrix}\right]$ or $\left[\begin{smallmatrix} A & \star \\ B^\top & C \end{smallmatrix}\right]$. For a left-continuous signal $f : \mathbb{R}_{\geq 0} \to \mathbb{R}^n$ and $t \in \mathbb{R}_{\geq 0}$, we use $f(t^+)$ to denote the limit $f(t^+) = \lim_{s \to t, s > t} f(s)$.

For X, Y Hilbert spaces with inner products $\langle \cdot, \cdot \rangle_X$ and $\langle \cdot, \cdot \rangle_Y$, respectively, a linear operator $U : X \to Y$ is called isometric if $\langle Ux_1, Ux_2 \rangle_Y = \langle x_1, x_2 \rangle_X$ for all $x_1, x_2 \in X$. We denote by $U^* : Y \to X$ the (Hilbert) adjoint operator that satisfies $\langle Ux, y \rangle_Y = \langle x, U^* y \rangle_X$ for all $x \in X$ and all $y \in Y$. The induced norm of U (provided it is finite) is denoted by $\|U\|_{X,Y} = \sup_{x \in X \setminus \{0\}} \frac{\|Ux\|_Y}{\|x\|_X}$. If the induced norm is finite we say that U is a bounded linear operator. If $X = Y$ we write $\|U\|_X$ and if X, Y are clear from the context we use the notation $\|U\|$. An operator $U : X \to X$ with X a Hilbert space is called self-adjoint if $U^* = U$. A self-adjoint operator $U : X \to X$ is called positive semi-definite if $\langle Ux, x \rangle \geq 0$ for all $x \in X$. Given a positive semi-definite U, we say that the bounded linear operator $A : X \to X$ is the square root of U if A is positive semi-definite and $A^2 = U$. This square root exists and is unique, see [31, Theorem 9.4-1]. We denote it by $U^{1/2}$.

To a Hilbert space X with inner product $\langle \cdot, \cdot \rangle_X$, we can associate the Hilbert space $\ell_2(X)$ consisting of infinite sequences $\tilde{x} = \{\tilde{x}_0, \tilde{x}_1, \tilde{x}_2, \ldots\}$ with $\tilde{x}_i \in X$, $i \in \mathbb{N}$, satisfying $\sum_{i=0}^{\infty} \|\tilde{x}_i\|_X^2 < \infty$, and the inner product $\langle \tilde{x}, \tilde{y} \rangle_{\ell_2(X)} = \sum_{i=0}^{\infty} \langle \tilde{x}_i, \tilde{y}_i \rangle_X$. We denote $\ell_2(\mathbb{R}^n)$ by ℓ_2 when $n \in \mathbb{N}_{\geq 1}$ is clear from the context. We also use the notation $\ell(X)$ to denote the set of all infinite sequences $\tilde{x} = \{\tilde{x}_0, \tilde{x}_1, \tilde{x}_2, \ldots\}$ with $\tilde{x}_i \in X$, $i \in \mathbb{N}$. Note that $\ell_2(X)$ can be considered a subspace of $\ell(X)$. As usual, we denote by $\mathbb{R}^n$ the standard n-dimensional Euclidean space with inner product $\langle x, y \rangle = x^\top y$ and norm $|x| = \sqrt{x^\top x}$ for $x, y \in \mathbb{R}^n$. $\mathscr{L}_2^n([0, \infty))$ denotes the set of square-integrable functions defined on $\mathbb{R}_{\geq 0} := [0, \infty)$ and taking values in $\mathbb{R}^n$ with $\mathscr{L}_2$-norm $\|x\|_{\mathscr{L}_2} = \sqrt{\int_0^\infty |x(t)|^2 dt}$ and inner product $\langle x, y \rangle_{\mathscr{L}_2} = \int_0^\infty x^\top(t) y(t) dt$ for $x, y \in \mathscr{L}_2^n([0, \infty))$. If n is clear from the context we also write $\mathscr{L}_2$. We also use square-integrable functions on subsets $[a, b]$ of $\mathbb{R}_{\geq 0}$ and then we write $\mathscr{L}_2^n([a, b])$ (or $\mathscr{L}_2([a, b])$ if n is clear from context) with the inner product and norm defined analogously. The set $\mathscr{L}_{2,e}^n([0, \infty))$ consists of all locally square-integrable functions, i.e., all functions x defined on $\mathbb{R}_{\geq 0}$, such that for each bounded domain $[a, b] \subset \mathbb{R}_{\geq 0}$ the restriction $x \mid_{[a,b]}$ is contained in $\mathscr{L}_2^n([a, b])$. We also will use the set of essentially bounded functions defined on $\mathbb{R}_{\geq 0}$ or $[a, b] \subset \mathbb{R}_{\geq 0}$, which are denoted by $\mathscr{L}_\infty^n([0, \infty))$ or $\mathscr{L}_\infty^n([a, b])$ with the norm given by the essential supremum denoted by $\|x\|_{\mathscr{L}_\infty}$ for an essentially bounded function x. A function $\beta : \mathbb{R}_{\geq 0} \to \mathbb{R}_{\geq 0}$ is called a $\mathscr{K}$-function if it is continuous, strictly increasing, and $\beta(0) = 0$.

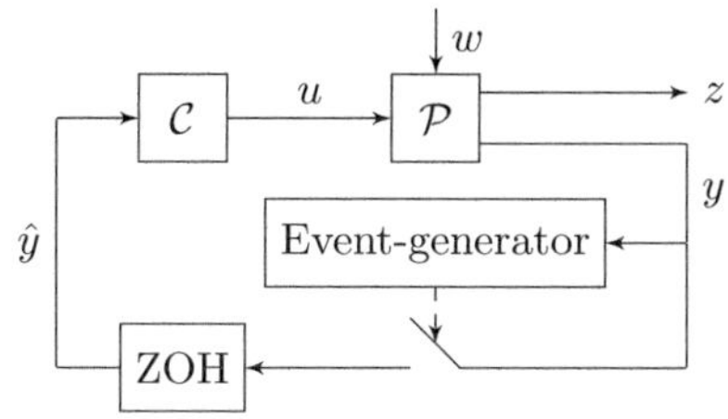

Fig. 7.1 Event-Triggered control setup

7.2 Event-Triggered Control Setup

In this paper, we consider the Event-Triggered control setup as shown in Fig. 7.1, in which the plant $\mathscr{P}$ is given by

$$\mathscr{P} : \begin{cases} \frac{\mathrm{d}}{\mathrm{d}t} x_p = A_p x_p + B_p u + B_{pw} w \\ y = C_y x_p + D_y u \\ z = C_z x_p + D_z u + D_{zw} w \end{cases} \tag{7.1}$$

and the controller $\mathscr{C}$ is given by

$$\mathscr{C} : \begin{cases} \frac{\mathrm{d}}{\mathrm{d}t} x_c = A_c x_c + B_c \hat{y} \\ u = C_u x_c + D_u \hat{y}. \end{cases} \tag{7.2}$$

For ease of exposition, we stick to the configuration of Fig. 7.1, although different control setups are possible as well, see, e.g., [25].

In (7.1) and (7.2), $x_p(t) \in \mathbb{R}^{n_{x_p}}$ denotes the state of the plant $\mathscr{P}$, $y(t) \in \mathbb{R}^{n_y}$ its measured output, $z(t) \in \mathbb{R}^{n_z}$ the performance output, and $w(t) \in \mathbb{R}^{n_w}$ a disturbance at time $t \in \mathbb{R}_{\geq 0}$. Furthermore, $x_c(t) \in \mathbb{R}^{n_{x_c}}$ denotes the state of the controller $\mathscr{C}$, $u(t) \in \mathbb{R}^{n_u}$ is the control input at time $t \in \mathbb{R}_{\geq 0}$, and $\hat{y}(t) \in \mathbb{R}^{n_y}$ denotes the output that is available at the controller, given by

$$\hat{y}(t) = y(t_k), \quad t \in (t_k, t_{k+1}], \tag{7.3}$$

where the sequence $\{t_k\}_{k \in \mathbb{N}}$ denotes the event (or transmission) times, which are generated by the *event-generator*.

In this chapter, we consider *periodic* event-generators, and *continuous* event-generators with *time-regularization*. We will provide their designs in Sects. 7.2.1 and 7.2.2, respectively. In order to do so, we will first define the state $\xi := (x_p, x_c, \hat{y}) \in \mathbb{R}^{n_\xi}$, with $n_\xi = n_{x_p} + n_{x_c} + n_y$, and the matrix $Y \in \mathbb{R}^{2n_y \times n_\xi}$ as

$$Y := \begin{bmatrix} C_y & D_y C_u & D_y D_u \\ O & O & I \end{bmatrix} \tag{7.4}$$

such that $\zeta := (y, \hat{y}) = Y\xi$.

7.2.1 Periodic Event-Triggered Control

In a periodic Event-Triggered control (PETC) setup, the plant output y is sampled periodically at fixed sample times $s_n = nh$, $n \in \mathbb{N}$, where $h \in \mathbb{R}_{>0}$ is the sample period. At each sample time s_n, $n \in \mathbb{N}$, the event-generator decides whether or not the measured output $y(s_n)$ should be transmitted to the controller. Hence, the sequence of event times $\{t_k\}_{k\in\mathbb{N}}$ is a subsequence of the sequence of sample times $\{s_n\}_{n\in\mathbb{N}}$.

In this work, we consider periodic event-generators of the form

$$t_0 = 0,\ t_{k+1} = \inf\{t > t_k \mid \zeta^\top(t) Q \zeta(t) > 0,\ t = nh,\ n \in \mathbb{N}\}, \tag{7.5}$$

where the scalar $h \in \mathbb{R}_{>0}$ and the matrix $Q \in \mathbb{R}^{2n_y \times 2n_y}$ are design parameters. A possible choice for Q is given by

$$Q = \begin{bmatrix} (1-\sigma^2)I & -I \\ -I & I \end{bmatrix} \tag{7.6}$$

with $\sigma \in (0, 1)$, such that (7.5) reduces to

$$t_0 = 0,\ t_{k+1} = \inf\{t > t_k \mid |\hat{y}(t) - y(t)|^2 > \sigma^2 |y(t)|^2,\ t = nh,\ n \in \mathbb{N}\},$$

which can be seen as the digital version of static continuous event-generators [41] of the type

$$t_0 = 0,\ t_{k+1} = \inf\{t \geq t_k \mid |\hat{y}(t) - y(t)|^2 > \sigma^2 |y(t)|^2,\ t \in \mathbb{R}_{\geq 0}\}.$$

Other control setups and other choices of Q are also possible, see, e.g., [25].

By introducing a timer variable $\tau \in [0, h]$, which keeps track of the time that has elapsed since the latest sample time, the closed-loop PETC system consisting of (7.1)–(7.3), and (7.5) can be written as the hybrid system

$$\frac{\mathrm{d}}{\mathrm{d}t}\begin{bmatrix} \xi \\ \tau \end{bmatrix} = \begin{bmatrix} A\xi + Bw \\ 1 \end{bmatrix}, \qquad \tau \in [0, h], \tag{7.7a}$$

$$\begin{bmatrix} \xi^+ \\ \tau^+ \end{bmatrix} = \begin{bmatrix} J\xi \\ 0 \end{bmatrix}, \qquad \tau = h \text{ and } \zeta^\top Q \zeta > 0, \tag{7.7b}$$

$$\begin{bmatrix} \xi^+ \\ \tau^+ \end{bmatrix} = \begin{bmatrix} \xi \\ 0 \end{bmatrix}, \qquad \tau = h \text{ and } \zeta^\top Q \zeta \leq 0, \tag{7.7c}$$

$$z = C\xi + Dw, \tag{7.7d}$$

where

$$A = \begin{bmatrix} A_p & B_pC_u & B_pD_u \\ O & A_c & B_c \\ O & O & O \end{bmatrix}, \ B = \begin{bmatrix} B_{pw} \\ O \\ O \end{bmatrix}, \ J = \begin{bmatrix} I & O & O \\ O & I & O \\ C_y & D_yC_u & D_yD_u \end{bmatrix}, \tag{7.8}$$
$$C = \begin{bmatrix} C_z & D_zC_u & D_zD_u \end{bmatrix}, \text{ and } D = D_{zw}.$$

At sample times $s_n = nh, n \in \mathbb{N}$, the reset (7.7b) occurs when an event is triggered by the event-generator, otherwise the state (ξ, τ) jumps according to (7.7c). In between the sample times, the system evolves according to the differential equation (7.7a), where $\big(\xi(s_n^+), \tau(s_n^+)\big)$ given by (7.7b) or (7.7c) denotes the starting point for the solution to (7.7a) in the interval $(s_n, s_{n+1}], n \in \mathbb{N}$. Hence, the solutions are considered to be left-continuous signals.

7.2.2 Time-Regularized Continuous Event-Triggered Control

In this chapter, we also consider *continuous* event-generators with *time-regularization*, of the form

$$t_{k+1} = \inf\{t \geq t_k + h \mid \zeta^\top(t)Q\zeta(t) > 0\}, \tag{7.9}$$

where now the scalar $h \in \mathbb{R}_{\geq 0}$ is a timer threshold (a waiting time), which enforces a MIET of (at least) h time units. If we again choose Q as in (7.6), then (7.9) constitutes the time-regularized version of (7.2.1). Note that the practical implementation of (7.9) requires continuous monitoring of the output y, which can be difficult to achieve on digital platforms.

The closed-loop CETC system consisting of (7.1)–(7.3) and (7.9) can be written as the hybrid system

$$\frac{\mathrm{d}}{\mathrm{d}t}\begin{bmatrix} \xi \\ \tau \end{bmatrix} = \begin{bmatrix} A\xi + Bw \\ 1 \end{bmatrix}, \qquad \tau \in [0, h] \text{ or } \zeta^\top Q\zeta \leq 0 \tag{7.10a}$$
$$\begin{bmatrix} \xi^+ \\ \tau^+ \end{bmatrix} = \begin{bmatrix} J\xi \\ 0 \end{bmatrix}, \qquad \tau \in [h, \infty) \text{ and } \zeta^\top Q\zeta > 0 \tag{7.10b}$$
$$z = C\xi + Dw, \tag{7.10c}$$

where the timer variable $\tau \in \mathbb{R}_{\geq 0}$ now keeps track of the time that has elapsed since the latest event time. The matrices A, B, C, D, and J are again given by (7.8).

7.2.3 Stability and Performance

As the objective of the paper is to study the $\mathscr{L}_2$-gain and internal stability of the systems (7.7) and (7.10), let us first provide rigorous definitions of these important concepts.

Definition 7.1 The hybrid system (7.7) or (7.10) is said to have an $\mathscr{L}_2$-gain from w to z smaller than γ if there exist a $\gamma_0 \in [0, \gamma)$ and a $\mathscr{K}$-function β such that, for any $w \in \mathscr{L}_2$ and any initial conditions $\xi(0) = \xi_0$ and $\tau(0) = h$, the corresponding solution to (7.7) or (7.10) satisfies $\|z\|_{\mathscr{L}_2} \leq \beta(|\xi_0|) + \gamma_0\|w\|_{\mathscr{L}_2}$. Sometimes, we also use the terminology γ-contractivity (in $\mathscr{L}_2$-sense) if this property holds. Moreover, 1-contractivity is also called contractivity (in $\mathscr{L}_2$-sense).

Definition 7.2 The hybrid system (7.7) or (7.10) is said to be internally stable if there exists a $\mathscr{K}$-function β such that, for any $w \in \mathscr{L}_2$ and any initial conditions $\xi(0) = \xi_0$ and $\tau(0) = h$, the corresponding solution to (7.7) or (7.10) satisfies $\|\xi\|_{\mathscr{L}_2} \leq \beta(\max(|\xi_0|, \|w\|_{\mathscr{L}_2}))$.

A few remarks are in order regarding this definition of internal stability. The requirement $\|\xi\|_{\mathscr{L}_2} \leq \beta(\max(|\xi_0|, \|w\|_{\mathscr{L}_2}))$ is rather natural in this context as we are working with $\mathscr{L}_2$-disturbances and investigate $\mathscr{L}_2$-gains. Indeed, just as in Definition 7.1, where a bound is required on the $\mathscr{L}_2$-norm of the output z (expressed in terms of a bound on $|\xi_0|$ and $\|w\|_{\mathscr{L}_2}$), we require in Definition 7.2 that a similar (though less strict) bound holds on the state trajectory ξ. Apart from internal stability, both design frameworks also lead to global attractivity of the origin (i.e., $\lim_{t\to\infty} \xi(t) = 0$ for all $w \in \mathscr{L}_2$, $\xi(0) = \xi_0$ and $\tau(0) = h$) and Lyapunov stability of the origin, see Proposition 7.1 for the lifting-based framework and [8] for the Riccati-based framework.

Remark 7.1 In this chapter, we focus on the contractivity of the systems (7.7) and (7.10) as γ-contractivity can be studied by proper scaling of the matrices C and D in (7.7), i.e., $C_{\text{scaled}} = \gamma^{-1}C$ and $D_{\text{scaled}} = \gamma^{-1}D$.

7.3 Lifting-Based Static PETC

In this section, we give an overview of our work [26], which provides a framework for the contractivity and internal stability analysis of the static PETC system (7.7) using ideas from lifting [4, 10, 16, 45, 46, 53]. To obtain necessary and sufficient conditions for internal stability and contractivity of (7.7), we use a procedure consisting of three main steps:

- In Sect. 7.3.2, we apply lifting-based techniques to (7.7) (having finite-dimensional input and output spaces) leading to a discrete-time system with infinite-dimensional input and output spaces (see (7.15) below). The internal stability and contractivity of both systems are equivalent.
- In Sect. 7.3.3, we apply a loop transformation to the infinite-dimensional system (7.15) in order to remove the feedthrough term, which is the only operator in the system description having both its domain and range being infinite dimensional. This transformation is constructed in such a manner that the internal stability and contractivity properties of the system are not changed. This step is crucial for

translating the infinite-dimensional system to a finite-dimensional system in the last step.

- In Sect. 7.3.4, the loop-transformed infinite-dimensional system is converted into a discrete-time *finite-dimensional* piecewise linear system (again without changing the stability and the contractivity properties of the system). Due to the finite dimensionality of the latter system, stability and contractivity in ℓ_2-sense can be analyzed, for instance, using well-known Lyapunov-based arguments. We elaborate on these computational aspects (which also exploit semi-definite programming) in Sect. 7.3.5.

These three steps lead to the main result as formulated in Theorem 7.2, which states that the internal stability and contractivity (in $\mathscr{L}_2$-sense) of (7.7) is equivalent to the internal stability and contractivity (in ℓ_2-sense) of a discrete-time finite-dimensional piecewise linear system. To facilitate the analysis, we first introduce the necessary preliminary definitions in Sect. 7.3.1.

7.3.1 Preliminaries

Consider the discrete-time system of the form

$$\xi_{k+1} = \chi(\xi_k, v_k) \tag{7.11a}$$
$$r_k = \psi(\xi_k, v_k) \tag{7.11b}$$

with $v_k \in V$, $r_k \in R$, $\xi_k \in \mathbb{R}^{n_\xi}$, $k \in \mathbb{N}$, with V and R Hilbert spaces, and $\chi : \mathbb{R}^{n_\xi} \times V \to \mathbb{R}^{n_\xi}$ and $\psi : \mathbb{R}^{n_\xi} \times V \to R$.

For this general discrete-time system, we also introduce ℓ_2-gain specifications and internal stability.

Definition 7.3 The discrete-time system (7.11) is said to have an ℓ_2-gain from v to r smaller than γ if there exist a $\gamma_0 \in [0, \gamma)$ and a $\mathscr{K}$-function β such that, for any $v \in \ell_2(V)$ and any initial state $\xi_0 \in \mathbb{R}^{n_\xi}$, the corresponding solution to (7.11) satisfies

$$\|r\|_{\ell_2(R)} \leq \beta(\|\xi_0\|) + \gamma_0 \|v\|_{\ell_2(V)}. \tag{7.12}$$

Sometimes, we also use the terminology γ-contractivity (in ℓ_2-sense) if this property holds. Moreover, 1-contractivity is also called contractivity (in ℓ_2-sense).

Definition 7.4 The discrete-time system (7.11) is said to be internally stable if there is a $\mathscr{K}$-function β such that, for any $v \in \ell_2(V)$ and any initial state $\xi_0 \in \mathbb{R}^{n_\xi}$, the corresponding solution ξ to (7.11) satisfies

$$\|\xi\|_{\ell_2} \leq \beta(\max(|\xi_0|, \|v\|_{\ell_2(V)}). \tag{7.13}$$

Note that this internal stability definition for the discrete-time system (7.11) parallels the continuous-time version in Definition 7.2. Moreover, since $\|\xi\|_{\ell_\infty} \leq \|\xi\|_{\ell_2}$ and $\|\xi\|_{\ell_2} < \infty$ implies $\lim_{k\to\infty} \xi_k = 0$, we also have global attractivity and Lyapunov stability properties of the origin when the discrete-time system is internally stable.

7.3.2 *Lifting the System*

To study contractivity, we introduce the lifting operator $W : \mathscr{L}_{2,e}[0,\infty) \to \ell(\mathscr{K})$ with $\mathscr{K} = \mathscr{L}_2[0,h]$ given for $w \in \mathscr{L}_{2,e}[0,\infty)$ by $W(w) = \tilde{w} = \{\tilde{w}_0, \tilde{w}_1, \tilde{w}_2, \ldots\}$ with

$$\tilde{w}_k(s) = w(kh+s) \text{ for } s \in [0,h] \tag{7.14}$$

for $k \in \mathbb{N}$. Using this lifting operator, we can rewrite the model in (7.7) as

$$\xi_{k+1} = \hat{A}\xi_k^+ + \hat{B}\tilde{w}_k \tag{7.15a}$$

$$\xi_k^+ = \begin{cases} J\xi_k, & \xi_k^\top Y^\top QY\xi_k > 0 \\ \xi_k, & \xi_k^\top Y^\top QY\xi_k \leq 0 \end{cases} \tag{7.15b}$$

$$\tilde{z}_k = \hat{C}\xi_k^+ + \hat{D}\tilde{w}_k \tag{7.15c}$$

in which ξ_0 is given and $\xi_k = \xi(kh)$, $k \in \mathbb{N}_{\geq 1}$, $\xi_k^+ = \xi(kh^+)$ (assuming that ξ is left-continuous) for $k \in \mathbb{N}$, and $\tilde{w} = \{\tilde{w}_0, \tilde{w}_1, \tilde{w}_2, \ldots\} = W(w) \in \ell_2(\mathscr{K})$ and $\tilde{z} = \{\tilde{z}_0, \tilde{z}_1, \tilde{z}_2, \ldots\} = W(z) \in \ell(\mathscr{K})$. Here we assume in line with Definition 7.1 that $\tau(0) = h$ in (7.7). Moreover,

$$\hat{A} : \mathbb{R}^{n_\xi} \to \mathbb{R}^{n_\xi}, \quad \hat{B} : \mathscr{K} \to \mathbb{R}^{n_\xi}, \quad \hat{C} : \mathbb{R}^{n_\xi} \to \mathscr{K}, \quad \text{and } \hat{D} : \mathscr{K} \to \mathscr{K}$$

are given for $x \in \mathbb{R}^{n_\xi}$ and $\omega \in \mathscr{K}$ by

$$\hat{A}x = e^{Ah}x \tag{7.16a}$$

$$\hat{B}\omega = \int_0^h e^{A(h-s)}B\omega(s)ds \tag{7.16b}$$

$$(\hat{C}x)(\theta) = Ce^{A\theta}\xi \tag{7.16c}$$

$$(\hat{D}\omega)(\theta) = \int_0^\theta Ce^{A(\theta-s)}B\omega(s)ds + D\omega(\theta), \tag{7.16d}$$

where $\theta \in [0,h]$.

It follows that (7.15) is contractive if and only if (7.7) is contractive. In fact, we have the following proposition

Proposition 7.1 [26] *The following statements hold:*

- *The hybrid system (7.7) is internally stable if and only if the discrete-time system (7.15) is internally stable.*
- *The hybrid system (7.7) is contractive if and only if the discrete-time system (7.15) is contractive.*
- *In case (7.7) is internally stable, it also holds that* $\lim_{t\to\infty}\xi(t) = 0$ *and* $\|\xi\|_{\mathscr{L}_\infty} \leq \beta(\max(|\xi_0|, \|w\|_{\mathscr{L}_2}))$ *for all* $w \in \mathscr{L}_2$, $\xi(0) = \xi_0$ *and* $\tau(0) = h$.

7.3.3 Removing the Feedthrough Term

Following [4], we aim at removing the feedthrough operator $\hat{D}$ as this is the only operator with both its domain and range being infinite dimensional. Removal can be accomplished by using an operator-valued version of Redheffer's lemma, see [4, Lemma 5]. The objective is to obtain a new system (without feedthrough term) and new disturbance inputs $\tilde{v}_k \in \mathscr{K}$, new state $\bar{\xi}_k \in \mathbb{R}^{n_\xi}$, and new performance output $\tilde{r}_k \in \mathscr{K}$, $k \in \mathbb{N}$, given by

$$\bar{\xi}_{k+1} = \bar{A}\bar{\xi}_k^+ + \bar{B}\tilde{v}_k \tag{7.17a}$$

$$\bar{\xi}_k^+ = \begin{cases} J\bar{\xi}_k, & \bar{\xi}_k^\top Y^\top Q Y \bar{\xi}_k > 0 \\ \bar{\xi}_k, & \bar{\xi}_k^\top Y^\top Q Y \bar{\xi}_k \leq 0 \end{cases} \tag{7.17b}$$

$$\tilde{r}_k = \bar{C}\bar{\xi}_k^+ \tag{7.17c}$$

such that (7.15) is internally stable and contractive if and only if (7.17) is internally stable and contractive. To do so, we first observe that a necessary condition for the contractivity (7.15) is that $\|\hat{D}\|_{\mathscr{K}} < 1$. Indeed, $\|\hat{D}\|_{\mathscr{K}} \geq 1$ would imply that for any $0 \leq \gamma_0 < 1$ there is a $\tilde{w}_0 \in \mathscr{K} \setminus \{0\}$ with $\|\hat{D}\tilde{w}_0\|_{\mathscr{K}} \geq \gamma_0\|\tilde{w}_0\|_{\mathscr{K}}$, which, in turn, would lead for the system (7.15) with $\xi_0 = 0$ and thus $\xi_0^+ = 0$ and disturbance sequence $\{\tilde{w}_0, 0, 0, \ldots\}$ to a contradiction with the contractivity of (7.15). We can now find an equivalent system of the form (7.17), with bounded linear operators

$$\bar{A} : \mathbb{R}^{n_\xi} \to \mathbb{R}^{n_\xi}, \quad \bar{B} : \mathscr{K} \to \mathbb{R}^{n_\xi}, \quad \text{and } \bar{C} : \mathbb{R}^{n_\xi} \to \mathscr{K}.$$

These operators are given by [26, Sect. IV.B]

$$\bar{A} = \hat{A} + \hat{B}\hat{D}^*(I - \hat{D}\hat{D}^*)^{-1}\hat{C}, \tag{7.18a}$$

$$\bar{B} = \hat{B}(I - \hat{D}^*\hat{D})^{-\frac{1}{2}}, \tag{7.18b}$$

$$\bar{C} = (I - \hat{D}\hat{D}^*)^{-\frac{1}{2}}\hat{C}. \tag{7.18c}$$

Hence, we establish the following result.

Theorem 7.1 [26] If $\|\hat{D}\|_{\mathscr{K}} < 1$, then internal stability and contractivity of system (7.15) with $\hat{A}$, $\hat{B}$, $\hat{C}$, and $\hat{D}$ as in (7.16) are equivalent to internal stability and contractivity of system (7.17) with $\bar{A}$, $\bar{B}$, and $\bar{C}$ as in (7.18).

7.3.4 From Infinite-Dimensional to Finite-Dimensional Systems

The system (7.17) is still an infinite-dimensional system, although the operators $\bar{A}$, $\bar{B}$, and $\bar{C}$ have finite rank and therefore have finite-dimensional matrix representations. Following (and slightly extending) [4], we now obtain the following result.

Theorem 7.2 [26] Consider system (7.7) and its lifted version (7.15) with $\|\hat{D}\|_{\mathscr{K}} < 1$. Define the discrete-time piecewise linear system

$$\xi_{k+1} = \begin{cases} A_1\xi_k + B_d v_k, & \xi_k^\top Y^\top Q Y \xi_k > 0 \\ A_2\xi_k + B_d v_k, & \xi_k^\top Y^\top Q Y \xi_k \leq 0 \end{cases} \tag{7.19a}$$

$$r_k = \begin{cases} C_1\xi_k, & \xi_k^\top Y^\top Q Y \xi_k > 0 \\ C_2\xi_k, & \xi_k^\top Y^\top Q Y \xi_k \leq 0, \end{cases} \tag{7.19b}$$

$k \in \mathbb{N}$, with $A_1 = A_d J$, $A_2 = A_d$, $C_1 = C_d J$, and $C_2 = C_d$, where A_d is defined by

$$A_d = \hat{A} + \hat{B}\hat{D}^*(I - \hat{D}\hat{D}^*)^{-1}\hat{C} \tag{7.20a}$$

and $B_d \in \mathbb{R}^{n_\xi \times n_v}$ and $C_d \in \mathbb{R}^{n_r \times n_\xi}$ are chosen such that

$$\begin{aligned} B_d B_d^\top &= \bar{B}\bar{B}^* = \hat{B}(I - \hat{D}^*\hat{D})^{-1}\hat{B}^* \text{ and} \\ C_d^\top C_d &= \bar{C}^*\bar{C} = \hat{C}^*(I - \hat{D}\hat{D}^*)^{-1}\hat{C}. \end{aligned} \tag{7.20b}$$

The system (7.7) is internally stable and contractive if and only if the system (7.19) is internally stable and contractive.

Hence, this theorem states that under the assumption $\|\hat{D}\|_{\mathscr{K}} < 1$ (which is a necessary condition for contractivity of (7.7)) the internal stability and contractivity (in $\mathscr{L}_2$-sense) of (7.7) is equivalent to the internal stability and contractivity (in ℓ_2-sense) of a discrete-time finite-dimensional piecewise linear system given by (7.19). In the next section, we will show how the matrices A_d, B_d, and C_d in (7.19) can be constructed, how the condition $\|\hat{D}\|_{\mathscr{K}} < 1$ can be tested, and how internal stability and contractivity can be tested for the system (7.19).

7.3.5 Computing the Discrete-Time Piecewise Linear System

To explicitly compute the discrete-time system (7.19) provided in Theorem 7.2, we need to determine the operators $\hat{B}\hat{D}^*(I - \hat{D}\hat{D}^*)^{-1}\hat{C}$, $\hat{B}(I - \hat{D}^*\hat{D})^{-1}\hat{B}^*$, and $\hat{C}^*(I - \hat{D}\hat{D}^*)^{-1}\hat{C}$ to obtain the triple (A_d, B_d, C_d) in (7.19). For the sake of self-containedness, we recall the procedure proposed in [9] to compute this triple, assuming throughout that $\|\hat{D}\|_{\mathcal{K}} < 1$.

First, we verify that $\|\hat{D}\|_{\mathcal{K}} < 1$, which is a necessary condition for the contractivity of (7.7). Define the Hamiltonian matrix

$$H := \begin{bmatrix} A+BMD^\top C & BMB^\top \\ -C^\top LC & -(A+BMD^\top C)^\top \end{bmatrix} \tag{7.21}$$

in which $L := (I - DD^\top)^{-1}$ and $M := (I - D^\top D)^{-1}$, and the matrix exponential

$$F(\tau) := e^{H\tau} = \begin{bmatrix} F_{11}(\tau) & F_{12}(\tau) \\ F_{21}(\tau) & F_{22}(\tau) \end{bmatrix}. \tag{7.22}$$

The condition $\|\hat{D}\|_{\mathcal{K}} < 1$ is equivalent to the following assumption [26].

Assumption 7.3 $\lambda_{\max}(D^\top D) < 1$ and $F_{11}(\tau)$ is invertible for all $\tau \in [0, h]$.

Invertibility of $F_{11}(\tau)$ for all $\tau \in [0, h]$ can always be achieved by choosing h sufficiently small, as $F_{11}(0) = I$ and F_{11} is a continuous function.

The procedure to find A_d, B_d, and C_d boils down to computing $F(h)$, which then leads to

$$A_d = \bar{F}_{11}^{-1}, \tag{7.23}$$

and

$$B_d B_d^\top = -\bar{F}_{11}^{-1}\bar{F}_{12}, \tag{7.24a}$$

$$C_d^\top C_d = \bar{F}_{21}\bar{F}_{11}^{-1}, \tag{7.24b}$$

where we used the notation $\bar{F}_{11} := F_{11}(h)$, $\bar{F}_{12} := F_{12}(h)$, $\bar{F}_{21} := F_{21}(h)$, and $\bar{F}_{22} := F_{22}(h)$.

This provides the matrices needed for explicitly determining the discrete-time piecewise linear system (7.19) for which the internal stability and contractivity tests need to be carried out.

To guarantee the internal stability and contractivity of a discrete-time piecewise linear system as in (7.19) (in order to guarantee these properties for the hybrid system (7.7) using Theorem 7.2), we aim at finding a Lyapunov function $V : \mathbb{R}^{n_\xi} \to \mathbb{R}_{\geqslant 0}$ that satisfies the dissipation inequality [47, 52]

$$V(\xi_{k+1}) - V(\xi_k) < -r_k^\top r_k + v_k^\top v_k, \ k \in \mathbb{N}, \tag{7.25}$$

and require that it holds along the trajectories of the system (7.19). An effective approach is to use versatile piecewise quadratic Lyapunov/storage functions [17, 30] of the form

$$V(\xi) = \begin{cases} \xi^\top P_1^p \xi & \text{with } p = \min\{q \in \{1, \ldots, N\} \mid \xi \in \Omega_q\} \text{ when } \xi^\top Y^\top QY\xi > 0 \\ \xi^\top P_2^p \xi & \text{with } p = \min\{q \in \{1, \ldots, N\} \mid \xi \in \Omega_q\} \text{ when } \xi^\top Y^\top QY\xi \le 0 \end{cases} \tag{7.26}$$

based on the regions

$$\Omega_p := \left\{\xi \in \mathbb{R}^{n_\xi} \,\middle|\, X_p\xi \ge 0\right\}, \quad p \in \{1, \ldots, N\} \tag{7.27}$$

in which the matrices X_p, $p \in \{1, \ldots, N\}$, are such that $\{\Omega_1, \Omega_2, \ldots, \Omega_N\}$ forms a partition of $\mathbb{R}^{n_\xi}$, i.e., $\cup_{p=1}^N \Omega_p = \mathbb{R}^{n_\xi}$ and the intersection of $\Omega_p \cap \Omega_q$ is of zero measure for all $p, q \in \{1, \ldots, N\}$ with $p \neq q$.

This translates into sufficient LMI-based conditions for stability and contractivity using three S-procedure relaxations [30], as formulated next.

Theorem 7.4 If there exist symmetric matrices $P_i^p \in \mathbb{R}^{n_\xi \times n_\xi}$, scalars $a_i^p, c_{ij}^{pq}, d_{ij}^{pq} \in \mathbb{R}_{>0}$, and symmetric matrices $E_i^p, U_{ij}^{pq}, W_{ij}^{pq} \in \mathbb{R}_{\ge 0}^{n_\xi \times n_\xi}$, with $i, j \in \{1, 2\}$, $p, q \in \{1, 2, \ldots, N\}$, such that

$$\left[\, P_i^p + (-1)^i a_i^p Y^\top QY - X_p^\top E_i^p X_p \,\right] \succ 0 \tag{7.28a}$$

and

$$\begin{aligned} &\begin{bmatrix} P_i^p - C_i^\top C_i - A_i^\top P_j^q A_i & -A_i^\top P_j^q B_d \\ -B_d P_j^q A_i & I - B_d^\top P_j^q B_d \end{bmatrix} \\ &\quad + \begin{bmatrix} (-1)^i c_{ij}^{pq} Y^\top QY + (-1)^j d_{ij}^{pq} A_i^\top Y^\top QY A_i & (-1)^j d_{ij}^{pq} A_i^\top Y^\top QY B_d \\ (-1)^j d_{ij}^{pq} B_d^\top Y^\top QY A_i & (-1)^j d_{ij}^{pq} B_d^\top Y^\top QY B_d \end{bmatrix} \\ &\qquad - \begin{bmatrix} X_p^\top U_{ij}^{pq} X_p + A_i^\top X_q^\top W_{ij}^{pq} X_q A_i & A_i^\top X_q^\top W_{ij}^{pq} X_q B_d \\ B_d^\top X_q^\top W_{ij}^{pq} X_q A_i & B_d^\top X_q^\top W_{ij}^{pq} X_q B_d \end{bmatrix} \prec 0 \end{aligned} \tag{7.28b}$$

hold for all $i, j \in \{1, 2\}$ and all $p, q \in \{1, 2, \ldots, N\}$, then the discrete-time piecewise linear system (7.19) is internally stable and contractive.

Two comments are in order regarding this theorem. First, note that due to the strictness of the LMIs (7.28), we guarantee that the ℓ_2-gain is strictly smaller than 1, which can be seen from appropriately including the strictness into the dissipativity inequality (7.25). Moreover, due to the strictness of the LMIs we also guarantee internal stability. Second, the LMI conditions of Theorem 7.4 are obtained by performing a contractivity analysis on the *discrete-time* piecewise linear system (7.19) using three S-procedure relaxations:

(i) require that $\xi^\top P_i^p \xi$ is positive only when $(-1)^i \xi^\top Y^\top Q Y \xi \leq 0$ and $X_p \xi \geq 0$ (this corresponds to the terms containing a_i^p and E_i^p in (7.28a), respectively);
(ii) use a relaxation related to the current time instant, i.e., if $V(\xi_k) = \xi_k^\top P_i^p \xi_k$, then it holds that $(-1)^i \xi_k^\top Y^\top Q Y \xi_k \leq 0$ and $X_p \xi_k \geq 0$ (this corresponds to the terms containing c_{ij}^{pq} and U_{ij}^{pq} in (7.28b), respectively);
(iii) use a relaxation related to the next time instant, i.e., if $V(\xi_{k+1}) = \xi_{k+1}^\top P_j^q \xi_{k+1}$, then it holds that $(-1)^j \xi_{k+1}^\top Y^\top Q Y \xi_{k+1} \leq 0$ and $X_q \xi_{k+1} \geq 0$ (this corresponds to the terms containing d_{ij}^{pq} and W_{ij}^{pq} in (7.28b), respectively).

Theorem 7.4 can be used to guarantee the internal stability and contractivity of (7.19) and hence, the internal stability and contractivity for the hybrid system (7.7). In the next section, we will rigorously show that these results form significant improvements with respect to the earlier conditions for contractivity of (7.7) presented in [11, 22, 25] and [48]. In Sect. 7.6, we also illustrate this improvement using two numerical examples.

7.4 Riccati-Based PETC

In this section, we recall the LMI-based conditions for analyzing the stability and contractivity analysis for the static PETC system (7.7) provided in [11, 25, 48], and show the relationship to the conditions obtained in Sect. 7.3.2. This also reveals that the conditions in Sect. 7.3.2 are (significantly) less conservative.

However, instead of reducing the conservatism in the stability and contractivity analyses of [11, 25, 48], we have shown in [7, 8] that we can also exploit this conservatism in order to reduce the amount of transmissions even further (with the same stability and performance guarantees as the static counterpart). This leads to the design of *dynamic* periodic event-generators, which we also cover in this section.

7.4.1 Static PETC

We follow here the setup discussed in [25], which is based on using a timer-dependent storage function $V : \mathbb{R}^{n_\xi} \times \mathbb{R}_{\geqslant 0} \to \mathbb{R}_{\geqslant 0}$, see [47], satisfying

$$\tfrac{\mathrm{d}}{\mathrm{d}t} V \leq -z^\top z + w^\top w, \tag{7.29}$$

during the flow (7.7a), and

$$V(J\xi, 0) < V(\xi, h), \text{ for all } \xi \text{ with } \xi^\top Y^\top Q Y \xi > 0, \tag{7.30a}$$

$$V(\xi, 0) < V(\xi, h), \text{ for all } \xi \text{ with } \xi^\top Y^\top Q Y \xi \leq 0, \tag{7.30b}$$

during the jumps (7.7b) and (7.7c). From these conditions, we can guarantee that the $\mathcal{L}_2$-gain from w to z is smaller than or equal to 1, see, e.g., [27].

In fact, in [25], $V(\xi, \tau)$ was chosen in the form

$$V(\xi, \tau) = \xi^\top P(\tau)\xi, \quad \tau \in [0, h], \tag{7.31}$$

where $P : [0, h] \to \mathbb{R}^{n_\xi \times n_\xi}$ is a continuously differentiable function with $P(\tau) \succ 0$ for $\tau \in [0, h]$. The function P will be chosen such that (7.31) becomes a storage function [47, 52] for the PETC system (7.7), (7.5) with the supply rate $\theta^{-2}z^\top z - w^\top w$. In order to do so, we select the function $P : [0, h] \to \mathbb{R}^{n_\xi \times n_\xi}$ to satisfy the Riccati differential equation (where we omitted τ for compactness of notation)

$$\tfrac{\mathrm{d}}{\mathrm{d}\tau} P = -A^\top P - PA - C^\top C - (PB + C^\top D)M(D^\top C + B^\top P). \tag{7.32}$$

Note that the solution to (7.32) exists under Assumption 7.3, see also [3, Lemma 9.2]. As shown in the proof of [25, Theorem III.2], this choice for the matrix function P implies the "flow condition" (7.29). The "jump condition" (7.30) is guaranteed in [25] by LMI-based conditions that lead to a proper choice of the boundary value $P_h := P(h)$.

To formulate the result of [25], we again consider the Hamiltonian matrix (7.21) and the matrix exponential (7.22). The function $P : [0, h] \to \mathbb{R}^{n_\xi \times n_\xi}$ is then explicitly defined for $\tau \in [0, h]$ by

$$P(\tau) = (F_{21}(h - \tau) + F_{22}(h - \tau)P(h))\,(F_{11}(h - \tau) + F_{12}(h - \tau)P(h))^{-1}, \tag{7.33}$$

provided that Assumption 7.3 holds.

Before stating the next theorem (which is a slight variation of [25, Theorem III.2]), let us introduce the notation $P_0 := P(0)$, $P_h := P(h)$, and a matrix $\bar{S}$ that satisfies $\bar{S}\bar{S}^\top := -\bar{F}_{11}^{-1}\bar{F}_{12}$. A matrix $\bar{S}$ exists under Assumption 7.3, because this assumption will guarantee that the matrix $-\bar{F}_{11}^{-1}\bar{F}_{12}$ is positive semi-definite.

Theorem 7.5 [7] If there exist matrices $N_T, N_N \in \mathbb{R}^{2n_y \times 2n_y}$ with $N_T, N_N \succeq 0$ and $P_h \in \mathbb{R}^{n_\xi \times n_\xi}$ with $P_h \succ 0$, and scalars $\beta, \mu \in \mathbb{R}_{\geq 0}$, such that

$$\begin{bmatrix} P_h - Y^\top (N_T + \mu Q)Y - J^\top \left(\bar{F}_{11}^{-\top} P_h \bar{F}_{11}^{-1} + \bar{F}_{21}\bar{F}_{11}^{-1}\right) J & J^\top \bar{F}_{11}^{-\top} P_h \bar{S} \\ \star & I - \bar{S}^\top P_h \bar{S} \end{bmatrix} \succ 0, \tag{7.34}$$

$$\begin{bmatrix} P_h - Y^\top (N_N - \beta Q)Y - \left(\bar{F}_{11}^{-\top} P_h \bar{F}_{11}^{-1} + \bar{F}_{21}\bar{F}_{11}^{-1}\right) & \bar{F}_{11}^{-\top} P_h \bar{S} \\ \star & I - \bar{S}^\top P_h \bar{S} \end{bmatrix} \succ 0, \tag{7.35}$$

and Assumption 7.3 hold, then the static PETC system (7.7) is internally stable and contractive.

Here, (7.29) is guaranteed by the choice of the function $P : [0, h] \to \mathbb{R}^{n_\xi \times n_\xi}$, (7.30a) is guaranteed by (7.34), and (7.30b) is guaranteed by (7.35).

In the spirit of Sect. 7.3.5, we can obtain that the LMI-based conditions in this proposition are equivalent to a *conservative* check of the ℓ_2-gain being smaller than or equal to 1 for the discrete-time piecewise linear system (7.19). In particular, the stability and contractivity tests in Theorem 7.5 use a *common* quadratic storage function (although extension towards a piecewise quadratic storage function is possible, see [8]) and only one of the S-procedure relaxations discussed in Sect. 7.3.5 (only (ii) is used). In addition to this new perspective on the results in [11, 22, 25], a strong link can be established between the existing LMI-based conditions described in Theorem 7.5 and the lifting-based conditions obtained in this section, as formalized next.

Theorem 7.6 [26] If the conditions of Theorem 7.5 hold and the regions in (7.27) are chosen such that for each $i = 1, 2, \dots, N$ there is a $\bar{\xi}_i \in \mathbb{R}^{n_\xi}$ such that $\bar{\xi}_i^\top X_i \bar{\xi}_i > 0$,[1] then $\|\hat{D}\|_{\mathscr{K}} < 1$ and the conditions of Theorem 7.4 hold.

This theorem reveals an intimate connection between the results obtained in [11, 22, 25] and the new lifting-based results obtained in the present paper. Indeed, as already mentioned, the LMI-based conditions in [11, 22, 25] as formulated in Theorem 7.5 boil down to an ℓ_2-gain analysis of a discrete-time piecewise linear system (7.19) based on a *quadratic storage function* using only a part of the S-procedure relaxations possible (only using (7.3.5), while the S-procedure relaxations (i) and (ii) mentioned at the end of Sect. 7.3.5 are not used). Moreover, Theorem 7.6 shows that the lifting-based results using Theorems 7.4 and 7.2 never provide worse estimates of the $\mathscr{L}_2$-gain of (7.7) than the results as formulated in Theorem 7.5. In fact, since the stability and contractivity conditions based on (7.19) can be carried out based on more versatile piecewise quadratic storage functions and more (S-procedure) relaxations (see Theorem 7.4), the conditions in Theorems 7.4 and 7.2 are typically significantly less conservative than the ones obtained in [11, 22, 25].

Remark 7.2 When Q is given by (7.6) with $\sigma = 0$, the static PETC system (7.7) reduces to a sampled-data system. Moreover, in this case the related discrete-time piecewise linear system reduces to a discrete-time LTI system, for which the l_2-gain conditions using a common quadratic Lyapunov/storage function are nonconservative (see [19, Lemma 5.1]). Hence, for sampled-data systems, Theorems 7.5 and 7.4 are equivalent and nonconservative.

7.4.2 Dynamic PETC

Although it is shown above that the stability and contractivity analysis in Theorem 7.5 is conservative, it does provide an explicit Lyapunov/storage function for the PETC

[1] This condition implies that each region has a non-empty interior thereby avoiding redundant regions of zero measure.

system (7.7), which the lifting-based approach does not. Moreover, that this conservatism can be exploited in order to further reduce the amount of communication in the system, while preserving the internal stability and contractivity guarantees [7, 8].

The idea is as follows. First, introduce the buffer variable $\eta \in \mathbb{R}$ (which will be included in the event-generator), and define the signal $\hat{o} : \mathbb{R}_{\geq 0} \to \mathbb{R}^{2n_y} \times [0, h] \times \mathbb{R}$ as

$$\hat{o}(t) := (\zeta(s_n), \tau(t), \eta(t)), \quad t \in (s_n, s_{n+1}], \ n \in \mathbb{N}, \tag{7.36}$$

which is the information that is available to the event-generator at time $t \in \mathbb{R}_{\geq 0}$.

The dynamic variable η will evolve according to

$$\tfrac{\mathrm{d}}{\mathrm{d}t}\eta = \Psi(\hat{o}), \qquad t \in (s_n, s_{n+1}), \ n \in \mathbb{N}, \tag{7.37a}$$
$$\eta^+ = \eta_T(\hat{o}), \qquad t \in \{t_k\}_{k \in \mathbb{N}}, \tag{7.37b}$$
$$\eta^+ = \eta_N(\hat{o}), \qquad t \in \{s_n\}_{n \in \mathbb{N}} \setminus \{t_k\}_{k \in \mathbb{N}}, \tag{7.37c}$$

where the functions $\Psi : \mathbb{R}^{2n_y} \times [0, h] \times \mathbb{R} \to \mathbb{R}$, $\eta_T : \mathbb{R}^{2n_y} \times [0, h] \times \mathbb{R} \to \mathbb{R}$ and $\eta_N : \mathbb{R}^{2n_y} \times [0, h] \times \mathbb{R} \to \mathbb{R}$ are to be designed. Note that at transmission times t_k, $k \in \mathbb{N}$, the variable η is updated differently than at the other sample times $s_n \neq t_k$, $n, k \in \mathbb{N}$, at which no transmission occurs.

The Lyapunov/storage function V given by (7.46) is often decreasing more than strictly necessary along jumps (7.7b) and (7.7c). To further reduce the amount of communication, we will store the "unnecessary" decrease of V as much as possible in a dynamic variable η, which acts as a buffer. For contractivity and internal stability, we need that the new Lyapunov/storage function $U(\xi, \tau, \eta) = V(\xi, \tau) + \eta$ satisfies

$$\tfrac{\mathrm{d}}{\mathrm{d}t}U(\xi, \tau, \eta) < w^\top w - z^\top z, \qquad \tau \in (0, h] \tag{7.38a}$$
$$U(\xi^+, \tau^+, \eta^+) \leq U(\xi, \tau, \eta), \qquad \tau = h. \tag{7.38b}$$

When a transmission is necessary according to the static event-generator (7.5), we might choose not to transmit at this sample time. As the state then jumps according to (7.7c), we can no longer guarantee that V does not increase along this jump. However, an increase of V can be compensated by reducing η, and hence we can defer the transmission until the buffer η is no longer large enough. The transmission only needs to occur if the buffer η would become negative otherwise.

First, we choose the flow dynamics (7.37a) of η as

$$\Psi(\hat{o}) = -\rho\eta, \ \text{for } \tau \in (0, h], \tag{7.39}$$

for any arbitrary decay rate $\rho \in \mathbb{R}_{>0}$. Together with (7.32), this choice of (7.39) implies that (7.38a) holds.

Remark 7.3 As Ψ is given by (7.39), it follows that $\eta(s_{n+1}) = e^{\rho h}\eta(s_n^+)$. Thus, since the event-generator only needs to know the value of η at sample times s_n, $n \in \mathbb{N}$, the variable η does not need to continuously evolve according to (7.39) in the event-generator. Instead, we can use the discrete-time dynamics just described.

For the functions η_T and η_N, we provide the following two designs. Together with the inequalities (7.34) and (7.35), both designs ensure that (7.38b) holds.

(1) State-based dynamic PETC:

$$\eta_T(\hat{o}) = \eta + \xi^\top (P_h - J^\top P_0 J)\xi, \tag{7.40a}$$

$$\eta_N(\hat{o}) = \eta + \xi^\top (P_h - P_0)\xi. \tag{7.40b}$$

(2) Output-based dynamic PETC:

$$\eta_T(\hat{o}) = \eta + \zeta^\top (N_T + \mu Q)\,\zeta, \tag{7.41a}$$

$$\eta_N(\hat{o}) = \eta + \zeta^\top (N_N - \beta Q)\,\zeta. \tag{7.41b}$$

Here, the scalars ρ, μ, and β, and the matrices N_T, N_N, P_0, and P_h follow from the stability analysis of the static PETC system in Theorem 7.5.

The first design requires that the full state $\xi(s_n)$ is known to the event-generator at sample time s_n, $n \in \mathbb{N}$. This is the case when $y = (x_p, x_c)$ (e.g., when $\mathscr{C}$ is a static state-feedback controller in which case $y = x_p$ and $n_{x_c} = 0$), as then $\zeta = \xi$. When $y = x_p$ and $n_{x_c} \neq 0$, a copy of the controller could be included in the event-generator in order to track the controller state x_c.

The second design is more conservative, but can also be used in case the event-generator does not have access to the complete vector (x_p, x_c), in which case $\zeta \neq \xi$. Hence, this choice can be used for output-based dynamic PETC.

Finally, from the definition of U it is clear that for all $\xi \in \mathbb{R}^{n_\xi}$, $\tau \in [0, h]$, and all $\eta \in \mathbb{R}_{\geq 0}$, it holds that

$$c_1|\xi|^2 + |\eta| \leq U(\xi, \tau, \eta) \leqslant c_2|\xi|^2 + |\eta|, \tag{7.42}$$

where c_1 and c_2 are defined by

$$c_1 = \min_{\tau\in[0,h]} \lambda_{min}(P(\tau)), \text{ and} \tag{7.43a}$$

$$c_2 = \max_{\tau\in[0,h]} \lambda_{max}(P(\tau)), \tag{7.43b}$$

and satisfy $c_2 \geq c_1 > 0$. In order to ensure that U is a proper storage function, we now only need to ensure that η does not become negative (i.e., that $\eta(t) \in \mathbb{R}_{\geq 0}$ for all $t \in \mathbb{R}_{\geq 0}$).

First, assume that we start with $\eta(0) \geq 0$. Next, note that in between jumps η evolves according to the differential equation (7.39). When after a jump at sample time $s_n, n \in \mathbb{N}$, we have that $\eta(s_n^+) \geq 0$, then due to (7.39) we have that $\eta(t) \geq 0$ for all $t \in (s_n, s_{n+1}]$. Hence, it only remains to show that η does not become negative due to the jumps at the sample times s_n, $n \in \mathbb{N}$. From (7.35), we know that $\eta_N(\hat{o}(s_n)) \geq 0$

when $\eta(s_n) \geq 0$ and $\zeta(s_n)^\top Q\zeta(s_n) \leq 0$. This implies that, as long as $\eta(s_n) \geq 0$, $\eta_N(\hat{o}(s_n))$ can only become negative when $\zeta(s_n)^\top Q\zeta(s_n) > 0$ (in which case the *static* periodic event-generator (7.5) would trigger a transmission). Moreover, in case $\zeta(s_n)^\top Q\zeta(s_n) > 0$, we know from (7.34) that $\eta_T(\hat{o}(s_n)) \geq 0$ when $\eta(s_n) \geq 0$. In other words, to ensure nonnegativity of η, we only need to trigger a transmission at the sample times s_n, $n \in \mathbb{N}$, at which $\eta_N(\hat{o}(s_n)) < 0$. Hence, we propose to generate the sequence of event/transmission times $\{t_k\}_{k\in\mathbb{N}}$ by a new *dynamic* periodic event-generator of the form

$$t_0 = 0,\ t_{k+1} = \inf\{t > t_k \mid \eta_N(\hat{o}(t)) < 0,\ t = nh,\ n \in \mathbb{N}\}. \tag{7.44}$$

Here, the scalar $h \in \mathbb{R}_{>0}$ and the matrix $Q \in \mathbb{R}^{2n_y \times 2n_y}$ are design parameters, in addition to the functions Ψ, η_T, and η_N. Note that the function η_N appears both in the update dynamics (7.37c), as well as in the triggering condition in (7.44).

The closed-loop dynamic PETC system consisting of (7.1)–(7.3), (7.37), and (7.44) can be written as the hybrid system

$$\frac{\mathrm{d}}{\mathrm{d}t}\begin{bmatrix}\xi\\ \tau\\ \eta\end{bmatrix} = \begin{bmatrix}A\xi + Bw\\ 1\\ \Psi(\hat{o})\end{bmatrix}, \qquad \tau \in [0, h] \tag{7.45a}$$

$$\begin{bmatrix}\xi^+\\ \tau^+\\ \eta^+\end{bmatrix} = \begin{bmatrix}J\xi\\ 0\\ \eta_T(\hat{o})\end{bmatrix}, \qquad \tau = h \text{ and } \eta_N(\hat{o}) < 0 \tag{7.45b}$$

$$\begin{bmatrix}\xi^+\\ \tau^+\\ \eta^+\end{bmatrix} = \begin{bmatrix}\xi\\ 0\\ \eta_N(\hat{o})\end{bmatrix}, \qquad \tau = h \text{ and } \eta_N(\hat{o}) \geq 0 \tag{7.45c}$$

$$z = C\xi + Dw. \tag{7.45d}$$

Theorem 7.7 [7] If $\eta(0) \geq 0$ and the conditions of Theorem 7.5 hold, then the dynamic PETC system (7.45) with (7.39) and (7.40a) or (7.41a) is internally stable and contractive.[2] Moreover, if the signal w is uniformly bounded, then also η is uniformly bounded.

While the static periodic event-generator (7.5) only has design parameters h and Q, the state-based dynamic event-generator (7.44) with (7.39) and (7.40a) has design parameters h, Q, ρ, P_0, and P_h, and the output-based dynamic event-generator (7.44) with (7.39) and (7.41a) has design parameters h, Q, ρ, N_T, N_N, μ, and β. However, for fixed h, Q, inequalities (7.34) and (7.35) are LMIs, in which case the parameters P_h, P_0, N_T, N_N, μ, and β can be synthesized (and optimized) numerically via semi-definite programming (e.g., using Yalmip/SeDuMi in MATLAB). Of course, manual tuning of one or more of these parameters is also possible, but can be difficult given the large design space.

[2] In the sense that Definitions 7.1 and 7.2 hold along solutions to the dynamic PETC system (7.45) with (7.39) and (7.40a) or (7.41a).

7.5 Riccati-Based Time-Regularized CETC

In the previous section, we analyzed internal stability and contractivity of the static PETC system (7.7) making use of matrix Riccati differential equations. Similar ideas can also be used to analyze internal stability and contractivity of the static CETC system with time-regularization as in (7.10), which we will discuss in this section. Moreover, just as in the previous section, this analysis also gives rise to a state-based and an output-based *dynamic* continuous event-generator design, which we will also provide here.

7.5.1 *Static CETC*

To analyze contractivity and stability of the system (7.10), we will now use a Lyapunov/storage function V of the form

$$V(\xi,\tau)=\begin{cases}\xi^\top P(\tau)\xi, & \text{when } \tau\in[0,h)\\ \xi^\top P(h)\xi, & \text{when } \tau\in[h,\infty),\end{cases}\tag{7.46}$$

where we select $P:[0,h]\to\mathbb{R}^{n_\xi\times n_\xi}$ to satisfy the Riccati differential equation (7.32), such that again $P:[0,h]\to\mathbb{R}^{n_\xi\times n_\xi}$ is a continuously differentiable function with $P(\tau)\succ 0$ for $\tau\in[0,h]$.

In order to guarantee contractivity and stability of the system (7.10), we need that

$$\tfrac{\mathrm{d}}{\mathrm{d}t}V\leq -z^\top z+w^\top w,\tag{7.47}$$

during flow (7.10a), and

$$V(J\xi,0)<V(\xi,h),\ \text{ for all } \xi \text{ with } \xi^\top Y^\top QY\xi>0,\tag{7.48}$$

during the jumps (7.10b).

Note that (7.48) and (7.30a) are identical, as well as (7.47) and (7.29) as long as $\tau\in[0,h]$. Hence, in contrast to Theorem 7.5, inequality (7.35) is not required, but is replaced by the condition

$$\tfrac{\mathrm{d}}{\mathrm{d}t}V\leq -z^\top z+w^\top w,\ \text{ for all } \xi \text{ with } \xi^\top Y^\top QY\xi\leq 0\tag{7.49}$$

when $\tau>h$. This leads to the following theorem.

Theorem 7.8 [6] Consider the CETC system (7.10) with (7.9), and $Q\in\mathbb{R}^{2n_y\times 2n_y}$. If there exist matrices $N_N, N_T\in\mathbb{R}^{2n_y\times 2n_y}$, $N_N, N_T\succeq 0$, and $P_h\in\mathbb{R}^{n_\xi\times n_\xi}$, $P_h\succ 0$, and scalars $\beta,\mu\in\mathbb{R}_{\geq 0}$, such that

$$\begin{bmatrix} A^\top P_h + P_h A + C^\top C + Y^\top (N_N - \beta Q) Y & \star \\ B^\top P_h + D^\top C & D^\top D - I \end{bmatrix} \prec 0, \tag{7.50}$$

(7.34) and Assumption 7.3 hold, then the system is internally stable and contractive.

Here, (7.47) is guaranteed by the choice of the function $P : [0, h] \to \mathbb{R}^{n_\xi \times n_\xi}$ when $\tau \in [0, h]$ and by (7.50) when $\tau > h$, and (7.48) is again guaranteed by (7.34).

7.5.2 Dynamic CETC

Similar to the PETC case, by adding a dynamic variable in the continuous event-generator, the conservatism in Theorem 7.8 can be exploited in order to further reduce the amount of communication in the system, while preserving the internal stability and contractivity guarantees.

Again, we introduce the buffer variable $\eta \in \mathbb{R}$ (which will be included in the event-generator). As in the CETC case, the output $y(t)$ can be measured continuously, now

$$o(t) := (\zeta(t), \tau(t), \eta(t)) \tag{7.51}$$

is the information that is available at the event-generator at time $t \in \mathbb{R}_{\geq 0}$,

The variable η will evolve according to

$$\frac{\mathrm{d}}{\mathrm{d}t}\eta = \Psi(o), \qquad t \in (t_k, t_{k+1}), \tag{7.52a}$$

$$\eta^+ = \eta_T(o), \qquad t = t_k, \tag{7.52b}$$

where $o(t) = (\zeta(t), \tau(t), \eta(t))$ is the information that is available at the event-generator at time $t \in \mathbb{R}_{\geq 0}$, and where the functions $\Psi : \mathbb{R}^{2n_y} \times \mathbb{R}^2_{\geq 0} \to \mathbb{R}$ and $\eta_T : \mathbb{R}^{2n_y} \times \mathbb{R}^2_{\geq 0} \to \mathbb{R}_{\geq 0}$ are to be designed.

Next, we design the dynamics (7.52) of the variable η with the goal of enlarging the (average) inter-event times compared to the static continuous event-generator (7.9), while maintaining the same stability and performance guarantees.

We now choose the flow dynamics (7.52a) of η as

$$\Psi(o) = \begin{cases} -2\rho\eta, & \text{when } \tau \in [0, h) \quad (7.53\text{a}) \\ -2\rho\eta + \zeta^\top (N_N - \beta Q)\zeta, & \text{when } \tau \in [h, \infty), \quad (7.53\text{b}) \end{cases}$$

for any arbitrary decay rate $\rho \in \mathbb{R}_{>0}$, and we again have two designs for the jump dynamics (7.52b) of η.

(1) State-based dynamic CETC:

$$\eta_T(o) = \eta + \xi^\top \left(P_h - J^\top P_0 J \right) \xi, \tag{7.54}$$

(2) Output-based dynamic CETC:

$$\eta_T(o) = \eta + \zeta^\top (N_T + \mu Q)\, \zeta \tag{7.55}$$

In order to ensure that $U(\xi, \tau, \eta) = V(\xi, \tau) + \eta$ is a proper storage function, we now only need to ensure that η does not become negative (i.e., that $\eta(t) \in \mathbb{R}_{\geq 0}$ for all $t \in \mathbb{R}_{\geq 0}$). Hence, we propose to generate the sequence of jump/event times $\{t_k\}_{k\in\mathbb{N}}$ by a dynamic *continuous* event-generator with time-regularization of the form

$$t_0 = 0,\ t_{k+1} = \inf\{t \geq t_k + h \mid \eta(t) < 0\}. \tag{7.56}$$

The closed-loop dynamic CETC system consisting of (7.1)–(7.3), (7.52), and (7.56) can be written as the hybrid system

$$\frac{\mathrm{d}}{\mathrm{d}t}\begin{bmatrix} \xi \\ \tau \\ \eta \end{bmatrix} = \begin{bmatrix} A\xi + Bw \\ 1 \\ \Psi(o) \end{bmatrix}, \qquad \tau \in [0, h] \text{ or } \eta \geq 0 \tag{7.57a}$$

$$\begin{bmatrix} \xi^+ \\ \tau^+ \\ \eta^+ \end{bmatrix} = \begin{bmatrix} J\xi \\ 0 \\ \eta_T(o) \end{bmatrix}, \qquad \tau > h \text{ and } \eta < 0 \tag{7.57b}$$

$$z = C\xi + Dw. \tag{7.57c}$$

Theorem 7.9 [6] If $\eta(0) \geq 0$ and the conditions of Theorem 7.8 hold, then the dynamic CETC system (7.57) with (7.53) and (7.54) or (7.55) is internally stable and contractive.[3] Moreover, if the signal w is uniformly bounded, then also η is uniformly bounded.

7.6 Numerical Example

Consider the unstable batch reactor of [27, 35, 50], with $n_{x_p} = 4$, $n_{x_c} = 2$, $n_y = n_w = n_u = n_z = 2$, and plant and controller dynamics given by (7.1) and (7.2) with

$$A_p = \begin{bmatrix} 1.3800 & -0.2077 & 6.7150 & -5.6760 \\ -0.5814 & -4.2900 & 0.0000 & 0.6750 \\ 1.0670 & 4.2730 & -6.6540 & 5.8930 \\ 0.0480 & 4.2730 & 1.3430 & -2.1040 \end{bmatrix}, \quad B_p = \begin{bmatrix} 0.0000 & 0.0000 \\ 5.6790 & 0.0000 \\ 1.1360 & -3.1460 \\ 1.1360 & 0.0000 \end{bmatrix},$$

$$B_{pw} = \begin{bmatrix} 10 & 0 & 10 & 0 \\ 0 & 5 & 0 & 5 \end{bmatrix}^\top, \quad C_y = C_z = \begin{bmatrix} 1 & 0 & 1 & -1 \\ 0 & 1 & 0 & 0 \end{bmatrix}, \quad D_y = D_z = D_{zw} = \begin{bmatrix} 0 & 0 \\ 0 & 0 \end{bmatrix},$$

$$A_c = \begin{bmatrix} 0 & 0 \\ 0 & 0 \end{bmatrix}, \quad B_c = \begin{bmatrix} 0 & 1 \\ 1 & 0 \end{bmatrix}, \quad C_u = \begin{bmatrix} -2 & 0 \\ 0 & 8 \end{bmatrix}, \quad D_u = \begin{bmatrix} 0 & -2 \\ 5 & 0 \end{bmatrix}.$$

[3] In the sense that Definitions 7.1 and 7.2 hold along solutions to the dynamic CETC system (7.57) with (7.53) and (7.54) or (7.55).

Note that for this system, the measured output y is not equal to the full plant/controller state (x_p, x_c), and thus we cannot use (7.40), (7.54), but we have to resort to (7.41) for the dynamic PETC case and to (7.55) for the dynamic CETC case.

We choose $h = 0.1$, $\rho = 10^{-3}$, and Q given by (7.6). For each choice of σ, we use a bisection algorithm to minimize the $\mathscr{L}_2$-gain γ (by appropriately scaling the matrices C and D, as discussed in Remark 7.1) based on Theorems 7.4, 7.5, and 7.8. For the lifting-based approach of Theorem 7.4, we use a single region $\Omega_1 = \mathbb{R}^{n_\xi}$, i.e., we choose $p = 1$ and $X_1 = O$. The matrices N_T and N_N and scalars β and μ follow from Theorem 7.5 for the dynamic PETC case, and from Theorem 7.8 for the dynamic time-regularized CETC case.

Figure 7.2a shows the guaranteed $\mathscr{L}_2$-gain γ as a function of σ for both PETC approaches and the time-regularized CETC approach. Here, we see that using the Riccati-based framework, a smaller $\mathscr{L}_2$-gain can be guaranteed by using a time-regularized CETC scheme than by using a PETC scheme. This makes sense intuitively, as after h time units have elapsed, a time-regularized continuous event-generator can trigger an event as soon as its event condition is violated, while a periodic event-generator can only do so at a sample time s_n, $n \in \mathbb{N}$. Figure 7.2a also shows that for a static PETC system, the lifting-based approach of Theorem 7.4 provides better $\mathscr{L}_2$-gain estimates than the Riccati-based approach of Theorem 7.5. However, the $\mathscr{L}_2$-gain estimate of Theorem 7.5 also holds for both *dynamic* PETC strategies, while the $\mathscr{L}_2$-gain estimate of Theorem 7.4 only holds for the *static* PETC strategy.

Figure 7.2b shows $\tau_{avg} =$ (total number of events)/(simulation time), the average inter-event times for the static and (output-based) dynamic event-generators, which have been obtained by simulating the systems for 100 time units with $\xi(0) = 0$, $\tau(0) = h$, and $\eta(0) = 0$, and disturbance w given by

$$w(t) = e^{-0.2t} \begin{bmatrix} 5\sin(3.5t) \\ -\cos(3t) \end{bmatrix}. \tag{7.58}$$

Finally, Fig. 7.2c shows the actual ratio $\|z\|_{\mathscr{L}_2}/\|w\|_{\mathscr{L}_2}$ for disturbance w given by (7.58), which has been obtained from the same simulations.

In Fig. 7.2c, we see that the dynamic event-generators exploit (part of) the conservatism in the $\mathscr{L}_2$-gain analysis of Theorems 7.8 and 7.5 to postpone the transmissions. This leads to higher ratios $\|z\|_{\mathscr{L}_2}/\|w\|_{\mathscr{L}_2}$ (but still below the guaranteed bounds in Fig. 7.2b), but also to consistently larger τ_{avg}, as can be seen in Fig. 7.2b.

Based on this example, we can conclude that for PETC systems, the lifting-based, and Riccati-based frameworks each has their own advantages. The lifting-based framework provides tighter $\mathscr{L}_2$-gain guarantees, while the Riccati-based framework allows to extend the transmission intervals by using a dynamic event-generator. For a fixed σ and a given desired performance, the lifting-based framework allows for larger h (hence, for larger *minimum* inter-event times) while the Riccati-based framework may lead to larger *average* inter-event times τ_{avg} by using a dynamic event-generator. Which framework is better thus depends on whether large minimum or average inter-event times are desired.

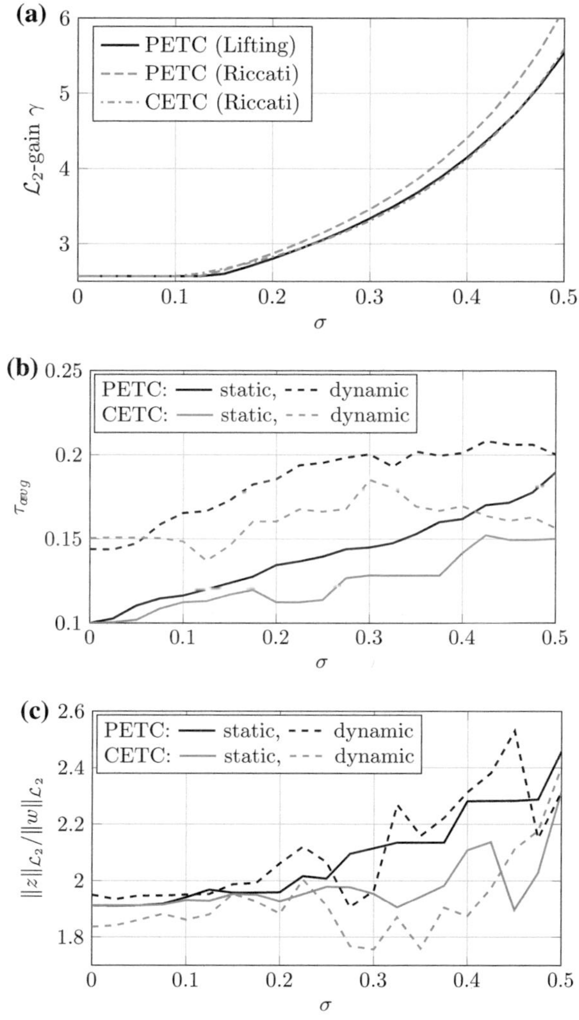

Fig. 7.2 Guaranteed $\mathscr{L}_2$-gain γ for varying σ (**a**), average inter-event times τ_{avg} for disturbance w given by (7.58) and different event-generators (**b**), and actual ratio $\|z\|_{\mathscr{L}_2}/\|w\|_{\mathscr{L}_2}$ for disturbance w given by (7.58) (**c**)

For the time-regularized CETC case, only the Riccati-based framework applies, which for this example yields (almost exactly) the same performance guarantees as the lifting-based PETC approach, with the same minimum inter-event time h, but often larger τ_{avg}. However, the designed continuous event-generator may be difficult to implement on a digital platform, as it requires continuous measuring of the output y.

To compare both frameworks with the (static or dynamic) time-regularized CETC solutions of [13, 14], note that for a given and $\mathscr{L}_2$-gain γ, the waiting time h (or τ_{MIET} in the terminology of [14]) of the continuous event-generator proposed in [13, 14] cannot exceed the maximally allowable transmission interval (MATI) of [27].

Moreover, for the same example in [27, Sect. IV], we can calculate that when using the sampled-data protocol, no notion of stability can be guaranteed for MATI larger than 0.063. In contrast, here we guarantee internal stability and $\mathcal{L}_2$-stability for $h = 0.1$. Hence, our frameworks tailored to linear systems are clearly much less conservative than our previous results for nonlinear systems in [13, 14]. See also [6] for a direct comparison between the static and dynamic continuous event-generators in Sect. 7.5 and the event-generators proposed in [13, 14].

7.7 Extensions

The results presented in this chapter can be extended in several ways.

First of all, the results in Sects. 7.4 and 7.5 can be extended toward the case with communication delays, as long as these delays are upper bounded by the sampling time or time threshold h, see [8]. In order to do so, for each possible delay $d \in [0, h]$, a function $P^d : [0, d] \to \mathbb{R}^{n_\xi \times n_\xi}$ satisfying the Riccati differential equation (7.32) needs to be synthesized. Hence, only a finite number of possible transmission delays can be considered using this approach. However, when the delays can have any value from a continuous interval, this situation can be effectively approximated by using a gridding approach, see also [8, Remark III.7]. Similar ideas can be used to extend the lifting-based approach in Sect. 7.3 toward delays. Moreover, certain Self-Triggered schemes (e.g., [23, 34, 49]) can also be captured in this lifting-based framework, see [40].

Second, our proposed frameworks can be extended toward decentralized setups in a similar manner as in [25, Sect. V] for the static PETC case. However, this requires that the clocks of all local event-generators are synchronized.

These extensions emphasize the usefulness of our new ETC solutions for linear systems, but also uncovers two potential drawbacks. In our earlier work [14], we considered decentralized CETC setups for *nonlinear* systems with transmission delays. These results can also be particularized to linear systems, giving rise to continuous event-generators with time-regularization that are similar (although more conservative) to those proposed in Sect. 7.5. However, the analysis proposed in [14] does not require clock synchronization for all local event-generators, and also directly allows that the transmission delays can have any value from a continuous interval. Hence, although the analysis in [14] (in the case of linear systems) provides less tight performance guarantees than our new results tailored to linear systems that we have proposed in this chapter, it does not suffer from the drawbacks that clocks need to be synchronized (in case of decentralized event-generators) and that only a finite number of possible transmission delays can be allowed.

7.8 Summary

In this chapter, we have provided an overview of our recent results in the design of time-regularized ETC and PETC schemes that are tailored to *linear systems* as provided in [8, 25, 26]. In particular, we have shown that stability and the contractivity in $\mathscr{L}_2$-sense (meaning that the $\mathscr{L}_2$-gain is smaller than 1) of static PETC closed-loop systems (which are hybrid systems) are equivalent to the stability and the contractivity in ℓ_2-sense (meaning that the ℓ_2-gain is smaller than 1) of an appropriate discrete-time piecewise linear system. These new insights are obtained by adopting a lifting-based perspective on this analysis problem, which led to computable ℓ_2-gain (and thus $\mathscr{L}_2$-gain) conditions, despite the fact that the linearity assumption, which is usually needed in the lifting literature, is not satisfied.

We have also reviewed the results in [8] that lead to the design of time-regularized CETC and PETC schemes based on Lyapunov/storage functions exploiting matrix Riccati differential equations. Moreover, we have identified the connections between the two approaches.

Additionally, we have discussed new designs of so-called (time-regularized and periodic) *dynamic* ETC strategies focused on linear systems. Interestingly, the inclusion of a dynamic variable in the event generator can lead to a significantly reduced consumption of communication and energy resources while leading to identical guarantees on stability and performance as their static counterparts.

Via a numerical example, we have demonstrated that a Riccati-based CETC design, a Riccati-based PETC design, and a lifting-based PETC design each has their own advantages. Hence, which choice of design framework is better depends on the system at hand.

Acknowledgements This work is part of the research programmes "Wireless control systems: A new frontier in automation" with project number 11382 and "Integrated design approach for safety-critical real-time automotive systems" with project number 12698, which are (partly) financed by the Netherlands Organisation for Scientific Research (NWO).

References

1. Abdelrahim, M., Postoyan, R., Daafouz, J., Nešić, D.: Input-to-state stabilization of nonlinear systems using event-triggered output feedback controllers. In: European Control Conference, pp. 2180–2185 (2015)
2. Abdelrahim, M., Postoyan, R., Daafouz, J., Nešić, D.: Stabilization of nonlinear systems using event-triggered output feedback controllers. IEEE Trans. Autom. Control. **61**(9), 2682–2687 (2016)
3. Başar, T., Bernhard, P.: H^∞-Optimal Control and Relaxed Minimax Design Problems: A Dynamic Game Approach, 2nd edn. Birkhäuser, Boston (1995)
4. Bamieh, B.A., Pearson, J.B.: A general framework for linear periodic systems with applications to H^∞/sampled-data control. IEEE Trans. Autom. Control. **37**(4), 418–435 (1992)
5. Borgers, D.P., Heemels, W.P.M.H.: Event-separation properties of event-triggered control systems. IEEE Trans. Autom. Control. **59**(10), 2644–2656 (2014)

6. Borgers, D.P., Dolk, V.S., Heemels, W.P.M.H.: Dynamic event-triggered control with time-regularization for linear systems. In: 55th IEEE Conference on Decision and Control, pp. 1352–1357 (2016)
7. Borgers, D.P., Dolk, V.S., Heemels, W.P.M.H.: Dynamic periodic event-triggered control for linear systems. In: Hybrid Systems: Computation and Control, pp. 179–186. Pittburgh, PA, USA (2017)
8. Borgers, D.P., Dolk, V.S., Heemels, W.P.M.H.: Riccati-based design of event-triggered controllers for linear systems with delays. IEEE Trans. Autom. Control. (2018). (To appear)
9. Cantoni, M.W., Glover, K.: H_∞ sampled-data synthesis and related numerical issues. Automatica **33**(12), 2233–2241 (1997)
10. Chen, T., Francis, B.A.: Optimal Sampled-Data Control Systems. Communications and Control Engineering, 1st edn. Springer, London (1995)
11. Dai, D., Hu, T., Teel, A.R., Zaccarian, L.: Output feedback synthesis for sampled-data system with input saturation. In: American Control Conference, pp. 1797–1802 (2010)
12. De Persis, C., Sailer, R., Wirth, F.: Parsimonious event-triggered distributed control: a Zeno free approach. Automatica **49**(7), 2116–2124 (2013)
13. Dolk, V.S., Borgers, D.P., Heemels, W.P.M.H.: Event-triggered control: tradeoffs between transmission intervals and performance. In: 53rd IEEE Conference on Decision and Control, pp. 2764–2769 (2014)
14. Dolk, V.S., Borgers, D.P., Heemels, W.P.M.H.: Output-based and decentralized dynamic event-triggered control with guaranteed $\mathcal{L}_p$-gain performance and Zeno-freeness. IEEE Trans. Autom. Control. **62**(1), 34–49 (2017)
15. Donkers, M.C.F., Heemels, W.P.M.H.: Output-based event-triggered control with guaranteed $\mathcal{L}_\infty$-gain and improved and decentralized event-triggering. IEEE Trans. Autom. Control. **57**(6), 1362–1376 (2012)
16. Dullerud, G.E., Lall, S.: Asynchronous hybrid systems with jumps - analysis and synthesis methods. Syst. Control. Lett. **37**(2), 61–69 (1999)
17. Ferrari-Trecate, G., Cuzzola, F.A., Mignone, D., Morari, M.: Analysis of discrete-time piecewise affine and hybrid systems. Automatica **38**(12), 2139–2146 (2002)
18. Forni, F., Galeani, S., Nešić, D., Zaccarian, L.: Event-triggered transmission for linear control over communication channels. Automatica **50**(2), 490–498 (2014)
19. Gahinet, P., Apkarian, P.: A linear matrix inequality approach to H_∞ control. Int. J. Robust Nonlinear Control. **4**(4), 421–448 (1994)
20. Garcia, E., Antsaklis, P.J.: Model-based event-triggered control for systems with quantization and time-varying network delays. IEEE Trans. Autom. Control. **58**(2), 422–434 (2013)
21. Girard, A.: Dynamic triggering mechanisms for event-triggered control. IEEE Trans. Autom. Control. **60**(7), 1992–1997 (2015)
22. Goebel, R., Sanfelice, R.G., Teel, A.R.: Hybrid dynamical systems. IEEE Control. Syst. Mag. **29**(2), 28–93 (2009)
23. Gommans, T.M.P., Antunes, D., Donkers, M.C.F., Tabuada, P., Heemels, W.P.M.H.: Self-triggered linear quadratic control. Automatica **50**(4), 1279–1287 (2014)
24. Heemels, W.P.M.H., Sandee, J.H., van den Bosch, P.P.J.: Analysis of event-driven controllers for linear systems. Int. J. Control. **81**(4), 571–590 (2008)
25. Heemels, W.P.M.H., Donkers, M.C.F., Teel, A.R.: Periodic event-triggered control for linear systems. IEEE Trans. Autom. Control. **58**(4), 847–861 (2013)
26. Heemels, W.P.M.H., Dullerud, G.E., Teel, A.R.: $\mathcal{L}_2$-gain analysis for a class of hybrid systems with applications to reset and event-triggered control: a lifting approach. IEEE Trans. Autom. Control. **61**(10), 2766–2781 (2016)
27. Heemels, W.P.M.H., Teel, A.R., van de Wouw, N., Nešić, D.: Networked control systems with communication constraints: tradeoffs between transmission intervals, delays and performance. IEEE Trans. Autom. Control. **55**(8), 1781–1796 (2010)
28. Heemels, W.P.M.H., Postoyan, R., Donkers, M.C.E., Teel, A.R., Anta, A., Tabuada, P., Nešić, D.: Periodic event-triggered control. In: Miskowicz, M. (ed.) Event-Based Control and Signal Processing. CRC Press/Taylor & Francis, Boca Raton (2015)

29. Henningsson, T., Johannesson, E., Cervin, A.: Sporadic event-based control of first-order linear stochastic systems. Automatica **44**(11), 2890–2895 (2008)
30. Johansson, M., Rantzer, A.: Computation of piecewise quadratic Lyapunov functions for hybrid systems. IEEE Trans. Autom. Control. **43**(4), 555–559 (1998)
31. Kreyszig, E.: Introductory Functional Analysis with Applications. Wiley, New York (1978)
32. Lehmann, D., Lunze, J.: Event-based control with communication delays and packet losses. Int. J. Control. **85**(5), 563–577 (2012)
33. Lunze, J., Lehmann, D.: A state-feedback approach to event-based control. Automatica **46**(1), 211–215 (2010)
34. Mazo Jr., M., Anta, A., Tabuada, P.: An ISS self-triggered implementation of linear controllers. Automatica **46**(8), 1310–1314 (2010)
35. Nešić, D., Teel, A.R.: Input-output stability properties of networked control systems. IEEE Trans. Autom. Control. **49**(10), 1650–1667 (2004)
36. Postoyan, R., Anta, A., Heemels, W.P.M.H., Tabuada, P., Nešić, D.: Periodic event-triggered control for nonlinear systems. In: 52nd IEEE Conference on Decision and Control, pp. 7397–7402 (2013)
37. Postoyan, R., Anta, A., Nešić, D., Tabuada, P.: A unifying Lyapunov-based framework for the event-triggered control of nonlinear systems. In: 50th IEEE Conference on Decision and Control and European Control Conference, pp. 2559–2564 (2011)
38. Postoyan, R., Tabuada, P., Nešić, D., Anta, A.: A framework for the event-triggered stabilization of nonlinear systems. IEEE Trans. Autom. Control. **60**(4), 982–996 (2015)
39. Selivanov, A., Fridman, E.: Event-triggered $\mathscr{H}_\infty$ control: a switching approach. IEEE Trans. Autom. Control. **61**(10), 3221–3226 (2016)
40. Strijbosch, N.W.A., Dullerud, G.E., Teel, A.R., Heemels, W.P.M.H.: $\mathscr{L}_2$-gain analysis of periodic event-triggered and self-triggered control systems with delays using lifting techniques. Submitted
41. Tabuada, P.: Event-triggered real-time scheduling of stabilizing control tasks. IEEE Trans. Autom. Control. **52**(9), 1680–1685 (2007)
42. Tallapragada, P., Chopra, N.: Event-triggered dynamic output feedback control for LTI systems. In: 51st IEEE Conference on Decision and Control, pp. 6597–6602 (2012)
43. Tallapragada, P., Chopra, N.: Decentralized event-triggering for control of nonlinear systems. IEEE Trans. Autom. Control. **59**(12), 3312–3324 (2014)
44. Tarbouriech, S., Seuret, A., Gomes da Silva Jr., J.M., Sbarbaro, D.: Observer-based event-triggered control co-design for linear systems. IET Control. Theory Appl. **10**(18), 2466–2473 (2016)
45. Toivonen, H.T.: Sampled-data control of continuous-time systems with an H_∞ optimality criterion. Automatica **28**(1), 45–54 (1992)
46. Toivonen, H.T., Sågfors, M.F.: The sampled-data H_∞ problem: a unified framework for discretization-based methods and Riccati equation solution. Int. J. Control. **66**(2), 289–310 (1997)
47. van der Schaft, A.: L_2-Gain and Passivity Techniques in Nonlinear Control. Lecture Notes in Control and Information Sciences, vol. 218. Springer, Berlin (1996)
48. van Loon, S.J.L.M., Heemels, W.P.M.H., Teel, A.R.: Improved $\mathscr{L}_2$-gain analysis for a class of hybrid systems with applications to reset and event-triggered control. In: 53rd IEEE Conference on Decision and Control, pp. 1221–1226 (2014)
49. Velasco, M., Fuertes, J.M., Marti, P.: The self triggered task model for real-time control systems. In: 24th IEEE Real-Time Systems Symposium, pp. 67–70 (2003)
50. Walsh, G.C., Ye, H., Bushnell, L.G.: Stability analysis of networked control systems. IEEE Trans. Control. Syst. Technol. **10**(3), 438–446 (2002)
51. Wang, X., Lemmon, M.D.: Event-triggering in distributed networked control systems. IEEE Trans. Autom. Control. **56**(3), 586–601 (2011)
52. Willems, J.C.: Dissipative dynamical systems part I: general theory. Arch. Ration. Mech. Anal. **45**(5), 321–351 (1972)
53. Yamamoto, Y.: New approach to sampled-data control systems-a function space method. In: 29th IEEE Conference on Decision and Control, vol. 3, pp. 1882–1887 (1990)

Chapter 8
Event-Triggered State-Feedback via Dynamic High-Gain Scaling for Nonlinearly Bounded Triangular Dynamics

J. Peralez, V. Andrieu, M. Nadri and U. Serres

Abstract This paper focuses on the construction of Event-Triggered state feedback laws. The approach followed is a high-gain approach. The event which triggers an update of the control law is based on a dynamical system in which the state is the high-gain parameter. This approach allows to design a control law ensuring convergence to the origin for nonlinear systems with triangular structure and a specific upper bound on the nonlinearities which is more general than a linear growth condition.

8.1 Introduction

The implementation of a control law on a process requires the use of an appropriate sampling scheme. In this regards, periodic control (with a constant sampling period) is the usual approach that is followed for practical implementation on digital platforms. Indeed, periodic control benefits from a huge literature, providing a mature theoretical background (see e.g., [2, 10, 12, 19, 20]) and numerous practical examples. The use of a constant sampling period makes easier the closed-loop analysis and the implementation, allowing solid theoretical results and a wide deployment in the

J. Peralez · V. Andrieu (✉) · M. Nadri · U. Serres
Univ. Lyon, Université Claude Bernard Lyon 1, CNRS, LAGEP UMR 5007,
43 bd du 11 novembre 1918, 69100 Villeurbanne, France
e-mail: vincent.andrieu@gmail.com

J. Peralez
e-mail: johan.peralez@gmail.com

M. Nadri
e-mail: madiha.nadri-wolf@univ-lyon1.fr

U. Serres
e-mail: ulysse.serres@univ-lyon1.fr

S. Tarbouriech et al. (eds.), *Control Subject to Computational and Communication Constraints*, Lecture Notes in Control and Information Sciences 475, https://doi.org/10.1007/978-3-319-78449-6_8

industry. However, the rate of control execution being fixed by a worst case analysis (the chosen period must guarantee the stability for all possible operating conditions), this may lead to an unnecessary fast sampling rate and then to an overconsumption of available resources.

The recent growth of shared networked control systems for which communication and energy resources are often limited goes with an increasing interest in aperiodic control design. This can be observed in the comprehensive overview on Event-Triggered and Self-Triggered control presented in [15]. Event-Triggered control strategies introduce a triggering condition assuming a continuous monitoring of the plant (that requires a dedicated hardware) while in Self-Triggered strategies, the control update time is based on predictions using previously received data. The main drawback of Self-Triggered control is the difficulty to guarantee an acceptable degree of robustness, especially in the case of uncertain systems.

Most of the existing results on Event-Triggered and Self-Triggered control for nonlinear systems are based on the input-to-state stability (ISS) assumption which implies the existence of a feedback control law ensuring an ISS property with respect to measurement errors [1, 9, 23, 27]. In this ISS framework, an emulation approach is followed: the knowledge of an existing robust feedback law in continuous time is assumed then some triggering conditions are proposed to preserve stability under sampling (see also the approach of [26]).

Another proposed approach consists in the redesign of a continuous-time stabilizing control. For instance, the authors of [18] adapted the original *universal formula* introduced by Sontag for nonlinear systems affine in the control. The relevance of this method was experimentally shown in [28] where the regulation of an omnidirectional mobile robot was addressed.

Although aperiodic control literature has proved an interesting potential, important fields still need to be further investigated to allow a wider practical deployment.

The high-gain approach is a very efficient tool to address the stabilizing control problem in the continuous-time case. It has the advantage to allow uncertainties in the model and to remain simple. Different approaches based on high-gain techniques have been followed in the literature to tackle the output feedback problem in the continuous-time case (see for instance [5, 6, 16]) and more recently for the (periodic) discrete-in-time case (see [25]). In the context of observer design, [8] proposed the design of a continuous discrete time observer, revisiting high-gain techniques in order to give an adaptive sampling stepsize.

In this work, we follow the strategy introduced in [8] and we address the Event-Triggered state feedback control. The results obtained extend the one of a conference paper which has been published in [21]. Compared to this one, the considered class of system has been extended. Moreover, all proofs are given. Note also that some of the results which are given here have been recently used in combination with an observer in the context of output feedback design (see [22]).

In high-gain designs, the asymptotic convergence is obtained by dominating the nonlinearities with high-gain techniques. In the proposed approach, the high-gain is dynamically adapted with respect to time-varying nonlinearities in order to allow an efficient trade-off between the high-gain parameter and the sampling step size. Moreover, the proposed strategy is shown to ensure the existence of a minimum inter-execution time.

The paper is organized as follows. The control problem and the class of considered systems is given in Sect. 8.2. In Sect. 8.3, some preliminary results concerning linear systems are given. The main result is stated in Sect. 8.4 and its proof is given in Sect. 8.5. Finally Sect. 8.6 contains an illustrative example.

8.2 Problem Statement

8.2.1 Class of Considered Systems

In this work, we consider the problem of designing a stabilizing Event-Triggered state feedback for the class of uncertain nonlinear systems described by the dynamical system

$$\dot{x}(t) = Ax(t) + Bu(t) + f(x(t), u(t)), \tag{8.1}$$

where the state x is in $\mathbb{R}^n$, $u : \mathbb{R} \to \mathbb{R}$ is the control signal in $\mathbb{L}^\infty(\mathbb{R}_+, \mathbb{R})$, where $A \in \mathbb{R}^{n\times n}$ is the upper shift matrix, $B = (0, \dots, 0, 1) \in \mathbb{R}^n$ and $f = (f_1, \dots, f_n)$ is a vector field on $\mathbb{R}^n$. We consider the case in which the vector field f satisfies the following assumption.

Assumption 8.1 (*Nonlinear bound*) There exist a nonnegative continuous function c, positive real numbers c_0, c_1 and q such that for all $x \in \mathbb{R}^n$, we have

$$|f_j(x, u)| \leqslant c(x) \left(|x_1| + |x_2| + \cdots + |x_j|\right), \tag{8.2}$$

with

$$c(x) = c_0 + \sum_{j=1}^{n} c_j |x_j|^{q_j}, \tag{8.3}$$

where $q_j < \frac{1}{j-1}$ for $j = 2, \dots, n$ and with no constraints on q_1.

Notice that Assumption 8.1 is more general than the incremental property introduced in [25] since the function c is not constant but depends on x. This bound can be related also to [16, 24] in which continuous-in-time output feedback law are designed. However, in these works the function c depends only on x_1 without any particular restriction (it may not be polynomial). Note moreover that in our context, we do not consider inverse dynamics.

This work is related to two other results which have been recently obtained in the context of Event-Triggered high-gain technics. The two other contexts which have been considered can be described as follows.

- **Observer design case in** [8]. For all $x \in \mathbb{R}^n$, e in $\mathbb{R}^n$ and u, we have

$$|f_j(x+e,u) - f_j(x,u)| \leqslant \Gamma(u)\left(|e_1| + |e_2| + \cdots + |e_j|\right), \tag{8.4}$$

  where Γ is any continuous function.
- **Output feedback case in** [22]. There exist positive real numbers c_0, c_1, q such that for all $x \in \mathbb{R}^n$, we have

$$|f_j(x,u)| \leqslant c(x_1)\left(|x_1| + |x_2| + \cdots + |x_j|\right), \tag{8.5}$$

  where c is function defined by

$$c(x_1) = c_0 + c_1|x_1|^q. \tag{8.6}$$

Note also that in a preliminary version of this work in [21], the bound (8.5) was also considered in the context of state-feedback.

8.2.2 *Updated Sampling Time Controller*

In the sequel, we restrict ourselves to a sample-and-hold implementation, i.e., the input is assumed to be constant between any two execution times. The control input u is defined through a sequence $(t_k, u_k)_{k\in\mathbb{N}}$ in $\mathbb{R}_+ \times \mathbb{R}$ in the following way:

$$u(t) = u_k, \quad \forall t \in [t_k, t_{k+1}), \quad k \in \mathbb{N}. \tag{8.7}$$

It can be noticed that for u to be well defined for all positive time, we need that $\lim_{k\to+\infty} t_k = +\infty$.

Our control objective is to design the sequence $(t_k, u_k)_{k\in\mathbb{N}}$ such that the origin of the obtained closed loop system is asymptotically stable. In addition to a feedback controller that computes the control input, Event-Triggered control systems need a *triggering mechanism* that determines when the control input has to be updated again. This rule is said to be *static* if it only involves the current state of the system, and *dynamic* if it uses an additional internal dynamic variable [13].

For simplicity, we also assume that the process of measurement, computing the control $u(t_k)$ and updating the actuators can be neglected. This assumption reflects that in many implementations, this time is much smaller than the time elapsed between the instants t_k and t_{k+1} [14].

8.2.3 *Notation*

We denote by $\langle \cdot, \cdot \rangle$ the canonical scalar product on $\mathbb{R}^n$ and by $\|\cdot\|$ the induced Euclidean norm; we use the same notation for the corresponding induced matrix norm. Also, we use the symbol $'$ to denote the transposition operation.

In the following, the notation $\xi(t^-)$ stands for $\lim\limits_{\substack{\tau \to t \\ \tau < t}} \xi(\tau)$. Also, to simplify the presentation, we introduce the notation $\xi_k = \xi(t_k)$ and $\xi_k^- = \xi(t_k^-)$.

8.3 Preliminary Results: The Linear Case

In high-gain designs, the idea is to consider the nonlinear terms (the f_i's) as disturbances. A first step consists in synthesizing a robust control for the linear part of the system, neglecting the effects of the nonlinearities. Then, the convergence and robustness are amplified through a high-gain parameter to deal with the nonlinearities.

Therefore, let us first focus on a general linear dynamical system

$$\dot{x}(t) = Ax(t) + Bu(t), \tag{8.8}$$

where the state x evolves in $\mathbb{R}^n$ and the control u is in $\mathbb{R}$. The matrix A is in $\mathbb{R}^{n\times n}$ and B is a column vector in $\mathbb{R}^n$.

In this preliminary case, we review a well-known result concerning periodic sampling approaches. Indeed, an emulation approach is adopted for the stabilization of the linear part: a feedback law is designed in continuous time and a triggering condition is chosen to preserve stability under sampling.

It is well known that if there exists a feedback control law (continuous-in-time) $u(t) = Kx(t)$ that asymptotically stabilizes the system then there exists a strictly positive inter-execution time $\delta_k = t_{k+1} - t_k$ such that the discrete-in-time control law $u(t) = Kx(t_k)$ for t in $[t_k, t_{k+1})$ renders the system asymptotically stable. This result is rephrased in Lemma 8.1 below whose proof is postponed in Appendix and for which we do not claim any originality.

Lemma 8.1 *Suppose the pair (A, B) is stabilizable, that is there exists a matrix K in $\mathbb{R}^n$ rendering $(A + BK)$ Hurwitz. Then there exists a positive real number δ^* such that for all δ in $[0, \delta^*)$ and the sequence $(t_k, u_k)_{k\in\mathbb{N}}$ defined as*

$$t_0 = 0, \quad t_{k+1} = t_k + \delta, \quad u_k = Kx_k, \quad \forall k \in \mathbb{N}, \tag{8.9}$$

makes the origin of the dynamical system (8.8) *a globally and asymptotically stable equilibrium.*

This result which is based on robustness is valid for general matrices A and B. The proof is based on the fact that if $A + BK$ is Hurwitz, the origin of the discrete time linear system defined for all k in $\mathbb{N}$ as

$$x_{k+1} = F_c(\delta)x_k, \tag{8.10}$$

where $F_c(\delta) = \exp(A\delta) + \int_0^\delta \exp(A(\delta - s))BKds$ is asymptotically stable for δ sufficiently small.

However, when we consider the particular case in which A and B satisfy the triangular form as in (8.1) (integrator chain), it is shown in the following theorem that the inter-execution time can be selected arbitrarily large as long as the control is modified.

Theorem 8.2 (Chain of integrator) *Suppose the matrix $A \in \mathbb{R}^{n\times n}$ is the upper shift matrix, and $B = (0, \ldots, 0, 1) \in \mathbb{R}^n$. Then, for all gain matrix K in $\mathbb{R}^n$ such that $A + BK$ is Hurwitz, there exists a positive real number α^* such that for all α in $[0, \alpha^*)$ and for all $\delta > 0$ the state feedback control law*

$$u(t) = K\mathscr{L}x(t_k), \qquad \forall t \in [t_k, t_{k+1}), \forall k \in \mathbb{N} \tag{8.11}$$

$$\mathscr{L} = diag(L^n, L^{n-1}, \ldots, L), \tag{8.12}$$

$$L = \frac{\alpha}{\delta}, \tag{8.13}$$

where the sequence $(t_k)_{k\in\mathbb{N}}$ defined as $t_0 = 0$, $t_{k+1} = t_k + \delta$ renders the origin of the dynamical system (8.8) a globally asymptotically stable equilibrium.

Before proving this theorem, we emphasize that in the particular case of the chain of integrator the sampling period time δ can be selected arbitrarily large.

Proof of Theorem 8.2: In order to analyze the behavior of the closed-loop system, let us mention the following algebraic properties of the matrix $\mathscr{L}$:

$$\mathscr{L}A = LA\mathscr{L}, \quad \mathscr{L}BK = LBK. \tag{8.14}$$

Let us introduce the following change of coordinates:

$$X = \frac{\mathscr{L}}{L^{n+1}}x = \begin{bmatrix} \frac{x_1}{L} & \frac{x_2}{L^2} & \cdots & \frac{x_n}{L^n} \end{bmatrix}'. \tag{8.15}$$

Employing (8.14), it yields that in the new coordinates the closed-loop dynamics are for all t in $[t_k, t_{k+1})$:

$$\dot{X}(t) = L(AX(t) + BKX_k). \tag{8.16}$$

By integrating the previous equality and employing (8.13) it yields for all k in $\mathbb{N}$:

$$\begin{aligned} X_{k+1} &= \left[\exp(AL\delta) + \int_0^\delta \exp(AL(\delta - s))LBKds\right]X_k \\ &= F_c(\alpha)X_k. \end{aligned}$$

In other word, this is the same discrete dynamics than the one given in (8.10) for system (8.8) in closed-loop with the state feedback KX_k. Consequently, according to Lemma 8.1, there exists a positive real number α^* such that $X = 0$ (and thus $x = 0$) is a GAS equilibrium for the system (8.16) provided that $L\delta$ is in $[0, \alpha^*)$. □

8.4 Main Result: The Nonlinear Case

We consider now the nonlinear system (8.1). Let K and α be chosen to stabilize the linear part of the system and consider the control

$$u_k = K\mathscr{L}_k x_k, \tag{8.17}$$

$$\mathscr{L}_k = diag\left(L_k^n, L_k^{n-1}, \ldots, L_k\right). \tag{8.18}$$

It remains to select the sequences L_k and δ_k to deal with the nonlinearities.

In the context of a linear growth condition (i.e., if the bound $c(x)$ defined in Assumption 8.1 is replaced by a constant), the authors of [25] have shown that a (well-chosen) constant parameter L_k can guarantee the global stability, provided that L_k is greater than a function of the bound. Here, we need to adapt the high-gain parameter to follow a function of the time-varying bound.

Following the idea presented in [8] in the context of observer design, we consider the following update law for the high-gain parameter:

$$\dot{L}(t) = a_2 L(t) M(t) c(x(t)), \qquad \forall t \in [t_k, t_k + \delta_k) \tag{8.19}$$

$$\dot{M}(t) = a_3 M(t) c(x(t)), \qquad \forall t \in [t_k, t_k + \delta_k) \tag{8.20}$$

$$L_k = L_k^-(1 - a_1\alpha) + a_1\alpha, \qquad \forall k \in \mathbb{N} \tag{8.21}$$

$$M_k = 1, \qquad \forall k \in \mathbb{N} \tag{8.22}$$

with $a_1\alpha < 1$, initial conditions $L(t_0) \geqslant 1$ and $M(t_0) = 1$, and where a_1, a_2, a_3 are positive real numbers to be chosen. For a justification of this type of high-gain update law, the interested reader may refer to [8] where is it shown that this update law is a continuous discrete version of the high-gain parameter update law introduced in [24].

Remark 8.1 Notice that the functions $L(\cdot)$ and $M(\cdot)$ are strictly increasing on any time interval $[t_k, t_k + \delta_k)$ and that $L_k \geqslant 1$ for all $k \in \mathbb{N}$.

Finally, the execution times t_k are given by the following relations:

$$t_0 = 0, \quad t_{k+1} = t_k + \delta_k, \tag{8.23}$$

$$\delta_k = \min\{s \in \mathbb{R}_+ \mid sL((t_k + s)^-) = \alpha\}. \tag{8.24}$$

Equations (8.23)–(8.24) constitute the triggering mechanism of the Event-Triggered strategy. This mechanism does not directly involve the state value x but the additional dynamic variable L and so can be referred as a dynamic triggering mechanism [13]. The relationship between L_k and δ_k comes from Eq. (8.13). It highlights the trade-off between high-gain value and inter-execution time (see [11, 25]).

We are now ready to state our main result which proof is given in Sect. 8.5.

Theorem 8.3 *Let Assumption 8.1 holds. Then there exist positive real numbers a_1, a_2, a_3, α^*, and a gain K such that, for all α in $[0, \alpha^*]$, there exists a positive real number $\ell_{\max}$ such that the set,*

$$\{x = 0, L \leqslant \ell_{\max}\} \subset \mathbb{R}^n \times \mathbb{R},$$

is GAS along the solution of system (8.1) *with the Event-Triggered feedback* (8.17)–(8.24). *More precisely, there exists a class $\mathscr{K}\mathscr{L}$ function β such that the solution $(x(\cdot), L(\cdot))$ initiated from $(x(0), L(0))$ with $L(0) \geqslant 1$ is defined for all $t \geqslant 0$ and satisfies*

$$|x(t)| + |\tilde{L}(t)| \leqslant \beta(|x(0)| + |\tilde{L}(0)|, t), \tag{8.25}$$

where $\tilde{L}(t) = \max\{L(t) - \ell_{\max}\}$. Moreover, there exists a positive real number $\delta_{\min}$ such that $\delta_k > \delta_{\min}$ for all k and so ensures the existence of a minimal inter-execution time.

8.5 Proof of Theorem 8.3

Let us introduce the following scaled coordinates along a trajectory of system (8.1) (compare with (8.15)). They will be used at different places in this paper.

$$X(t) = \mathscr{S}(t)x(t), \tag{8.26}$$

$$\mathscr{S}(t) = diag\left(\frac{1}{L(t)^b}, \cdots, \frac{1}{L(t)^{n+b-1}}\right) = \frac{\mathscr{L}(t)}{L(t)^{n+b}}, \tag{8.27}$$

where $1 > b > 0$ is such that

$$(j + b - 1)q_j < 1\ ,\ \forall 1 \leqslant j \leqslant n, \tag{8.28}$$

with q_j given in Assumption 8.1. Note that since by assumption we have $q_j < \frac{1}{j-1}$ for $j > 1$, it is always possible to find such b. Note that the matrix-valued function $\mathscr{L}(\cdot)$ satisfies:

$$\mathscr{L}(t)A = L(t)A\mathscr{L}(t), \tag{8.29}$$

$$\mathscr{L}(t)\exp(At) = \exp(L(t)At)\mathscr{L}(t), \tag{8.30}$$

$$\mathscr{L}(t)BK = L(t)BK. \tag{8.31}$$

8.5.1 Selection of the Gain Matrix K

Let D be the diagonal matrix in $\mathbb{R}^{n\times n}$ defined by $D = diag(b, 1+b, \ldots, n+b-1)$. Let P be a symmetric positive definite matrix and K a vector in $\mathbb{R}^n$ such that (always possible, see [7]) (8.32), (8.33) and

$$P(A+BK)+(A+BK)'P \leqslant -I, \tag{8.32}$$

$$p_1 I \leqslant P \leqslant p_2 I, \tag{8.33}$$

$$p_3 P \leqslant PD+DP \leqslant p_4 P, \tag{8.34}$$

with $p_1, \ldots, p_4$ positive real numbers.

With the matrix K selected, it remains to select the parameters a_1, a_2, a_3, and α^*. This is done in Propositions 8.1 and 8.2. Proposition 8.1 focuses on the existence of (x_k, L_k) for all k in $\mathbb{N}$, whereas, based on a Lyapunov analysis, Proposition 8.2 shows that a sequence of quadratic function of scaled coordinates is decreasing. Based on these two propositions, the proof of Theorem 8.3 is given in Sect. 8.5.4 where it is shown that the time function L is bounded.

8.5.2 Existence of the Sequence $(t_k, x_k, L_k)_{k\in\mathbb{N}}$

The first step of the proof is to show that the sequence $(x_k, L_k)_{k\in\mathbb{N}} = (x(t_k), L(t_k))_{k\in\mathbb{N}}$ is well defined. Note that it does not imply that $x(t)$ is defined for all t since for the time being it has not been shown that the sequence t_k is unbounded. This will be obtained in Sect. 8.5.4 when proving Theorem 8.3.

Proposition 8.1 (Existence of the sequence) *Let a_1, a_3, and α be positive, and $a_2 \geqslant \frac{2n}{p_3}$. Then, the sequence $(t_k, x_k, L_k)_{k\in\mathbb{N}}$ is well defined.*

Proof of Proposition 8.1: We proceed by contradiction. Assume that $k \in \mathbb{N}$ is such that (t_k, x_k, L_k) is well defined but $(t_{k+1}, x_{k+1}, L_{k+1})$ is not. This means that there exists a time $t^* > t_k$ such that $x(\cdot)$ and $L(\cdot)$ are well defined for all t in $[t_k, t^*)$ and such that

$$\lim_{t\to t^*} \big(|x(t)| + |L(t)|\big) = +\infty. \tag{8.35}$$

Since $L(\cdot)$ is increasing and, in addition, for all t in $[t_k, t^*)$ we have (according to (8.24)) $L(t) \leqslant \frac{\alpha}{(t-t_k)}$, we get:

$$L^* = \lim_{t\to t^*} L(t) \leqslant \frac{\alpha}{(t^*-t_k)} < +\infty. \tag{8.36}$$

Consequently, $\lim_{t\to t^*}|x(t)| = +\infty$, which together with (8.26) and (8.27) yields

$$\lim_{t\to t^*}|X(t)| = +\infty. \tag{8.37}$$

On the other hand, denoting $V(X(t)) = X(t)'PX(t)$, we have along the solution of (8.1) and for all t in $[t_k, t^*)$

$$\dot{\overline{V(X(t))}} = \dot{X}(t)'PX(t) + X(t)'P\dot{X}(t), \tag{8.38}$$

where

$$\begin{aligned}
\dot{X}(t) &= \dot{\mathscr{S}}(t)x(t) + \mathscr{S}(t)\dot{x}(t)\\
&= -\frac{\dot{L}(t)}{L(t)}D\mathscr{S}(t)x(t) + \mathscr{S}(t)\left[Ax(t) + BK\mathscr{L}_k x_k + f(x(t))\right]\\
&= -\frac{\dot{L}(t)}{L(t)}DX(t) + L(t)AX(t) + L(t)BKX_k + \mathscr{S}(t)f(x(t)).
\end{aligned}$$

With the previous equality, (8.38) becomes for all t in $[t_k, t^*)$

$$\begin{aligned}
\dot{\overline{V(X(t))}} = &-\frac{\dot{L}(t)}{L(t)}X(t)'(PD + DP)X(t) \\
&+L(t)[X(t)'(A'P + PA)X(t) + 2X(t)'PBKX_k] + 2X(t)'P\mathscr{S}(t)f(x(t)).
\end{aligned} \tag{8.39}$$

Since $M \geqslant 1$, we have with (8.19) and (8.34) for all t in $[t_k, t^*)$

$$\begin{aligned}
-\frac{\dot{L}(t)}{L(t)}X(t)'(PD + DP)X(t) &\leqslant -\, p_3\frac{\dot{L}(t)}{L(t)}X(t)'PX(t)\\
&= -\, p_3a_2M(t)c(x_1(t))V(X(t))\\
&\leqslant -\, p_3a_2c(x_1(t))V(X(t)).
\end{aligned}$$

Moreover, using Young's inequality, we get

$$2X(t)'PBKX_k \leqslant X(t)'PX(t) + X_k(K'B'P + PBK)X_k.$$

Hence, we have, for all t in $[t_k, t^*)$

$$\begin{aligned}
\dot{\overline{V(X(t))}} &\leqslant -p_3a_2c(x_1(t))V(X(t))\\
&\quad +L[X(t)'(A'P + PA)X(t) + X_k'(K'B'P + PBK)X_k] + 2nc(x_1(t))V(X(t))\\
&\leqslant \left(-p_3a_2c(x_1(t)) + L(t)\lambda_1 + 2nc(x_1(t))\right)V(X(t)) + L(t)\lambda_2V_k\,,
\end{aligned}$$

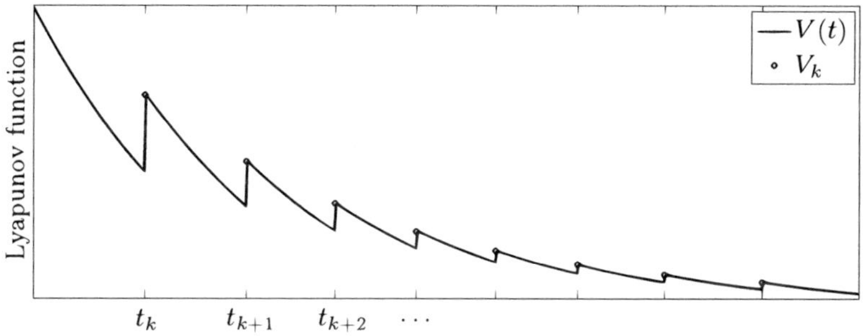

Fig. 8.1 Time evolution of Lyapunov function V

where[1] $\lambda_1 = \max\{0, \frac{\lambda_{\max}(A'P+PA)}{\lambda_{\min}(P)}\}$ and $\lambda_2 = \max\{0, \frac{\lambda_{\max}(K'B'P+PBK)}{\lambda_{\min}(P)}\}$. Bearing in mind that $L(t) \leqslant L^*$ for all t in $[t_k, t^*)$ and since $a_2 \geqslant \frac{2n}{p_3}$, the previous inequality becomes

$$\overset{\cdot}{\overline{V(X(t))}} \leqslant L^*\lambda_1 V(X(t)) + L^*\lambda_2 V_k.$$

This gives for all t in $[t_k, t^*)$

$$V(t) \leqslant \exp(\lambda_1 L^*(t-t_k))V_k + \int_0^{t-t_k} \exp\left(\lambda_1 L^*(t-t_k-s)\right)\lambda_2 V_k ds \tag{8.40}$$

$$\leqslant \left[\exp(\lambda_1 \alpha) + (\exp(\lambda_1 \alpha) - 1)\frac{\lambda_2}{\lambda_1}\right] V_k. \tag{8.41}$$

Hence, $\lim_{t\to t^*} |X(t)| < +\infty$ which contradicts (8.37) and thus, ends the proof. □

8.5.3 *Lyapunov Analysis*

This section is devoted to the Lyapunov analysis. It is shown that a good choice of the parameters a_1, a_2, and a_3 in the high-gain update law (8.19)–(8.22) yields the decrease of the sequence $(V(X_k))_{k\in\mathbb{N}}$ (see Fig. 8.1).

Remark 8.2 Drawing on the results obtained in [24] on lower triangular systems, the dynamic scaling (8.27) includes a number b. Although the decrease of $V(X_k)$ can be obtained with $b = 1$, it will be required that $bq < 1$ in order to ensure the boundedness of $L(\cdot)$ (see Eq. (8.65) in Sect. 8.5.4).

The aim of this subsection is to show the following intermediate result.

[1] If Z is a symmetric matrix, $\lambda_{\max}(Z)$ and $\lambda_{\min}(Z)$ denote its largest and its smallest eigenvalue, respectively.

Proposition 8.2 (Decrease of scaled coordinates) *There exist positive real numbers a_1 (sufficiently small), a_2 (sufficiently large), and α^* such that for $a_3 = 2n$ and for all α in $[0, \alpha^*]$ the following property is satisfied:*

$$V_{k+1} - V_k \leqslant -\left(\frac{\alpha}{p_2}\right)^2 V_k \tag{8.42}$$

Proof of Proposition 8.2: Let $a_2 \geqslant \frac{2n}{p_3}$. Then, according to Proposition 8.1, the sequence $(t_k, x_k, L_k)_{k\in\mathbb{N}}$ is well defined. Let k be in $\mathbb{N}$. The nonlinear system (8.1) with control (8.17) gives the closed-loop dynamics

$$\dot{x}(t) = Ax(t) + BK\mathscr{L}_k x_k + f(x(t)), \quad \forall t \in [t_k, t_k + \delta_k).$$

Integrating the preceding equality between t_k and t_{k+1} yields

$$\begin{aligned} x_{k+1} = \exp(A\delta_k)x_k &+ \int_0^{\delta_k} \exp(A(\delta_k - s))BK\mathscr{L}_k x_k ds \\ &+ \int_0^{\delta_k} \exp(A(\delta_k - s)) f(x(t_k + s))ds. \end{aligned}$$

Employing the algebraic properties (8.29)–(8.31) and (8.26) we get,

$$\mathscr{S}_k\left(\exp(A\delta_k)x_k + \int_0^{\delta_k} \exp(A(\delta_k - s))BK\mathscr{L}_k x_k ds\right) = F_c(\alpha_k)X_k, \tag{8.43}$$

where $\alpha_k = \delta_k L_k$ and F_c is defined in (8.10). Hence,

$$x_{k+1} = (\mathscr{S}_k)^{-1}F_c(\delta_k L_k) + \int_0^{\delta_k} \exp(A(\delta_k - s)) f(x(t_k + s))ds. \tag{8.44}$$

Employing the algebraic properties (8.29)–(8.31) we get, when left multiplying (8.44) by $\mathscr{S}_{k+1}^{-}$,

$$\mathscr{S}_{k+1}^{-} x_{k+1} = R + \mathscr{S}_{k+1}^{-}(\mathscr{S}_k)^{-1}F_c(\alpha_k)X_k, \tag{8.45}$$

where

$$R = \int_0^{\delta_k} \exp(L_{k+1}^{-} A(\delta_k - s))\mathscr{S}_{k+1}^{-} f(x(t_k + s))ds. \tag{8.46}$$

Note that, since we have $X_{k+1} = \Psi\mathscr{S}_{k+1}^{-}x_{k+1}$ with $\Psi = \mathscr{S}_{k+1}(\mathscr{S}_{k+1}^{-})^{-1}$, (8.45) yields

$$V(X_{k+1}) = (\Psi\mathscr{S}_{k+1}^{-}x_{k+1})'P\Psi\mathscr{S}_{k+1}^{-}x_{k+1} = V(X_k) + T_1 + T_2,$$

with

$$T_1 = X_k' F_c(\alpha_k)' \mathscr{S}_k^{-1} \mathscr{S}_{k+1}^{-} \Psi P \Psi \mathscr{S}_{k+1}^{-} \mathscr{S}_k^{-1} F_c(\alpha_k) X_k - V(X_k),$$
$$T_2 = 2X_k' F_c(\alpha_k)' \mathscr{S}_k^{-1} \mathscr{S}_{k+1}^{-} \Psi P \Psi R + R' \Psi P \Psi R.$$

The next two lemmas provide upper bounds for T_1 and T_2. The term T_1, which will be shown to be negative, guarantees that the Lyapunov function decreases, whereas the term T_2 is handled by robustness. Let β be defined by

$$\beta = n \int_0^{\delta_k} c(x_1(t_k + s))ds. \tag{8.47}$$

Lemma 8.2 *Let $a_1 \leqslant \frac{2}{p_4 p_2}$ and $a_3 = 2n$. Then, there exists $\alpha^* > 0$ sufficiently small such that for all α in $[0, \alpha^*)$*

$$T_1 \leqslant -\left(\frac{\alpha}{p_2}\right)^2 V(X_k) - \|\mathscr{S}_{k+1}^{-} x_k\|^2 (e^{2\beta} - 1) \frac{p_3 p_1 a_2}{2n}. \tag{8.48}$$

Lemma 8.3 *There exists a positive continuous real-valued function N such that the following inequality holds:*

$$T_2 \leqslant \|\mathscr{S}_{k+1}^{-} x_k\|^2 (e^{2\beta} - 1) N(\alpha).$$

The proofs of Lemmas 8.2 and 8.3 are postponed in Appendix.

With the two bounds obtained for T_1 and T_2, we get

$$V(X_{k+1}) - V(X_k) \leqslant -\left(\frac{\alpha}{p_2}\right)^2 V(X_k) + \left\|\mathscr{S}_{k+1}^{-} x_k\right\|^2 (e^{2\beta} - 1)\left[-\frac{p_3 p_1 a_2}{2n} + N(\alpha)\right].$$

For $a_2 \geqslant 2n \dfrac{N(\alpha)}{p_3 p_1}$ the result follows. □

8.5.4 Boundedness of L and Proof of Theorem 8.3

Although the construction of the updated law for the high-gain parameter (8.19)–(8.22) follows the idea developed in [8], the study of the behavior of the high-gain parameter is more involved. Indeed, in the context of observer design of [8], the nonlinear function c was assumed to be essentially bounded while in the present work, c is depending on x. This implies that the interconnection structure between state and high-gain dynamics must be further investigated.

Proof of Theorem 8.3 Assume a_1, a_2, a_3, and α^* meet the conditions of Propositions 8.1 and 8.2. Consider solutions $(x(\cdot), L(\cdot), M(\cdot))$ for system (8.1) with the Event-Triggered state feedback with initial condition $x(0)$ in $\mathbb{R}^n$, $L(0)) \geqslant 1$ and $M(0) = 1$. With Proposition 8.1, the sequence $(t_k, x_k, L_k)_{k\in\mathbb{N}}$ is well defined.

The existence of a strictly positive dwell time is obtained from the following proposition whose proof of this proposition is given in Appendix.

Proposition 8.3 *There exist* $\ell_{\max} > 0$*, a class* $\mathcal{K}$ *function* γ *and a nondecreasing function in both argument* ρ *such that*

$$\tilde{L}_{k+1} \leqslant \left(1 - \frac{a_1\alpha}{2}\right)\tilde{L}_k + \gamma(V_k), \quad \forall k \in \mathbb{N}, \tag{8.49}$$

with $\tilde{L}_k = \max\{L_k - \ell_{\max}, 0\}$, $\gamma(s) = 0$ *for all* s *in* $[0, 1]$*, and for all* t *for which* $L(t)$ *exists*

$$1 \leqslant L(t) \leqslant \rho(\tilde{L}_0, V_0). \tag{8.50}$$

With this proposition in hand, note that it yields for all k in $\mathbb{N}$, $\delta_k \geqslant \frac{\alpha}{\rho(\tilde{L}_0, V_0)} > 0$. Consequently, there is a dwell time and the solutions are complete. (i.e., $\sum_k \delta_k = +\infty$). Moreover, for all k in $\mathbb{N}$, $\frac{L_k}{L_{k+1}^-} \geqslant \frac{1}{\rho(\tilde{L}_0, V_0)}$. Consequently, inequality (8.42) becomes

$$V_{k+1} \leqslant \left(1 - \sigma(\tilde{L}_0, V_0)\right) V_k,$$

where $\sigma(\tilde{L}_0, V_0) = \frac{\alpha N(\alpha)}{\rho(L_0, V_0)^{2(n-1+b)}}$ is a decreasing function of both arguments. This gives $V_k \leqslant (1 - \sigma(\tilde{L}_0, V_0))^k V_0$, for all k in $\mathbb{N}$. With (8.49), it yields $\tilde{L}_k \leqslant \tilde{\beta}(\tilde{L}_0 + V_0, k)$ where

$$\tilde{\beta}(s, k) = \left(1 - \frac{a_1\alpha}{2}\right)^k \left(s + \sum_{j=1}^{k} \frac{\gamma\left((1 - \sigma(s, s))^{k-j} s\right)}{\left(1 - \frac{a_1\alpha}{2}\right)^{k-j}}\right).$$

The function $\tilde{\beta}$ is of class $\mathcal{K}$ in s. Moreover, since $\gamma(s) = 0$ for $s \leqslant 1$, this implies that there exists $k^*(s)$ such that the mapping $k \mapsto \tilde{\beta}(s, k)$ is decreasing for all $k \geqslant k^*(s)$. Moreover, we have $\lim_{k\to\infty} \tilde{\beta}(s, k) = 0$. On another hand, since $\delta_k \leqslant \alpha$, it implies that $k \leqslant \frac{t}{\alpha}$ for all t in $[t_k, t_{k+1})$.

$$\tilde{L}(t) \leqslant \frac{\tilde{L}_{k+1}}{1 - a_1\alpha} \leqslant \frac{\tilde{\beta}(\tilde{L}_0 + V_0, k+1)}{1 - a_1\alpha}. \tag{8.51}$$

Finally, with (8.41), it yields

$$V(t) \leqslant \xi(\alpha)(1 - \sigma(\tilde{L}_0, V_0))^{\frac{t}{\alpha}} V_0. \tag{8.52}$$

With the right-hand side of (8.50) and the definition of the Lyapunov function V, we have

$$\frac{p_1 + \mu q_1}{2\rho(\tilde{L}_0, V_0)^{2(n-1+b)}} \|x(t)\|^2 \leqslant V(t), \tag{8.53}$$

Moreover, we also have

$$V_0 \leqslant 2(p_2 + \mu q_2) \|x(0)\|^2 . \tag{8.54}$$

From inequalities (8.51)–(8.54) and the properties of the function $\tilde{\beta}$, it yields readily that there exists a class $\mathscr{K}\mathscr{L}$ function β such that inequality (8.25) holds.

8.6 Illustrative Example

We apply our approach to the following uncertain third-order system proposed in [16]:

$$\begin{cases} \dot{x}_1 = x_2 \\ \dot{x}_2 = x_3 \\ \dot{x}_3 = \theta x_1^2 x_3 + u, \end{cases} \tag{8.55}$$

where θ is a constant parameter in which only a magnitude bound $\theta_{\max}$ is known. The stabilization of this problem is not trivial even in the case of a continuous-in-time controller. The difficulties arise from the nonlinear term $x_1^2 x_3$ that makes the x_3 dynamic not globally Lipschitz, and from the uncertainty on the θ value, preventing the use of a feedback to cancel the nonlinearity.

However, system (8.55) belongs to the class of systems (8.1) and Assumption 8.1 is satisfied with $c(x_1) = \theta_{\max} x_1^2$. Hence, by Theorem 8.3, a Event-Triggered feedback controller (8.17)–(8.24) can be constructed. Simulations were conducted with a gain matrix K and a coefficient α selected as

$$K = \begin{bmatrix} -1 & -3 & -3 \end{bmatrix}', \quad \alpha = 0.4$$

to stabilize the linear part of the system (8.55). Parameters a_1, a_2, and a_3 were then selected through a trial and error procedure as follows:

$$a_1 = 1, \quad a_2 = 1, \quad a_3 = 1.$$

Simulation results are given in Figs. 8.2 and 8.3. The evolution of the control and state trajectories are displayed in Fig. 8.2. The corresponding evolution of the Lyapunov function V and the high-gain L are shown in Fig. 8.2. We can see how the inter-execution times δ_k adapts to the nonlinearity. Interestingly, it allows a significant increase of δ_k when the state is close to the origin: $L(t)$ then goes to 1 and consequently δ_k increases toward value α ($\alpha = 0.4$ in this simulation).

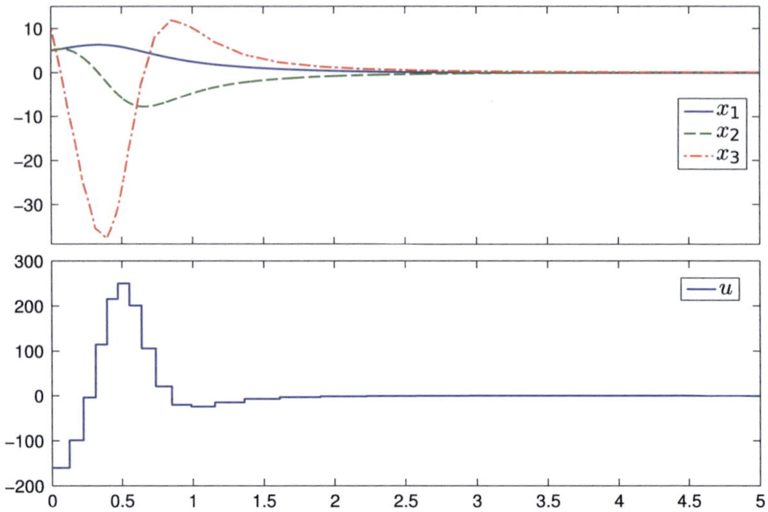

Fig. 8.2 Control signal and state trajectories of (8.55) with $(x_1, x_2, x_3) = (5, 5, 10)$ as initial conditions

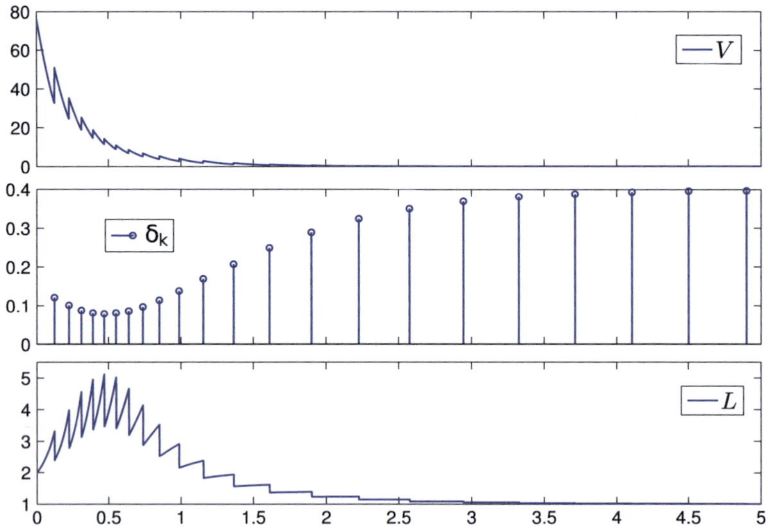

Fig. 8.3 Simulation results

8.7 Conclusion

In this paper, a novel Event-Triggered state feedback law has been given. This law is based on a high-gain methodology. The event which triggers an update of the control law is based on a dynamical system in which state is the high-gain parameter.

This approach allows to design control laws ensuring convergence to the origin for nonlinear systems with triangular structure and a specific upper bound on the nonlinearities. Current research line focuses on the design of a Event-Triggered output feedback (see [4]).

8.8 Proofs of Lemmas

Proof of Lemma 8.1

The proof of Lemma 8.1 is based on this Lemma.

Lemma 8.4 *Let P be a positive definite matrix such that* (8.32) *and* (8.33) *hold then there exists δ_m such that for all $\delta \leqslant \delta_m$, we have*

$$PF_c(\delta) + F_c(\delta)'P - P \leqslant -\frac{\delta}{2p_2}P. \tag{8.56}$$

Proof Let v in $\mathbb{R}^n$ be such that $\|v\| = 1$. Consider the mapping

$$\nu(\delta) = v'\left(PF_c(\delta) + F_c(\delta)'P - P\right)v.$$

Note that $\nu(0) = 0$. Moreover, we have

$$\frac{d\nu}{d\delta}(0) = v'\left(P(A+BK) - (A+BK)'P\right)v \leqslant -\|v\|^2.$$

This yields the existence of a positive real number δ_m such that for all $\delta \leqslant \delta_m$, we have

$$\nu(\delta) \leqslant -\frac{\delta}{2}\|v\|^2 \leqslant -\frac{\delta}{2p_2}v'Pv.$$

This property being true for every v in S^{n-1}, we have

$$F_c(\delta)'PF_c(\delta) \leqslant \left(1 - \frac{\delta}{2p_2}\right)P.$$

To prove Lemma 8.1, let $\delta \leqslant \delta_m$ and P be a positive definite matrix such that (8.32) and (8.33) hold and consider $V(x) = x'Px$. We have for all t in $[t_k, t_{k+1})$

$$V(x(t)) \leqslant \left(1 - \frac{\delta}{2p_2}\right)^k \left(1 - \frac{t - t_k}{2p_2}\right) V(x_0).$$

Hence, this yields that the origin is globally and asymptotically stable.

Proof of Lemma 8.2

In order to prove Lemma 8.2, we need the following lemma which will be proved in the next section.

Lemma 8.5 *Let* $\Psi = \mathscr{S}_{k+1}(\mathscr{S}_{k+1}^{-})^{-1}$. *The matrix* P *satisfies the following property for all* a_1 *and* α *such that* $a_1\alpha < 1$*:*

$$\Psi P \Psi \leqslant \left(1 + \alpha \frac{a_1 p_4}{2}\right) P. \tag{8.57}$$

Applying Lemma 8.5 to T_1 yields the following inequality:

$$T_1 \leqslant \left(1 + \alpha \frac{a_1 p_4}{2}\right) V\left(\mathscr{S}_{k+1}^{-} \mathscr{S}_k^{-1} F_c(\alpha_k) X_k\right) - V(X_k).$$

On another hand, we have, for all v in $\mathbb{R}^n$

$$v' \mathscr{S}_{k+1}^{-} P \mathscr{S}_{k+1}^{-} v - v \mathscr{S}_k P \mathscr{S}_k v = v' \left(\int_{t_k}^{t_{k+1}} \frac{d\mathscr{S}(s)}{ds} P \mathscr{S}(s) + \mathscr{S}(s) P \frac{d\mathscr{S}(s)}{ds} ds \right) v.$$

However, we have for all s in $[t_k, t_{k+1})$

$$\frac{d\mathscr{S}}{ds}(s) = -\frac{\dot{L}(s)}{L(s)} D \mathscr{S}(s).$$

Consequently, it yields

$$v' \mathscr{S}_{k+1}^{-} P \mathscr{S}_{k+1}^{-} v - v \mathscr{S}_k P \mathscr{S}_k v = v' \left(\int_{t_k}^{t_{k+1}} -\frac{\dot{L}(s)}{L(s)} \mathscr{S}(s)[DP + PD] \mathscr{S}(s) ds \right) v.$$

Note that since $L(0) > 1$, it yields that $L(t) > 1$ on the time of existence of the solution. Moreover, we have also $\dot{L} \geqslant 0$ and taking into account the bounds on P in (8.33) and on $DP + PD$ in (8.34), we get

$$\begin{aligned}
& v' \mathscr{S}_{k+1}^{-} P \mathscr{S}_{k+1}^{-} v - v \mathscr{S}_k P \mathscr{S}_k v \\
& \quad \leqslant v' \left(p_3 \int_{t_k}^{t_{k+1}} -\frac{\dot{L}(s)}{L(s)} \mathscr{S}(s) P \mathscr{S}(s) ds \right) v \\
& \quad = v' \left(p_3 \int_{t_k}^{t_{k+1}} -a_2 M(s) c(s) \mathscr{S}(s) P \mathscr{S}(s) ds \right) v \\
& \quad = v' \left(p_3 \int_{t_k}^{t_{k+1}} -a_2 \exp\left(a_3 \int_{t_k}^{t_{k+1}} c(r) dr \right) c(s) \mathscr{S}(s) P \mathscr{S}(s) ds \right) v \\
& \quad \leqslant -p_3 p_1 a_2 v' \left(\int_{t_k}^{t_{k+1}} \exp\left(a_3 \int_{t_k}^{t_{k+1}} c(r) dr \right) c(s) \|\mathscr{S}(s)\|^2 ds \right) v
\end{aligned}$$

Note that since $L_k \leqslant L_{k+1}^-$, we finally get

$$
\begin{aligned}
& v'\mathscr{S}_{k+1}^- P \mathscr{S}_{k+1}^- v - v\mathscr{S}_k P \mathscr{S}_k v \\
& \leqslant -p_3 p_1 a_2 v' \left(\int_{t_k}^{t_{k+1}} \exp\left(a_3 \int_{t_k}^{t_{k+1}} c(r)dr \right) c(s) \|\mathscr{S}_{k+1}^-\|^2 ds \right) v \\
& = -\frac{p_3 p_1 a_2}{a_3} v' \left(\exp\left(a_3 \int_{t_k}^{t_{k+1}} c(r)dr \right) - 1 \right) \|\mathscr{S}_{k+1}^-\|^2 v \\
& \leqslant -\frac{p_3 p_1 a_2}{a_3} \left(\exp\left(a_3 \int_{t_k}^{t_{k+1}} c(r)dr \right) - 1 \right) \|\mathscr{S}_{k+1}^- v\|^2 .
\end{aligned}
$$

The previous inequality with $v = \mathscr{S}_k^{-1} F_c(\alpha_k) X_k$, $a_3 = 2n$ and the notation (8.47) yield

$$
\begin{aligned}
T_1 \leqslant & \left(1 + \alpha \frac{a_1 p_4}{2}\right) V\left(F_c(\alpha_k) X_k\right) - V(X_k) \\
& - \frac{p_3 p_1 a_2}{2n} \left(e^{2\beta} - 1\right) \|\mathscr{S}_{k+1}^- \mathscr{S}_k^{-1} F_c(\alpha_k) X_k\|^2 .
\end{aligned}
$$

Note that $\alpha_k \leqslant \alpha$. Consequently, with Lemma 8.4 and α sufficiently small, this yields

$$
\begin{aligned}
T_1 \leqslant & \left[\left(1 + \alpha \frac{a_1 p_4}{2}\right) \left(1 - \frac{\alpha}{p_2}\right) - 1 \right] V(X_k) \\
& - \frac{p_3 p_1 a_2}{2n} \left(e^{2\beta} - 1\right) \|\mathscr{S}_{k+1}^- \mathscr{S}_k^{-1} F_c(\alpha_k) X_k\|^2 .
\end{aligned}
$$

With $a_1 \leqslant \frac{2}{p_4 p_2}$ this yields

$$
T_1 \leqslant -\left(\frac{\alpha}{p_2}\right)^2 V(X_k) - \frac{p_3 p_1 a_2}{2n} \left(e^{2\beta} - 1\right) \|\mathscr{S}_{k+1}^- \mathscr{S}_k^{-1} F_c(\alpha_k) X_k\|^2 .
$$

However, we have

$$
\mathscr{S}_{k+1}^- (\mathscr{S}_k)^{-1} F_c(\alpha_k) X_k = \left[\exp(A\alpha) + R_c(\alpha) G(L_k, L_{k+1}^-)\right] \mathscr{S}_{k+1}^- x_k, \qquad (8.58)
$$

where

$$
R_c(\alpha) = \int_0^{\alpha} \exp(A(\alpha - s)) ds B K_c,
$$

$$
G(L_k, L_{k+1}^-) = \left(\frac{L_k}{L_{k+1}^-}\right)^{n+1} \mathscr{S}_k (\mathscr{S}_{k+1}^-)^{-1} .
$$

Now, we have

$$\begin{aligned}&[\exp(A\alpha) + R_c(\alpha)G(L_k, L_{k+1}^-)]' \left[\exp(A\alpha) + R_c(\alpha)G(L_k, L_{k+1}^-)\right]\\&\quad= \exp((A + A')\alpha) + \exp(A'\alpha)R_c(\alpha)G(L_k, L_{k+1}^-)\\&\qquad + G(L_k, L_{k+1}^-)R_c(\alpha)' \exp(A\alpha) + R_c(\alpha)'R_c(\alpha)G(L_k, L_{k+1}^-)^2.\end{aligned}$$

Note that $L_{k+1}^- \geqslant L_k$. Hence,

$$\left\|G(L_k, L_{k+1}^-)\right\| \leqslant 1. \tag{8.59}$$

Moreover, for all $\varepsilon > 0$, employing the continuity of the mapping $|R(\cdot)|$ and $|\exp(A'\cdot)|$ and the fact that $|R(0)| = 0$, we can find sufficiently small α, such that we have

$$\|R_c(\alpha)\| \leqslant \varepsilon\ ,\ \left\|\exp(A'\alpha)\right\| \leqslant 1 + \varepsilon\ ,\ \|\exp(A\alpha)\| \leqslant 1 + \varepsilon,$$

and

$$\exp((A + A')\alpha) \geqslant (1 - \varepsilon)I.$$

Hence,

$$\left[\exp(A\alpha) + R_c(\alpha)G(L_k, L_{k+1}^-)\right]' \left[\exp(A\alpha) + R_c(\alpha)G(L_k, L_{k+1}^-)\right] \geqslant (1 - 3\varepsilon - 3\varepsilon^2)I$$

So, select ε such that $(1 - 3\varepsilon - 3\varepsilon^2) = \frac{1}{2}$ (for instance) yields

$$T_1 \leqslant -\left(\frac{\alpha}{p_2}\right)^2 V(X_k) - \frac{p_3p_1a_2}{2n}\left(e^{2\beta} - 1\right)\|\mathscr{S}_{k+1}^- x_k\|^2 .$$

Proof of Lemma 8.5

In order to prove Lemma 8.5, we need the following lemma which will be proved in the next section.

Lemma 8.6 *The matrix P satisfies the following property for all a_1 and α such that $a_1\alpha < 1$*

$$\Psi P\Psi \leqslant \psi_0(\alpha)P\psi_0(\alpha),$$

where

$$\psi_0(\alpha) = diag\left(\frac{1}{(1 - a_1\alpha)^b}, \ldots, \frac{1}{(1 - a_1\alpha)^{n+b-1}}\right).$$

Given v in $S^{n-1} = \{v \in \mathbb{R}^n \mid \|v\| = 1\}$, consider the function

$$\nu(\alpha, v) = v'\psi_0(\alpha)P\psi_0(\alpha)v.$$

We have

$$\psi_0(0) = I\ , \ \frac{\partial \psi_0}{\partial \alpha}(0) = a_1 D,$$

then

$$\nu(0, v) = v' P v, \ \frac{\partial \nu}{\partial \alpha}(0, v) = a_1 v' \left[PD + DP\right] v.$$

So using the inequalities in (8.32)–(8.34)

$$\frac{\partial \nu}{\partial \alpha}(0, v) \leqslant a_1 p_4 v' P v.$$

Now, we can write

$$\nu(\alpha, v) = v' P v + \alpha \frac{\partial \nu}{\partial \alpha}(0, v) + \rho(\alpha, v),$$

with $\lim_{\alpha \to 0} \frac{\rho(\alpha, v)}{\alpha} = 0$. This equality implies that

$$\nu(\alpha, v) \leqslant v' P v \left[1 + \alpha a_1 p_4\right] + \rho(\alpha, v).$$

The vector v being in a compact set and the function r being continuous, there exists α^* such that for all α in $[0, \alpha^*)$ we have $\rho(\alpha, v) \leqslant \alpha \frac{a_1 p_4}{2} v' P v$ for all v. This gives

$$\nu(\alpha, v) \leqslant v' P v \left[1 + \alpha \frac{a_1 p_4}{2}\right], \forall\, \alpha \in [0, \alpha^*), \forall\, v \in S^{n-1}.$$

This property being true for every v, this ends the proof of Lemma 8.5.

Proof of Lemma 8.3

First, we seek for an upper bound of the norm of $\mathscr{S}_{k+1}^{-} f(x(t_k + s))$. We have

$$\begin{aligned}
\|\mathscr{S}_{k+1}^{-} f(x(t_k + s))\|^2 &= \sum_{j=1}^{n} \left((L_{k+1}^{-})^{-b-j+1} f_j(x(t_k + s))\right)^2 \\
&\leqslant \sum_{j=1}^{n} (L_{k+1}^{-})^{2(-b-j+1)} \left(\sum_{i=1}^{j} c(t_k + s)|x_i(t_k + s)|\right)^2 \\
&= c(t_k + s)^2 \sum_{j=1}^{n} \left(\sum_{i=1}^{j} (L_{k+1}^{-})^{-b-j+1} |x_i(t_k + s)|\right)^2 .
\end{aligned}$$

Since $L_{k+1}^{-} \geqslant 1$, we have $(L_{k+1}^{-})^{-b-j+1} \leqslant (L_{k+1}^{-})^{-b-i+1}$ whenever $1 \leqslant i \leqslant j$. It yields

$$
\begin{aligned}
\|\mathscr{S}_{k+1}^{-} f(x(t_k+s))\|^2 &\leqslant c(t_k+s)^2 \sum_{j=1}^{n} \left(\sum_{i=1}^{n} (L_{k+1}^{-})^{-b-i+1} |x_i(t_k+s)| \right)^2 \\
&\leqslant c(t_k+s)^2 \sum_{j=1}^{n} n \|\mathscr{S}_{k+1}^{-} x(t_k+s)\|^2 \\
&= n^2 c(t_k+s)^2 \|\mathscr{S}_{k+1}^{-} x(t_k+s)\|^2. \qquad (8.60)
\end{aligned}
$$

Hence, from (8.46) and (8.60), we get

$$
\begin{aligned}
\|R\| &\leqslant \int_0^{\delta_k} \exp\left(L_{k+1}^{-} \|A\| (\delta_k - s)\right) nc(t_k+s) \|\mathscr{S}_{k+1}^{-} x(t_k+s)\| ds \\
&= \exp(\|A\|\,\alpha) \int_0^{\delta_k} \exp\left(-L_{k+1}^{-} \|A\| s\right) nc(t_k+s) \|\mathscr{S}_{k+1}^{-} x(t_k+s)\| ds. \qquad (8.61)
\end{aligned}
$$

Moreover, we have for all s in $[0; \delta_k)$

$$
\mathscr{S}_{k+1}^{-} \dot{x}(t_k+s) = \mathscr{S}_{k+1}^{-} A x(t_k+s) + \mathscr{S}_{k+1}^{-} BK \mathscr{L}_k x_k + \mathscr{S}_{k+1}^{-} f(x(t_k+s)).
$$

Denoting by $w(s)$ the expression $\mathscr{S}_{k+1}^{-} x(t_k+s)$, this gives

$$
\begin{aligned}
\frac{d}{ds} \|w(s)\| &= \frac{\langle \dot{w}(s), w(s) \rangle}{\|w(s)\|} \\
&\leqslant \|\dot{w}(s)\| \\
&\leqslant \left\| L_{k+1}^{-} A w(s) \right\| + \left\| \mathscr{S}_{k+1}^{-} BK \mathscr{L}_k x_k \right\| + \left\| \mathscr{S}_{k+1}^{-} f(x(t_k+s) \right\| \\
&\leqslant (L_{k+1}^{-} \|A\| + nc(t_k+s)) \|w(s)\| + \left\| BK (L_{k+1}^{-})^{-b-n+1} \mathscr{L}_k x_k \right\|, \text{ by (8.60)} \\
&\leqslant (L_{k+1}^{-} \|A\| + nc(t_k+s)) \|w(s)\| + L_{k+1}^{-} \|BK\| \|w(0)\|.
\end{aligned}
$$

Hence, integrating the previous inequality, we obtain

$$
\|w(s)\| \leqslant \int_0^s (L_{k+1}^{-} \|A\| + nc(t_k+r)) \|w(r)\| dr + \|BK\|\| \|w(0)\| L_{k+1}^{-} s + \|w(0)\|.
$$

Since $(L_{k+1}^{-} \|A\| + nc(t_k+s))$ is a continuous nonnegative function and $(\|BK\| L_{k+1}^{-} s + 1) \|w(0)\|$ is nondecreasing, applying a variant of the Gronwall–Bellman inequality (see [3, Theorem 1.3.1]), it becomes

$$
\|w(s)\| \leqslant (\|BK\| L_{k+1}^{-} s + 1) \|w(0)\| \times \exp\left(\int_0^s (L_{k+1}^{-} \|A\| + nc(t_k+r) dr \right),
$$

and we have

$$
\begin{aligned}
&\|\mathscr{S}_{k+1}^{-}x(t_k+s)\| \\
&\qquad \leqslant (\|BK\|\, L_{k+1}^{-}s+1)\exp\left(\int_0^s L_{k+1}^{-}\|A\|+nc(t_k+r)dr\right)\left\|\mathscr{S}_{k+1}^{-}x_k\right\| \\
&\qquad = (\|BK\|\, L_{k+1}^{-}s+1)\exp\left(L_{k+1}^{-}\|A\|s\right)\exp\left(\int_0^s nc(t_k+r)dr\right)\left\|\mathscr{S}_{k+1}^{-}x_k\right\|.
\end{aligned}
\tag{8.62}
$$

Consequently, according to (8.61) and (8.62), we get

$$
\begin{aligned}
&\|R\| \leqslant \\
&\exp(\|A\|\,\alpha)\int_0^{\delta_k} nc(t_k+s)(\|BK\|\, L_{k+1}^{-}s+1)\exp\left(\int_0^s (nc(t_k+r)dr\right)\left\|\mathscr{S}_{k+1}^{-}x_k\right\| ds \\
&\leqslant \exp(\|A\|\,\alpha)\int_0^{\delta_k} nc(t_k+s)(\|BK\|\,\alpha+1)\exp\left(\int_0^s (nc(t_k+r)dr\right)\left\|\mathscr{S}_{k+1}^{-}x_k\right\| ds \\
&\leqslant \exp(\|A\|\,\alpha)(\alpha\,\|BK\|+1)\int_0^{\delta_k} nc(t_k+s)\exp\left(\int_0^s (nc(t_k+r)dr\right) ds\left\|\mathscr{S}_{k+1}^{-}x_k\right\| \\
&= \exp(\|A\|\,\alpha)(\alpha\,\|BK\|+1)\left[\exp\left(\int_0^{\delta_k} (nc(t_k+r)dr\right)-1\right]\left\|\mathscr{S}_{k+1}^{-}x_k\right\|.
\end{aligned}
$$

On another hand, employing (8.58), we have

$$
\left\|\mathscr{S}_{k+1}^{-}(\mathscr{S}_k)^{-1}F_c(\alpha_k)X_k\right\| \leqslant \left[\|\exp(A\alpha)\|+\|R_c(\alpha)\|\,\left\|G(L_k,L_{k+1}^{-}\right\|)\right]\left\|\mathscr{S}_{k+1}^{-}x_k\right\|.
$$

Hence, employing Lemma 8.6 and Eq. (8.59), this gives the existence of two continuous function N_1 and N_2 such that

$$
\begin{aligned}
T_2 =& R'\Psi P\Psi R+2X_k'F_c(\alpha_k)'\mathscr{S}_k^{-1}\mathscr{S}_{k+1}^{-}\Psi P\Psi R, \\
\leqslant& \|\mathscr{S}_{k+1}^{-}x_k\|^2 N_1(\alpha)\left[\exp\left(n\int_0^{\delta_k} c(t_k+r)dr\right)-1\right]^2 \\
&+\left\|\mathscr{S}_{k+1}^{-}x_k\right\|^2 N_2(\alpha)\left[\exp\left(n\int_0^{\delta_k} c(t_k+r)dr\right)-1\right],
\end{aligned}
$$

where

$$
N_1(\alpha)=\exp(2\|A\|\alpha)(\alpha\,\|BK\|+1)^2\frac{\|P\|}{(1-a_1\alpha)^{2(n-b+1)}},
$$

$$N_2(\alpha) = 2\exp(\|A\|\alpha)(\alpha\,\|BK\| + 1)\frac{(\|\exp(A\alpha)\| + \|R_c(\alpha)\|)\,\|P\|}{(1 - a_1\alpha)^{2(n-b+1)}}.$$

Proof of Lemma 8.6

Consider the matrix function defined as

$$\mathscr{P}(s) = diag(s^b, \ldots, s^{n+b-1})P diag(s^b, \ldots, s^{n+b-1}).$$

Note that for all v in $\mathbb{R}^n$

$$\frac{d}{ds}v'\mathscr{P}(s)v = \frac{1}{s}v' diag(s^b, \ldots, s^{n+b-1})(D'P + PD) diag(s^b, \ldots, s^{n+b-1})v > 0.$$

Hence, $\mathscr{P}$ is an increasing function. Furthermore, we have

$$\begin{aligned}
\Psi P \Psi =& \mathscr{S}_{k+1}(\mathscr{S}_{k+1}^{-})^{-1} P \mathscr{S}_{k+1}(\mathscr{S}_{k+1}^{-})^{-1}\\
=& diag\left(\left(\frac{L_{k+1}^{-}}{L_{k+1}}\right)^b, \ldots, \left(\frac{L_{k+1}^{-}}{L_{k+1}}\right)^{n+b-1}\right) P\\
&\times diag\left(\left(\frac{L_{k+1}^{-}}{L_{k+1}}\right)^b, \ldots, \left(\frac{L_{k+1}^{-}}{L_{k+1}}\right)^{n+b-1}\right)\\
=& \mathscr{P}\left(\frac{L_{k+1}^{-}}{L_{k+1}^{-}(1 - a_1\alpha) + a_1\alpha}\right),
\end{aligned}$$

Hence, as

$$\frac{L_{k+1}^{-}}{L_{k+1}^{-}(1 - a_1\alpha) + a_1\alpha} \leqslant \frac{1}{1 - a_1\alpha},$$

we get the inequality of Lemma 8.6, i.e., $\Psi P \Psi \leqslant \mathscr{P}\left(\frac{1}{1-a_1\alpha}\right)$.

Proof of Proposition 8.3

Inequality (8.42) of Proposition 8.2 implies that $(V_k)_{k\in\mathbb{N}}$ is a nonincreasing sequence. Consequently, being nonnegative, $(V_k)_{k\in\mathbb{N}}$ is bounded. One infers, using inequality (8.41), that $V(t)$ is bounded. Hence, by the left parts in inequality (8.33), we get that, on the time T_x $(= \sum \delta_k)$ of existence of the solution, $X(t)$ is bounded. Then we get that $\frac{x_j(t)}{L(t)^{b+j-1}}$ for $j = 1, \ldots n$ are bounded. Summing up, there exists a class $\mathscr{K}$ function γ such that

$$\frac{|x_j(t)|}{L(t)^{b+j-1}} \leqslant \gamma(V_k) \leqslant \gamma(V_0), \quad \forall (j, t, k) \in \{1, \ldots n\} \times [t_k, T_x) \times \mathbb{N}. \tag{8.63}$$

With this result in hand, let us analyze the high-gain dynamics. According to equations (8.19) and (8.20), we have, for all k and all t in $[t_k, t_{k+1})$, $\dot{L}(t) = \frac{a_2}{a_3} L(t)\dot{M}(t)$, which implies that

$$L(t) = \exp\left(\frac{a_2}{a_3}\int_{t_k}^{t}\dot{M}(s)ds\right)L_k$$
$$= \exp\left(\frac{a_2}{a_3}M(t) - \frac{a_2}{a_3}\right)L_k, \quad \forall t \in [t_k, t_{k+1}), k \in \mathbb{N}. \tag{8.64}$$

Consequently, from (8.21)

$$L_{k+1} = \exp\left(\frac{a_2}{a_3}(M_{k+1}^{-} - 1)\right)L_k(1 - a_1\alpha) + a_1\alpha.$$

Moreover, we have

$$\begin{aligned}\dot{M}(t) &= a_3 M(t)c(x(t)) \\ &= a_3 M(t)\left(c_0 + \sum_{j=1}^{n} c_j|x_j|^{q_j}\right) \\ &\leqslant a_3 M(t)\left(c_0 + \sum_{j=1}^{n} c_j\gamma(V_k)^{q_j}L(t)^{(b+j-1)q_j}\right) \qquad \text{(by (8.63))}.\end{aligned}$$

Picking $\mathfrak{b} = \max_j\{(b + j - 1)q_j\}$, it yields,

$$\begin{aligned}\dot{M}(t) &\leqslant a_3\left(c_0 + \sum_{j=1}^{n} c_j\gamma(V_k)^{q_j}\right)M(t)L(t)^{\mathfrak{b}} \qquad \text{(since } L(t) \geqslant 1) \\ &\leqslant \tilde{c}(\gamma(V_k))M(t)\exp\left(\frac{a_2}{a_3}\mathfrak{b}(M(t) - 1)\right)L_k^{\mathfrak{b}}, \qquad \text{(by (8.64))}\end{aligned}$$

where $\tilde{c}(\gamma(V_k)) = a_3(c_0 + \sum_{j=1}^{n} c_j\gamma(V_k)^{q_j})$. Let $\psi(t)$ be the solution to the scalar dynamical system

$$\dot{\psi}(t) = \psi(t)\exp\left(\frac{a_2}{a_3}\mathfrak{b}(\psi(t) - 1)\right), \quad \psi(0) = 1.$$

$\psi(\cdot)$ is defined on $[0, T_\psi)$ where T_ψ is a positive real number possibly equal to $+\infty$. Note that we have (see, e.g., [17, Theorem 1.10.1]) that for all t such that $0 \leqslant \tilde{c}(\gamma(V_k))(t - t_k)L_k^{\mathfrak{b}} < T_\psi$

$$M(t) \leqslant \psi\left(\tilde{c}(\gamma(V_k))(t - t_k)L_k^{\mathfrak{b}}\right).$$

Consequently, for all k such that $\tilde{c}(\gamma(V_k))\alpha L_k^{\mathfrak{b}-1} < T_\psi$

$$M_{k+1}^{-} \leqslant \psi\left(\tilde{c}(\gamma(V_k))\delta_k L_k^{\mathfrak{b}}\right) \leqslant \psi\left(\tilde{c}(\gamma(V_k))\alpha L_k^{\mathfrak{b}-1}\right),$$

where the last inequality follows from the fact that $L(\cdot)$ is nondecreasing on $[t_k, t_{k+1})$. It follows, employing (8.8) that, for all k such that $\tilde{c}(\gamma(V_k))\alpha L_k^{\mathfrak{b}} < T_\psi$,

$$L_{k+1} \leqslant F_k(L_k), \tag{8.65}$$

where

$$F_k(L) = \exp\left(\psi\left(\tilde{c}(\gamma(V_k))\alpha L^{\mathfrak{b}-1}\right) - 1\right) L(1 - a_1\alpha) + a_1\alpha.$$

Note that, since $\mathfrak{b} < 1$, $\lim_{L\to+\infty} L^{\mathfrak{b}-1} = 0$. Thus,

$$\lim_{L\to+\infty} \frac{F_k(L)}{L} = 1 - a_1\alpha < 1.$$

Consequently, there exists an increasing function ℓ_1 such that for all $L > \ell_1(V_k)$

$$\tilde{c}(\gamma(V_k))\alpha L^{\mathfrak{b}-1} < T_\psi, \quad F_k(L) < \left(1 - \frac{a_1\alpha}{2}\right) L. \tag{8.66}$$

On the other hand, consider the following nonlinear system with input χ in $\mathbb{R}^n$:

$$\begin{cases} \dot{L}(t) = a_2 L(t) M(t) \left(c_0 + \sum_{j=1}^{n} c_j \chi_j(t)^{q_j} L(t)^{\mathfrak{b}}\right) \\ \dot{M}(t) = a_3 M(t) \left(c_0 + \sum_{j=1}^{n} c_j \chi_j(t)^{q_j} L(t)^{\mathfrak{b}}\right), \end{cases} \tag{8.67}$$

We assume that the norm of the input signal satisfies the bound

$$|\chi_j(\cdot)| \leqslant \gamma(v), \tag{8.68}$$

where v is a given positive real number. Notice that the couple (L, M) which satisfies Eqs. (8.19) and (8.20) between $[t_k, t_{k+1})$ is also a solution of the previous nonlinear system with input $\chi_j(t) = X_j(t)$ which satisfies (8.68) with $v = V_k$. Let $\phi_{s,t}$ denotes the flow of (8.67) issued from s, i.e., $\phi_{s,t}(a, b)$ is the solution of (8.67) that takes value (a, b) at $t = s$. Let C_1, C_2, be the two compact subsets of $\mathbb{R}^2$ defined by

$$C_1 = \{1 \leqslant L \leqslant \ell_1(v), M = 1\}, \ C_2 = \{1 \leqslant L \leqslant 2\ell_1(v), 0 \leqslant M \leqslant 2\}.$$

The set C_1 is included in the interior of C_2, and we have the following Lemma.

Lemma 8.7 *There exists a nonincreasing function d such that for all input function χ which satisfies the bound* (8.68) *the following holds.*

$$\forall k \in \mathbb{N}, \quad \forall t \leqslant d(v), \quad \phi_{t_k, t_k+t}(C_1) \subset C_2. \tag{8.69}$$

The proof of Lemma 8.7 is given in Appendix. Let

$$\ell_2(v) = \max\left\{2\ell_1(v), \frac{\alpha}{d(v)}\right\}.$$

Note that L_k satisfies the following properties:

$$L_k > \ell_1(V_k) \Longrightarrow L_{k+1} \leqslant \left(1 - \frac{a_1\alpha}{2}\right) L_k \tag{8.70}$$

$$L_k \leqslant \ell_1(V_k) \Longrightarrow L_{k+1} \leqslant \ell_2(V_k) \tag{8.71}$$

Equation (8.70) follows immediately from (8.65) and (8.66). We now prove (8.71). Notice that, because $L_{k+1}^- \geqslant 1$ and $a_1\alpha < 1$, (8.21) implies

$$L_{k+1} \leqslant L_{k+1}^-. \tag{8.72}$$

Suppose first that $\delta_k \leqslant d(V_k)$. In that case, $(L_k, M_k) \in C_1$. Then (8.72) and (8.69) with $v = V_k$ yield

$$L_{k+1} \leqslant L_{k+1}^- = L\left((t_k + \delta_k)^-\right) \leqslant 2\ell_1(V_k) \leqslant \ell_2(V_k).$$

Suppose now that $\delta_k > d(V_k)$. Since $\delta_k L_{k+1}^- = \alpha$, it follows that

$$L_{k+1} \leqslant \frac{\alpha}{\delta_k} \leqslant \frac{\alpha}{d(V_k)} \leqslant \ell_2(V_k).$$

Note that properties (8.70) and (8.71) in combination with the fact that the sequence $(V_k)_{k\in\mathbb{N}}$ is decreasing imply that

$$L_k \leqslant \max\{L_0, \ell_2(V_0)\}, \quad k \in \mathbb{N}.$$

Moreover, for all k in $\mathbb{N}$ and all t in $[t_k, t_{k+1})$

$$\begin{aligned} L(t) &\leqslant L_{k+1}^- && \text{(since } \dot{L}(t) \geqslant 0) \\ &= \frac{L_{k+1} - a_1\alpha}{1 - a_1\alpha} && \text{by (8.21)} \\ &\leqslant \frac{\max\{L_0, \ell_2(V_0)\}}{1 - a_1\alpha}, \end{aligned}$$

and the result holds with $\rho(L_0, V_0) = \frac{\max\{\tilde{L}_0 + \ell_2(1), \ell_2(V_0)\}}{1 - a_1\alpha}$.

Proof of Lemma 8.7

Let Ω_L and Ω_M be the increasing functions

$$\Omega_L(v) = 4a_2\ell_1(v)\left(c_0 + \sum_{j=1}^{n} c_j\gamma(v)^{q_j}(2\ell_1(v))^{\mathfrak{b}}\right),$$

$$\Omega_M(v) = 2a_3\left(c_0 + \sum_{j=1}^{n} c_j\gamma(v)^{q_j}(2\ell_1(v))^{\mathfrak{b}}\right).$$

Note that if $(L(t), M(t))$ is in C_2 and $\chi(t)$ satisfies the bound (8.68), we have

$$\dot{L}(t) \leqslant \Omega_L(v), \quad \dot{M}(t) \leqslant \Omega_M(v). \tag{8.73}$$

We claim that the function $d = \min\left\{\frac{1}{\Omega_L}, \frac{1}{\Omega_M}\right\}$ satisfies the properties of Lemma 8.7. Assume this is not the case. Hence, there exists $M(t_k)$, $L(t_k)$ in C_1, χ which satisfies the bound (8.68) and $t^* \leqslant d(v)$ such that $(L(t_k + t^*), M(t_k + t^*)) \notin C_2$. Let s^* be the time at which the solution leaves C_2. More precisely, let $s^* = \inf\{s \mid t_k \leqslant s \leqslant t_k + t^*, (L(s), M(s)) \notin C_2\}$. Note that $(L(s^*), M(s^*))$ is at the border of C_2 and $t_k < s^* < t_k + d(v)$. Moreover, with (8.73), it yields

$$M(s^*) \leqslant 1 + (s^* - t_k)\Omega_M(v) < 1 + d(v)\Omega_M(v) \leqslant 2.$$

Similarly, we have

$$L(s^*) < L(t_k) + d(v)\Omega_L \leqslant L(t_k) + 1 \leqslant 2\ell_1(v),$$

where the last inequality is obtained since $\ell_1(v) \geqslant 1$. This implies that $(L(s^*), M(s^*))$ is not at the border of C_2 which contradicts the existence of t^*.

References

1. Abdelrahim, M., Postoyan, R., Daafouz, J., Nesic, D.: Input-to-state stabilization of nonlinear systems using event-triggered output feedback controllers. In: 14th European Control Conference, ECC'15 (2015)
2. Alur, R., Arzen, K.-E., Baillieul, J., Henzinger, T.A., Hristu-Varsakelis, D., Levine, W.S.: Handbook of Networked and Embedded Control Systems. Springer Science & Business Media (2007)
3. Ames, W.F., Pachpatte, B.G.: Inequalities for Differential and Integral Equations, vol. 197. Academic press, San Diego (1997)
4. Andrieu, V., Praly, L.: A unifying point of view on output feedback designs. In: 7th IFAC Symposium on Nonlinear Control Systems, pp. 8–19 (2007)
5. Andrieu, V., Tarbouriech, S.: Global asymptotic stabilization for a class of bilinear systems by hybrid output feedback. IEEE Trans. Autom. Control. **58**(6), 1602–1608 (2013)

6. Andrieu, V., Praly, L., Astolfi, A.: Asymptotic tracking of a reference trajectory by output-feedback for a class of non linear systems. Syst. Control. Lett. **58**(9), 652–663 (2009)
7. Andrieu, V., Praly, L., Astolfi, A.: High gain observers with updated gain and homogeneous correction terms. Automatica **45**(2), 422–428 (2009)
8. Andrieu, V., Nadri, M., Serres, U., Vivalda, J.-C.: Self-triggered continuous-discrete observer with updated sampling period. Automatica **62**, 106–113 (2015)
9. Anta, A., Tabuada, P.: To sample or not to sample: self-triggered control for nonlinear systems. IEEE Trans. Autom. Control. **55**(9), 2030–2042 (2010)
10. Aström, K.J., Wittenmark, B.: Computer-controlled systems. Prentice Hall, Englewood Cliffs (1997)
11. Dabroom, A.M., Khalil, H.K.: Output feedback sampled-data control of nonlinear systems using high-gain observers. IEEE Trans. Autom. Control. **46**(11), 1712–1725 (2001)
12. Dinh, T.-N., Andrieu, V., Nadri, M., Serres, U.: Continuous-discrete time observer design for lipschitz systems with sampled measurements. IEEE Trans. Autom. Control. **60**(3), 787–792 (2015)
13. Girard, A.: Dynamic triggering mechanisms for event-triggered control. IEEE Trans. Autom. Control. **60**(7), 1992–1997 (2015)
14. Heemels, W.P.M.H., Johansson, K.-H., Tabuada, P.: An introduction to event-triggered and self-triggered control. In: 2012 IEEE 51st Annual Conference on Decision and Control (CDC), pp. 3270–3285 (2012)
15. Heemels, W.P.M.H., Johansson, K.-H., Tabuada, P.: Event-triggered and self-triggered control. In: Baillieul, J., Samad, T. (eds.) Encyclopedia of Systems and Control, pp. 1–10. Springer, London (2014)
16. Krishnamurthy, P., Khorrami, F.: Dynamic high-gain scaling: state and output feedback with application to systems with iss appended dynamics driven by all states. IEEE Trans. Autom. Control. **49**(12), 2219–2239 (2004)
17. Lakshmikantham, V., Leela, S.: Differential and Integral Inequalities: Theory and applications. Vol. I: Ordinary Differential Equations. Mathematics in Science and Engineering, vol. 55-I. Academic Press, New York (1969)
18. Marchand, N., Durand, S., Castellanos, J.F.G.: A general formula for event-based stabilization of nonlinear systems. IEEE Trans. Autom. Control. **58**(5), 1332–1337 (2013)
19. Mazenc, F., Andrieu, V., Malisoff, M.: Design of continuous-discrete observers for time-varying nonlinear systems. Automatica **57**, 135–144 (2015)
20. Nesic, D., Teel, A.R., Kokotović, P.V.: Sufficient conditions for stabilization of sampled-data nonlinear systems via discrete-time approximations. Syst. Control. Lett. **38**(45), 259–270 (1999)
21. Peralez, J., Andrieu, V., Nadri, M., Serres, U.: Self-triggered control via dynamic high-gain scaling. In: *IFAC proceedings Volumes (IFAC-PapersOnLine)*, vol. 10 (2016)
22. Peralez, J., Andrieu, V., Nadri, M., Serres, U.: Event-triggered output feedback stabilization via dynamic high-gain scaling. IEEE Trans. Autom. Control. (2018)
23. Postoyan, R., Tabuada, P., Nesic, D., Anta, A.: A framework for the event-triggered stabilization of nonlinear systems. IEEE Trans. Autom. Control. **60**(4), 982–996 (2015)
24. Praly, L.: Asymptotic stabilization via output feedback for lower triangular systems with output dependent incremental rate. IEEE Trans. Autom. Control. **48**(6), 1103–1108 (2003)
25. Qian, C., Du, H.: Global output feedback stabilization of a class of nonlinear systems via linear sampled-data control. IEEE Trans. Autom. Control. **57**(11), 2934–2939 (2012)
26. Seuret, A., Prieur, C., Marchand, N.: Stability of non-linear systems by means of event-triggered sampling algorithms. IMA J. Math. Control. Inf. (2013)
27. Tabuada, P.: Event-triggered real-time scheduling of stabilizing control tasks. IEEE Trans. Autom. Control. **52**(9), 1680–1685 (2007)
28. Villarreal-Cervantes, M.G., Guerrero-Castellanos, J.F., Ramrez-Martnez, S., Snchez-Santana, P.: Stabilization of a (3,0) mobile robot by means of an event-triggered control. ISA Trans. **58**, 605–613 (2015)

Chapter 9
Insights on Event-Triggered Control for Linear Systems Subject to Norm-Bounded Uncertainty

S. Tarbouriech, A. Seuret, C. Prieur and L. Zaccarian

Abstract The chapter deals with the design of Event-Triggered rules to stabilize a class of uncertain linear control systems where the uncertainty affecting the plant is assumed to be norm-bounded. The event-triggering rule uses only local information, namely the control updates are generated only by the output signals available to the controller. The proposed approach combines a hybrid framework to describe the closed-loop system with techniques based on looped functionals. The suggested design conditions are formulated in terms of linear matrix inequalities (LMIs), ensuring global robust asymptotic stability of the closed-loop system. A tunable parameter allows guaranteeing an adjustable dwell-time property of the solutions. The effectiveness of the approach is evaluated on an example taken from the literature.

9.1 Introduction

In recent years, sampled-data control designs for linear or nonlinear plants have been studied in several works. In particular, robust stability analysis with respect to aperiodic sampling has been widely studied (see, for example, [7, 17, 23] and references therein), where variations on the sampling intervals are seen as a disturbance,

S. Tarbouriech (✉)
LAAS-CNRS, Université de Toulouse, Toulouse, France
e-mail: tarbour@laas.fr

C. Prieur
University Grenoble Alpes, CNRS, Grenoble INP, GIPSA-lab, France
e-mail: christophe.prieur@gipsa-lab.fr

L. Zaccarian · A. Seuret
LAAS-CNRS, Université de Toulouse, CNRS, Toulouse, France
e-mail: zaccarian@laas.fr

A. Seuret
e-mail: aseuret@laas.fr

L. Zaccarian
Dipartimento di Ingegneria Industriale, University of Trento, Trento, Italy

S. Tarbouriech et al. (eds.), *Control Subject to Computational and Communication Constraints*, Lecture Notes in Control and Information Sciences 475, https://doi.org/10.1007/978-3-319-78449-6_9

an undesired perturbation of the periodic case. The objective is then to provide an analysis of such systems using the discrete-time approach [8, 17], the input delay approach [12, 31], or the impulsive systems approach [22]. Furthermore, an alternative and interesting vision of sampled-data systems has been proposed in [3, 5], suggesting to adapt the sampling sequence to certain events related to the state evolution (see, for example, [4, 15, 18, 20, 34, 38]). This is called "Event-Triggered sampling", which naturally mixes continuous and discrete-time dynamics. Thus, the Event-Triggered algorithm design can be rewritten as the stability study of a hybrid dynamical system, which has been carried out in different contexts in [13, 14, 27, 29].

In the context of Event-Triggered control, two objectives can be pursued: (1) the controller is a priori designed and only the Event-Triggered rules have to be designed, or (2) the joint design of the control law and the event-triggering conditions has to be performed. The first case is called the emulation approach, whereas the second one corresponds to the co-design problem. A large part of the existing works is dedicated to the design of efficient event-triggering rules, that is the design done by emulation: see, for example, [2, 16, 26, 35, 37] and references therein. Moreover, most of the results on Event-Triggered control consider that the full state is available, which can be unrealistic from a practical point of view. Hence, it is interesting to address the design of Event-Triggered controllers by using only measured signals. Some works have addressed this challenge as, for example, in [36] where the dynamic controller is an observer-based one, [1], where the co-design of the output feedback law and the event-triggering conditions is addressed by using the hybrid framework.

The results proposed in the current chapter take place in the context of the emulation approach, when the predesigned controller is issued from a hybrid dynamic output feedback controller, with the aim at using only the available signals. The controller under consideration is a continuous controller possibly including some reset loop as in [28] (see also [10, 11, 32] for more details on reset control systems). Actually, the approach proposed combines a hybrid framework to describe the sampled-data system with Lyapunov-based techniques. Constructive conditions, in the sense that linear matrix inequality (LMI) conditions are associated to a convex optimization scheme, are proposed to design the Event-Triggered rule ensuring asymptotic stability of the closed-loop system. Differently from [1], a condition involving the allowable maximal sampling period T can be deduced by solving a set of LMIs proposed using a similar approach to the one in [21]. Furthermore, complementary to most of the results in the literature, uncertainty affecting the continuous plant is considered in our approach. The results of this chapter are complementary to those in [33] where polytopic uncertainties (rather than norm bounded ones) are considered with similar design approaches.

The chapter is organized as follows. In Sect. 9.2, the system under consideration is defined, together with the sampled-data architecture. The problem that we intend to solve is also formally stated in Sect. 9.2, describing the associated hybrid formulation. Section 9.3 is dedicated to presenting the main conditions, allowing to design the event-triggering rules in both the nominal and the uncertain cases. The condition to design the associated dwell-time is also derived. We point out that the contribution is twofold. On the first hand, we provide a new Event-Triggered algorithm yielding robust stability controllers for the closed-loop system. Secondly, the stability conditions depend on a dwell-time T, which appears as an explicit tuning parameter for the selection of the control law. Section 9.4 illustrates the results and compares them with some existing approach. Finally, in Sect. 9.5, some concluding remarks end the chapter.

Notation. The sets $\mathbb{N}$, $\mathbb{R}^+$, $\mathbb{R}^n$, $\mathbb{R}^{n\times n}$ and $\mathbb{S}^n$ denote respectively the sets of positive integers, positive scalars, n-dimensional vectors, $n \times n$ matrices and symmetric matrices in $\mathbb{R}^{n\times n}$. If a matrix P in $\mathbb{S}^n_+$, it means that P is symmetric positive definite. The superscript "⊤" stands for matrix transposition, and the notation $\mathrm{He}(P)$ stands for $P + P^\top$. The Euclidean norm is denoted $|\cdot|$. Given a compact set $\mathscr{A}$, the notation $|x|_{\mathscr{A}} := \min\{|x - y|,\ y \in \mathscr{A}\}$ indicates the distance of the vector x from the set $\mathscr{A}$. The symbols I and 0 represent the identity and the zero matrices of appropriate dimensions.

9.2 Problem Formulation

The chapter deals with linear systems fed by an output feedback sampled-data control described by the following hybrid dynamical system

$$\begin{cases} \dot{x} = Ax + Bu, & \\ \dot{u} = 0, & (x, u, \sigma) \in \mathscr{C}, \\ \dot{\sigma} \in g_T(\sigma), & \end{cases}$$
$$\begin{cases} x^+ = x, & \\ u^+ = KCx, & (x, u, \sigma) \in \mathscr{D}, \\ \sigma^+ = 0, & \end{cases} \tag{9.1}$$

where $x \in \mathbb{R}^n$ represents the state of the system and $u \in \mathbb{R}^m$ represents the zero order holder of the system input since the last sampling time. The output y of the system is given by

$$y = Cx \in \mathbb{R}^p. \tag{9.2}$$

System (9.1)–(9.2) can appear when connecting, for instance, a linear continuous-time plant with a dynamic output feedback controller.

Remark 9.1 In [28], a reset controller is considered, which corresponds to modifying system (9.1)–(9.2) by considering $x^+ = Jx$, where J is a matrix of appropriate dimensions.

To study stability properties for (9.1), the hybrid formalism of [13, 27, 30] can be used. Matrices A, B, C characterize the system dynamics and matrix K corresponds to the controller gain. While C is assumed to be constant and known, let us assume that matrices A and B are constant but uncertain (see, for example, [24, 39]) and expressed by

$$\left[A \ B \right] = \left[A_0 \ B_0 \right] + DF \left[E_1 \ E_2 \right], \tag{9.3}$$

with A_0, B_0, D, E_1 and E_2 constant and known matrices. These matrices define the structure of the uncertainty. A_0 and B_0 define the nominal case and the uncertainty parameter is F, which is supposed to be constant and belongs to the set:

$$\mathbb{F} = \left\{ F \in \mathbb{R}^{f \times f};\ F^\top F \leqslant I \right\}. \tag{9.4}$$

Timer $\sigma \in [0, 2T]$ flows by keeping track of the elapsed time since the last sample (where it was reset to zero) according to the following set-valued dynamics:

$$g_T(\sigma) := \begin{cases} 1 & \sigma \leqslant 2T \\ [0, 1] & \sigma = 2T, \end{cases} \tag{9.5}$$

whose rationale is that whenever $\sigma < 2T$, its value exactly represents the elapsed time since the last sample, moreover $\sigma \in [T, 2T]$ implies that at least T seconds have elapsed since the last sample.

Remark 9.2 The use of a set-valued map for the right hand side g_T of the flow equation for σ enables us to confine the timer σ to a compact set $[0, 2T]$. Note also that with the selection in (9.5), the regularity conditions in [14, As. 6.5] and the desirable robustness properties of stability of compact attractors established in [14, Chap. 7] are satisfied.

In (9.1), the so-called flow and jump sets $\mathscr{C}$ and $\mathscr{D}$ must be suitably selected to induce a desirable behavior of the sampled-data system, and are the available degrees of freedom in the design of the Event-Triggered algorithm addressed here. In particular, the problem that we intend to solve in this chapter can be summarized as follows.

Problem 9.1 Given an uncertain linear plant and a hybrid controller defined by (9.1)–(9.5), design an event-triggering rule, with a prescribed dwell-time T that makes the closed loop globally asymptotically stable to a compact set wherein $x = 0$ and $u = 0$.

Problem 9.1 corresponds to an emulation problem (see, for example [16, 26, 35, 37] and the references therein) since we assume that the controller gain K is given.

9.3 Event-Triggered Design

In order to address Problem 9.1, we focus on hybrid dynamics (9.1) for suitably selecting the flow and jump sets $\mathscr{C}$ and $\mathscr{D}$, whose role is precisely to rule when a sampling should happen, based on the available signals to the controller, namely output $y = Cx$, the last sampled input u and timer σ. Then, we select the following sets $\mathscr{C}$ and $\mathscr{D}$:

$$\mathscr{C} := (\mathscr{F} \times [0, 2T]) \cup \left(\mathbb{R}^{n+m} \times [0, T]\right) \quad (9.6a)$$
$$\mathscr{D} := \left(\mathscr{J} \times [0, 2T]\right) \cap \left(\mathbb{R}^{n+m} \times [T, 2T]\right), \quad (9.6b)$$

where sets $\mathscr{F}$ and $\mathscr{J}$ are selected as

$$\mathscr{F} := \left\{(x, u);\ \begin{bmatrix} y \\ u - Ky \end{bmatrix}^\top M \begin{bmatrix} y \\ u - Ky \end{bmatrix} \leqslant 0\right\}, \quad (9.6c)$$

$$\mathscr{J} := \left\{(x, u);\ \begin{bmatrix} y \\ u - Ky \end{bmatrix}^\top M \begin{bmatrix} y \\ u - Ky \end{bmatrix} \geqslant 0\right\}, \quad (9.6d)$$

where matrix $M = \begin{bmatrix} M_1 & M_2 \\ M_2^\top & M_3 \end{bmatrix} \in \mathbb{R}^{(p+m)\times(p+m)}$ has to be designed, and y is defined in (9.2). Solution (9.6) to the considered Event-Triggered problem is parametrized by M and T.

Note that the jump set selection in (9.6b) ensures that all solutions satisfy a dwell-time constraint corresponding to T. Moreover the definition of the flow and jump sets provided in (9.6) meets the one provided in the recent paper [25]. The novelty of this definition, which is also used in [33], relies on the consideration of a general matrix M. For example, selecting $M_2 = 0$ leads to the definition of the flow and jump sets usually employed in the literature, issued from an Input-to-State (or Input-to-Output) analysis. See [25] for more details.

9.3.1 *Nominal Case*

In this section, the design is addressed for the nominal case, namely $A = A_0$ and $B = B_0$ (which corresponds to $F = 0$ in (9.3)).

Theorem 9.1 *Assume that, for a given $T > 0$, there exist matrices $P \in \mathbb{S}^n_+$, $M = \begin{bmatrix} M_1 & M_2 \\ M_2^\top & M_3 \end{bmatrix} \in \mathbb{S}^{p+m}$ satisfying*

$$\Psi_{\mathscr{C}}(A_0, B_0) := \begin{bmatrix} \mathrm{He}(P(A_0 + B_0KC)) - C^\top M_1 C & \star \\ B_0^\top P - M_2^\top C & -M_3 \end{bmatrix} < 0,$$

$$\Psi_{\mathscr{D}}(A_0, B_0, T) := \begin{bmatrix} I \\ KC \end{bmatrix}^\top \Lambda(A_0, B_0, T)^\top P \Lambda(A_0, B_0, T) \begin{bmatrix} I \\ KC \end{bmatrix} - P < 0, \quad (9.7)$$

with

$$\Lambda(A_0, B_0, T) := \begin{bmatrix} I & 0 \end{bmatrix} e^{\begin{bmatrix} A_0 & B_0 \\ 0 & 0 \end{bmatrix} T} \in \mathbb{R}^{n\times(n+m)}. \tag{9.8}$$

Then the compact attractor

$$\mathscr{A} := \{(x, u, \sigma) : \ x = 0, u = 0, \sigma \in [0, 2T]\} \tag{9.9}$$

is globally asymptotically stable for the nominal closed-loop dynamics (9.1), (9.6) with $\begin{bmatrix} A & B \end{bmatrix} = \begin{bmatrix} A_0 & B_0 \end{bmatrix}$.

Proof To prove this result one uses a non-smooth Lyapunov function and the hybrid invariance principle given in [33]. In particular, the following function is considered, with the standard notation $|z|_P^2 := z^\top P z$:

$$V(x, u, \sigma) := \underbrace{e^{-\rho \min\{\sigma, T\}} \left| \Lambda(T - \min\{\sigma, T\}) \begin{bmatrix} x \\ u \end{bmatrix} \right|_P^2}_{=:V_0(x,u,\sigma)} + \underbrace{\eta |u|^2}_{=:V_u(u)}, \tag{9.10}$$

with Λ given in (9.8), and where ρ and η are sufficiently small positive scalars.

Let us first denote $\xi := (x, u, \sigma)$. We can also notice that in (9.6a) the flow set is the union of two sets, and then one can split the analysis in three cases:

- Case 1: $\sigma \in [0, T)$;
- Case 2: $\sigma = T$;
- Case 3: $(x, u) \in \mathscr{F}$ and $\sigma \in [T, 2T]$.

Then along flowing solutions one gets:

- In Case 1, after some simplifications (as done in [33]) we get:

$$\dot{V}(\xi) = -\rho e^{-\rho \min\{\sigma, T\}} \left| \Lambda(T - \min\{\sigma, T\}) \begin{bmatrix} x \\ u \end{bmatrix} \right|_P^2 = -\rho V_0(\xi) \leqslant 0,$$

- In Case 3 (which also addresses Case 2, because no flowing is possible for a solution from $\sigma = T$, unless $(x, u) \in \mathscr{F}$), from inequality $\Psi_{\mathscr{C}}(A_0, B_0) < 0$ in (9.7), there exists a sufficiently small $\varepsilon > 0$ such that $\Psi_{\mathscr{C}}(A_0, B_0) < -\varepsilon I$, and then one obtains the following strict decrease property:

$$\dot{V}(\xi) \leqslant -\varepsilon \left\| \begin{bmatrix} x \\ u - Ky \end{bmatrix} \right\|^2, \text{ if } (x, u) \in \mathscr{F} \text{ and } \sigma \geqslant T. \tag{9.11}$$

Therefore, the Lyapunov function V is strictly decreasing along flows.

Along jumps, it is easy to verify that, for all $\xi \in \mathscr{D}$, after some calculations (see again [33]), the condition $\Psi_{\mathscr{D}}(A_0, B_0, T) < 0$ ensures that

$$V^+(\xi) = e^{-\rho T} V_0(\xi) \leqslant e^{-\rho T} V(\xi), \tag{9.12}$$

which proves the strict decrease of the Lyapunov function, across any jump from a point outside $\mathscr{A}$.

One can finally show that no "bad" complete solution exists, which keeps V constant and nonzero. If any such "bad" complete solution existed, then it would start outside $\mathscr{A}$ (where $V(\xi) \neq 0$) and it could not jump, because otherwise, from (9.12), a decrease of V would be experienced across the jump. However, any solution flowing forever outside $\mathscr{A}$ would eventually reach a point where $\sigma > T$, and (9.11) would imply again some decrease of V. The proof is then completed by applying the invariance principle in [33]. □

Remark 9.3 Let us provide some comments on the conditions of Theorem 9.1.

- The satisfaction of $\Psi_{\mathscr{C}}(A_0, B_0) < 0$ imposes that the Lyapunov function V in (9.10) is decreasing while flowing with $\sigma \geqslant T$ (which requires $(x, u) \in \mathscr{F}$).
- The satisfaction of $\Psi_{\mathscr{D}}(A_0, B_0, T) < 0$ can be interpreted as an asymptotic stability criterion for system (9.1) when the control updates are performed periodically with a period T, which motivates the union and intersection in (9.6a) and (9.6b).

Remark 9.4 The interest of the proposed approach with respect to the literature, where the dwell time is computed a posteriori, resides in the fact that Theorem 9.1 includes a guaranteed dwell-time T as a tuning parameter. In particular, if one can find a solution to the LMI conditions (9.7) for a given parameter T, then this same T can be employed in the definition of the flow and jumps sets (9.6) and is a guaranteed dwell time for all solutions of (9.1), (9.6). This method can be compared to [35] or [1] where a similar triggering rule includes a dwell time constraint, but in the current case, the dwell time T is a parameter for the design of event triggering algorithm.

Remark 9.5 Theorem 9.1 can be stated when an additional reset control component, as mentioned in Remark 9.1, is included in the jump dynamics of system (9.1)–(9.6). Preliminary results in this direction are provided in [28].

9.3.2 Uncertain Case

Let us address now the case where matrices A and B are uncertain as defined in (9.3)–(9.4). In this case, it is difficult to verify the inequality $\Psi_{\mathscr{D}}(A, B, T) < 0$, which will depend nonlinearly on the uncertain parts. Nevertheless, it is possible to adapt the result developed in [31, Theorem 1], which are based on the recent developments arising from stability analysis of periodic sampled-data systems (see also [19]). Then, the following lemma can be proven by using the looped-functional approach developed in [6, 31].

Lemma 9.1 *For a given positive scalar T and matrices A, B, K, C as defined in (9.3)–(9.4), if there exist $P, Z \in \mathbb{S}^n_+$, $Q, X \in \mathbb{S}^n$, $R \in \mathbb{R}^{n\times n}$, $Y \in \mathbb{R}^{2n\times n}$ such that inequalities*

$$\begin{aligned} \Psi_{\mathscr{D}1}(A, B, T) &:= F_0(A, B, T) + T F_1(A, B) < 0, \\ \Psi_{\mathscr{D}2}(A, B, T) &:= \begin{bmatrix} F_0(A, B, T) & TY \\ \star & -TZ \end{bmatrix} < 0, \end{aligned} \tag{9.13}$$

hold for all pairs (A, B) satisfying (9.3)–(9.4), *where*

$$\begin{aligned} F_0(A, B, T) &:= \mathrm{He}(e_S^\top P e_1 - e_{12}^\top R e_2 - Y e_{12}) - e_{12}^\top Q e_{12} - T e_2^\top X e_2, \\ F_1(A, B) &:= \mathrm{He}(e_S^\top Q e_{12} + e_S^\top R e_2) + e_S^\top Z e_S + 2 e_2^\top X e_2, \end{aligned}$$

and $e_S := \begin{bmatrix} A & BKC \end{bmatrix}$, $e_1 := \begin{bmatrix} I_n & 0 \end{bmatrix}$, $e_2 := \begin{bmatrix} 0 & I_n \end{bmatrix}$ and $e_{12} := \begin{bmatrix} I_n & -I_n \end{bmatrix}$, then inequality $\Psi_{\mathscr{D}}(A, B, T) < 0$ in (9.7) *holds for any pair (A, B) satisfying (9.3)–(9.4).*

By adapting Lemma 9.1 to the case of norm-bounded uncertainty, we can derive the following theorem solving Problem 9.1. It is based on the non-smooth hybrid Lyapunov function introduced in (9.10), which is weak in the sense that it does not provide a strict decrease both during flow and across jumps (samplings) of the proposed Event-Triggered sampled-data system. The proof then relies on the non-smooth invariance principle presented in [33].

Theorem 9.2 *Given positive scalar $T > 0$ and matrices A_0, B_0, D, E_1, E_2 as defined in (9.3)–(9.4). Assume that there exist matrices $P \in \mathbb{S}^n_+$, $M = \begin{bmatrix} M_1 & M_2 \\ M_2^\top & M_3 \end{bmatrix} \in \mathbb{S}^{p+m}$, and matrices $Z \in \mathbb{S}^n_+$, $Q, U \in \mathbb{S}^n$, $R \in \mathbb{R}^{n\times n}$ and $Y \in \mathbb{R}^{2n\times n}$ and positive scalars ε_i, $i = 0, 1, 2, 3$ satisfying conditions*

$$\Theta_{\mathscr{C}} := \begin{bmatrix} \Psi_{\mathscr{C}}(A_0, B_0) + \varepsilon_0 e_{\mathscr{C}}^\top e_{\mathscr{C}} & \star \\ \begin{bmatrix} D^\top P & 0 \end{bmatrix} & -\varepsilon_0 I \end{bmatrix} < 0, \tag{9.14}$$

$$\Theta_{\mathscr{D}1} := \begin{bmatrix} F_0(A_0, B_0, T) + T F_1(A_0, B_0) + (\varepsilon_1 + T\varepsilon_3) e_{\mathscr{D}}^\top e_{\mathscr{D}} & \star \\ D^\top (P e_1 + T(Q e_{12} + R e_2 + Z e_0)) & -\varepsilon_1 I \end{bmatrix} < 0, \tag{9.15}$$

$$\Theta_{\mathscr{D}2} := \begin{bmatrix} F_0(A_0, B_0, T) + \varepsilon_2 e_{\mathscr{D}}^\top e_{\mathscr{D}} & \star & \star \\ TY^\top & -TZ & \star \\ D^\top P e_1 & 0 & -\varepsilon_2 I \end{bmatrix} < 0, \tag{9.16}$$

$$\Theta_{\mathscr{D}3} := \varepsilon_3 I - D^\top Z D > 0, \tag{9.17}$$

with

$$\begin{aligned} F_0(A_0, B_0, T) &:= \mathrm{He}(e_0^\top P e_1 - e_{12}^\top R e_2 - Y e_{12}) - e_{12}^\top Q e_{12} - T e_2^\top X e_2 \\ F_1(A_0, B_0) &:= \mathrm{He}(e_0^\top Q e_{12} + e_0^\top R e_2) + e_0^\top Z e_0 + 2 e_2^\top X e_2 \end{aligned}$$

and $e_0 := \begin{bmatrix} A_0 & B_0KC \end{bmatrix}$, $e_{\mathscr{C}} = \begin{bmatrix} E_1 + E_2KC & E_2 \end{bmatrix}$, $e_{\mathscr{D}} = \begin{bmatrix} E_1 & E_2KC \end{bmatrix}$, $e_1 := \begin{bmatrix} I_n & 0 \end{bmatrix}$, $e_2 := \begin{bmatrix} 0 & I_n \end{bmatrix}$ *and* $e_{12} := \begin{bmatrix} I_n & -I_n \end{bmatrix}$. *Then the compact attractor* $\mathscr{A}$ *in* (9.9) *is globally asymptotically stable for the uncertain closed-loop dynamics (9.1)–(9.6) that is for each pair* (A, B) *satisfying (9.3)–(9.4).*

Proof The proof of the theorem follows by showing that (9.14)–(9.17) imply the two conditions in (9.7) (with $[A_0\ B_0]$ replaced by $[A\ B]$), and then the proof can be completed by following the same exact steps as those in the proof of Theorem 9.1.

First condition in (9.7). Condition $\Psi_{\mathscr{C}}(A_0, B_0) < 0$ of (9.7) has to be replaced by $\Psi_{\mathscr{C}}(A, B) < 0$ where (A, B) have the expression in (9.3). After some calculations, this substitution gives:

$$\Psi_{\mathscr{C}}(A, B) = \Psi_{\mathscr{C}}(A_0, B_0) + \mathrm{He}\left(\begin{bmatrix} PD \\ 0 \end{bmatrix} F \underbrace{\begin{bmatrix} E_1 + E_2KC & E_2 \end{bmatrix}}_{=e_{\mathscr{C}}}\right) < 0 \tag{9.18}$$

Using the fact that $F^\top F \leqslant I$, this expression can be upper-bounded as follows:

$$\Psi_{\mathscr{C}}(A, B) \leqslant \Psi_{\mathscr{C}}(A_0, B_0) + \varepsilon_0 e_{\mathscr{C}}^\top e_{\mathscr{C}} + \varepsilon_0^{-1} \begin{bmatrix} PD \\ 0 \end{bmatrix} \begin{bmatrix} PD \\ 0 \end{bmatrix}^\top \tag{9.19}$$

for any positive scalar ε_0. By using a Schur complement on the rightmost term of (9.19), one obtains the matrix $\Theta_{\mathscr{C}}$ in (9.14). Then its follows that inequality (9.14) implies that $\Psi_{\mathscr{C}}(A, B) < 0$.

Second condition in (9.7). To prove this second condition, we use Lemma 9.1 and show that (9.15)–(9.17) imply (9.13). In particular, the expressions of $F_0(A, B, T)$ and $F_1(A, B)$ in Lemma 9.1 are developed by substituting $e_S = e_0 + DFe_{\mathscr{D}}$, as defined in Theorem 9.2. Indeed, we note that

$$\begin{aligned} \Psi_{\mathscr{D}1}(A, B, T) &= \Psi_{\mathscr{D}1}(A_0, B_0, T) + \mathrm{He}\left(e_{\mathscr{D}}^\top F^\top D^\top (Pe_1 + TQe_{12} + TRe_2 + TZe_0)\right) \\ &\quad + Te_{\mathscr{D}}^\top F^\top D^\top ZDFe_{\mathscr{D}}, \\ \Psi_{\mathscr{D}2}(A, B, T) &= \Psi_{\mathscr{D}2}(A_0, B_0, T) + \begin{bmatrix} \mathrm{He}\left(e_{\mathscr{D}}^\top F^\top D^\top Pe_1\right) & 0 \\ 0 & 0 \end{bmatrix}, \end{aligned}$$

where we recall that $e_{\mathscr{D}} = \begin{bmatrix} E_1 & E_2KC \end{bmatrix}$. First we note that, since $\Theta_{\mathscr{D}3} > 0$ and $F^\top F \leqslant I$, the last term of $\Psi_{\mathscr{D}1}(A, B, T)$ can be upper-bounded by $\varepsilon_3 Te_{\mathscr{D}}^\top e_{\mathscr{D}}$. Following the same procedure as for $\Theta_{\mathscr{C}}$ in (9.19), and using $F^\top F \leqslant I$, for any positive selection of $\varepsilon_1, \varepsilon_2$ we have,

$$\begin{aligned}\Psi_{\mathscr{D}1}(A,B,T) &\leqslant \Psi_{\mathscr{D}1}(A_0,B_0,T) + (\varepsilon_1 + T\varepsilon_3)e_{\mathscr{D}}^\top e_{\mathscr{D}} \\ &\quad + \varepsilon_1^{-1}(Pe_1 + T(Qe_{12} + Re_2 + Ze_0))^\top DD^\top (Pe_1 + T(Qe_{12} + Re_2 + Ze_0)) \\ \Psi_{\mathscr{D}2}(A,B,T) &\leqslant \Psi_{\mathscr{D}2}(A_0,B_0,T) + \begin{bmatrix} \varepsilon_2 e_{\mathscr{D}}^\top e_{\mathscr{D}} + \varepsilon_2^{-1} e_1^\top PDD^\top Pe_1 & 0 \\ 0 & 0 \end{bmatrix}.\end{aligned}$$

Finally the expression of $\Theta_{\mathscr{D}1}$ and $\Theta_{\mathscr{D}2}$ are retrieved by application of the Schur complement. This shows that (9.15)–(9.17) imply the two conditions in (9.13), and also the second inequality in (9.7) is proven to hold, which completes the proof. □

9.3.3 Optimization and Computational Issues

Note first that conditions of Theorems 9.1 and 9.2 are linear in all the decision variables provided that K and T are given, as classically in an emulation problem. It is important to note that if matrix $A_0 + B_0KC$ is Hurwitz, there always exists a small enough positive scalar T such that the conditions of Theorems 9.1 or 9.2 are feasible. Indeed, the Event-Triggered rule is defined through two design parameters, which are matrix M and dwell-time T. The implicit objective is to reduce the number of control updates.

Let us observe that in the LMI $\Theta_{\mathscr{C}} < 0$ in (9.14) (see also (9.7)), the blocks $\mathrm{He}(P(A_0 + B_0KC)) - C^\top M_1 C$ and $-M_3$ are required to be negative definite. A natural optimization procedure could then consist in minimizing the effect of the off-diagonal term $PB_0 - C^\top M_2$, which could be performed by minimizing the size of the positive definite matrix M_3 appearing on the diagonal. Obtaining small values of the diagonal term $-M_3$ will indeed reduce also the off-diagonal term in (9.7). This optimization problem can be formulated in terms of an LMI optimization as follows

$$\begin{aligned} &\min_{P,M,Z,Q,U,R,Y,\varepsilon_0,\varepsilon_1,\varepsilon_2,\varepsilon_3} \quad Trace(M_3), \\ &\text{subject to:} \\ &\Theta_{\mathscr{C}} < 0, \Theta_{\mathscr{D}1} < 0, \Theta_{\mathscr{D}2} < 0, \Theta_{\mathscr{D}3} > 0 \\ &P > I,\ M_1 < 0, \end{aligned} \tag{9.20}$$

In the optimization problem (9.20), the additional constraint $P > I$ is imposed for well conditioning the LMI constraints. In addition, constraint $M_1 < 0$ has been included in order to obtain negative definiteness of $\mathrm{He}(P(A_0 + B_0KC))$ in (9.7), which avoids exponentially unstable continuous dynamics, thereby giving more graceful inter-sample transients. Furthermore, the fact of minimizing the trace of M_3 aims at increasing the negativity of matrix M_3, which leads to larger flow sets (see equation (9.6)). Since the jump set is the closed complement of the flow set, it is expected that solutions will flow longer and jump less in light of larger flow sets. Moreover, the dwell-time T being also a design parameter, whose role is connected to the expected average sampling rate of the Event-Triggered implementation, one can seek for maximizing its value through problem (9.20) by iteratively increasing T and testing the conditions.

9.4 Illustrative Example

In this example, we consider that system (9.1) is issued from the connection of a linear plant with a dynamical output feedback controller, inspired from [1, 9]. Furthermore, this example can be viewed as a complementary example to that one tackled in [33], where polytopic uncertainties were considered. Hence, we consider the following plant:

$$\begin{cases} \dot{x}_p = A_p x_p + B_p u_p = (A_{p0} + D_p F E_{p1}) x_p + (B_{p0} + D_p F E_{p2}) u_p, \\ y_p = C_p x_p, \end{cases} \tag{9.21}$$

for which the nominal data is given by

$$A_{p0} = \begin{bmatrix} 0 & 1 \\ -2 & 3 \end{bmatrix}, \ B_{p0} = \begin{bmatrix} 0 \\ 1 \end{bmatrix}, \ C_p = \begin{bmatrix} -1 & 4 \end{bmatrix} \tag{9.22}$$

and the matrices describing the norm-bounded uncertainty are defined by:

$$D_p = \begin{bmatrix} 0 \\ 1 \end{bmatrix}, E_{p1} = \begin{bmatrix} 0 & \omega_0 \end{bmatrix}, E_{p2} = 0.1\omega_0 \tag{9.23}$$

with a positive constant ω_0. The following controller obtained using an optimization process provided in [1] is also considered:

$$\begin{cases} \dot{x}_c = A_c x_c + B_c y_p, \\ u_p = C_c x_c + D_c y_p, \end{cases} \tag{9.24}$$

With plant (9.21) interconnected to the controller (9.24), denote the overall state by $x := \begin{bmatrix} x_p \\ x_c \end{bmatrix}$. The complete closed-loop system under consideration in this chapter can be reformulated as system (9.1) with

$$\left[\begin{array}{c|c} A & B \\ \hline K & C \end{array}\right] = \left[\begin{array}{cc|cc} A_p & 0 & B_p & 0 \\ 0 & A_c & 0 & B_c \\ \hline D_c & C_c & C_p & 0 \\ I & 0 & 0 & I \end{array}\right] \tag{9.25}$$

Similarly, from (9.3) one can define the nominal and uncertain parts as follows:

$$A_0 = \begin{bmatrix} A_{p0} & 0 \\ 0 & A_c \end{bmatrix}; \; B_0 = \begin{bmatrix} B_{p0} & 0 \\ 0 & B_c \end{bmatrix} \; D = \begin{bmatrix} D_p \\ 0 \end{bmatrix}; \; E_1 = \begin{bmatrix} E_{p1} & 0 \end{bmatrix}; \; E_2 = \begin{bmatrix} E_{p2} & 0 \end{bmatrix} \tag{9.26}$$

9.4.1 Nominal Case

The nominal case corresponds to setting $\omega_0 = 0$. Let us note that in [1], some improvements with respect to the literature (for instance with respect to [9]) have been reported in the nominal case. More precisely, the authors of [1] obtained a dwell-time $T = 0.0114$ s, whereas by using the conditions of Theorem 9.1, one can verify that there exist values of the design parameter T up to 0.113 s providing feasible designs. This corresponds to a parameter T ten times larger than the solution provided in [1], which well illustrates the potential of the proposed method. Moreover, it is worth pointing out that the numerical results obtained from the application of Theorem 9.1 (which is only applicable to the nominal case) are very similar to the maximal guaranteed dwell-time obtained by Theorem 9.2 specialized to the nominal case by setting $E_1 = 0$ and $E_2 = 0$. This means that the conservatism introduced by Theorem 9.2, to be able to provide an Event-Triggered algorithm for the system, is quite limited compared to Theorem 9.1.

Figure 9.1 shows several simulation results of the nominal system obtained for four dwell-time parameters T selected in Theorem 9.1. The caption of the figure also shows the number of control updates (N_u) that have been required by each Event-Triggered simulation. While increasing T leads apparently to a notable reduction of the number of control updates, it can also be seen in Fig. 9.1d that the selection of a too large guaranteed dwell-time has several drawbacks. First of all, a similar number of control updates N_u are required for simulations (c) and (d). The sampling algorithm in (c) is still able to often trigger the sampling well after $T = 0.1$ times after the previous update, and the inter-sampling time may reach up to 0.8 s. The sampling algorithm employed in (d) results in a periodic implementation of the control law. As another consequence, the simulation provided in (d) shows some undesirable oscillatory behavior that makes this emulation rule not effective with respect to some performance index.

Apart from that, the three simulations depicted in Fig. 9.1a, b, c are quite similar if one only regards the x state response. The main difference can be seen in the triggering rule and in the number of control updates N_u. Indeed, a trend can be seen in these simulations, which consists in noting that increasing the dwell-time parameter T allows to notably reduce the number of control updates, while obtaining similar responses in the x variable. Of course, regarding the previous remarks on Fig. 9.1d, increasing too much T up to the maximal feasibility value of the LMI conditions of Theorem 9.2 (or Theorem 9.1), is not a good option to obtain effective emulation algorithms.

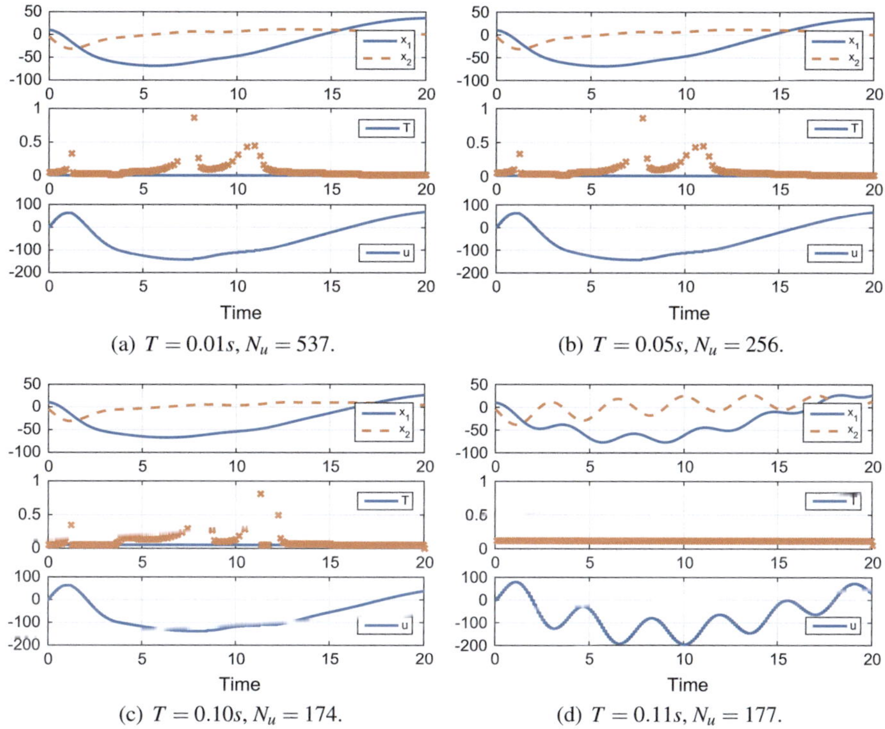

Fig. 9.1 Figure representing the state of the plant x_p, the inter-sampling times (with the dwell time T) and the control input u issued from Theorem 9.2 for several values or T for the same initial condition $x_p(0) = [10 \ -5]^\top$, $x_c(0) = [0 \ 0]^\top$

9.4.2 Uncertain Case

In order to illustrate the uncertain case, corresponding to the situation where ω_0 is not equal to 0 anymore, we have conducted the following test. For five values of ω_0 taken in the interval $[0, 0.1]$, the average number of control updates N_u obtained over 60 different initial conditions is computed for several values of T in $[0, 0.1]$. Figure 9.2 shows these simulations. One can first note that the maximal value of ω_0, for which a solution to the conditions of Theorem 9.2 can be found, depends on the dwell-time parameter T. More precisely, increasing T reduces the maximal allowable uncertainty range ω_0. For $T = 0.01$, solutions to the conditions of Theorem 9.2 can be found up to $\omega_0 = 0.1$ while for $T = 0.1$, solutions can be found up to $\omega_0 = 0.025$.

In addition, the decreasing trends shown in the figure reveal that the expected control updates, suitably averaged over the 60 initial conditions, are a decreasing function of the dwell-time parameter T.

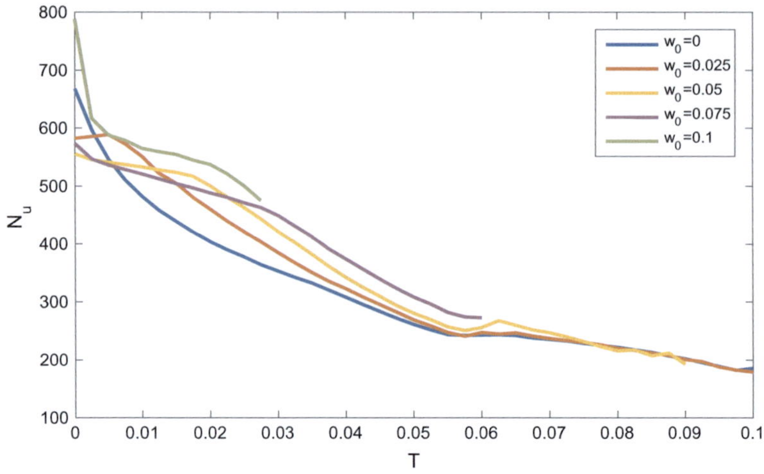

Fig. 9.2 Evolution of the average number of control updates N_u with respect to the dwell time parameter T for several values of ω_0

9.5 Conclusion

In this chapter, we have presented a method to provide efficient output-feedback Event-Triggered controls for linear systems subject to norm-bounded uncertainties. Based on an existing control law, which ensures, a priori, the stability of the associated continuous-time closed-loop system, the chapter presents several constructive theorems providing an efficient Event-Triggered sampling algorithm dedicated to the nominal and the uncertain cases. The conditions are expressed in terms of LMIs where a guaranteed dwell-time appears as a tunable parameter. The method is then evaluated on an example taken from the literature, which demonstrates the potential of the proposed solutions.

References

1. Abdelrahim, M., Postoyan, R., Daafouz, J., Nešić, D. Co-design of output feedback laws and event-triggering conditions for linear systems. In: 53rd IEEE Conference on Decision and Control, (CDC 2014), pp. 3560–3565. Los Angeles (CA), USA (2014)
2. Abdelrahim, M., Postoyan, R., Daafouz, J., Nešić, D. Stabilization of nonlinear systems using event-triggered output feedback laws. In: 21st International Symposium on Mathematical Theory of Networks and Systems. Groningen, The Netherlands (2014)
3. Årzén, K.-E. A simple event-based PID controller. In: Proceedings of the 14th IFAC World Congress, vol. 18, pp. 423–428. Beijing, China (1999)
4. Aström, K.J. Event based control. In: Analysis and Design of Nonlinear Control Systems, pp. 127–147. Springer, Berlin (2008)

5. Åström, K.J., Bernhardsson, B.: Comparison of periodic and event based sampling for first-order stochastic systems. In: Proceeding 14th IFAC World Congress, vol. 11, pp. 301–306, Beijing, China (1999)
6. Briat, C., Seuret, A.: A looped-functional approach for robust stability analysis of linear impulsive systems. Syst. Control Lett. **61**(10), 980–988 (2012)
7. Chen, T., Francis, B.A.: Optimal Sampled-data Control Systems. Springer, Berlin, Germany (1995)
8. Cloosterman, M.B.G., Hetel, L., Van De Wouw, N., Heemels, W.P.M.H., Daafouz, J., Nijmeijer, H.: Controller synthesis for networked control systems. Automatica **46**(10), 1584–1594 (2010)
9. Donkers, M.C.F., Heemels, W.P.M.H.: Output-based event-triggered control with guaranteed-gain and improved and decentralized event-triggering. IEEE Trans. Autom. Control **57**(6), 1362–1376 (2012)
10. Fichera, F., Prieur, C., Tarbouriech, S., Zaccarian, L.: A Convex Hybrid $\mathscr{H}_\infty$ synthesis with guaranteed convergence rate. In: 51st IEEE Conference on Decision and Control, pp. 4217–4222. Maui (HI), USA (2012)
11. Fichera, F., Prieur, C., Tarbouriech, S., Zaccarian, L.: Using Luenberger observers and Dwell-time logic for feedback hybrid loops in continuous-time control systems. Int. J. Robust Nonlinear Control **23**, 1065–1086 (2013)
12. Fridman, E., Seuret, A., Richard, J.-P.: Robust sampled-data stabilization of linear systems. an input delay approach. Automatica **40**(8), 1141–1446 (2004)
13. Goebel, R., Sanfelice, R., Teel, A.R.: Hybrid dynamical systems. IEEE Control Syst. Mag. **29**(2), 28–93 (2009)
14. Goebel, R., Sanfelice, R.G., Teel, A.R.: Hybrid Dynamical Systems: Modeling, Stability, and Robustness. Princeton University Press, Princeton (2012)
15. Heemels, W.P.M.H., Donkers, M.C.F., Teel, A.R.: Periodic event-triggered control for linear systems. IEEE Trans. Autom. Control **58**(4), 847–861 (2013)
16. Heemels, W.P.M.H., Johansson, K.H., Tabuada, P.: An introduction to event-triggered and self-triggered control. In: 51st IEEE Conference on Decision and Control, pp. 3270–3285. Maui (HI), USA (2012)
17. Heemels, W.P.M.H., Teel, A.R., van de Wouw, N., Nešić, D.: Networked control systems with communication constraints: tradeoffs between transmission intervals, delays and performance. IEEE Trans. Autom. Control **55**(8), 1781–1796 (2010)
18. Hespanha, J.P., Naghshtabrizi, P., Xu, Y.: A survey of recent results in networked control systems. Proc. IEEE **95**(1), 138–162 (2007)
19. Hetel, L., Fiter, C., Omran, H., Seuret, A., Fridman, E., Richard, J.-P., Niculescu, S.-I.: Recent developments on the stability of systems with aperiodic sampling: an overview. Automatica **76**, 309–335 (2017)
20. Lunze, J., Lehmann, D.: A state-feedback approach to event-based control. Automatica **46**(1), 211–215 (2010)
21. Mazo, M., Anta, A., Tabuada, P.: An ISS self-triggered implementation of linear controllers. Automatica **46**(8), 1310–1314 (2010)
22. Naghshtabrizi, P., Hespanha, J.P., Teel, A.R.: Exponential stability of impulsive systems with application to uncertain sampled-data systems. Syst. Control Lett. **57**(5), 378–385 (2008)
23. Nešić, D., Teel, A.R.: A framework for stabilization of nonlinear sampled-data systems based on their approximate discrete-time models. IEEE Trans. Autom. Control **49**(7), 1103–1122 (2004)
24. Petersen, I.R.: A stabilization algorithm for a class of uncertain linear systems. Syst. Control Lett. **8**, 351–356 (1987)
25. Postoyan, R., Girard, A.: Triggering mechanism using freely selected sensors for linear time-invariant systems. In: 54th IEEE Conference on Decision and Control, pp. 4812–4817. Osaka, Japan (2015)
26. Postoyan, R., Tabuada, P., Nešić, D., Anta, A.: Event-triggered and self-triggered stabilization of distributed networked control systems. In: 50th IEEE Conference on Decision and Control and European Control Conference (CDC-ECC), pp. 2565–2570. Orlando (FL), USA (2011)

27. Prieur, C., Goebel, R., Teel, A.R.: Hybrid feedback control and robust stabilization of nonlinear systems. IEEE Trans. Autom. Control **52**(11), 2103–2117 (2007)
28. Prieur, C., Seuret, A., Tarbouriech, S., Zaccarian, L.: Event-triggered control via reset control systems framework. In: IFAC Symposium on Nonlinear Control Systems (NOLCOS). Monterey, CA, USA (2016)
29. Prieur, C., Tarbouriech, S., Zaccarian, L.: Guaranteed stability for nonlinear systems by means of a hybrid loop. In: IFAC Symposium on Nonlinear Control Systems (NOLCOS), pp. 72–77. Bologna, Italy (2010)
30. Prieur, C., Tarbouriech, S., Zaccarian, L.: Lyapunov-based hybrid loops for stability and performance of continuous-time control systems. Automatica **49**(2), 577–584 (2013)
31. Seuret, A.: A novel stability analysis of sampled-data systems with applications to multi-rate sampling and packet loss. Automatica **48**(1), 177–182 (2012)
32. Seuret, A., Prieur, C.: Event-based sampling algorithms based on a Lyapunov function. In: 50th IEEE Conference on Decision and Control and European Control Conference (CDC-ECC), pp. 6128–6133. Orlando (FL), USA (2011)
33. Seuret, A., Prieur, C., Tarbouriech, S., Zaccarian, L., Teel, A.R.: A nonsmooth hybrid invariance principle applied to robust event-triggered design. Available: https://hal.laas.fr/hal-01526331 (2017) [Online]
34. Tabuada, P.: Event-triggered real-time scheduling of stabilizing control tasks. IEEE Trans. Autom. Control **52**(9), 1680–1685 (2007)
35. Tallapragada, P., Chopra, N.: Event-triggered dynamic output feedback control for LTI systems. In: 51st IEEE Conference on Decision and Control, pp. 6597–6602. Maui (HI), USA (2012)
36. Tarbouriech, S., Seuret, A., Gomes da Silva Jr., J.M., Sbarbaro, D. Observer-based event-triggered control co-design for linear systems. IET Control Theory Appl. **10**(18), 2466–2473 (2016)
37. Wang, X., Lemmon, M.D.: Event design in event-triggered feedback control systems. In: 47th IEEE Conference on Decision and Control, pp. 2105–2110. Cancun, Mexico (2008)
38. Zampieri, S.: A survey of recent results in networked control systems. In: Proceedings of the 17th IFAC World Congress, pp. 2886–2894. Seoul, Korea (2008)
39. Zhou, K., Doyle, J.C., Glover, K.: Robust and Optimal Control. Prentice Hall Inc., Upper Saddle River, New Jersey (1996)

Chapter 10
Abstracted Models for Scheduling of Event-Triggered Control Data Traffic

M. Mazo Jr., A. Sharifi-Kolarijani, D. Adzkiya and C. Hop

Abstract Event-Triggered control (ETC) implementations have been proposed to overcome the inefficiencies of periodic (time-triggered) controller designs, namely the over-exploitation of the computing and communication infrastructure. However, the potential of aperiodic Event-Triggered techniques to reuse the freed bandwidth, and to reduce energy consumption on wireless settings, has not yet been truly reached. The main limitation to fully exploit ETC's great traffic reductions lies on the difficulty to predict the occurrence of controller updates, forcing the use of conservative scheduling approaches in practice. Having a model of the timing behaviour of ETC is of paramount importance to enable the construction of model-based schedulers for such systems. Furthermore, on wireless control systems these schedulers allow to tightly schedule listening times, thus reducing energy consumption. In this chapter we describe an approach to model ETC traffic employing ideas from the symbolic abstractions literature. The resulting models of traffic are timed-automata. We also discuss briefly how these models can be employed to automatically synthesize schedulers.

M. Mazo Jr. (✉) · A. Sharifi-Kolarijani · C. Hop
Delft University of Technology, Delft, The Netherlands
e-mail: m.mazo@tudelft.nl

A. Sharifi-Kolarijani
e-mail: a.sharifikolarijani@tudelft.nl

C. Hop
e-mail: c.m.j.hop@student.tudelft.nl

D. Adzkiya
Institut Teknologi Sepuluh Nopember, Surabaya, Indonesia
e-mail: dieky@matematika.its.ac.id

S. Tarbouriech et al. (eds.), *Control Subject to Computational and Communication Constraints*, Lecture Notes in Control and Information Sciences 475, https://doi.org/10.1007/978-3-319-78449-6_10

10.1 Introduction

A surge of Event-Triggered control (ETC) implementation strategies has appeared in the last decade promising to alleviate the inefficiencies of periodic (time-triggered) controller designs. Periodic controller implementations abuse the computing and communication infrastructures employing periodic feedback independently of the current state of the system. In ETC these inefficiencies are mitigated by letting the sensors decide (employing their limited computation capabilities) whether a measurement is worth transmitting for the computation of corrective actions. This results in aperiodic transmissions of measurements and computation of corrective actions. The amount of transmissions is often orders of magnitude smaller than the amount required to achieve a similar performance with periodic control.

However, this potential of aperiodic Event-Triggered techniques, to release channel capacity and reduce energy consumption on wireless settings, has not yet been truly reached. See e.g., [1] in which a tenfold traffic reduction resulted in barely a 57% energy consumption reduction of a wireless sensing infrastructure. The main limitation to fully exploit ETC's great traffic reductions lies on the difficulty to predict the occurrence of controller updates. Having a model of the timing behaviour of ETC is therefore of paramount importance to enable the construction of model-based schedulers for such systems, and to tightly schedule listening times on wireless communications to reduce energy consumption.

Different from controller/scheduler co-design approaches, see e.g., [2–4], we suggest to retain the separation of concerns between controller design and scheduling by defining a proper interface between these two realms. We propose a formal approach to derive models that capture the timing behaviour (of controller updates) of a family of Event-Triggered strategies for Linear Time-Invariant systems [5]. The constructed models provide an over-approximation of all the updates' timing behaviours generated by the aperiodic ETC system. Then, techniques from games over timed automata (TA) can be leveraged to synthesise schedulers [6].

Inspired by the state-dependent sampling proposed in [7], we employ a two-step approach to compute sampling intervals associated to states: first, the state space is partitioned (abstracted) into a finite number of convex polyhedral cones (pointed at the origin); then, for each conic region the time interval in which events can be originated is computed using a convex embedding approach [8] and Linear Matrix Inequalities derived from Lyapunov conditions. In the resulting timing models, transitions among discrete states (associated to the conic regions) are derived through reachability analysis over the sampling intervals computed earlier, see e.g., [9].

Furthermore, we show that the resulting models of traffic can be alternatively encoded as TA [10]. This allows us to additionally address the problem of scheduling the access to a shared resource by multiple ETC systems by solving games over TA [6, 11]. To this end, we enrich the constructed models of ETC traffic with actions that trade communication traffic for control performance. Then, a scheduler is synthesised as a strategy providing actions (for one player) that prevents the set of ETC tasks and shared resource (the other player) from entering in a conflict situation. Useful classes of games for such a synthesis are readily solvable in tools like UPPAAL-TIGA [12].

10.2 Mathematical Preliminaries

We start by introducing some common notation employed throughout the chapter, and the basic theoretical notions of finite-state abstractions, timed automata, and Event-Triggered control that support the remainder of the chapter.

10.2.1 Notation

We use $\mathbb{R}^n$ to denote the n-dimensional Euclidean space, $\mathbb{R}^+$ and $\mathbb{R}_0^+$ to denote the positive and nonnegative reals, respectively, $\mathbb{N}$ is the set of positive integers, and $\mathbb{IR}^+$ is the set of all closed intervals $[a, b]$ such that $a, b \in \mathbb{R}^+$ and $a \leqslant b$. For any set S, we denote by 2^S the set of all subsets of S, i.e., the power set of S. The sets of all $m \times n$ real-valued matrices and the set of all $n \times n$ real-valued symmetric matrices are denoted by $\mathcal{M}_{m\times n}$ and $\mathcal{M}_n$, respectively. Given a matrix M, $M \preceq 0$ (or $M \succeq 0$) indicates that M is a negative (or positive) semidefinite matrix and $M \prec 0$ (or $M \succ 0$) denotes M is a negative (positive) definite matrix. For a given matrix M, we denote by $[M]_{(i,j)}$ its i-th row, j-th column entry. The largest integer not greater than $x \in \mathbb{R}$ is denoted by $\lfloor x \rfloor$ and $|y|$ denotes the Euclidean norm of a vector $y \in \mathbb{R}^n$. Given two sets Z_a and Z_b, every relation $Q \subseteq Z_a \times Z_b$ admits $Q^{-1} = \{(z_b, z_a) \in Z_b \times Z_a | (z_a, z_b) \in Q\}$ as its inverse relation. For $Q \subseteq Z \times Z$, an equivalence relation on a set Z, $[z]$ denotes the equivalence class of $z \in Z$ and Z/Q denotes the set of all equivalence classes. For a set $A \subseteq \mathbb{R}^n$ we denote its Lebesgue measure by $\mu(A)$.

Given an ordinary differential equation of the form $\dot{\xi}(t) = f(\xi(t))$, admitting a unique solution, we denote by $\xi_x : \mathbb{R}_0^+ \to \mathbb{R}^n$ the solution to the initial value problem with $\xi_x(0) = x$. Finally, we also employ the notion of *flow pipe*:

Definition 10.1 (*Flow Pipe* [9]) The set of reachable states, or flow pipe, from an initial set $X_{0,s}$ in the time interval $[\underline{\tau}_s, \bar{\tau}_s]$ is denoted by:

$$\mathscr{X}_{[\underline{\tau}_s, \bar{\tau}_s]}(X_{0,s}) = \bigcup_{t \in [\underline{\tau}_s, \bar{\tau}_s]} \mathscr{X}_t(X_{0,s}) = \bigcup_{t \in [\underline{\tau}_s, \bar{\tau}_s]} \{\xi_{x_0}(t) \,|\, x_0 \in X_{0,s}\}. \tag{10.1}$$

10.2.2 Symbolic Abstractions

We revise in the following the framework from [13] to relate different models of a system.

Definition 10.2 (*Generalized Transition System* [13])
A system S is a tuple $(X, X_0, U, \longrightarrow, Y, H)$ consisting of:

- a set of states X;
- a set of initial states $X_0 \subseteq X$;
- a set of inputs U;
- a transition relation $\longrightarrow \subseteq X \times U \times X$;
- a set of outputs Y;
- an output map $H : X \to Y$.

We say a system is finite-state (or infinite-state) when X is a finite (or infinite) set, and that the system S is metric if the output set Y is a metric space (with some appropriately defined metric).

Definition 10.3 (*Approximate Simulation Relation* [13]) Consider two metric systems S_a and S_b with $Y_a = Y_b$, and let $\varepsilon \in \mathbb{R}_0^+$. A relation $R \subseteq X_a \times X_b$ is an ε-approximate simulation relation from S_a to S_b if the following three conditions are satisfied:

1. $\forall x_{a0} \in X_{a0}, \exists x_{b0} \in X_{b0}$ such that $(x_{a0}, x_{b0}) \in R$;
2. $\forall (x_a, x_b) \in R$ we have $d(H_a(x_a), H_b(x_b)) \leqslant \varepsilon$;
3. $\forall (x_a, x_b) \in R$ for all $(x_a, u_a, x_a') \in \underset{a}{\longrightarrow}$, $\exists (x_b, u_b, x_b') \in \underset{b}{\longrightarrow}$ satisfying $(x_a', x_b') \in R$.

Whenever an ε-approximate simulation relation from S_a to S_b exists we write $S_a \preceq^\varepsilon S_b$, and say that S_b ε-approximately simulates S_a. Intuitively, under some technical conditions, $S_a \preceq^\varepsilon S_b$ implies that all possible output sequences that S_a can produce are contained in the set of output sequences that S_b can generate.

Let us also introduce the following alternative notion of *quotient system* (see e.g., [13] for the traditional definition):

Definition 10.4 (*Power Quotient System* [5]) Let $S = (X, X_0, U, \longrightarrow, Y, H)$ be a system and R be an equivalence relation on X. The power quotient of S by R, denoted by $S_{/R}$, is the system $(X_{/R}, X_{/R,0}, U_{/R}, \underset{/R}{\longrightarrow}, Y_{/R}, H_{/R})$ consisting of:

- $X_{/R} = X/R$;
- $X_{/R,0} = \{x_{/R} \in X_{/R} | x_{/R} \cap X_0 \neq \varnothing\}$;
- $U_{/R} = U$;
- $(x_{/R}, u, x'_{/R}) \in \underset{/R}{\rightarrow}$ if $\exists (x, u, x') \in \rightarrow$ in S with $x \in x_{/R}$ and $x' \in x'_{/R}$;
- $Y_{/R} \subset 2^Y$;
- $H_{/R}(x_{/R}) = \bigcup_{x \in x_{/R}} H(x)$.

In the case considered in this chapter we rarely are able to compute such power quotient systems. In particular, the transition relation and output maps in general need to be over-approximated. We introduce the following relaxed version of the previous definition, followed by a Lemma establishing the relation between such quotient systems and the original concrete system.

Definition 10.5 (*Approximate Power Quotient System* [5])
Let $S = (X, X_0, U, \longrightarrow, Y, H)$ be a system, R be an equivalence relation on X, and $S_{/R} = (X_{/R}, X_{/R,0}, U_{/R}, \underset{/R}{\longrightarrow}, Y_{/R}, H_{/R})$ be the power quotient of S by R.

An approximate power quotient of S by R, denoted by $\bar{S}_{/R}$, is a system $(X_{/R}, X_{/R,0}, U_{/R}, \underset{\bar{/R}}{\longrightarrow}, \bar{Y}_{/R}, \bar{H}_{/R})$ such that:

- $\underset{\bar{/R}}{\rightarrow} \supseteq \underset{/R}{\rightarrow}$,
- $\bar{Y}_{/R} \supseteq Y_{/R}$, and
- $\bar{H}_{/R}(x_{/R}) \supseteq H_{/R}(x_{/R}), \forall x_{/R} \in X_{/R}$,

Lemma 10.1 [5] *Let S be a metric system, R be an equivalence relation on X, and let the metric system $\bar{S}_{/R}$ be the approximate power quotient system of S by R. For any*

$$\varepsilon \geqslant \max_{\substack{x \in x_{/R} \\ x_{/R} \in X_{/R}}} d(H(x), \bar{H}_{/R}(x_{/R})),$$

with d the Hausdorff distance over the set 2^Y, $\bar{S}_{/R}$ ε-approximately simulates S, i.e., $S \preceq_S^{\varepsilon} \bar{S}_{/R}$.

For any set Y, $Y \in 2^Y$, which allows us to employ the Hausdorff distance [14] as a common metric for output sets of the power quotient and the original system.

10.2.3 Timed Safety and Timed Game Automata

The abstraction methodology we propose results in models semantically equivalent to Timed Safety Autamata (TSA) [15]. TSA are a simplified version of the classical *timed automata* [10] (TA). While TA employ *Büchi-acceptance* conditions to specify progress properties, in TSA local invariant conditions are employed to this same end (see [16, Sect. 2] for a detailed discussion). Here, we just recall briefly the definition of TSA from [16]. Let Σ be a finite alphabet of actions, and $\mathscr{C}$ a set of finitely many real-valued variables employed to represent clocks. Consider $\sim \in \{>, \geqslant, <, \leqslant\}$, a clock constraint δ is a conjunctive formula of atomic constraints $c_1 \sim k$ or $c_1 - c_2 \sim k$ for $c_1, c_2 \in \mathscr{C}$, and $k \in \mathbb{N}$. We employ $\mathscr{B}(\mathscr{C})$ to denote the set of all possible clock constraints.

Definition 10.6 (*Timed Safety Automata* [16]) A *timed safety automata* is a tuple $\mathscr{A} = (L, L_0, \Sigma, \mathscr{C}, E, I)$ where

- L is a finite set of locations (or discrete states);
- $L_0 \subseteq L$ is a set of start locations;
- Σ is the set of actions;
- $\mathscr{C}$ is the set of clocks;
- $E \subseteq L \times \mathscr{B}(\mathscr{C}) \times \Sigma \times 2^{\mathscr{C}} \times L$ is the set of transitions.
- $I : L \longrightarrow \mathscr{B}(\mathscr{C})$ assigns invariants to locations.

The shorthand notation $l \xrightarrow{g,a,r} l'$ is used to denote $(l, g, a, r, l') \in E$, i.e., a transition from state l to state l' under input symbol a, with $r \subseteq \mathscr{C}$ the set of clocks reset when this transition is taken, and a clock constraint g over $\mathscr{C}$ as the guard enabling this transition.

Definition 10.7 (*Operational Semantics* [16]) The semantics of a timed safety automaton is a transition system (also known as timed transition system) where states are pairs (l, u), with $l \in L$ and u a clock valuation, and transitions are defined by the rules:

- $(l, u) \xrightarrow{d} (l, u + d)$ if $u \models I(l)$ and $(u + d) \in I(l)$ for a scalar $d \in \mathbb{R}^+$;
- $(l, u) \xrightarrow{a} (l', u')$ if $l \xrightarrow{g,a,r} l'$, $u \models g$, $u' = [r \to \mathbf{0}]u$ and $u' \models I(l')$.

Remark 10.1 In a TSA, the *guards* and *invariants* assert necessary and sufficient conditions respectively for transitions to take place. The sufficient conditions established by *invariants* must not be violated by letting time advance. Therefore, invariants establish upper bounds for the time to take the next transition [15].

Remark 10.2 Note that a timed automaton is a particular class of hybrid automata in which the only allowed continuous dynamics are of the form $\dot{c} = 1$, and in which *guard* and *invariant* sets are in the form of clock constraints.

TSA evolve over uncountable state spaces, due to its clock variables. Nonetheless, it has been shown that its reachability analysis is decidable [10]. This decidability allows the development of powerful tools for verification and synthesis [10, 17], which can be used to generate schedulers for real-time systems, whose timing is modelled as TSAs [6].

Timed Game Automata (TGA) are an extension of TSA where the set of actions is partitioned into controllable actions (activated by the controller) and uncontrollable actions (activated by the environment or an opponent).

Definition 10.8 (*Timed Game Automaton* [18]) A timed game automaton is a tuple $\mathscr{G} = (L, L_0, \Sigma_c, \Sigma_u, \mathscr{C}, E, I)$ where

- $(L, L_0, \Sigma_c \cup \Sigma_u, \mathscr{C}, E, I)$ is a timed safety automaton;
- Σ_c is a set of controllable actions;
- Σ_u is a set of uncontrollable actions;
- $\Sigma_c \cap \Sigma_u = \emptyset$.

A feature of TGA is its modularity. Constructing a TGA for complex systems can be done by constructing a TGA for each part and then combining (or composing) them. The resulting object is denoted a Network of TGA (NTGA), and is constructed through a synchronized parallel composition in which uncontrollable inputs of a TGA are linked to outputs of another TGA. For a more detailed discussion of such composition we refer the reader to [19].

10.2.4 Event-Triggered Control for LTI Systems

We describe next a basic framework for Event-Triggered control in the case of Linear Time Invariant (LTI) control systems with state-feedback. The techniques we present in the remainder of the paper only focus on this class of control systems. Many extensions to this simple framework have been proposed, see for instance the rest of the papers in Part 2 of this book and references therein. See the conclusion of the chapter for a brief discussion on generalizing the current results.

Consider an LTI system without disturbances:

$$\dot{\xi}(t) = A\xi(t) + B\upsilon(t),\ \ \xi(t) \in \mathbb{R}^n,\ \upsilon(t) \in \mathbb{R}^m \tag{10.2}$$

and a linear state-feedback controller implemented in a sample-and-hold fashion:

$$\upsilon(t) = \upsilon(t_k) = K\xi(t_k),\ \ \forall t \in [t_k, t_{k+1}),\ \ k \in \mathbb{N}. \tag{10.3}$$

The following quadratic triggering mechanism:

$$t_{k+1} := \inf\{t > t_k |\ |\xi(t_k) - \xi(t)|^2 \geqslant \alpha|\xi(t)|^2\}, \tag{10.4}$$

with $\alpha \in \mathbb{R}^+$ a design parameter properly selected, renders the closed-loop system asymptotically stable [20]. Let us denote the inter-sample time associated to a state by:

$$\tau(x) := t_{k+1} - t_k, \quad \text{with } x = \xi(t_k). \tag{10.5}$$

For LTI systems, the solutions ξ in some time interval $[t_k, t_k + \sigma]$ can be easily expressed in terms of the initial condition:

$$\xi(t_k + \sigma) = \Lambda(\sigma)\xi(t_k), \tag{10.6}$$

$$\Lambda(\sigma) = [I + \textstyle\int_0^\sigma e^{Ar} dr(A + BK)]. \tag{10.7}$$

Thus, the state-dependent inter-sampling times can be rewritten as:

$$\tau(x) = \inf\{\sigma > 0 |\ x^T \Phi(\sigma) x \geqslant 0\}, \tag{10.8}$$

$$\Phi(\sigma) = [I - \Lambda^T(\sigma)][I - \Lambda(\sigma)] - \alpha \Lambda^T(\sigma)\Lambda(\sigma). \tag{10.9}$$

10.3 Timing Abstractions of Event-Triggered Control Systems

We are interested in obtaining models capturing the evolution over time of the inter-sample times generated by an ETC loop. Such dynamics are actually provided by the following system:

$$S = (X, X_0, U, \longrightarrow, Y, H)$$

where

- $X = \mathbb{R}^n$;
- $X_0 \subseteq \mathbb{R}^n$;
- $U = \varnothing$, i.e., the system is autonomous;
- $\longrightarrow \in X \times U \times X$ such that $\forall x, x' \in X : (x, x') \in \longrightarrow$ iff $\xi_x(\tau(x)) = x'$;
- $Y \subset \mathbb{R}^+$;
- $H : \mathbb{R}^n \to \mathbb{R}^+$ where $H(x) = \tau(x)$.

The system S generates as output sequences all possible sequences of inter-sampling intervals that a given ETC loop can exhibit. However, S is an infinite-state system and the map H (a copy of τ) is not an explicit function.

Problem 10.1 We seek to construct finite-state systems capturing, up to some computable precision, all the possible traffic patterns of an ETC system, i.e., all possible sequences $\{\tau(\xi(t_k))\}_{k \in \mathbb{N}}$.

In order to solve this problem, we propose to abstract the system S by a power quotient system $S_{/R}$ as follows:

$$S_{/R} = (X_{/R}, X_{0/R}, U_{/R}, \underset{/R}{\longrightarrow}, Y_{/R}, H_{/R})$$

where

- $X_{/R} = \mathbb{R}^n_{/R} := \{\mathscr{R}_1, \ldots, \mathscr{R}_q\}$;
- $X_{/R,0} = \{\mathscr{R}_i \mid X_0 \cap \mathscr{R}_i \neq \emptyset\}$;
- $U_{/R} = \varnothing$, i.e., the system is autonomous;
- $(x_{/R}, x'_{/R}) \in \underset{/R}{\longrightarrow}$ if $\exists x \in x_{/R}, \exists x' \in x'_{/R}$ such that $\xi_x(H(x)) = x'$;
- $Y_{/R} \subset 2^Y \subset \mathrm{I\!R}^+$;
- $H_{/R}(x_{/R}) = [\inf_{x \in x_{/R}} H(x),\ \sup_{x \in x_{/R}} H(x)] := [\underline{\tau}_{x_{/R}}, \bar{\tau}_{x_{/R}}]$.

In general, constructing such a power quotient system is not possible, thus we focus on constructing an approximate power quotient system $\bar{S}_{/R}$.

Remark 10.3 By Lemma 10.1 we know that $S \preceq_S^{\varepsilon} \bar{S}_{/R}$. This can be interpreted as the (approximate) power quotient system producing output sequences $\{T_k\}_{k\in\mathbb{N}}$, $T_k \in \mathbb{IR}^+$, such that $\tau(\xi(t_k)) \in T_k$. The fact that all the possible *timing sequences* $\{t_{k+1} - t_k\}_{k\in\mathbb{N}}$ of the ETC system, i.e., output sequences of the infinite system S, are captured by $\bar{S}_{/R}$ allows us to employ these abstractions to synthesize *schedulers*, cf. Sect. 10.4.

In the remaining of this section we describe how to select an appropriate equivalence relation R, compute the intervals $[\underline{\tau}_{x_{/R}}, \bar{\tau}_{x_{/R}}]$, and determine the transition relation $\underset{/\bar{R}}{\longrightarrow}$.

10.3.1 State Set

In the traditional construction of quotient systems, one bundles together states that produce the same output. In our proposed construction of power quotient systems this is no longer the case, but the precision ε achieved depends on how close are the outputs of states bundled together, cf. Lemma 10.1. In the case of LTI Event-Triggered systems one can easily characterize states that produce the same output, i.e., states x, x' such that $\tau(x) = \tau(x')$, see e.g., [7, 21]:

Proposition 10.1 *States lying on the same ray crossing the origin have the same inter-sample time, i.e., $\tau(x) = \tau(\lambda x)$, $\forall \lambda \neq 0, x \neq 0$.*

Hence, to construct a quotient system of S, the abstract states need to be rays in the state space at hand. But there is an infinite number of rays, thus, in order to obtain a finite state abstraction, we suggest to take as abstract states unions of an infinite number of such rays. In particular, polyhedral cones pointed at the origin are a choice which makes the construction of finite state-space partitions relatively easy. We denote such cones by $\mathcal{R}_s$ where $s \in \{1, \ldots, q\}$ and $\bigcup_{s=1}^{q} \mathcal{R}_s = \mathbb{R}^n$ (see Fig. 10.1 for an example in $\mathbb{R}^2$).

In order to construct such a partition of the state-space, we use a so called *isotropic covering*. Consider first the case of partitioning $\mathbb{R}^2$ via cones pointed at the origin. This is easily achieved by first splitting the interval $\Theta = [-\frac{\pi}{2}, \frac{\pi}{2})$ uniformly in a number of sub-intervals $\Theta_s = [\underline{\theta}_s, \overline{\theta}_s)$. Then for each of those intervals one can construct the corresponding cone as:

$$\mathcal{R}_s = \{x \in \mathbb{R}^2 \mid x^T Q_s x \geqslant 0\}, \; Q_s = \frac{1}{2}\begin{bmatrix} -2\sin\underline{\theta}_s \sin\overline{\theta}_s & \sin(\underline{\theta}_s + \overline{\theta}_s) \\ \sin(\underline{\theta}_s + \overline{\theta}_s) & -2\cos\underline{\theta}_s \cos\overline{\theta}_s \end{bmatrix}. \tag{10.10}$$

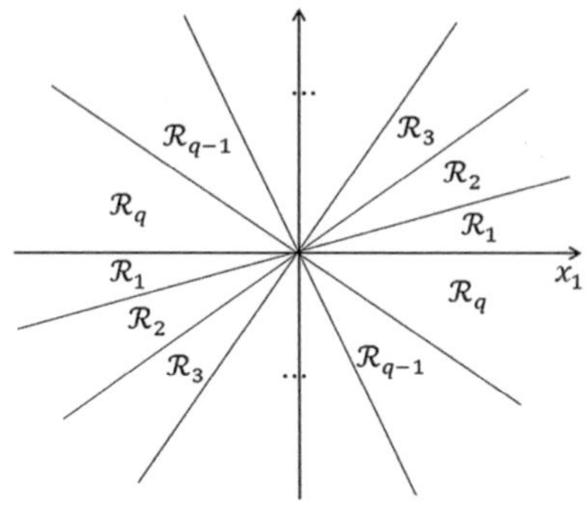

Fig. 10.1 Example of a state-space partitioning with polyhedral cones in $\mathbb{R}^2$ [5]

Remark 10.4 Note that even though it may look like we only partition in this form half of the space (as we only ranged in the polar coordinates between $[-\frac{\pi}{2}, \frac{\pi}{2})$ radians), in reality the other half space is covered by this same $\mathscr{R}_s$ sets. To see this, just observe that a point defined in polar coordinates by the pair (r, θ), i.e., $x_1 = r\cos\theta,\ x_2 = r\sin\theta$, and the point $(-r, \theta)$ (or alternatively $(r, \theta + \pi)$) belong to the same set $\mathscr{R}_s$ as $x^T Q_s x = (-x)^T Q_s(-x)$. Furthermore, this poses no problem in terms of the times associated to the set as $\tau(x) = \tau(-x)$ from Proposition 10.1.

We can generalize this partitioning approach to cover arbitrary higher dimensions as follows. Consider a point $x = [x_1,\ x_2, \dots x_n]^T \in \mathbb{R}^n$, and define the projection of that point on its $i - j$ coordinates as $(x)_{(i,j)} = (x_i, x_j)$. Now, let the sets defining the partition of the state-space to be defined as:

$$\mathscr{R}_{(s_1, s_2, \dots s_{n-1})} = \left\{ x \in \mathbb{R}^n \mid \bigwedge_{i=1}^{n-1} (x)^T_{(i,i+1)} Q_{s_i} (x)_{(i,i+1)} \geqslant 0 \right\}. \tag{10.11}$$

By ranging over all possible indices $s = (s_1, s_2, \dots s_{n-1}) \in \{1, 2, \dots, m\}^{n-1}$ the whole state space can be covered. Here m denotes the number of intervals employed to subdivide $[-\frac{\pi}{2}, \frac{\pi}{2})$ in constructing the Q_s matrices.

The equivalence relation $R \subseteq \mathbb{R}^n \times \mathbb{R}^n$ in Definition 10.5 is thus given by $(x, x') \in R \Leftrightarrow x, x' \in \mathscr{R}_s$, for some s.

10.3.2 Output Map

Constructing the output map $\bar{H}_{/R}$ (and the associated output set $\bar{Y}_{/R}$) boils down to computing the time intervals $[\underline{\tau}_s, \bar{\tau}_s]$ such that $\forall x \in \mathscr{R}_s : \tau(x) \in [\underline{\tau}_s, \bar{\tau}_s]$. In other words, we need to compute lower and upper bounds on the inter-sample times that can be observed for different states (among the infinite number) in a region $\mathscr{R}_s$.

Employing the state transition matrix in (10.9), one can express a necessary condition for $\underline{\tau}$ to be a lower bound of $\tau(x)$ as:

$$x^T \Phi(\sigma)x \leqslant 0, \forall \sigma \in [0, \underline{\tau}] \Rightarrow \underline{\tau}_s \leqslant \tau(x).$$

Note that this condition involves a matrix functional $\Phi(\sigma)$ and thus it cannot be directly checked. To address this issue, we employ the approach from Hetel et al. [8] to construct a convex polytope (in the space of matrices) containing $\Phi(\sigma)$. Then, employing convexity, one can replace the condition involving an infinite number of matrices $\Phi(\sigma)$ by a finite (and thus computable) set of inequalities involving only a finite set of matrices $\underline{\Phi}_{\kappa}$, with $\kappa \in \mathcal{K}$, i.e., :

$$(x^T \underline{\Phi}_{\kappa} x \leqslant 0, \forall \kappa \in \mathcal{K}) \Longrightarrow (x^T \Phi(\sigma)x \leqslant 0, \forall \sigma \in [0, \underline{\tau}_s]). \tag{10.12}$$

Assumption 10.1 Assume that a scalar $\bar{\sigma} > 0$ exists such that $x^T \Phi(\bar{\sigma})x \geqslant 0, \forall x \in \mathbb{R}^n$.

Remark 10.5 This constant $\bar{\sigma}$ is a global upper bound for the inter-sample times. It can be computed through a line search until the matrix $\Phi(\bar{\sigma})$ becomes positive definite. In general, such an upper-bound may not exist. Think e.g., of a stable real eigenvector v of the open-loop system and a controller setting the control action for $u = Kv = 0$, in this case the system may not trigger new controller updates. A simple solution to this issue is to modify the triggering condition by fixing a certain upper bound for the triggering times $\bar{\sigma}$:

$$\tau(x) = \min\{\bar{\sigma}, \sigma > 0 |\ x^T \Phi(\sigma)x \geqslant 0\}. \tag{10.13}$$

The following Lemma provides the construction of a finite number of LMIs to numerically check condition (10.12). Consider a positive integer $N_{conv} \geqslant 0$ such that $N_{conv} + 1$ is the number of vertices of the polytope employed to cover $\Phi(\sigma)$ in the time interval $[0, \bar{\sigma}]$, and an integer number $l \geqslant 1$ for the number of intervals in which to divide the cover, see Fig. 10.2 for an intuitive illustration.

Lemma 10.2 [5] *Consider a time bound* $\underline{\tau} \in (0, \bar{\sigma}]$. *If* $x^T \underline{\Phi}_{(i,j),\underline{\tau}} x \leqslant 0$ *holds* $\forall (i, j) \in \mathcal{K}_{\underline{\tau}} = (\{0, \ldots, N_{conv}\} \times \{0, \ldots, \lfloor \frac{\underline{\tau} l}{\bar{\sigma}} \rfloor\})$, *then:*

$$x^T \Phi(\sigma)x \leqslant 0, \quad \forall \sigma \in [0, \underline{\tau}]$$

with Φ *defined in* (10.9) *and*

$$\underline{\Phi}_{(i,j),\underline{\tau}} = \underline{\hat{\Phi}}_{(i,j),\underline{\tau}} + \underline{\nu} I,$$

$$\underline{\hat{\Phi}}_{(i,j),\underline{\tau}} = \begin{cases} \sum_{k=0}^{i} L_{k,j} (\frac{\bar{\sigma}}{l})^k & \text{if } j < \lfloor \frac{\underline{\tau} l}{\bar{\sigma}} \rfloor, \\ \sum_{k=0}^{i} L_{k,j} (\underline{\tau} - \frac{j\bar{\sigma}}{l})^k & \text{if } j = \lfloor \frac{\underline{\tau} l}{\bar{\sigma}} \rfloor, \end{cases}$$

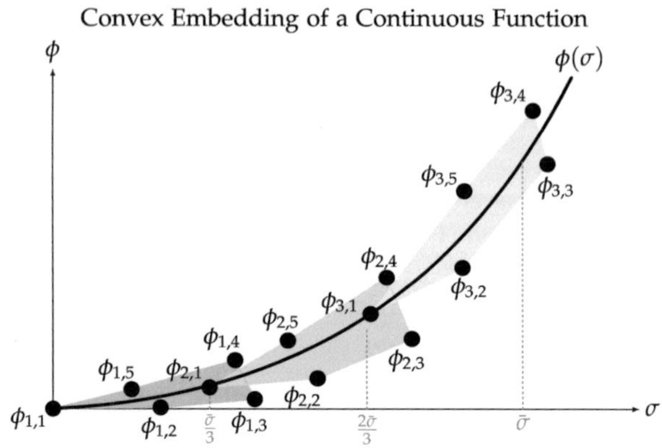

Fig. 10.2 Polytopic bounding of a (scalar) exponential function, with $N_{conv} = 4$ and $l = 3$

$$\begin{cases} L_{0,j} &= I - \Pi_{1,j} - \Pi_{1,j}^T + (1-\alpha)\Pi_{1,j}^T \Pi_{1,j}, \\ L_{1,j} &= [(1-\alpha)\Pi_{1,j}^T - I]\Pi_{2,j} \\ & \quad +\Pi_{2,j}^T[(1-\alpha)\Pi_{1,j} - I], \\ L_{k\geqslant 2,j} &= [(1-\alpha)\Pi_{1,j}^T - I]\frac{A^{k-1}}{k!}\Pi_{2,j} \\ & \quad +\Pi_{2,j}^T \frac{(A^{k-1})^T}{k!}[(1-\alpha)\Pi_{1,j} - I] \\ & \quad +(1-\alpha)\Pi_{2,j}^T(\sum_{i=1}^{k-1} \frac{(A^{i-1})^T}{i!} \frac{A^{k-i-1}}{(k-i)!})\Pi_{2,j}, \end{cases} \tag{10.14}$$

$$\begin{cases} \Pi_{1,j} = I + M_j(A+BK), & M_j = \int_0^{j\frac{\bar{\sigma}}{l}} e^{As} ds, \\ \Pi_{2,j} = N_j(A+BK), & N_j = AM_j + I, \end{cases} \tag{10.15}$$

$$\underline{\nu} \geqslant \max_{\substack{\sigma' \in [0, \frac{\bar{\sigma}}{l}] \\ r \in \{0,\dots,l-1\}}} \lambda_{\max}(\Phi(\sigma' + r\tfrac{\bar{\sigma}}{l}) - \tilde{\Phi}_{N_{conv},r}(\sigma')), \tag{10.16}$$

$$\tilde{\Phi}_{N_{conv},r}(\sigma) = \textstyle\sum_{k=0}^{N_{conv}} L_{k,r}\sigma^k. \tag{10.17}$$

For a given state x we can now employ this result to compute a lower bound on $\tau(x)$. In order to compute a lower bound $\underline{\tau}_s$ for the bundle of states defined by a conic region $\mathscr{R}_s$, one can leverage the S-procedure as in the following theorem. Before stating the result, we need to define some new set of matrices $\tilde{Q}_s^{(i,j)}$ as:

$$\tilde{Q}_s^{(i,j)} \in \mathscr{M}_n, \text{ such that } \begin{cases} [\tilde{Q}_s^{(i,j)}]_{(i,i)} = [Q_s]_{(1,1)} \\ [\tilde{Q}_s^{(i,j)}]_{(i,j)} = [Q_s]_{(1,2)} \\ [\tilde{Q}_s^{(i,j)}]_{(j,i)} = [Q_s]_{(2,1)} \\ [\tilde{Q}_s^{(i,j)}]_{(j,j)} = [Q_s]_{(2,2)} \\ [\tilde{Q}_s^{(i,j)}]_{(k,l)} = 0 & \text{otherwise} \end{cases} \tag{10.18}$$

where $Q_s \in \mathscr{M}_2$ are as defined in (10.10).

Theorem 10.2 (Regional Lower Bound Approximation) *Consider a scalar* $\underline{\tau}_s \in (0, \bar{\sigma}]$ *and matrices* $\underline{\Phi}_{\kappa,\underline{\tau}_s}$, $\kappa = (i, j) \in \mathcal{K}_{\underline{\tau}_s}$, *defined as in Lemma 10.2. If there exist scalars* $\underline{\varepsilon}_{\kappa,\underline{\tau}_s} \geqslant 0$ *such that for all* $\kappa \in \mathcal{K}_{\underline{\tau}_s}$ *the following LMIs hold:*

$$\underline{\Phi}_{\kappa,\underline{\tau}_s} + \sum_{i=1}^{n-1} \underline{\varepsilon}_{\kappa,s_i} \tilde{Q}_{s_i}^{(i,i+1)} \preceq 0$$

the inter-sample time (10.4) *of the system* (10.2)–(10.3) *is regionally bounded from below by* $\underline{\tau}_s$, $\forall x \in \mathcal{R}_s$.

Proof The proof is verbatim the proof on [5], replacing the linear representation of conic partitions in dimensions higher than two, by the alternative quadratic representation of (10.11).

Similarly, one can compute upper bounds $\bar{\tau}_s$ of the inter-sample time for a conic region, employing Lemma 10.3 and Theorem 10.3.

Lemma 10.3 [5] *Consider a time bound* $\bar{\tau} \in [\underline{\tau}, \sigma]$. *If* $x^T \bar{\Phi}_{(i,j),\bar{\tau}} x \geqslant 0$ *holds* $\forall (i, j) \in \mathcal{K}_{\bar{\tau}} = (\{0, \ldots, N_{conv}\} \times \{\lfloor \frac{\bar{\tau} l}{\bar{\sigma}} \rfloor, \ldots, l-1\})$, *then:*

$$x^T \Phi(\sigma) x \geqslant 0, \quad \forall \sigma \in [\bar{\tau}, \bar{\sigma}]$$

with Φ *defined in* (10.9) *and:*

$$\bar{\Phi}_{(i,j),\bar{\tau}} = \bar{\hat{\Phi}}_{(i,j),\bar{\tau}} + \bar{\nu} I,$$

$$\bar{\hat{\Phi}}_{(i,j),\bar{\tau}} = \begin{cases} \sum_{k=0}^{i} L_{k,j} (\frac{(j+1)\bar{\sigma}}{l} - \bar{\tau})^k & \text{if } j = \lfloor \frac{\bar{\tau} l}{\bar{\sigma}} \rfloor, \\ \sum_{k=0}^{i} L_{k,j} (\frac{\bar{\sigma}}{l})^k & \text{if } j > \lfloor \frac{\bar{\tau} l}{\bar{\sigma}} \rfloor, \end{cases}$$

$$\bar{\nu} \leqslant \max_{\substack{\sigma' \in [0, \frac{\bar{\sigma}}{l}] \\ r \in \{0,\ldots,l-1\}}} \lambda_{\min}(\Phi(\sigma' + r\tfrac{\bar{\sigma}}{l}) - \tilde{\Phi}_{N_{conv},r}(\sigma')), \tag{10.19}$$

where $L_{k,j}$ *and* $\tilde{\Phi}_{N_{conv},r}$ *are given by* (10.14) *and* (10.17), *respectively.*

Theorem 10.3 (Regional Upper Bound Approximation) *Consider a scalar* $\bar{\tau}_s \in (0, \bar{\sigma}]$ *and matrices* $\bar{\Phi}_{\kappa,\bar{\tau}_s}$, $\kappa = (i, j) \in \mathcal{K}_{\bar{\tau}_s}$, *defined as in Lemma 10.2. If there exist scalars* $\bar{\varepsilon}_{\kappa,\bar{\tau}_s} \geqslant 0$ *such that for all* $\kappa \in \mathcal{K}_{\bar{\tau}_s}$ *the following LMIs hold:*

$$\bar{\Phi}_{\kappa,\bar{\tau}_s} + \sum_{i=1}^{n-1} \bar{\varepsilon}_{\kappa,s_i} \tilde{Q}_{s_i}^{(i,i+1)} \preceq 0$$

the inter-sample time (10.4) *of the system* (10.2)–(10.3) *is regionally bounded from above by* $\bar{\tau}_s$, $\forall x \in \mathcal{R}_s$.

Proof Analogous to the proof of Theorem 10.2.

Remark 10.6 In order to employ Theorem 10.2 (Theorem 10.3) to compute lower (upper) bounds for each of the regions, one needs to apply a line search over $\underline{\tau}_s \in [0, \bar{\sigma}]$ ($\bar{\tau}_s \in [\underline{\tau}_s, \bar{\sigma}]$). This requires checking the feasibility of the LMIs at each step of the line search.

10.3.3 Transition Relation

Finally, the transition relation $\underset{/R}{\longrightarrow}$ of the abstraction is given by:

$$(x_{/R}, x'_{/R}) \in \underset{/R}{\longrightarrow} \Leftrightarrow \mu(\mathscr{X}_{[\underline{\tau}_s, \bar{\tau}_s]}(x_{/R}) \cap x'_{/R}) > 0. \tag{10.20}$$

Remark 10.7 Equation (10.20) explicitly enforces that the intersection between the sets needs to be strictly larger than just the trivial coincidence in the origin, or one facet of the sets, by requiring that such intersection has non-zero measure.

In other words, to construct the relation one needs to compute which sets $\mathscr{R}_{s'}$ are (non-trivially) intersected by $\mathscr{X}_{[\underline{\tau}_s, \bar{\tau}_s]}(\mathscr{R}_s)$: the reachable set from $\mathscr{R}_s$ in the time interval $[\underline{\tau}_s, \bar{\tau}_s]$. In practice, we can only compute approximations of this reachable set. Nevertheless, in order to construct an approximate abstraction $\bar{S}_{/R}$ it suffices to compute the intersection with outer approximations i.e., $\hat{\mathscr{X}}_{[\underline{\tau}_s, \bar{\tau}_s]}(\mathscr{R}_s)$ such that $\mathscr{X}_{[\underline{\tau}_s, \bar{\tau}_s]}(\mathscr{R}_s) \subseteq \hat{\mathscr{X}}_{[\underline{\tau}_s, \bar{\tau}_s]}(\mathscr{R}_s)$.

Remark 10.8 Employing an outer approximation of the reachable sets can potentially introduce *spurious* transitions, i.e., $\underset{/R}{\longrightarrow} \subseteq \underset{\bar{/R}}{\longrightarrow}$ but as stated in Lemma 10.1 the desired approximate simulation relation is retained.

Note that $\mathscr{R}_s$ are not compact sets (they are unbounded cones). However, all of those cones share the origin which is an invariant point of the state space. Therefore, it is sufficient to compute the reachable set of $\underline{\mathscr{R}}_s := \mathscr{R}_s \cap \mathscr{E}_s$ for some affine hyperplane $\mathscr{E}_s = \{x | e^T x + c \leqslant 0\}$ with $e \in \mathbb{R}^n$ and $c \neq 0 \in \mathbb{R}$. Then from the convex hull of this polytope in $\mathbb{R}^{n-1}$ and the origin one can construct a non-empty convex subset $\hat{\mathscr{R}}_s$ of $\mathscr{R}_s$ as follows:

$$\hat{\mathscr{R}}_s = \{\lambda x_e \mid \lambda \in [0, 1],\ x_e \in \underline{\mathscr{R}}_s\}. \tag{10.21}$$

Now observe that, thanks to the linearity of (10.6) and the fact that all sets are pointed at the origin, the condition (10.20) can be replaced by:

$$\mu(\mathscr{X}_{[\underline{\tau}_s, \bar{\tau}_s]}(\mathscr{R}_s) \cap \mathscr{R}_{s'}) > 0 \Leftrightarrow \mu(\mathscr{X}_{[\underline{\tau}_s, \bar{\tau}_s]}(\hat{\mathscr{R}}_s) \cap \hat{\mathscr{R}}_{s'}) > 0 \tag{10.22}$$

There are many techniques available to compute polytopic outer approximations of reachable sets of polytopes as the set $\hat{\mathscr{X}}_{[\underline{\tau}_s, \bar{\tau}_s]}(\mathscr{R}_s)$. In particular, we employ in our implementations the approach from [9]. Similarly, there are many tools that enable the computation of intersection of polytopic sets and check that such sets have no empty interior, e.g., [22, 23].

10.3.4 Increasing the Precision of the Abstractions

The precision of the abstractions obtained can be considered from both: the distance of the output traces, i.e., ε in Lemma 10.1, and in terms of the amount of spurious transitions introduced.

The conservatism introduced through the polytopic embedding, cf. Sect. 10.3.2 can be reduced by increasing N_{conv} and l in Lemmas 10.2 and 10.3. This results in more LMI constraints, but in general leads to a smaller ε by reducing $|\bar{\tau}_s - \underline{\tau}_s|$ for each $\mathscr{R}_s$. Having tighter bounds for the inter-sample times also reduces the conservatism introduced in computing the reachable sets in Sect. 10.3.3, which in turn reduces the amount of spurious transitions. Similarly, one can employ more precise or tight outer approximations of the reachable sets to reduce spurious transitions.

Finally, one can also refine the conic regions of an abstraction $S_{/R}$ into more regions $\mathscr{R}_s$. As long as in the new abstraction $S_{/R'}$ the equivalence classes are subsets of the classes in the first abstraction $S_{/R}$, the precision of the inter-sample bounds cannot decrease, i.e., $|\bar{\tau}_s - \underline{\tau}_s|$ cannot increase. Formally:

$$\big(\forall (x, x') \in R' \Rightarrow (x, x') \in R\big) \Rightarrow \varepsilon' \leqslant \varepsilon, \tag{10.23}$$

where $S \preceq_S^{\varepsilon} \bar{S}_{/R}$, and $S \preceq_S^{\varepsilon'} \bar{S}_{/R'}$. Note that this does not need to hold if the partition defined by R' is not a refinement of the original partition determined by R.

10.4 Timed Automata and Scheduling

In this section we briefly show that the abstractions $\bar{S}_{/R}$, whose construction is described in the previous sections, are in fact semantically equivalent to TSA. Then, we illustrate a few possibilities to enrich the obtained abstractions with *controllable actions*, which may be employed to design schedulers for ETC systems on shared resources.

Let us first interpret the semantics of the proposed abstractions $\bar{S}_{/R}$. Note that the system $\bar{S}_{/R}$ only captures discrete events. However, just like the concrete system S, the connection with actual time is established through the outputs produced by these models. The abstraction $\bar{S}_{/R}$ is a finite-state dynamical system, but with an infinite output set $\bar{Y}_{/R}$ capturing time intervals. When the last transmitted measurement x satisfies $x \in x_{/R}$, the output $y_{/R} = \bar{H}(x_{/R})$ indicates that the original control system:

1. does not trigger updates during the interval $[0, \underline{\tau}_{x_{/R}})$;
2. may trigger a controller update during the time interval $[\underline{\tau}_{x_{/R}}, \bar{\tau}_{x_{/R}})$; and
3. must trigger an update if $\bar{\tau}_{x_{/R}}$ seconds have elapsed since the last transmission.

In the model $\bar{S}_{/R}$ a controller update of the ETC system (cf. Sect. 10.2.4) is captured by a transition between states $x_{/R} \rightarrow x'_{/R}$ of the abstraction $\bar{S}_{/R}$. As described in Sect. 10.2.3, one can capture the same type of semantics with a TSA. In particular, the TSA $\bar{S}_{TSA} = (L, L_0, \Sigma, \mathscr{C}, E, I)$ has the same semantics as the abstraction $\bar{S}_{/R}$, where:

- the set of locations $L := X_{/R} = \{l_1, \ldots, l_q\}$;
- the set of initial locations $L_0 := X_{/R,0}$;
- the set of actions $\Sigma = \{*\}$ is an arbitrary labeling of discrete transitions (or edges);
- the clock set $\mathscr{C} = \{c\}$ contains a single clock;
- the set of edges E is such that $(l_s, g, a, r, l_{s'}) \in E$ iff $l_s \underset{/R}{\longrightarrow} l_{s'}, g = \{\underline{\tau}_s \leqslant c \leqslant \bar{\tau}_s\}$, $a = *$, and $r = \{c := 0\}$;
- the invariant map $I(l_s) := \{0 \leqslant c \leqslant \bar{\tau}_s\}, \forall s \in \{1, \ldots, q\}$.

10.4.1 Automatic Synthesis of Schedulers

One may think of a scheduler as a coordinating controller that prevents several systems from entering into a conflict configuration. We consider a set of ETC systems that share a common resource, e.g., a computing or communication platform, and propose to use an NTGA-based approach to synthesize schedulers in this set-up. After each update the shared resource is unavailable for some predefined time interval, and to prevent a conflict (a control loop requesting access while the resource is being used) the scheduler decides which control loop shall be updated next by selecting an update mechanism for each control loop (see next paragraph). The process to automatically synthesize schedulers consists of three steps: first, construct an NTGA associated with the set of NCSs; then, define the set of bad states (representing conflicts); and finally, employ a tool like UPPAAL-Tiga [12] to obtain a safe strategy avoiding the bad states. The NTGA derived from the set of NCSs: $\mathscr{G}^{NCSs}$, is a parallel synchronized composition of the TGA associated with the network: $\mathscr{G}^{net}$, and the TGA associated with the control loops $\mathscr{G}^{cli}$ for all $i \in \{1, \ldots, N\}$.

In order to design schedulers, the models of the systems to schedule need to expose variables enabling the control of their dynamics. More precisely, the TSA $\bar{S}_{TSA}$ needs to be enriched to contain more than one *action* in the set Σ, resulting in the TGA $\mathscr{G}^{cli}$. Several update mechanisms, providing different controllable actions for the scheduler can be considered:

- the update time is based on a triggering mechanism, where a triggering coefficient is selected from the finite set $\{\alpha_1, \ldots, \alpha_p\}$. In TGA $\mathscr{G}^{cli}$ for each $s \in \{1, \ldots, q\}$, we introduce additional locations $l_s^{\alpha_1}, \ldots, l_s^{\alpha_p}$ representing the choice of the triggering coefficient $\alpha_1, \ldots, \alpha_p$ selected at state $x_{/R} = \mathscr{R}_s$. For each $s \in \{1, \ldots, q\}$, the edges from l_s to $l_s^{\alpha_1}, \ldots, l_s^{\alpha_p}$ are controllable enabling the scheduler to choose the triggering coefficient.
- the update time is forced at a predefined time, which is earlier than the minimum inter-sample time of the active triggering condition. In TGA $\mathscr{G}^{cli}$ for each $s \in$

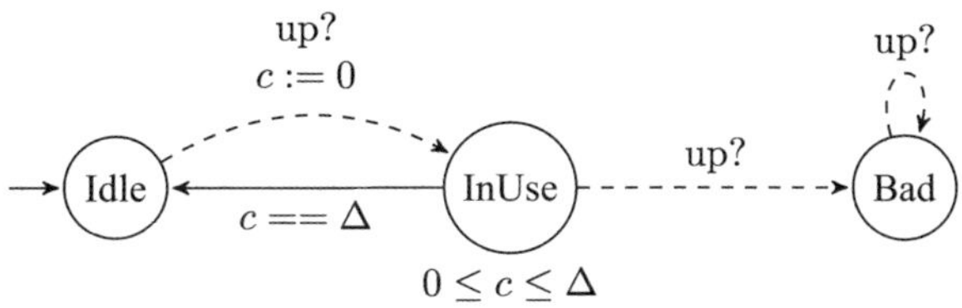

Fig. 10.3 TGA of a shared resource

$\{1, \ldots, q\}$, we introduce controllable edges originated from l_s that represent earlier controller updates.

- the update time is based on a triggering mechanism but delayed a predefined amount of time to be selected from some set $\{\tau_1^d, \ldots, \tau_r^d\}$. Note that ETC naturally tolerates a maximum amount of delay $\bar{\Delta}$ [20], thus one must select these delays smaller than such $\bar{\Delta}$. In TGA $\mathcal{G}^{cli}$ for each $s \in \{1, \ldots, q\}$, we introduce locations $l_s^{\tau_1^d}, \ldots, l_s^{\tau_r^d}$ that represent the sampled state is in $\mathcal{R}_s$ and the chosen delay is $\tau_1^d, \ldots, \tau_r^d$, respectively. Again, these new edges from l_s to $l_s^{\tau_1^d}, \ldots, l_s^{\tau_r^d}$ are controllable for the scheduler.

Each of these mechanisms can be employed on their own or combined to provide more control handles to the scheduler. The status of the shared resource – available, unavailable and conflict – and the possible transitions among them are modeled by TGA $\mathcal{G}^{net}$ depicted in Fig. 10.3. Thus, the bad states of the NTGA are defined as the set of states such that the location of $\mathcal{G}^{net}$ is *Bad*. For a more detailed treatment of this procedure we refer the reader to the report [19].

10.5 Illustrative Examples

In what follows, we illustrate the described abstraction construction on two examples. First, we consider a simple academic two-dimensional LTI system:

$$\begin{aligned}\dot{\xi}_1(t) &= \begin{bmatrix} -14 & 10 \\ -24 & 17 \end{bmatrix} \xi_1(t) + \begin{bmatrix} 1 \\ 2 \end{bmatrix} \nu_1(t), \\ \nu_1(t) &= [9 \quad -6.5]\, \xi_1(t).\end{aligned} \tag{10.24}$$

We employ the following values for the abstraction parameters: the triggering coefficient $\alpha = 0.05$, the upper bound of the inter-sample interval $\bar{\sigma} = 1$ s, the order of polynomial approximation $N_{\text{conv}} = 5$, the number of polytopic subdivisions $l = 100$ and the total number of state space partitions $q = 20$.

The resulting abstraction of the closed-loop system (10.24) is provided in Fig. 10.4, depicting $\underline{\tau}_s$ and $\bar{\tau}_s$; $\underline{\tau}_s$ and $\bar{\tau}_s$ in a radial manner, and a representation of the discrete transitions in the resulting TSA. The achieved precision of the abstraction is $\varepsilon = 0.284$ s. Figure 10.5 illustrates the validity of the theoretical bounds that we

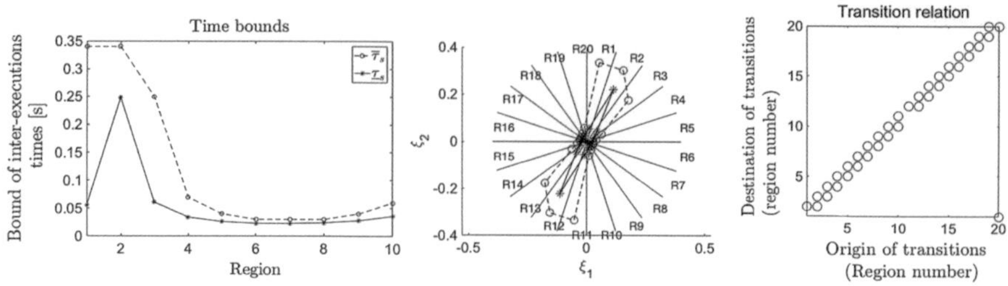

Fig. 10.4 System (10.24) lower and upper bounds of inter-sample times depicted by solid and dashed curves, respectively (left panel). Spider-web representation of times (note the symmetry) (center panel). Graphic representation of the transition relation (right panel)

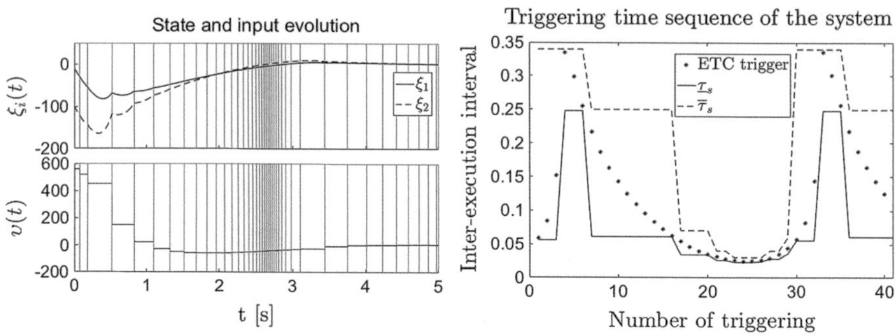

Fig. 10.5 System (10.24) states and input trajectories (left panel) and triggering times (right panel) of a simulation of the ETC system. Time between triggering (asterisks), predicted lower bound (solid line) and upper bound (dashed line)

found for $\underline{\tau}_s$ (solid line) and $\bar{\tau}_s$ (dashed line). The asterisks represent the inter-sample times sequence during 5 s simulation of the ETC system.

The second example is a somewhat more realistic system: an intelligent vehicle headway controller [24]:

$$\begin{bmatrix} \dot{E}_r(t) \\ \dot{E}_v(t) \\ \dot{a}(t) \end{bmatrix} = \begin{bmatrix} 0 & 1 & 0 \\ 0 & 0 & 1 \\ 0 & -1.43 & -2.149 \end{bmatrix} \begin{bmatrix} E_r(t) \\ E_v(t) \\ a(t) \end{bmatrix} + \begin{bmatrix} 0 \\ 0 \\ 0.01077 \end{bmatrix} u(t), \tag{10.25}$$

$$u(t) = -\begin{bmatrix} 40 \; 55.78 \; 24.45 \end{bmatrix} [E_r(t) \quad E_v(t) \quad a(t)]^T, \tag{10.26}$$

where $E_r = R_h - R$, $E_v = V - V_p$, with R_h and R the desired and actual headway, V_p and V the preceding and host vehicle velocities, and a the host velocity acceleration. The controller is implemented with a triggering coefficient $\alpha = 0.05$. In the abstraction we select $m = 10$, the number of subdivisions for each angular coordinate in the interval $[-\frac{\pi}{2}, \frac{\pi}{2})$, which results in $q = 2 \times m^{(n-1)} = 2 \times 10^2 = 200$ states of the abstraction, i.e., regions in which the state space is divided. The rest of the parameters are selected as $\bar{\sigma} = 2\,\text{s}$, $N_{conv} = 5$, and $l = 100$.

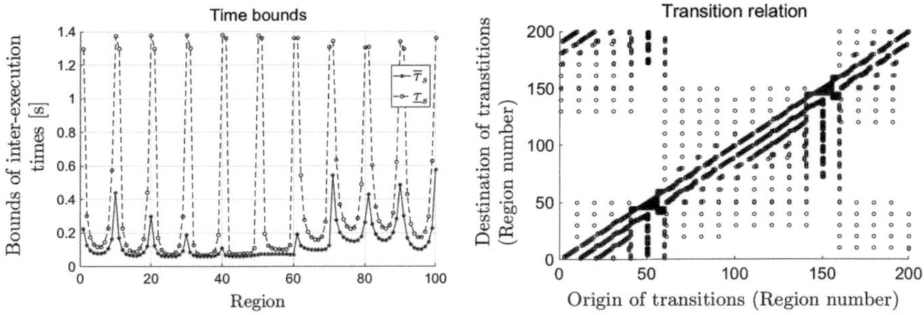

Fig. 10.6 System (10.25) lower and upper bounds of inter-sample times depicted by solid and dashed curves, respectively (left panel). Graphic representation of the transition relation (right panel)

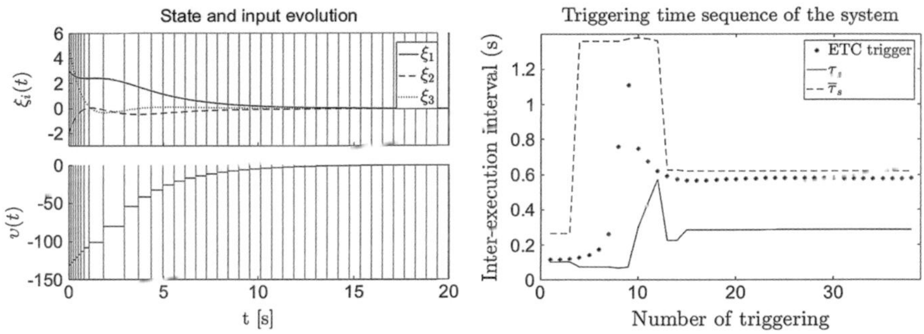

Fig. 10.7 System (10.25) states and input trajectories (left panel) and triggering times (right panel) of a simulation of the ETC system. Time between triggering (asterisks), predicted lower bound (solid line) and upper bound (dashed line)

Figure 10.6 shows the resulting abstraction of the closed-loop system (10.25). Note that times for only half of the state space (100 regions) are plotted, as the symmetric half of the state space results in identical bounds. The precision of the constructed abstractions is $\varepsilon = 1.3$ s. The validity of the theoretical bounds that we found on a simulation, with initial condition $x_0 = [3 \quad -2 \quad 5]^T$, is visualized in Fig. 10.7.

10.6 Conclusion

We have presented a methodology to construct models describing the timing patterns of updates in Event-Triggered control systems. The resulting models can be recast as timed automata, for which a large body of literature and tools are available. In particular, one can employ techniques from the literature on timed automata to automatically synthesize schedulers arbitrating the access to shared resources between ETC loops and possibly other (real-time) tasks.

An apparent drawback of the proposed approach is the amount of computation required to construct the models. In the construction of each abstract state one needs to solve several LMI feasibility problems to construct the output set. Then a reachability analysis must be run for each of these states. Fortunately, this is a procedure that can be easily parallelized and that is only run offline. However, the amount of abstract states that is required (assuming a uniform partitioning) scales exponentially with the dimensionality of the system. A couple of promising approaches to address the challenge of scalability are the use of compositional ideas, as in e.g., [25], and the use of model order reduction techniques, as in e.g., Chap. 1 of this book.

The versatility of timed automata to model the traffic of model-based aperiodic controllers has been also demonstrated in e.g., [26], or Chap. 6 of this book. Extensions to other types of event-based controller implementations, like periodic ETC with dynamic controllers [27], or to the non-linear context [20] can be constructed similarly, provided that: (i) the reachability of the considered systems is possible, and (ii) one can construct computable triggering checks dependent solely on the last sampled state. Many approaches are available for the reachability of non-linear systems, see e.g., [28–30]. The second condition is closely related to the idea of Self-Triggered control, for which large classes of non-linear systems have been studied in e.g., [21, 31]

Future work shall investigate if other applications of these sort of abstractions can be found in the real-time control context. An interesting possibility is the study of security, where timed automata may serve to characterize resilient traffic flows (or conversely attacker patterns) as in the context of Chap. 11 in this book.

References

1. Araújo, J., Mazo Jr., M., Anta, A., Tabuada, P., Johansson, K.H.: System architectures, protocols and algorithms for aperiodic wireless control systems. IEEE Trans. Ind. Inf. **10**(1), 175–184 (2014)
2. Caccamo, M., Buttazzo, G., Sha, L.: Elastic feedback control. In: Proceedings 12th Euromicro Conference on Real-Time Systems, pp. 121–128. (2000)
3. Bhattacharya, R., Balas, G.: Anytime control algorithm: model reduction approach. J. Guid. Control Dyn. **27**(5), 767–776 (2004)
4. Al-Areqi, S., Gorges, D., Reimann, S., Liu, S.: Event-based control and scheduling codesign of networked embedded control systems, pp. 5299–5304. (2013)
5. Kolarijani, A.S., Mazo Jr., M.: A formal traffic characterization of LTI event-triggered control systems. IEEE Trans. Control Netw. Syst. **5**(1), 274–283. (2018)
6. Abdeddaïm, Y., Asarin, E., Maler, O.: Scheduling with timed automata. Theor. Comput. Sci. **354**(2), 272–300 (2006)
7. Fiter, C., Hetel, L., Perruquetti, W., Richard, J.-P.: A state dependent sampling for linear state feedback. Automatica **48**(8), 1860–1867 (2012)
8. Hetel, L., Daafouz, J., Lung, C.: LMI control design for a class of exponential uncertain systems with application to network controlled switched systems. In: Proceedings of the American Control Conference, pp. 1401–1406. (2007)
9. Chutinan, A., Krogh, B.: Computing polyhedral approximations to flow pipes for dynamic systems. In: Proceedings 37th IEEE Conference on Decision and Control, vol. 2, pp. 2089–2094. (1998)

10. Alur, R., Dill, D.: A theory of timed automata. Theor. Comput. Sci. **126**(2), 183–235 (1994)
11. Maler, O., Pnueli, A., Sifakis, J.: On the synthesis of discrete controllers for timed systems. In: Mayr E., Puech C. (eds.) Proceedings of the 12th Symposium on Theoretical Aspects of Computer Science, vol. 900, pp.229–242 (1995)
12. Uppaal tiga. [Online]. Available: http://people.cs.aau.dk/~adavid/tiga/
13. Tabuada, P.: Verification and Control of Hybrid Systems: A Symbolic Approach. Springer, London, Limited (2009)
14. Ewald, G.: Graduate texts in mathematics. In: Combinatorial Convexity And Algebraic Geometry. Springer, New York (1996)
15. Henzinger, T., Nicollin, X., Sifakis, J., Yovine, S.: Symbolic model checking for real-time systems. Inf. Comput. **111**(2), 193–244 (1994)
16. Bengtsson, J., Yi, W.: Timed automata: semantics, algorithms and tools. Lect. Concurr. Petri Nets **3098**, 87–124 (2004)
17. Behrmann, G., David, A., Larsen, K.: A tutorial on Uppaal. Form. Methods Des. Real-Time Syst. **3185**, 200–236 (2004)
18. Chatain, T., David, A., Larsen, K.G.: Playing games with timed games. IFAC Proc. Vol. **42**(17), 238–243 (2009)
19. Adzkiya, D., Mazo Jr., M.: Scheduling of event-triggered networked control systems using timed game automata. arXiv:1610.03729 (2016)
20. Tabuada, P.: Event-triggered real-time scheduling of stabilizing control tasks. IEEE Trans. Autom. Control **52**(9), 1680–1685 (2007)
21. Anta, A., Tabuada, P.: To sample or not to sample: self-triggered control for nonlinear systems. IEEE Trans. Autom. Control **55**, 2030–2042 (2010)
22. Kvasnica, M., Grieder, P., Baotic, M., Morari, M.: Multi-parametric toolbox (MPT). In: HSCC, pp. 448–462. Springer, Berlin (2004)
23. Bagnara, R., Hill, P.M., Zaffanella, E.: The parma polyhedra library: toward a complete set of numerical abstractions for the analysis and verification of hardware and software systems. Sci. Comput. Program. **72**(1), 3–21 (2008)
24. Bin, L., Rongben, W., Jiangwei, C.: A new optimal controller for intelligent vehicle headway distance. In: IEEE Intelligent Vehicle Symposium, vol. 2, pp. 387–392 (2002)
25. Wang, X., Lemmon, M.D.: Event-triggering in distributed networked control systems. IEEE Trans. Autom. Control **56**(3), 586–601 (2011)
26. Zamani, M., Dey, S., Mohamed, S., Dasgupta, P., Mazo Jr., M.: Scheduling of controllers' update-rates for residual bandwidth utilization. In: 14th International Conference on Formal Modeling and Analysis of Timed Systems, LNCS, vol. 9884, pp. 85–101 (2016)
27. Heemels, W.H., Donkers, M., Teel, A.R.: Periodic event-triggered control for linear systems. IEEE Trans. Autom. Control **58**(4), 847–861 (2013)
28. Balluchi, A., Casagrande, A., Collins, P., Ferrari, A., Villa, T., Sangiovanni-Vincentelli, A.L.: Ariadne: a framework for reachability analysis of hybrid automata. In: Proceedings of the International Syposium on Mathematical Theory of Networks and Systems (2006)
29. Frehse, G., Le Guernic, C., Donzé, A., Cotton, S., Ray, R., Lebeltel, O., Ripado, R., Girard, A., Dang, T., Maler, O.: Spaceex: scalable verification of hybrid systems. Comput. Aided Verif. 379–395 (2011)
30. Althoff, M.: An introduction to CORA 2015. In: ARCH@ CPSWeek, pp. 120–151 (2015)
31. Di Benedetto, M.D., Di Gennaro, S., D'Innocenzo, A.: Digital self-triggered robust control of nonlinear systems. Int. J. Control **86**(9), 1664–1672 (2013)

Chapter 11
Resilient Self-Triggered Network Synchronization

D. Senejohnny, P. Tesi and C. De Persis

Abstract In this chapter, we investigate Self-Triggered synchronization of linear oscillators in the presence of communication failures caused by denial-of-Service (DoS). A general framework is considered in which network links can fail independent of each other. A characterization of DoS frequency and duration to preserve network synchronization is provided, along with an explicit characterization of the effect of DoS on the time required to achieve synchronization. A numerical example is given to substantiate the analysis.

11.1 Introduction

Cyber-physical systems (CPSs) exhibit a tight conjoining of computational and physical components. The fact that any breach in the cyberspace can have a tangible effect on the physical world has recently triggered attention toward cybersecurity also within the engineering community [1, 2]. In CPSs, attacks to the cyber-layer are mainly categorized as either denial-of-service (DoS) attacks or deception attacks. The latter affects the reliability of data by manipulating the transmitted packets over network; see [3, 4]. On the other hand, DoS attacks are primarily intended to affect the timeliness of the information exchange, i.e., to cause packet losses; see for instance [5, 6] for an introduction to the topic. This chapter aims at considering the effect of DoS attacks.

D. Senejohnny (✉) · P. Tesi · C. De Persis
ENTEG and Jan C. Willems Center for Systems and Control,
University of Groningen,
9747 AG Groningen, The Netherlands
e-mail: d.senejohnny@rug.nl

C. De Persis
e-mail: c.de.persis@rug.nl

P. Tesi
DINFO, Università degli Studi di Firenze, 50139 Firenze, Italy
e-mail: pietro.tesi@unifi.it; p.tesi@rug.nl

S. Tarbouriech et al. (eds.), *Control Subject to Computational and Communication Constraints*, Lecture Notes in Control and Information Sciences 475, https://doi.org/10.1007/978-3-319-78449-6_11

In the literature, the issue of resilience against DoS has been mostly investigated in *centralized* settings [7–14]. Very recently, [15, 16] explored this problem in a *distributed* setting with emphasis on consensus-like networks. The main goal of this chapter is to address the issue of resilience against DoS for network coordination problems in which node dynamics are more general than simple integrators. Specifically, we study *synchronization* networks of the same type as in [17]. Inspired by [18] and [19], we consider a *Self-Triggered* coordination scheme, in which the available information to each agent is used to update local controls and to specify the next update time. We consider Self-Triggered coordination schemes since they are of major interest when synchronization has to be achieved in spite of possibly severe communication constraints. In this respect, a remarkable feature of Self-Triggered coordination lies in the possibility of ensuring coordination properties in the absence of any global information on the graph topology and with no need to resort to synchronous communication.

The primary step in the analysis of distributed coordination problems in the presence of DoS pertains to the modeling of DoS itself. In [12, 13], a general model is considered that only constrains DoS patterns in terms of their average frequency and duration. This makes it possible to describe a wide range of DoS-generating signals, e.g., trivial, periodic, random, and protocol-aware *jamming* [5, 6, 20, 21]. The occurrence of DoS has a different effect on the communication, depending on the network architecture. For networks operating through a single access point, in the so-called "infrastructure" mode, DoS may cause all the network links to fail simultaneously [15]. In this chapter, we consider instead a more general scenario in which the network links can fail independent of each other, thus extending the analysis to "ad-hoc" (peer-to-peer) network architectures. In this respect, a main contribution of this chapter is an explicit characterization of the frequency and duration of DoS at the various network links under which coordination can be preserved. In addition to extending the results of [19] to independent polling of neighbors, we also provide an explicit characterization of the effects of DoS on the coordination time. A preliminary and incomplete account of this work without the relevant proofs has appeared in [22].

The problem of network coordination under communication failures can be viewed as a coordination problem in the presence of switching topologies. For purely continuous-time systems, this problem has been thoroughly investigated under assumptions such as, point-wise, period-wise, and joint connectivity [23–25]. In CPSs, however, due to the presence of a digital communication layer, the situation is drastically different. In fact, the presence of a digital communication layer implies that the time span between any two consecutive transmissions cannot be arbitrarily small. As a consequence, the classic connectivity notions developed for purely continuous-time systems are not directly applicable to a digital setting as the one considered here. In this respect, we introduce a notion of *persistency-of-communication* (PoC), which requires graph (link) connectivity be be satisfied over periods of time that are consistent with the constraints imposed by the communication medium [15, 16].

The remainder of this chapter is organized as follows. In Sect. 11.2, we formulate the problem of interest and provide the results for Self-Triggered synchronization. In Sect. 11.3, we describe the considered class of DoS patterns. The main results are provided in Sect. 11.4. A numerical example is given in Sect. 11.5. Finally, Sect. 11.6 ends the chapter with concluding remarks.

Notation: The following notation is used throughout this chapter. The stacking of N column vectors $x_1, x_2, \ldots, x_n$ is denoted by x, i.e., $x = \left[x_1^\top \; x_2^\top \; \ldots \; x_n^\top\right]^\top$. The N- dimensional identity matrix is denoted by I_N. Vectors of all ones and zeros are denoted by $\mathbf{1}$ and $\mathbf{0}$, respectively. The ℓth component of vector x is denoted by x_ℓ or, interchangeably, by $[x]_\ell$.

11.2 Self-Triggered Synchronization

11.2.1 System Definition

We consider a connected and undirected graph $\mathscr{G} = (\mathscr{I}, \mathscr{E})$, where $\mathscr{I} := \{1, 2, \cdots, N\}$ is the set of nodes and $\mathscr{E} \subseteq \mathscr{I} \times \mathscr{I}$ is the set of links (edges). Given a node $i \in \mathscr{I}$, we shall denote by $\mathscr{N}_i = \{j \in \mathscr{I} : (i, j) \in \mathscr{E}\}$ the set of its neighbors, i.e., the set of nodes that exchange information with node i, and by $d^i = |\mathscr{N}_i|$, i.e., the cardinality of $\mathscr{N}_i$. Notice that the order of the elements i and j in (i, j) is irrelevant since the graph is assumed undirected. Throughout the chapter, we shall refer to $\mathscr{G}$ as the "nominal" network (the network configuration when communication is allowed for every link).

We assume that each network node is a dynamical system consisting of a linear oscillator with dynamics

$$\dot{x}^i = Ax^i + Bu^i \tag{11.1}$$

where (A, B) is a stabilizable pair and all eigenvalues of A lie on imaginary axis with unitary geometric multiplicity; $x^i, u^i \in \mathbb{R}^n$ represent node state and control variables. The network nodes exchange information according to the configuration described by the links of $\mathscr{G}$. To achieve synchronization with constrained flow of information, we employ a hybrid controller with state variables $(x, \eta, \xi, \theta) \in \mathbb{R}^{n\times N} \times \mathbb{R}^{n\times N} \times \mathbb{R}^{n\times d} \times \mathbb{R}^{n\times d}$, where $d := \sum_{i=1}^N d^i$. The controller also makes use of a quantization function.

The specific quantizer of choice is $\text{sign}_\varepsilon : \mathbb{R} \to \{-1, 0, 1\}$, which is given by

$$\text{sign}_\varepsilon(z) := \begin{cases} \text{sign}(z) & \text{if } |z| \geq \varepsilon \\ 0 & \text{otherwise} \end{cases} \tag{11.2}$$

where $\varepsilon > 0$ is a sensitivity parameter, which is selected at the design stage to trade-off between synchronization accuracy and communication frequency. The flow dynamics are given by

$$\dot{\eta}^i = (A + BK)\eta^i + \sum_{j \in \mathcal{N}_i} \xi^{ij} \tag{11.3a}$$

$$\dot{\xi}^{ij} = A\xi^{ij} \tag{11.3b}$$

$$\dot{\theta}^{ij} = -\mathbf{1} \tag{11.3c}$$

$$u^i = K\eta^i, \tag{11.3d}$$

where $A + KB$ is Hurwitz; $\eta^i \in \mathbb{R}^n$ and $\xi^{ij} \in \mathbb{R}^n$ are controller states, and $\theta^{ij} \in \mathbb{R}^n$ is the local clock over the link $(i, j) \in \mathcal{E}$, where $\theta^{ij}(0) = 0$. As it will become clear in the sequel, the superscript "ij" appearing in ξ and θ indicates that these variables are common to nodes i and j. The continuous evolution of the edge-based controller dynamic holds as long as the set

$$\mathcal{S}(\theta, t) := \{(i, j, \ell) \in \mathcal{I} \times \mathcal{I} \times \mathcal{L} : \theta_\ell^{ij}(t^-) = 0\} \tag{11.4}$$

is nonempty, where $s(t^-)$ denotes the limit from below of a signal $s(t)$, i.e., $s(t^-) = \lim_{\tau \nearrow t} s(\tau)$, and where $\ell \in \mathcal{L} := \{1, 2, \ldots, n\}$. At these time instants, in the "nominal" operating mode, a discrete transition (jump) occurs, which is given by

$$\begin{aligned}
x_\ell^i(t) &= x_\ell^i(t^-) \\
\eta_\ell^i(t) &= \eta_\ell^i(t^-) \\
\xi_\ell^{ij}(t) &= \begin{cases} [e^{At}\,\mathrm{sign}_\varepsilon(e^{-At}\mathcal{D}^{ij}(\eta(t) - x(t)))]_\ell & \text{if } (i, j, \ell) \in \mathcal{S}(\theta, t) \\ \xi_\ell^{ij}(t^-) & \text{otherwise} \end{cases} \\
\theta_\ell^{ij}(t) &= \begin{cases} f_\ell^{ij}(t) & \text{if } (i, j, \ell) \in \mathcal{S}(\theta, t) \\ \theta_\ell^{ij}(t^-) & \text{otherwise} \end{cases}
\end{aligned} \tag{11.5}$$

for every $i \in \mathcal{I}$, $j \in \mathcal{N}_i$, and $\ell \in \mathcal{L}$.

Here, $\mathcal{D}^{ij}(\alpha(t)) = \alpha^j(t) - \alpha^i(t)$ and $f_\ell^{ij} : \mathbb{R}^n \to \mathbb{R}_{>0}$ is given by

$$f_\ell^{ij}(x) = \max\left\{ \frac{\left|\left[e^{-At}\mathcal{D}^{ij}(\eta(t) - x(t))\right]_\ell\right|}{2(d^i + d^j)}, \frac{\varepsilon}{2(d^i + d^j)} \right\}. \tag{11.6}$$

Note that for all $(i, j) \in \mathcal{E}$ we have $\theta^{ij}(t) = \theta^{ji}(t)$ and $\xi^{ij}(t) = -\xi^{ji}(t)$ for all $t \in \mathbb{R}_{\geqslant 0}$. As such, (11.1)–(11.5) can be regarded as an edge-based synchronization protocol. Here, the term "Self-Triggered", first adopted in the context of real-time systems [26], expresses the property that the data exchange between nodes is driven by local clocks, which avoids the need for a common global clock.

A few comments are in order.

Remark 11.1 (*Controller structure*) The controller emulates the node dynamics (11.1), with an extra coupling term as done in [17]. The coupling is through the variable ξ^{ij}, which is updated at discrete times and emulates the open-loop behavior of (11.1) during its the controller continuous evolution [19]. Slightly different from [17], the coupling term ξ^{ij} takes into account the discrepancy between node and controller states. This choice of coupling is due to the use of the quantizer (11.2) which triggers at discrete instances. ■

Remark 11.2 (*Clock variable* θ_ℓ^{ij}) Each clock variable θ_ℓ^{ij} plans ahead the update time of component ℓ of controller state ξ^{ij}. Whenever θ_ℓ^{ij} reaches zero, the ℓth component of the controller state and clock variables is updated. In order to avoid arbitrarily fast sampling (Zeno phenomena), we use the threshold ε in the update of the function f^{ij} in (11.6). In particular, this implies that for every edge $(i, j) \in \mathscr{E}$ and for any time $\mathscr{T}$, no more than $n\lfloor\frac{2(d^i+d^j)\mathscr{T}}{\varepsilon} + 1\rfloor$ number of updates can occur over an interval of length $\mathscr{T}$. ■

11.2.2 Practical Self-Triggered Synchronization

Inspired by [17], we analyze (11.1)–(11.5) using the change of coordinates

$$\begin{aligned} x^i(t) &= x^i(t) \\ \mathscr{X}^i(t) &= e^{-At}(\eta^i(t) - x^i(t)) \\ \mathscr{U}^{ij}(t) &= e^{-At}\xi^{ij}(t) \\ \theta^{ij}(t) &= \theta^{ij}(t). \end{aligned} \tag{11.7}$$

Accordingly, the network-state variables become $(x, \mathscr{X}, \mathscr{U}, \theta) \in \mathbb{R}^{n\times N} \times \mathbb{R}^{n\times N} \times \mathbb{R}^{n\times d} \times \mathbb{R}^{n\times d}$ with corresponding flow dynamics

$$\dot{x}^i(t) = (A + BK)x^i(t) + BKe^{At}\mathscr{X}^i(t) \tag{11.8a}$$

$$\begin{aligned} \dot{\mathscr{X}}^i(t) &= \sum_{j\in\mathscr{N}_i} \mathscr{U}^{ij} \\ \dot{\mathscr{U}}^{ij}(t) &= \mathbf{0} \\ \dot{\theta}^{ij}(t) &= -\mathbf{1} \end{aligned} \tag{11.8b}$$

and discrete transitions (jumps)

$$x_\ell^i(t) = x_\ell^i(t^-) \tag{11.9a}$$

$$\begin{aligned}
\mathscr{X}_\ell^i(t) &= \mathscr{X}_\ell^i(t^-) \\
\mathscr{U}_\ell^{ij}(t) &= \begin{cases} \operatorname{sign}_\varepsilon(\mathscr{D}_\ell^{ij}(\mathscr{X}(t)) & \text{if } (i,j,\ell) \in \mathscr{S}(\theta,t) \\ \mathscr{U}_\ell^{ij}(t^-) & \text{otherwise} \end{cases} \\
\theta_\ell^{ij}(t) &= \begin{cases} g_\ell^{ij}(\mathscr{X}(t)) & \text{if } (i,j,\ell) \in \mathscr{S}(\theta,t) \\ \theta_\ell^{ij}(t^-) & \text{otherwise} \end{cases}
\end{aligned} \tag{11.9b}$$

where $(i, j, \ell) \in \mathscr{I} \times \mathscr{I} \times \mathscr{L}$ and

$$g_\ell^{ij}(\mathscr{X}(t)) = \max\left\{ \frac{\left|\mathscr{D}_\ell^{ij}(\mathscr{X}(t))\right|}{2(d^i + d^j)}, \frac{\varepsilon}{2(d^i + d^j)} \right\}. \tag{11.10}$$

Notice that the notion of local time in both coordinates is the same. The reason for considering this change of coordinates is to transform the original synchronization problem into a consensus problem that involves integrator variables $\mathscr{X}^i$.

The result which follows is the main result of this section.

Theorem 11.1 (Practical Synchronization) *Let all the eigenvalues of A lie on the imaginary axis with geometric multiplicity equal to one. Let $(x, \mathscr{X}, \mathscr{U}, \theta)$ be the solution to system (11.8) and (11.9). Then, there exist a finite time T such that $\mathscr{X}$ converges within the time T to a point $\mathscr{X}_* = [\mathscr{X}_*^{1\top}, \ldots, \mathscr{X}_*^{N\top}]^\top$ in the set*

$$\mathscr{E} := \left\{ \mathscr{X} \in \mathbb{R}^{nN} : |\mathscr{D}_\ell^{ij}(\mathscr{X})| < \delta \;\; \forall (i,j,\ell) \in \mathscr{I} \times \mathscr{I} \times \mathscr{L} \right\}, \tag{11.11}$$

where $\delta = \varepsilon(N-1)$, and $\mathscr{U}(t) = \mathbf{0}$ for all $t \geq T$. Moreover, for any arbitrary small $\varepsilon_c \in \mathbb{R}_{>0}$ there exist a time $T_c(\varepsilon_c) \geqslant T$ such that

$$\left|x_\ell^i(t) - x_\ell^j(t)\right| < 2\varepsilon_c + \sqrt{n}\,\delta \;\; \forall (i,j,\ell) \in \mathscr{I} \times \mathscr{I} \times \mathscr{L} \tag{11.12}$$

for all $t \geqslant T_c(\varepsilon_c)$, where n is the dimension of the vector x.

Proof See the appendix. ∎

Equations (11.11) and (11.12) involve a notion of "practical" synchronization. This amounts to saying that the solutions eventually synchronize up to an error, which can be made as small as desired by reducing ε (at the expense of an increase in the communication cost since, in view of (11.6), the minimum inter-transmission time decreases with ε). Theorem 11.1 will be used as a reference frame for the analysis of Sect. 11.4. The case of asymptotic synchronization can be pursued along the lines of [18].

11.3 Network Denial-of-Service

We shall refer to denial-of-service (DoS, in short) as the phenomenon by which communication between the network nodes is interrupted. We shall consider the very general scenario in which the network communication links can fail independent of each other. From the perspective of modeling, this amounts to considering multiple DoS signals, one for each network communication link.

11.3.1 DoS Characterization

Let $\{h_n^{ij}\}_{n\in\mathbb{Z}_{\geq 0}}$ with $h_0^{ij}\geq 0$ denote the sequence of DoS off/on transitions affecting the link (i, j), namely the sequence of time instants at which the DoS status on the link (i, j) exhibits a transition from zero (communication is possible) to one (communication is interrupted). Then

$$H_n^{ij} := \{h_n^{ij}\} \cup \left[h_n^{ij}, h_n^{ij} + \tau_n^{ij}\right[\tag{11.13}$$

represents the nth DoS time-interval, of a length $\tau_n^{ij} \in \mathbb{R}_{\geqslant 0}$, during which communication on the link (i, j) is not possible.

Given $t, \tau \in \mathbb{R}_{\geq 0}$, with $t \geq \tau$, let

$$\Xi^{ij}(\tau, t) := \bigcup_{n\in\mathbb{Z}_{\geq 0}} H_n^{ij} \bigcap [\tau, t] \tag{11.14}$$

and

$$\Theta^{ij}(\tau, t) := [\tau, t] \setminus \Xi^{ij}(\tau, t) \tag{11.15}$$

where $\setminus$ denotes relative complement. In words, for each interval $[\tau, t]$, $\Xi^{ij}(\tau, t)$ and $\Theta^{ij}(\tau, t)$ represent the sets of time instants where communication on the link (i, j) is denied and allowed, respectively.

The first question to be addressed is that of determining a suitable modeling framework for DoS. Following [13], we consider a general model that only constrains DoS attacks in terms of their average frequency and duration. Let $n^{ij}(\tau, t)$ denote the number of DoS off/on transitions on the link (i, j) occurring on the interval $[\tau, t]$.

Assumption 11.2 (*DoS frequency*) For each $(i, j) \in \mathscr{E}$, there exist $\eta^{ij} \in \mathbb{R}_{\geq 0}$ and $\tau_f^{ij} \in \mathbb{R}_{>0}$ such that

$$n^{ij}(\tau, t) \leq \eta^{ij} + \frac{t-\tau}{\tau_f^{ij}} \tag{11.16}$$

for all $t, \tau \in \mathbb{R}_{\geq 0}$ with $t \geq \tau$. ■

Assumption 11.3 (*DoS duration*) For each $(i, j) \in \mathscr{E}$, there exist $\kappa^{ij} \in \mathbb{R}_{\geq 0}$ and $\tau_d^{ij} \in \mathbb{R}_{>1}$ such that

$$|\Xi^{ij}(\tau, t)| \leq \kappa^{ij} + \frac{t-\tau}{\tau_d^{ij}} \tag{11.17}$$

for all $t, \tau \in \mathbb{R}_{\geq 0}$ with $t \geq \tau$. ■

In Assumption 11.2, the term "frequency" stems from the fact that τ_f^{ij} provides a measure of the "dwell time" between any two consecutive DoS intervals on the link (i, j). The quantity η^{ij} is needed to render (11.16) self-consistent when $t = \tau = h_n^{ij}$ for some $n \in \mathbb{Z}_{\geqslant 0}$, in which case $n^{ij}(\tau, t) = 1$. Likewise, in Assumption 11.3, the term "duration" is motivated by the fact that τ_d^{ij} provides a measure of the fraction of time ($\tau_d^{ij} > 1$) the link (i, j) is under DoS. Like η^{ij}, the constant κ^{ij} plays the role of a regularization term. It is needed because during a DoS interval, one has $|\Xi(h_n^{ij}, h_n^{ij} + \tau_n^{ij})| = \tau_n^{ij} \geqslant \tau_n^{ij}/\tau_d^{ij}$ since $\tau_d^{ij} > 1$, with $\tau_n^{ij} = \tau_n^{ij}/\tau_d^{ij}$ if and only if $\tau_n^{ij} = 0$. Hence, κ^{ij} serves to make (11.17) self-consistent. Thanks to the quantities η^{ij} and κ^{ij}, DoS frequency and duration are both average quantities.

11.3.2 Discussion

The considered assumptions only pose limitations on the frequency of the DoS status and its duration. As such, this characterization can capture many different scenarios, including trivial, periodic, random and protocol-aware jamming [5, 6, 20, 21]. For the sake of simplicity, we limit our discussion to the case of radio frequency (RF) jammers, although similar considerations can be made with respect to spoofing-like threats [27].

Consider for instance the case of *constant jamming*, which is one of the most common threats that may occur in a wireless network [5, 28]. By continuously emitting RF signals on the wireless medium, this type of jamming can lower the packet send ratio (PSR) for transmitters employing carrier sensing as a medium access policy as well as lower the packet delivery ratio (PDR) by corrupting packets at the receiver. In general, the percentage of packet losses caused by this type of jammer depends on the jamming-to-signal ratio and can be difficult to quantify as it depends, among many things, on the type of anti-jamming devices, the possibility to adapt the signal strength threshold for carrier sensing, and the interference signal power, which may vary with time. In fact, there are several provisions that can be taken in order to *mitigate* DoS attacks, including spreading techniques, high-pass filtering, and encoding [21, 29]. These provisions decrease the chance that a DoS attack will be successful, and, as such, limit in practice the frequency and duration of the time intervals over which communication is effectively denied. This is nicely captured by the considered formulation.

As another example, consider the case of *reactive jamming* [5, 28]. By exploiting the knowledge of the 802.1i MAC layer protocols, a jammer may restrict the RF signal to the packet transmissions. The collision period need not be long since with many CRC error checks a single-bit error can corrupt an entire frame. Accordingly, jamming takes the form of a (high-power) burst of noise, whose duration is determined by the length of the symbols to corrupt [29, 30]. Also, this case can be nicely accounted for via the considered assumptions.

11.4 Main Result

11.4.1 Resilient Self-Triggered Synchronization

When DoS disrupts link communications, the former controller state ξ_ℓ^{ij} is not available any more. In order to compensate for the communication failures, the control action is suitably modified as follows during the controller discrete updates,

$$
\begin{aligned}
x_\ell^i(t) &= x_\ell^i(t^-) \\
\mathscr{X}_\ell^i(t) &= \mathscr{X}_\ell^i(t^-) \\
\mathscr{U}_\ell^{ij}(t) &= \begin{cases} \mathrm{sign}_\varepsilon(\mathscr{D}_\ell^{ij}(\mathscr{X})) & \text{if } (i,j,\ell) \in \mathscr{S}(\theta,t) \wedge t \in \Theta^{ij}(0,t) \\ 0 & \text{if } (i,j,\ell) \in \mathscr{S}(\theta,t) \wedge t \in \Xi^{ij}(0,t) \\ \mathscr{U}_\ell^{ij}(t^-) & \text{otherwise} \end{cases} \\
\theta_\ell^{ij}(t) &= \begin{cases} g_\ell^{ij}(t) & \text{if } (i,j,\ell) \in \mathscr{S}(\theta,t) \wedge t \in \Theta^{ij}(0,t) \\ \dfrac{\varepsilon}{2(d^i+d^j)} & \text{if } (i,j,\ell) \in \mathscr{S}(\theta,t) \wedge t \in \Xi^{ij}(0,t) \\ \theta_\ell^{ij}(t^-) & \text{otherwise} \end{cases}
\end{aligned}
\tag{11.18}
$$

In words, the control action $\mathscr{U}^{ij}$ is reset to zero whenever the link (i, j) is in DoS status.[1] In addition to $\mathscr{U}$, also the local clocks are modified upon DoS, yielding a *two-mode* sampling logic. Let $\{t_{\ell_k}^{ij}\}_{\ell_k \in \mathbb{Z}_{\geq 0}}$ denote the sequence of transmission attempts for ℓth component of ξ^{ij} over the link $(i, j) \in \mathscr{E}$. Then, when a communication

[1]Notice that this requires that the nodes are able to detect the occurrence of DoS. This is the case, for instance, with transmitters employing carrier sensing as medium access policy. Another example is when transceivers use TCP-like protocols.

attempt is successful $t^{ij}_{\ell_{k+1}} = t^{ij}_{\ell_k} + g^{ij}_\ell(t)$, and when it is unsuccessful $t^{ij}_{\ell_{k+1}} = t^{ij}_{\ell_k} + \varepsilon/(2(d^i + d^j))$.

In order to characterize the overall network behavior in the presence of DoS. The analysis is subdivided into two main steps: (i) we first prove that all the edge-based controllers eventually stop updating their local controls; and (ii) we then provide conditions on the DoS frequency and duration such that synchronization, in the sense of (11.12), is preserved. This is achieved by resorting to a notion of persistency-of-Communication (PoC), which naturally extends the PoE condition [25] to a digital networked setting by requiring graph connectivity over periods of time that are consistent with the constraints imposed by the communication medium.

As for (i), we have the following result.

Proposition 11.1 (Convergence of the solutions) *Let $(x, \mathscr{X}, \mathscr{U}, \theta)$ be the solutions to (11.8) and (11.18). Then, there exists a finite time T_* such that, for any $(i, j) \in \mathscr{E}$, it holds that $\mathscr{U}^{ij}_\ell(t) = 0$ for all $\ell \in \mathscr{L}$ and for all $t \geqslant T_*$.*

Proof See the appendix. ■

The above result does not allow one to conclude anything about the final disagreement vector in the sense that given a pair of nodes (i, j), the asymptotic value of $|\mathscr{X}^j_\ell(t) - \mathscr{X}^i_\ell(t)|$ and/or $|x^j_\ell(t) - x^i_\ell(t)|$ can be arbitrarily large. As an example, if node i is never allowed to communicate then $\mathscr{X}^i(t) = \mathscr{X}^i(0)$ and the oscillator state $x^i(t)$ satisfies $\dot{x}^i(t) = Ax^i(t)$ with initial condition $-\mathscr{X}^i(0)$ for all $t \in \mathbb{R}_{\geqslant 0}$. In order to recover the same conclusions as in Theorem 11.1, bounds on DoS frequency and duration have to be enforced. The result which follows provides one such characterization. Let $(i, j) \in \mathscr{E}$ be a generic network link, and consider a DoS sequence on (i, j), which satisfies Assumptions 11.2 and 11.3. Define

$$\alpha^{ij} := \frac{1}{\tau^{ij}_d} + \frac{\Delta^{ij}_*}{\tau^{ij}_f} \tag{11.19}$$

where

$$\Delta^{ij}_* := \frac{\varepsilon}{2(d^i + d^j)}. \tag{11.20}$$

As for (ii), we have the following result.

Proposition 11.2 (Persistency-of-communication (PoC)) *Consider any link $(i, j) \in \mathscr{E}$ employing the transmission protocol* (11.18). *Also consider any DoS sequence on (i, j), which satisfies Assumptions 11.2 and 11.3 with η^{ij} and κ^{ij} arbitrary, and τ^{ij}_d and τ^{ij}_f such that $\alpha^{ij} < 1$. Let*

$$\Phi^{ij} := \frac{\kappa^{ij} + (\eta^{ij} + 1)\Delta^{ij}_*}{1 - \alpha^{ij}}. \tag{11.21}$$

Then, for any given unsuccessful transmission attempt $t_{\ell_k}^{ij}$, at least one successful transmission occurs over the link (i, j) within the interval $[t_{\ell_k}^{ij}, t_{\ell_k}^{ij} + \Phi^{ij}]$.

Proof See the appendix. ∎

The following result extends the conclusions of Theorem 11.1 to the presence of DoS.

Theorem 11.4 *Let $(x, \mathscr{X}, \mathscr{U}, \theta)$ be the solution to (11.8) and* (11.18). *For each $(i, j) \in \mathscr{E}$, consider any DoS sequence that satisfies Assumptions 11.2 and 11.3 with η^{ij} and κ^{ij} arbitrary, and τ_d^{ij} and τ_f^{ij} such that $\alpha^{ij} < 1$. Then, $\mathscr{X}$ converges in a finite time T_* to a point $\mathscr{X}^*$ in (11.11), and $\mathscr{U}(t) = \mathbf{0}$ for all $t \geq T_*$. Moreover, for every $\varepsilon_c \in \mathbb{R}_{>0}$ there exists a time $T_c(\varepsilon_c) \geqslant T_*$ such that* (11.12) *is satisfied for all $t \geqslant T_c(\varepsilon_c)$.*

Proof By Proposition 11.1, all the local controls become zero in a finite time T_*. In turn, Proposition 11.2 excludes that this is due to the persistence of a DoS status. Then the result follows along the same lines as in Theorem 11.1. ∎

Remark 11.3 One main reason for considering DoS comes from studying network coordination problems in the presence of possibly malicious attacks. In fact, the proposed modeling framework allows to consider DoS patterns that need not follow a given class of probability distribution, which is instead a common hypothesis when dealing with "genuine" DoS phenomena such as network congestion or communication errors due to low-quality channels. In this respect, [16] discusses how genuine DoS can be incorporated into this modeling framework. ∎

11.4.2 Effect of DoS on the Synchronization Time

By Theorem 11.4, $\dot{\mathscr{X}}$ becomes zero in a finite time T_* after which the network states x exponentially synchronize. Thus, it is of interest to characterize T_*, which amounts to characterizing the effect of DoS on the time needed to achieve synchronization.

Lemma 11.1 (Bound on the convergence time) *Consider the same assumptions as in Theorem 11.4. Then,*

$$T_* \leqslant \left[\frac{1}{\varepsilon} + \frac{d_{\max}}{\varepsilon d_{\min}} + \frac{4 d_{\max}}{\varepsilon^2}\Phi\right] \sum_{i \in \mathscr{I}} \sum_{\ell \in \mathscr{L}} (\eta_\ell^i(0) - x_\ell^i(0))^2, \tag{11.22}$$

where $d_{\min} := \min_{i \in \mathscr{I}} d^i$ and $\Phi := \max_{(i,j) \in \mathscr{E}} \Phi^{ij}$.

Proof Consider the same Lyapunov function V as in the proof of Theorem 11.1. Notice that, by construction of the control law and the scheduling policy, for every successful transmission $t^{ij}_{\ell_k}$ characterized by $|\mathscr{D}^{ij}_\ell(\mathscr{X}(t^{ij}_{\ell_k})| \geqslant \varepsilon$, the function V decreases with rate not less than $\varepsilon/2$ for at least $\varepsilon/(4d_{\max})$ units of time, in which case V decreases by at least $\varepsilon^2/(8d_{\max}) =: \varepsilon_*$. Considering all the network links, such transmissions are in total no more than $\lfloor V(0)/\varepsilon_* \rfloor$ since, otherwise, the function V would become negative. Hence, it only remains to compute the time needed to have $\lfloor V(0)/\varepsilon_* \rfloor$ of such transmissions. In this respect, pick any $t^*_\ell \geqslant 0$ such that consensus has still not been reached on the ℓth component of $\mathscr{X}$. Note that we can have $\mathscr{U}^{ij}_\ell(t^*_\ell) = 0$ for all $(i, j) \in \mathscr{E}$. However, this condition can last only for a limited amount of time. In fact, if $\mathscr{U}^{ij}_\ell(t^*_\ell) = 0$ then the next transmission attempt, say l^{ij}_ℓ, over the link (i, j) and component-ℓ will necessarily occur at a time less than or equal to $t^*_\ell + \Delta^{ij}_*$ with $\Delta^{ij}_* \leqslant \varepsilon/(4d_{\min})$. Let $\mathscr{Q} := [t^*_\ell, t^*_\ell + \Delta^{ij}_*]$, and suppose that over $\mathscr{Q}$ some of the controls $\mathscr{U}^{ij}_\ell$ have remained equal to zero. This implies that for some $(i, j) \in \mathscr{E}$ we necessarily have that l^{ij}_ℓ is unsuccessful. This is because if $\mathscr{U}^{ij}_\ell(t) = 0$ for all $(i, j) \in \mathscr{E}$ and all $t \in \mathscr{Q}$ then $\mathscr{X}^i_\ell(t) = \mathscr{X}^i_\ell(t^*_\ell)$ for all $i \in \mathscr{I}$ and all $t \in \mathscr{Q}$. Hence, if all the l^{ij}_ℓ were successful, we should also have $\mathscr{U}^{ij}_\ell(l^{ij}_\ell) \neq 0$ for some $(i, j) \in \mathscr{E}$ since, by hypothesis, consensus is not reached at time t^*_ℓ. Hence, applying Proposition 11.2 we conclude that at least one of the controls $\mathscr{U}^{ij}_\ell$ will become nonzero before $l^{ij}_\ell + \Phi^{ij}$. As each vector component ℓ has the same Δ^{ij}_*, at least one of the control vectors $\mathscr{U}^{ij}$ will become nonzero before the same amount of time. Overall, this implies that at least one control will become nonzero before $\varepsilon/(4d_{\min}) + \Phi$ units of time have elapsed. Since t^*_ℓ is generic, we conclude that V decreases by at least ε_* every $\varepsilon/(4d_{\max}) + \varepsilon/(4d_{\min}) + \Phi$ units of time, which implies that

$$T_* \leqslant \left[\frac{\varepsilon}{4d_{\max}} + \frac{\varepsilon}{4d_{\min}} + \Phi\right]\frac{V(0)}{\varepsilon_*}. \tag{11.23}$$

The thesis follows by recalling that $V(0)$ can be rewritten as

$$V(0) = \frac{1}{2}\sum_{i\in\mathscr{I}}\sum_{\ell\in\mathscr{L}}(\mathscr{X}^i_\ell(0))^2. \tag{11.24}$$

■

11.5 A Numerical Example

We consider a random (connected) undirected graph with $N = 6$ nodes and with $d^i = 2$ for all $i \in \mathscr{I}$. Each node has harmonic oscillator dynamics of the form

Table 11.1 DoS average duty cycle over links

Link (i, j)	Duty cycle (%)	Link (i, j)	Duty cycle (%)
$\{1, 2\}$	56.07 %	$\{1, 4\}$	55.12 %
$\{2, 3\}$	55.2 %	$\{3, 6\}$	56.3 %
$\{4, 5\}$	66.06 %	$\{5, 6\}$	59.72 %

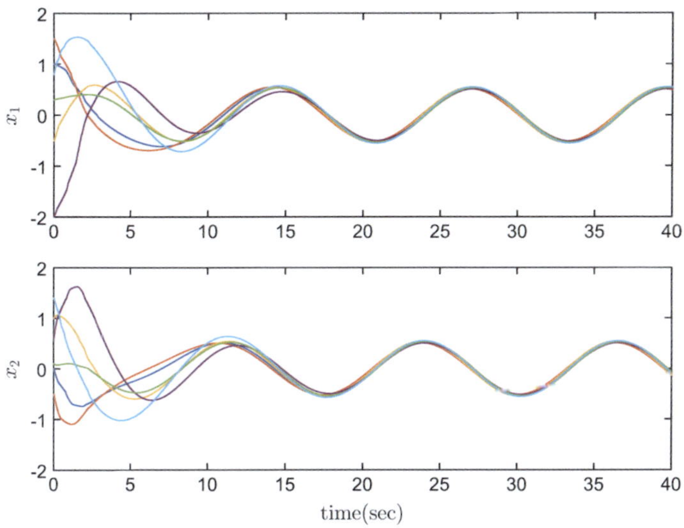

Fig. 11.1 Evolution of x, corresponding to the solution to (11.1)–(11.3) and (11.18) for a random graph with $N = 6$ nodes in the presence of DoS

$$\dot{x}^i(t) = \begin{bmatrix} 0 & 1 \\ -1 & 0 \end{bmatrix} x^i(t) + \begin{bmatrix} 0 \\ 1 \end{bmatrix} u^i(t). \tag{11.25}$$

The nodes initial values are randomly within interval $[-2, 2]$ and $(\eta(0), \xi(0), \theta(0)) = (\mathbf{0}, \mathbf{0}, \mathbf{0})$.

In the simulations, we considered DoS attacks which affect each of the network links independently. For each link, the corresponding DoS pattern takes the form of a pulse-width-modulated signal with variable period and duty cycle (maximum period of 0.4sec and maximum duty cycle equal to 55%), both generated randomly. These patterns are reported in Table 11.1 for each network link.

The evolution of x, corresponding to the solutions to (11.1)–(11.3) and (11.18) with $\varepsilon = 0.04$ is depicted in Fig. 11.1. One sees that x exhibits a quite smooth response. In fact, the impact of loss of information can be better appreciated by looking at the controller dynamics, which are reported in Figs. 11.2 and 11.3. This can be explained simply by noting that the controller state ξ is affected by DoS directly while x is affected by DoS indirectly since ξ enters the node dynamics after being filtered twice.

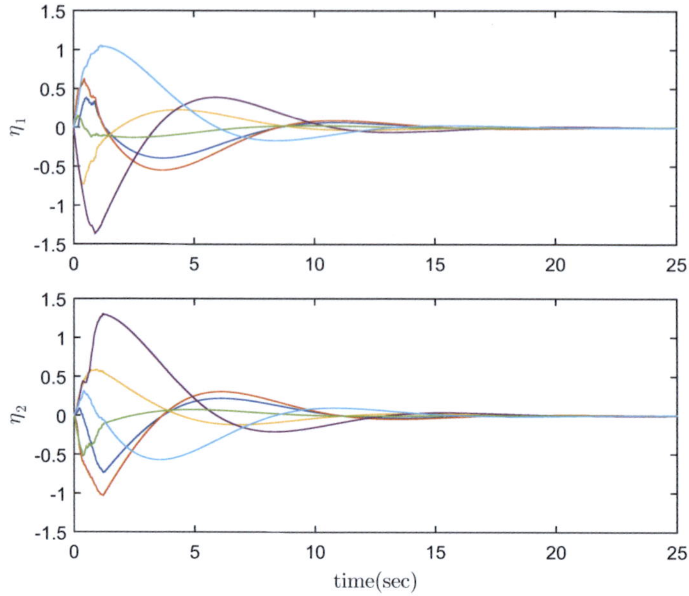

Fig. 11.2 Evolution of the controller state η in the absence of DoS

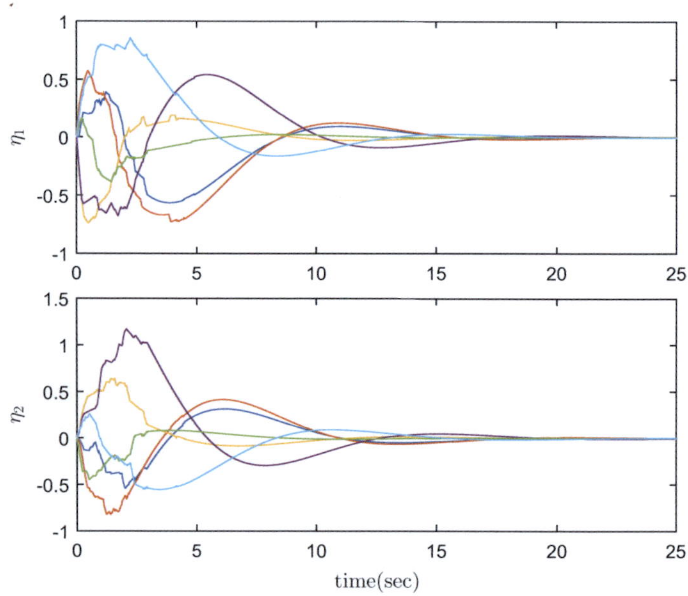

Fig. 11.3 Evolution of the controller state η in the presence of DoS

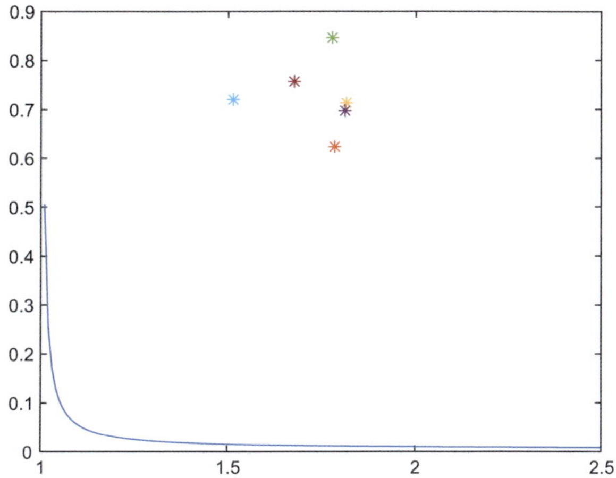

Fig. 11.4 Locus of the points $1/\tau_d + \Delta_*/\tau_f^{ij} = 1$ as a function of (τ_d, τ_f) with $\Delta_* = 0.05$ (blue solid line). The horizontal axis represents τ_d and the vertical axis represents τ_f. Notice that $\Delta_* = \Delta_*^{ij}$ for all $(i, j) \in \mathscr{E}$, so that the locus of point does not vary with (i, j). The various "$*$" represent the values of $(\tau_d^{ij}, \tau_f^{ij})$ for the network links

As a final comment, note that for each DoS pattern one can compute corresponding values for $(\eta^{ij}, \kappa^{ij}, \tau_f^{ij}, \tau_d^{ij})$. They can be determined by computing $n^{ij}(\tau, t)$ and $|\Xi^{ij}(\tau, t)|$ of each DoS pattern (cf. Assumptions 11.2 and 11.3) over the considered simulation horizon. Figure 11.4 depicts the values obtained for τ_f^{ij} and τ_d^{ij} for each $(i, j) \in \mathscr{E}$. One sees that these values are consistent with the requirements imposed by the PoC condition.

11.6 Conclusions

In this chapter, we have investigated Self-Triggered synchronization of group of harmonic oscillators in presence of denial-of-service at communication links. In the considered framework each of the network links fail independently, which is relevant for peer-to-peer networks architectures. A characterization of DoS frequency and duration is provided under which network synchronization is preserved, along with an explicit estimate of the effect of DoS on the time required to achieve synchronization.

Appendix

Proof of Theorem 11.1 As a first step, we analyze the consensus of subsystem $(\mathscr{X}, \mathscr{U}, \theta)$. Afterward, we will investigate the synchronization of the states x^i throughout the relation $\mathscr{X}^i(t) = e^{-At}(\eta^i(t) - x^i(t))$.

Consider the Lyapunov function $V(\mathscr{X}) = \frac{1}{2}\mathscr{X}^\top\mathscr{X}$, and let $t_{\ell_k}^{ij} := \max\{t_l^{ij} : t_l^{ij} \leqslant t, l \in \mathbb{Z}_{\geqslant 0}\}$. The derivative of V along the solutions to (11.8) satisfies

$$\begin{aligned}\dot{V}(\mathscr{X}(t)) &= \sum_{i=1}^{N} \mathscr{X}^{i\top}(t)\,\dot{\mathscr{X}}^{i}(t)\\ &= -\sum_{(i,j)\in\mathscr{E}} (\mathscr{X}^{j}(t)-\mathscr{X}^{i}(t))^{\top}\mathscr{U}^{ij}(t^{ij}_{\ell_k})\\ &= -\sum_{(i,j)\in\mathscr{E}}\sum_{\ell=1}^{n} \mathscr{D}^{ij}_{\ell}(\mathscr{X}(t))\,\mathrm{sign}_{\varepsilon}(\mathscr{D}^{ij}_{\ell}(\mathscr{X}(t^{ij}_{\ell_k}))).\end{aligned} \tag{11.26}$$

During the continuous evolution $|\dot{\mathscr{D}}^{ij}_{\ell}(\mathscr{X}(t))| \le d^i + d^j$ for $t \in [t^i_k, t^i_{k+1}[$, where $\mathscr{D}^{ij}(\mathscr{X}(t)) = \mathscr{X}^j(t) - \mathscr{X}^i(t)$. Exploiting this fact and recalling the definition of $g^{ij}_{\ell}(\mathscr{X}(t))$ in (11.10), it holds that if $|\mathscr{D}^{ij}_{\ell}(\mathscr{X}(t^{ij}_{\ell_k}))| \ge \varepsilon$ then

$$\begin{aligned}|\mathscr{D}^{ij}_{\ell}(\mathscr{X}(t))| &\ge |\mathscr{D}^{ij}_{\ell}(\mathscr{X}(t^{ij}_{\ell_k}))| - (d^i + d^j)(t - t^{ij}_{\ell_k})\\ &\ge \frac{|\mathscr{D}^{ij}_{\ell}(\mathscr{X}(t^{ij}_{\ell_k}))|}{2}\end{aligned} \tag{11.27}$$

and

$$\mathrm{sign}_{\varepsilon}(\mathscr{D}^{ij}_{\ell}(\mathscr{X}(t))) = \mathrm{sign}_{\varepsilon}(\mathscr{D}^{ij}_{\ell}(\mathscr{X}(t^{ij}_{\ell_k}))). \tag{11.28}$$

Using (11.27) and (11.28) we conclude that

$$\dot{V}(\mathscr{X}(t)) \le -\sum_{(i,j)\in\mathscr{E}} \sum_{\substack{\ell\in\mathscr{L}:\\ |\mathscr{D}^{ij}_{\ell}(\mathscr{X}(t^{ij}_{\ell_k}))|\ge\varepsilon}} \frac{|\mathscr{D}^{ij}_{\ell}(\mathscr{X}(t^{ij}_{\ell_k}))|}{2} \tag{11.29}$$

In view of (11.29), there must exist a finite time T such that, for every $(i, j) \in \mathscr{E}$ and every k, ℓ with $t^{ij}_{\ell_k} \geqslant T$, it holds that $|\mathscr{D}^{ij}_{\ell}(\mathscr{X}(t^{ij}_{\ell_k}))| < \varepsilon$. This is because, otherwise, V would become negative. The inequality in (11.11) follows by recalling that, in a graph with N nodes the graph diameter is $N-1$. This shows that $\mathscr{X}$ converges in a finite time T to a point $\mathscr{X}_*$ in the set $\mathscr{E}$.

We now focus on x. In view of (11.2), $\mathscr{U}$ converges to zero in a finite time. Moreover, in view of (11.7), we have that $\eta^i(t) - x^i(t)$ converges to $e^{At}\mathscr{X}^i_*$ and ξ to $\mathbf{0}$ in a finite time. As for η, recall that η^i has flow and jump dynamics given by

$$\begin{aligned}\dot{\eta}^i(t) &= (A + BK)\eta^i(t) + \sum_{j\in\mathscr{N}_i} \xi^{ij}(t)\\ \eta^i(t) &= \eta^i(t^-).\end{aligned} \tag{11.30}$$

Hence, η converges exponentially to the origin since ξ converges to $\mathbf{0}$ is a finite time and $A + BK$ is Hurwitz. Combining this fact with the property that $\eta^i(t) - x^i(t)$ convergence asymptotically to $e^{At}\mathscr{X}^i_*$, we have that $x^i(t)$ convergence asymptotically to

$-e^{At}\mathscr{X}_*^i$. This implies that for any node $i \in \mathscr{I}$ and any $\varepsilon_c \in \mathbb{R}_{>0}$, there exists a time $T_c(\varepsilon_c)$ after which $\|x^i(t) + e^{At}\mathscr{X}_*^i\| \leq \varepsilon_c$, where $\|\cdot\|$ stands for Euclidean norm.

Notice that, in general, $\mathscr{X}_*^i \neq \mathscr{X}_*^j$ for $i \neq j$ in accordance with the practical consensus property (11.11). Therefore, the solutions x^i and x^j for all $(i, j) \in \mathscr{I} \times \mathscr{I}$ will achieve practical consensus as well. In particular, an upper bound on their disagreement level can be estimated as

$$\begin{aligned} \|x^i(t) - x^j(t)\| &\leq \|x^i(t) + e^{At}\mathscr{X}_*^i\| + \|x^j(t) + e^{At}\mathscr{X}_*^i\| \\ &\leqslant \|x^i(t) + e^{At}\mathscr{X}_*^i\| + \|x^j(t) + e^{At}\mathscr{X}_*^j\| + \|e^{At}\mathscr{X}_*^i - e^{At}\mathscr{X}_*^j\| \\ &\leq 2\varepsilon_c + \|e^{At}(\mathscr{X}_*^j - \mathscr{X}_*^i)\| \\ &\leq 2\varepsilon_c + \sqrt{n}\,\delta \end{aligned} \tag{11.31}$$

where the last inequality is obtained from (11.11) and the fact that A has purely imaginary eigenvalues by hypothesis. This concludes the proof. ■

Proof of Proposition 1 Reasoning as in the proof of Theorem 11.1, it is an easy matter to see that in the presence of DoS (11.29) modifies into

$$\dot{V}(\mathscr{X}(t)) \leq - \sum_{(i,j)\in\mathscr{E}} \sum_{\substack{\ell\in\mathscr{L}: \\ |\mathscr{D}_\ell^{ij}(\mathscr{X}(t_{\ell_k}^{ij}))|\geq\varepsilon \,\wedge \\ t_{\ell_k}^{ij}\in\Theta^{ij}(0,t)}} \frac{|\mathscr{D}_\ell^{ij}(\mathscr{X}(t_{\ell_k}^{ij}))|}{2}. \tag{11.32}$$

In words, the derivative of V decreases whenever, for some $(i, j) \in \mathscr{E}$, $\ell \in \mathscr{L}$, two conditions are met: (i) $|\mathscr{D}_\ell^{ij}(\mathscr{X}(t_{\ell_k}^{ij}))| \geqslant \varepsilon$, which means that i and j are not component-wise ε-close; and (ii) communication on the link that connects i and j is possible.

From (11.32) there must exist a finite time T_* such that, for every $\{i, j, \ell\} \in \mathscr{E} \times \mathscr{L}$ and every k with $t_{\ell_k}^{ij} \geqslant T_*$, it holds that $|\mathscr{D}_\ell^{ij}(\mathscr{X}(t_{\ell_k}^{ij}))| < \varepsilon$ or $t_{\ell_k}^{ij} \in \Xi^{ij}(0, t)$. This is because, otherwise, V would become negative. The proof follows by recalling that in both the cases $|\mathscr{D}_\ell^{ij}(\mathscr{X}(t_{\ell_k}^{ij}))| < \varepsilon$ and $t_{\ell_k}^{ij} \in \Xi^{ij}(0, t)$ the control $\mathscr{U}_\ell^{ij}(t)$ is set equal to zero. ■

Proof of Proposition 11.2 Consider any link $(i, j) \in \mathscr{E}$, and suppose that a certain transmission attempt $t_{\ell_k}^{ij}$ is unsuccessful. We claim that a successful transmission over the link (i, j) does always occur within $[t_{\ell_k}^{ij}, t_{\ell_k}^{ij} + \Phi^{ij}]$. We prove the claim by contradiction. To this end, we first introduce a number of auxiliary quantities. Denote by $\bar{H}_n^{ij} := \{h_n^{ij}\} \cup [h_n^{ij}, h_n^{ij} + \tau_n^{ij} + \Delta_*^{ij}$ [the nth DoS interval over the link (i, j) prolonged by Δ_*^{ij} units of time. Also, let

$$\bar{\Xi}^{ij}(\tau, t) := \bigcup_{n\in\mathbb{Z}_{\geq 0}} \bar{H}_n^{ij} \bigcap [\tau, t] \tag{11.33}$$

$$\bar{\Theta}^{ij}(\tau, t) := [\tau, t] \setminus \bar{\Xi}^{ij}(\tau, t). \tag{11.34}$$

Suppose then that the claim is false, and let $t_\ell^\star$ denote the last transmission attempt over $[t_{\ell_k}^{ij}, t_{\ell_k}^{ij} + \Phi^{ij}]$. Notice that this necessarily implies $|\bar{\Theta}^{ij}(t_{\ell_k}^{ij}, t_\ell^\star)| = 0$. To see this, first note that, in accordance with (11.18), the inter-sampling time over the interval $[t_{\ell_k}^{ij}, t_\ell^\star]$ is equal to $\varepsilon/(2(d^i + d^j)) = \Delta_*^{ij}$. Hence, we cannot have $|\bar{\Theta}^{ij}(t_{\ell_k}^{ij}, t_\ell^\star)| > 0$ since this would imply the existence of a DoS-free interval within $[t_{\ell_k}^{ij}, t_\ell^\star]$ of length greater than Δ_*^{ij}, which is not possible since, by hypothesis, no successful transmission attempt occurs within $[t_{\ell_k}^{ij}, t_\ell^\star]$. Thus $|\bar{\Theta}^{ij}(t_{\ell_k}^{ij}, t_\ell^\star)| = 0$. Moreover, since $t_\ell^\star$ is unsuccessful, it must be contained in a DoS interval, say H_q^{ij}. This implies $[t_\ell^\star, t_\ell^\star + \Delta_*^{ij}[\subseteq \bar{H}_q^{ij}$. Hence, we have

$$
\begin{aligned}
|\bar{\Theta}^{ij}(t_{\ell_k}^{ij}, t_\ell^\star + \Delta_*^{ij})| &= |\bar{\Theta}^{ij}(t_{\ell_k}^{ij}, t_\ell^\star)| + |\bar{\Theta}^{ij}(t_\ell^\star, t_\ell^\star + \Delta_*^{ij})| \\
&= 0
\end{aligned}
\tag{11.35}
$$

However, condition $|\bar{\Theta}(t_{\ell_k}^{ij}, t_\ell^\star + \Delta_*^{ij})| = 0$ is not possible. To see this, notice that

$$
\begin{aligned}
|\bar{\Theta}^{ij}(t_{\ell_k}^{ij}, t)| &= t - t_{\ell_k}^{ij} - |\bar{\Xi}^{ij}(t_{\ell_k}^{ij}, t)| \\
&\geqslant t - t_{\ell_k}^{ij} - |\Xi^{ij}(t_{\ell_k}^{ij}, t)| - (n(t_{\ell_k}^{ij}, t) + 1)\Delta_*^{ij} \\
&\geqslant (t - t_{\ell_k}^{ij})(1 - \alpha^{ij}) - \kappa^{ij} - (\eta^{ij} + 1)\Delta_*^{ij}
\end{aligned}
\tag{11.36}
$$

for all $t \geqslant t_{\ell_k}^{ij}$ where the first inequality follows from the definition of the set $\bar{\Xi}^{ij}(\tau, t)$ while the second one follows from Assumptions 11.2 and 11.3. Hence, by (11.36), we have $|\bar{\Theta}^{ij}(t_{\ell_k}^{ij}, t)| > 0$ for all $t > t_{\ell_k}^{ij} + (1 - \alpha^{ij})^{-1}(\kappa^{ij} + (\eta^{ij} + 1)\Delta_*^{ij}) = t_{\ell_k}^{ij} + \Phi^{ij}$. Accordingly, $|\bar{\Theta}(t_{\ell_k}^{ij}, t_\ell^\star + \Delta_*^{ij})| = 0$ cannot occur because $t_\ell^\star + \Delta_*^{ij} > t_{\ell_k}^{ij} + \Phi^{ij}$. In fact, by hypothesis, $t_\ell^\star$ is defined as the last unsuccessful transmission attempt within $[t_{\ell_k}^{ij}, t_{\ell_k}^{ij} + \Phi^{ij}]$, and, by (11.18), the next transmission attempt after $t_\ell^\star$ occurs at time $t_\ell^\star + \Delta_*^{ij}$. This concludes the proof. ■

References

1. Sandberg, H., Amin, S., Johansson, K.: Cyberphysical security in networked control systems: an introduction to the issue. IEEE Control Syst. **35**(1), 20–23 (2015)
2. Cardenas, A.A., Amin, S., Sastry, S.: Secure control: towards survivable cyber-physical systems. In: The 28th International Conference on Distributed Computing Systems Workshops, pp. 495–500 (2008)
3. Fawzi, H., Tabuada, P., Diggavi, S.: Secure state-estimation for dynamical systems under active adversaries. In: 2011 49th Annual Allerton Conference on Communication, Control, and Computing (Allerton), pp. 337–344 (2011)
4. Pasqualetti, F., Dorfler, F., Bullo, F.: Control-theoretic methods for cyberphysical security: geometric principles for optimal cross-layer resilient control systems. IEEE Control Syst. **35**(1), 110–127 (2015)

5. Xu, W., Ma, K., Trappe, W., Zhang, Y.: Jamming sensor networks: attack and defense strategies. IEEE Netw. **20**(3), 41–47 (2006)
6. Thuente, D., Acharya, M.: Intelligent jamming in wireless networks with applications to 802.11 b and other networks. In: Proceedings of the 25th IEEE Communications Society Military Communications Conference (MILCOM06), pp. 1–7. Washington, DC (2006)
7. Amin, S., Càrdenas, A., Sastry, S.: Safe and secure networked control systems under denial of-service attacks. In: Hybrid Systems: Computation and Control, pp. 31–45 (2009)
8. Gupta, A., Langbort, C., Basar, T.: Optimal control in the presence of an intelligent jammer with limited actions. In: Proceedings of the IEEE Conference on Decision and Control, pp. 1096–1101 (2010)
9. Befekadu, G., Gupta, V., Antsaklis, P.: Risk-sensitive control under a class denial-of-service attack models. In: 2011 American Control Conference. CA, USA, San Francisco (2011)
10. Teixeira, A., Shames, I., Sandberg, H., Johansson, K.H.: A secure control framework for resource-limited adversaries. Automatica **51**, 135–148 (2015)
11. Foroush, H.S., Martínez, S.: On event-triggered control of linear systems under periodic denial-of-service jamming attacks. In: Proceedings of the IEEE Conference on Decision and Control, pp. 2551–2556 (2012)
12. De Persis, C., Tesi, P.: Resilient control under denial-of-service. In: Proceedings of the IFAC World Conference. Cape Town, South Africa, pp. 134–139 (2013)
13. De Persis, C., Tesi, P.: Input-to-state stabilizing control under denial-of-service. IEEE Trans. Autom. Control **60**, 2930–2944 (2015)
14. De Persis, C., Tesi, P.: Networked control of nonlinear systems under denial-of-service. Syst. Control Lett. **96**, 124–131 (2016)
15. Senejohnny, D., Tesi, P., De Persis, C.: Self-triggered coordination over a shared network under denial-of-service. In: Proceedings of the IEEE Conference on Decision and Control, pp. 3469–3474. Osaka, Japan (2015)
16. Senejohnny, D., Tesi, P., De Persis, C.: A jamming-resilient algorithm for self-triggered network coordination. IEEE Trans. Control Netw. Syst. **PP**, 1–1 (2017). Inpress
17. Scardovi, L., Sepulchre, R.: Synchronization in networks of identical linear systems. Automatica **45**(11), 2557–2562 (2009)
18. De Persis, C., Frasca, P.: Robust self-triggered coordination with ternary controllers. IEEE Trans. Autom. Control **58**(12), 3024–3038 (2013)
19. De Persis, C.: On self-triggered synchronization of linear systems. Estim. Control Networked Syst. **4**(1), 247–252 (2013)
20. Xu, W., Trappe, W., Zhang, Y., Wood, T.: The feasibility of launching and detecting jamming attacks in wireless networks. In: Proceedings of the 6th ACM International Symposium on Mobile Ad Hoc Networking and Computing, pp. 46–57. ACM (2005)
21. Tague, P., Li, M., Poovendran, R.: Mitigation of control channel jamming under node capture attacks. IEEE Trans. Mob. Comput. **8**(9), 1221–1234 (2009)
22. Senejohnny, D., Tesi, P., De Persis, C.: Resilient self-triggered network synchronization. In: Proceedings of the IEEE Conference on Decision and Control, pp. 489–494. Las Vegas, USA (2016)
23. Olfati-Saber, R., Murray, R.M.: Consensus problems in networks of agents with switching topology and time-delays. IEEE Trans. Autom. Control **49**(9), 1520–1533 (2004)
24. Jadababaie, A., Lin, J., Morse, A.: Coordination of groups of mobile autonomous agents using nearest neighbour rules. IEEE Trans. Autom. Control **48**(6), 988–1001 (2003)
25. Arcak, M.: Passivity as a design tool for group coordination. IEEE Trans. Autom. Control **52**(8), 1380–1390 (2007)
26. Velasco, P.M.M., Fuertes, J.: The self-triggered task model for real-time control systems. In: Proceedings of 24th IEEE Real-Time Systems Symposium, Work-in-Progress Session (2003)
27. Bellardo, J., Savage, S.: 802.11 denial-of-service attacks: real vulnerabilities and practical solutions. In: USENIX security, pp. 15–28 (2003)
28. Pelechrinis, K., Iliofotou, M., Krishnamurthy, S.V.: Denial of service attacks in wireless networks: the case of jammers. IEEE Commun. Surv. Tutor. **13**(2), 245–257 (2011)

29. DeBruhl, B., Tague, P.: Digital filter design for jamming mitigation in 802.15.4 communication. In: 2011 Proceedings of 20th International Conference on Computer Communications and Networks (ICCCN), pp. 1–6 (2011)
30. Wood, A.D., Stankovic, J., et al.: Denial of service in sensor networks. Computer **35**(10), 54–62 (2002)

Part III
Distributed Control of Cyber-Physical Systems

Chapter 12
Distributed Hybrid Control Synthesis for Multi-Agent Systems from High-Level Specifications

M. Guo, D. Boskos, J. Tumova and D. V. Dimarogonas

Abstract Current control applications necessitate in many cases the consideration of systems with multiple interconnected components. These components/agents may need to fulfill high-level tasks at a discrete planning layer and also coupled constraints at the continuous control layer. Toward this end, the need for combined decentralized control at the continuous layer and planning at the discrete layer becomes apparent. While there are approaches that handle the problem in a top-down centralized manner, decentralized bottom-up approaches have not been pursued to the same extent. We present here some of our results for the problem of combined, hybrid control and task planning from high-level specifications for multi-agent systems in a bottom-up manner. In the first part, we present some initial results on extending the necessary notion of abstractions to multi-agent systems in a distributed fashion. We then consider a setup where agents are assigned individual tasks in the form of linear temporal logic (LTL) formulas and derive local task planning strategies for each agent. In the last part, the problem of combined distributed task planning and control under coupled continuous constraints is further considered.

M. Guo · D. Boskos · D. V. Dimarogonas (✉)
ACCESS Linnaeus Center and Center for Autonomous Systems, KTH Royal Institute of Technology, 100 44, Stockholm, Sweden
e-mail: dimos@kth.se

M. Guo
e-mail: mengg@kth.se

D. Boskos
e-mail: boskos@kth.se

J. Tumova
Robotics, Perception, and Learning Department, KTH Royal Institute of Technology, 100 44, Stockholm, Sweden
e-mail: tumova@kth.se

S. Tarbouriech et al. (eds.), *Control Subject to Computational and Communication Constraints*, Lecture Notes in Control and Information Sciences 475, https://doi.org/10.1007/978-3-319-78449-6_12

12.1 Introduction

We consider multi-agent systems that need to fulfill high-level tasks, e.g., to periodically reach a given subset of states, or to reach a certain set of states in a particular order, and also undergo dynamically coupled constraints, e.g., to maintain connectivity, or avoid collisions. Toward this end, the need for combined decentralized control at the continuous layer and planning at the discrete layer becomes apparent. While there are approaches that handle the problem in a top-down centralized manner, decentralized bottom-up approaches have not been pursued to the same extent. We present here some of our results for the problem of hybrid control synthesis of multi-agent systems under high-level specifications in a bottom-up manner. This approach can enhance various system properties, such as reducing the computational complexity induced by the number of agents, improving modularity in terms of new agents entering the system, and incorporating fault tolerance, robustness, and adaptability to changes in the workspace.

The next part focuses on distributed abstractions of the multi-agent system. We assume that the agents' dynamics consist of feedback interconnection terms, which represent dynamically coupled constraints of the multi-agent system, and additional bounded input terms, which we call free inputs and provide the ability for motion planning under the coupled constraints. For the derivation of the symbolic models, we quantify admissible space-time discretizations in order to capture reachability properties of the original system. We provide sufficient conditions which establish that the abstraction of our original system is well posed, in the sense that the finite transition system which serves as an abstract model for the motion capabilities of each agent has at least one outgoing transition for every discrete state. Each agent's abstract model is based on the knowledge of its neighbors' discrete positions and the transitions are performed through the selection of appropriate hybrid control laws in place of the agent's free input, which enables the manipulation of the coupling terms and can drive the agent to its possible successor states. In addition, the derived discretizations include parameters whose tuning enables multiple transitions and provides quantifiable motion planning capabilities for the system. Finally, the corresponding results are generalized by allowing for a varying degree of decentralization i.e., by building each agent's abstract model based on the knowledge of its neighbors' discrete positions up to a tunable distance in the communication graph.

In the next part, we deal with dependent temporal logic specifications at the discrete planning level. Namely, the agents' behaviors are limited by mutually independent temporal logic constraints, allowing to express safety, surveillance, sequencing, or reachability properties of their traces, and, at the same time, a part of the specification expresses the agents' tasks in terms of the services to be provided along the trace. These may impose requests for the other agent's collaborations. We propose a two-phase solution based on automata-based model checking, in which the planning procedure for the two types of specifications is systematically decoupled. While this procedure significantly reduces the cost in the case of sparse dependencies, it meets the complexity the centralized solution at worst case. We introduce an additional iterative limited horizon planning technique as a complementary technique.

We then tackle the multi-agent control problem under local temporal logic tasks and continuous-time constraints. The local tasks are dependent due to collaborative services, while at the same time the agents are subject to dynamic constraints with their neighboring agents. Thus, integration of the continuous motion control with the high-level discrete network structure control is essential. Particularly, the agents are subject to relative-distance constraints which relate to the need of maintaining connectivity of the overall network. The local tasks capture the temporal requirements on the agent's actions, while the relative-distance constraints impose requirements on the collective motion of the whole team. Our approach to the problem involves an offline and an online step. In the offline step, we synthesize a high-level plan in the form of a sequence of services for each of the agents. In the online step, we dynamically switch between the high-level plans through leader election and choose the associated continuous controllers. The whole team then follows the leader toward until its next service is provided and then a new leader is selected. It is guaranteed that each agent's local task will be accomplished and the communication network remains connected at all time.

12.2 Decentralized Abstractions

12.2.1 Introduction

In this section, we focus on multi-agent systems with continuous dynamics consisting of feedback terms, which induce coupling constraints, and bounded additive inputs, which provide the agents' control capabilities. The feedback interconnection between the agents can represent internal dynamics of the system, or alternatively, a control design guaranteeing certain system properties (e.g., network connectivity or collision avoidance), which appears often in the multi-agent literature. The results are based on our recent works [3] and [4], which provide sufficient conditions for the existence of distributed discrete models for multi-agent systems with coupled dynamics. In particular, our main goal is to obtain a partition of the workspace into cells and select a transition time step, in order to derive for each agent an abstract discrete model with at least one outgoing transition from each discrete state. Compositional approaches for symbolic models of interconnected systems have been also studied in the recent works [7, 17, 19, 20], and [21], and are primarily focused on the discrete time case.

12.2.2 Problem Formulation

We consider multi-agent systems of the form

$$\dot{x}_i = f_i(x_i, \mathbf{x}_j) + v_i, x_i \in \mathbf{R}^n, i \in \mathcal{N}, \tag{12.1}$$

where $\mathcal{N} := \{1, \ldots, N\}$ stands for the agents' set. Each agent is assumed to have a fixed number N_i of neighbors $j_1, \ldots, j_{N_i}$. The dynamics in (12.1) are decentralized and consist for each $i \in \mathcal{N}$ of a feedback term $f_i(\cdot)$, which depends on i's state x_i and the states of its neighbors, which are compactly denoted by $\mathbf{x}_j (= \mathbf{x}_{j(i)}) := (x_{j_1}, \ldots, x_{j_{N_i}})$, and an additional input term v_i, which we call free input. We assume that the feedback terms $f_i(\cdot)$ are globally bounded, namely, there exists a constant $M > 0$ such that

$$|f_i(x_i, \mathbf{x}_j)| \leqslant M, \forall (x_i, \mathbf{x}_j) \in \mathbf{R}^{(N_i+1)n} \tag{12.2}$$

and that they are globally Lipschitz. Thus, there exist constants $L_1, L_2 > 0$, such that

$$|f_i(x_i, \mathbf{x}_j) - f_i(x_i, \mathbf{y}_j)| \leqslant L_1 |(x_i, \mathbf{x}_j) - (x_i, \mathbf{y}_j)|, \tag{12.3}$$

$$|f_i(x_i, \mathbf{x}_j) - f_i(y_i, \mathbf{x}_j)| \leqslant L_2 |(x_i, \mathbf{x}_j) - (y_i, \mathbf{x}_j)|, \tag{12.4}$$

for all $x_i, y_i \in \mathbf{R}^n$, $\mathbf{x}_j, \mathbf{y}_j \in \mathbf{R}^{N_i n}$ and $i \in \mathcal{N}$. Furthermore, we consider piecewise continuous free inputs v_i that satisfy the bound

$$|v_i(t)| \leqslant v_{\max}, \forall t \geqslant 0, i \in \mathcal{N}. \tag{12.5}$$

The coupling terms $f_i(x_i, \mathbf{x}_j)$ are encountered in a large set of multi-agent protocols [16], including consensus, connectivity maintenance, collision avoidance, and formation control. In addition, (12.1) may represent internal dynamics of the system as for instance in the case of smart buildings (see e.g., [1]). It is also assumed that the maximum magnitude of the feedback terms is higher than that of the free inputs, namely, that $v_{\max} < M$. This assumption is in part motivated by the fact that we are primarily interested in maintaining the property that the feedback is designed for, and secondarily, in exploiting the free inputs in order to accomplish high-level tasks. A class of multi-agent systems of the form (12.1) which justifies this assumption has been studied in our companion work [5], which is focused on robust network connectivity maintenance by means of bounded feedback laws. It is worthwhile mentioning that all these assumptions are removed in [6], where the discrete models are built online over a bounded time horizon, and require only forward completeness of the system's trajectories.

In what follows, we consider a cell decomposition $\mathbf{S} = \{S_l\}_{l \in \mathcal{I}}$ of the state space $\mathbf{R}^n$, which can be regarded as a partition of $\mathbf{R}^n$, and a time step $\delta t > 0$. We will refer to this selection as a space and time discretization. Given the indices $\mathcal{I}$ of a decomposition, we use the notation $\mathbf{l}_i = (l_i, l_{j_1}, \ldots, l_{j_{N_i}}) \in \mathcal{I}^{N_i+1}$ to denote the indices of the cells where agent i and its neighbors belong and call it the cell configuration of i. Our goal is to build an individual transition system of each agent i with state set the cells of the decomposition, actions determined through the possible cells of its neighbors, and transition relation specified as follows. Given the initial cells of agent i and its neighbors, it is possible for i to perform a transition to a final cell, *if for all states in its initial cell there exists a free input, such that its trajectory will reach the*

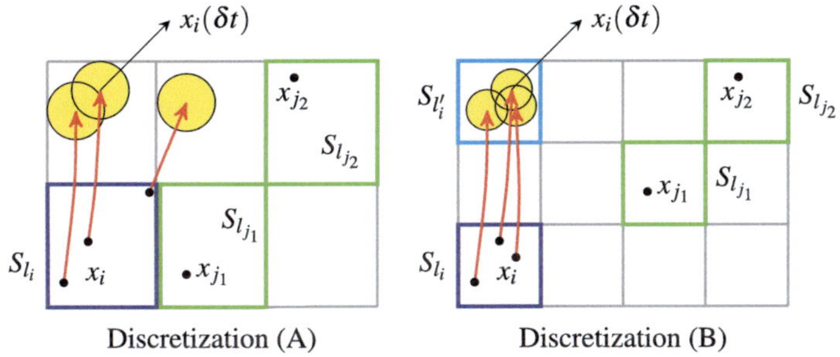

Fig. 12.1 Illustration of a non-well posed (A) and a well posed (B) discretization

final cell at time δt, for all possible initial states of its neighbors in their cells, and their corresponding free inputs.

For the synthesis of high-level plans, we require the discretization to be well posed, in the sense that for each agent and any initial cell it is possible to perform a transition to at least one final cell. In order to illustrate the concept of a well-posed space-time discretization, consider the cell decompositions depicted in Fig. 12.1 and a time step δt. For both decompositions in the figure we depict a cell configuration of agent i and represent the endpoints of agent's i trajectories at time δt through the tips of the arrows. In the left decomposition, we select three distinct initial conditions of i and observe that the corresponding reachable sets at δt lie in different cells. Thus, given this cell configuration of i it is not possible to find a cell in the decomposition which is reachable from every point in the initial cell, and we conclude that the discretization is not well posed for the system. In the right figure, we observe however that for the three distinct initial positions in cell S_{l_i}, it is possible to drive agent i to cell $S_{l'_i}$ at time δt. We assume that this is possible for all initial conditions in this cell and irrespectively of the initial conditions of i's neighbors in their cells and the inputs they choose. By additionally assuming this property for all configurations of the agents, we establish a well posed discretization for the system.

12.2.3 *Derivation of Well-Posed Discretizations*

In order to enable the desired transitions of each agent in the presence of the coupling terms $f_i(\cdot)$, we assign hybrid control laws to the free inputs v_i. We next provide the specific feedback laws that are utilized therefore. Consider first a cell decomposition $\mathbf{S} = \{S_l\}_{l\in\mathcal{I}}$ of $\mathbf{R}^n$ and a time step δt. For each agent $i \in \mathcal{N}$ and cell configuration $\mathbf{l}_i = (l_i, l_{j_1}, \ldots, l_{j_{N_i}})$ of i select an $N_i + 1$-tuple of reference points $(x_{i,G}, \mathbf{x}_{j,G}) \in S_{l_i} \times (S_{l_{j_1}} \times \cdots \times S_{l_{j_{N_i}}})$ and define $F_{i,\mathbf{l}_i}(x_i) := f_i(x_i, \mathbf{x}_{j,G})$, $x_i \in \mathbf{R}^n$. Also, let $z_i(\cdot)$ be the solution of the initial value problem $\dot{z}_i = F_{i,\mathbf{l}_i}(z_i)$, $z_i(0) = x_{i,G}$, which we call the reference trajectory of i. This trajectory is obtained by "freezing" agent i's neighbors at their corresponding reference points through the feedback term

$$k_{i,\mathbf{l}_i,1}(t, x_i, \mathbf{x}_j) := f_i(z_i(t), \mathbf{x}_{j,G}) - f_i(x_i, \mathbf{x}_j), \tag{12.6}$$

in place of the agent's free input v_i. Also, by selecting a vector w_i from the set

$$W := B(\lambda v_{\max}), \lambda \in (0, 1), \tag{12.7}$$

(the ball with radius $\lambda v_{\max}$ in $\mathbf{R}^n$) and assuming that we can superpose to the reference trajectory the motion of i with constant speed w_i, namely, move along the curve $\bar{x}_i(\cdot)$ defined as $\bar{x}_i(t) := z_i(t) + tw_i$, $t \geqslant 0$, we can reach the point x inside the depicted ball in Fig. 12.2 at time δt from the reference point $x_{i,G}$. The parameter λ in (12.7) stands for the part of the free input that is used to increase the transition choices from the given cell configuration. In a similar way, it is possible to reach any point inside the ball by a different selection of w_i. This ball has radius

$$r := \lambda v_{\max} \delta t, \tag{12.8}$$

namely, the distance that the agent can cross in time δt by exploiting the part of the free input that is available for reachability purposes. *For the abstraction, we require the ability to perform a transition to each cell which has nonempty intersection with* $B(z_i(\delta t); r)$. These transitions are enabled via the feedback laws

$$k_{i,\mathbf{l}_i}(t, x_i, \mathbf{x}_j; x_{i0}, w_i) := k_{i,\mathbf{l}_i,1}(t, x_i, \mathbf{x}_j) + k_{i,\mathbf{l}_i,2}(x_{i0}) + k_{i,\mathbf{l}_i,3}(w_i), \tag{12.9}$$

parameterized by $x_{i0} \in S_{l_i}$, $w_i \in W$, where $k_{i,\mathbf{l}_i,1}(\cdot)$ is given in (12.6) and with

$$k_{i,\mathbf{l}_i,2}(x_{i0}) := \frac{1}{\delta t}(x_{i,G} - x_{i0}), \quad k_{i,\mathbf{l}_i,3}(w_i) := w_i, \quad x_{i0} \in S_{l_i}, w_i \in W. \tag{12.10}$$

In order to perform for instance a transition to the cell where the point x in Fig. 12.2 belongs, we require that the feedback law $k_{i,\mathbf{l}_i}(\cdot)$ drives agent i to the endpoint of the curve $\bar{x}_i(\cdot)$ from each initial condition in S_{l_i}. This is accomplished by exploiting the extra terms $k_{i,\mathbf{l}_i,2}(\cdot)$ and $k_{i,\mathbf{l}_i,3}(\cdot)$. The derivation of well-posed discretizations is additionally based on the choice of cell decompositions and associated time steps δt which ensure that the magnitude of the feedback law apart from the term w_i in $k_{i,\mathbf{l}_i,3}(\cdot)$ does not exceed $(1 - \lambda)v_{\max}$. Thus, due to (12.7), which implies that $|w_i| \leqslant \lambda v_{\max}$, it follows that the total magnitude of the applied control law will be consistent with assumption (12.5) on the free inputs' bound. Notice also that due to the assumption $v_{\max} < M$, it is in principle not possible to cancel the interconnection terms. Furthermore, the control laws $k_{i,\mathbf{l}_i}(\cdot)$ are decentralized, since they only use information of agent i's neighbors states and they depend on the cell configuration $\mathbf{l}_i$, through the reference points $(x_{i,G}, \mathbf{x}_{j,G})$ which are involved in (12.6) and (12.10).

Based on the control laws in (12.9) and assuming given a space-time discretization $\mathbf{S} - \delta t$, we derive the transition system $TS_i := (Q_i, Act_i, \longrightarrow_i)$ of each agent, where Q_i is a set of states, Act_i is a set of actions, and $\longrightarrow_i$ is a transition relation with $\longrightarrow_i \subset Q_i \times Act_i \times Q_i$. In particular, TS_i is given by $Q_i := \mathscr{I}$, i.e., the

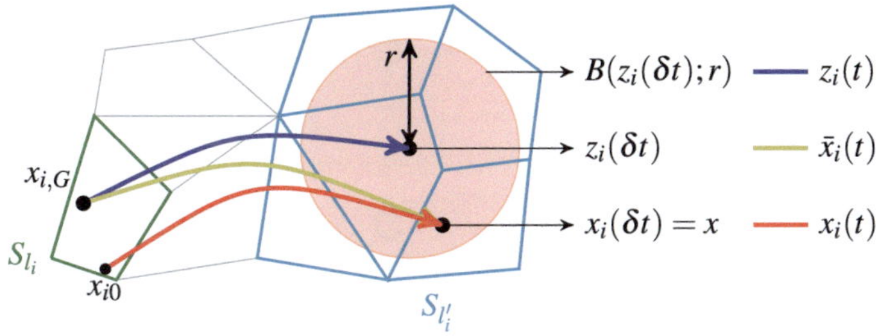

Fig. 12.2 Consider any point x inside the ball with center $z_i(\delta t)$. Then, the control law $k_{i,\mathbf{l}_i}(\cdot)$ ensures that for each initial condition $x_{i0} \in S_{l_i}$, agent's i trajectory $x_i(\cdot)$ will reach x at time δt. Thus, since $x \in S_{l_i'}$, we obtain a transition to $S_{l_i'}$

cells of the decomposition, $Act_i := \mathscr{I}^{N_i+1}$ and the transition relation $\longrightarrow_i$ defined as follows: given $l_i \in \mathscr{I}, \mathbf{l}_i = (l_i, l_{j_1}, \ldots, l_{j_{N_i}})$ and $l_i' \in \mathscr{I}$, we enable the transition $(l_i, \mathbf{l}_i, l_i') \in \longrightarrow_i$, if there exists a parameter w_i, such that the control law $k_{i,\mathbf{l}_i}(\cdot)$ in (12.9) guarantees that the agent will reach cell $S_{l_i'}$ at δt, from any initial condition in its cell, based only on the fact that its neighbors belong to the corresponding cells in $\mathbf{l}_i$. By denoting $\text{Post}_i(l_i; \mathbf{l}_i) := \{l_i' \in \mathscr{I} : (l_i, \mathbf{l}_i, l_i') \in \longrightarrow_i\}$, it follows that the discretization is well posed iff for each agent $i \in \mathscr{N}$ and cell configuration $\mathbf{l}_i = (l_i, l_{j_1}, \ldots, l_{j_{N_i}})$ it holds $\text{Post}_i(l_i; \mathbf{l}_i) \neq \emptyset$. In particular we have the following result.

Theorem 12.1 *Consider a cell decomposition* $\mathbf{S}$ *of* $\mathbf{R}^n$ *with diameter* $d_{\max}$*, a time step* δt*, the parameter* $\lambda \in (0, 1)$ *and define* $L := \max\{3L_2 + 4L_1\sqrt{N_i}, i \in \mathscr{N}\}$*, with* L_1 *and* L_2 *as given in (12.3) and (12.4). We assume that* $d_{\max} \in \left(0, \frac{(1-\lambda)^2 v_{\max}^2}{4ML}\right]$ *and* $\delta t \in \left[\frac{(1-\lambda)v_{\max} - \sqrt{(1-\lambda)^2 v_{\max}^2 - 4MLd_{\max}}}{2ML}, \frac{(1-\lambda)v_{\max} + \sqrt{(1-\lambda)^2 v_{\max}^2 - 4MLd_{\max}}}{2ML}\right]$*. Then, the space-time discretization is well posed for (12.1). In particular, for each agent* $i \in \mathscr{N}$ *and cell configuration* $\mathbf{l}_i = (l_i, l_{j_1}, \ldots, l_{j_{N_i}})$*, it holds* $\text{Post}_i(l_i; \mathbf{l}_i) = \{l \in \mathscr{I} : S_l \cap B(z_i(\delta t); r) \neq \emptyset\}$*, with* $z_i(\cdot)$ *denoting the corresponding reference trajectory of* i *and* r *as in (12.8).*

12.2.4 Abstractions of Varying Decentralization Degree

We next present a generalization of the previous approach, where each agent's abstract model has been based on the knowledge of the discrete positions of its neighbors, by allowing the agent to have this information for all members of the network up to a certain distance in the communication graph. The latter provides an improved estimate of the potential evolution of its neighbors and allows for more accurate discrete agent models, due to the reduction of the control magnitude which is required for the manipulation of the coupling terms. Therefore we introduce also some extra notation. Given $m \geqslant 1$, we denote by $\mathscr{N}_i^m$ the set of agents from which i is

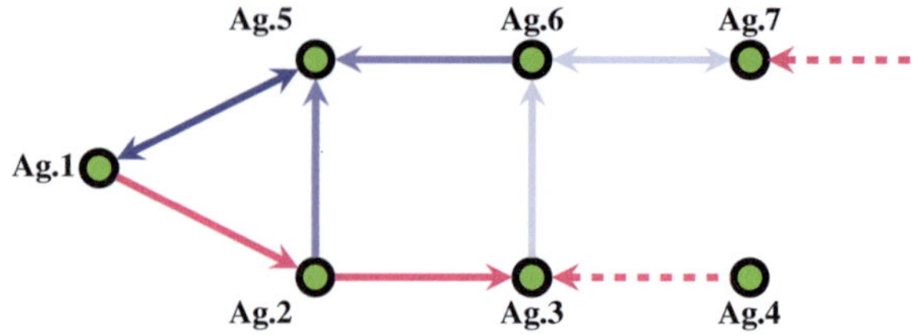

Fig. 12.3 This figure illustrates 7 agents of a network. The sets $\mathcal{N}_1^m$, $\bar{\mathcal{N}}_1^m$ of agent 1 up to paths of length $m = 3$ are: $\bar{\mathcal{N}}_1^1 = \{1, 5\}$, $\mathcal{N}_1^1 = \{5\}$; $\bar{\mathcal{N}}_1^2 = \{1, 2, 5, 6\}$, $\mathcal{N}_1^2 = \{2, 6\}$; $\bar{\mathcal{N}}_1^3 = \{1, 2, 3, 5, 6, 7\}$, $\mathcal{N}_1^3 = \{3, 7\}$

reachable through a path of length m and not by a shorter one, excluding also the possibility to reach itself through a cycle. We also define the set $\bar{\mathcal{N}}_i^m := \bigcup_{\ell=1}^m \mathcal{N}_i^\ell \cup \{i\}$, namely, the set of all agents from which i is reachable by a path of length at most m, including i, and call it the *m-neighbor set* of i (see Fig. 12.3).

The derivation of the discrete models is based as previously on the design of appropriate hybrid feedback laws in place of the v_i's, which enable the desired transitions. We, therefore, provide a modification of the control law (12.6) which is based on a more accurate estimation for the evolution of agent i's neighbors. In particular, select an ordered tuple of indices $\mathbf{l}_i$ corresponding to the cells where the agents in i's m-neighbor set belong, and assume without any loss of generality that $\mathcal{N}_i^{m+1} \neq \emptyset$. Also, choose a reference point $x_{\ell,G}$ from the cell of each agent in $\bar{\mathcal{N}}_i^m$ and consider the initial value problem (IVP) $\dot{z}_\ell(t) = f_\ell(z_\ell(t), z_{j(\ell)_1}(t), \ldots, z_{j(\ell)_{N_\ell}}(t))$, $t \geqslant 0$, $\ell \in \bar{\mathcal{N}}_i^{m-1}$, $z_\ell(0) = x_{\ell,G}$, for all $\ell \in \bar{\mathcal{N}}_i^{m-1}$, with the terms $z_\ell(\cdot)$, $\ell \in \mathcal{N}_i^m$ defined as $z_\ell(t) := x_{\ell,G}$, for all $t \geqslant 0$, $\ell \in \mathcal{N}_i^m$. This IVP provides a solution of the unforced, i.e., without free inputs subsystem formed by the m-neighbor set of agent i. In addition, the agents are initiated from their reference points in their cells and the neighbors precisely m hops away are considered fixed at their corresponding reference points for all times. In analogy to the previous section, we will call the ith component $z_i(\cdot)$ of the solution to the IVP the reference trajectory of i. We also compactly denote as $\mathbf{z}_j(\cdot) := (z_{j_1}(\cdot), \ldots, z_{j_{N_i}}(\cdot))$ the corresponding components of i's neighbors. The key part in this modification is that the latter provide a more accurate estimate of the neighbors' possible evolution over the time interval $[0, \delta t]$. Thus, by replacing the feedback component in (12.6) by $k_{i,\mathbf{l}_i,1}(t, x_i, \mathbf{x}_j) := f_i(z_i(t), \mathbf{z}_j(t)) - f_i(x_i, \mathbf{x}_j)$, $t \in [0, \infty)$, $(x_i, \mathbf{x}_j) \in \mathbf{R}^{(N_i+1)n}$, we can exploit the control law in (12.9) to obtain analogous transition capabilities as in the previous section. It is noted that this selection reduces the control effort which is required to compensate for the evolution of i's neighbors and leads to improved discretizations. Sufficient conditions for the derivation of well posed discretizations along the lines of Theorem 12.1 can be found in [4].

12.3 Multi-agent Plan Synthesis

In this section, we focus on task and motion planning for a multi-agent system that has already been abstracted as a discrete, finite, state-transition system using the technique introduced in Sect. 12.2, or similar. In contrast to Sect. 12.4, the agents here are dependent on each other in terms of their LTL specifications that capture potentially collaborative tasks, whereas we assume that they can communicate in a limited way and they are not subject to relative-distance constraints. Next to the task specifications, each agent is subject to an independent motion specification, also given in LTL. We tackle high-computational demands associated with centralized planning via introducing a two-phase procedure that largely decouples task planning and motion planning. Moreover, we discuss that the solution can benefit further from utilizing receding horizon planning approach. This section, thus, overviews results presented in [23] and [22] and introduces their integration into a single task and motion planning technique. An interested reader is referred to [23] and [22] for full technical details on task and motion planning decomposition and receding horizon planning, respectively.

12.3.1 Problem Formulation

Similarly as in the previous sections, we consider a team of N possibly heterogeneous, autonomous agents with unique identities (IDs) $i \in \mathcal{N} = \{1, \ldots, N\}$. However, here the agent i's capabilities are modeled through *finite transition systems (TS)*

$$T_i = (S_i, s_{init,i}, A_i, \rightarrow_i, \Pi_i, L_i, \Sigma_i, \mathcal{L}_i, Sync_i),$$

where the set of states S_i of the TS represent discrete states of the agent i (e.g., the location of the agent in the environment that is partitioned into a finite number of cells), and $s_{init,i} \in S_i$ is the agent i's initial state (e.g., its initial cell). The actions A_i abstract the agent's low-level controllers, and a transition $s \xrightarrow{\alpha}_i s'$ from $s \in S_i$ to $s' \in S_i$ correspond to the agent's capability to execute the action $\alpha \in A_i$ (e.g., to move between two cells of the environment). We note that a transition duration is arbitrary and unknown prior its execution. Atomic propositions Π_i together with the labeling function $L_i : S_i \rightarrow 2^{\Pi_i}$ are used to mark interesting properties of the system states (e.g., a cell is safe). Labeling function $\mathcal{L}_i : Act \rightarrow 2^{\Sigma_i} \cup \mathcal{E}_i$ associates each action $\alpha \in A_i$ with a set of services of interest $\sigma \in 2^{\Sigma_i}$ that are provided upon its execution (e.g., an object pick-up), or with a special silent service set $\mathcal{E}_i = \{\varepsilon_i\}$, $\varepsilon_i \notin \Sigma_i$, indicating that no service of interest is provided upon the execution of action α. Traces of the transition system are infinite alternating sequences of states and actions that start in the initial state and follow the transition function. Intuitively, they provide abstractions of the agent's long-term behaviors (e.g., the agent's trajectories in the environment). A trace $\tau_i = s_{i,1}\alpha_{i,1}s_{i,2}\alpha_{i,2}\ldots$ produces words $w(\tau_i) = L(s_{i,1})L(s_{i,2})\ldots$, and

$\omega(\tau_i) = \mathscr{L}(\alpha_{i,1})\mathscr{L}(\alpha_{i,2})\ldots$ representing the sequences of state properties that hold true, and services that are provided, respectively.

The agents can communicate, and in particular they all follow this synchronization protocol: An agent i can send a synchronization request $sync_i(I)$ to a subset of agents $\{i\} \subseteq I \subseteq \mathscr{N}$ notifying that it is ready to synchronize. Then, before proceeding with execution of any action $\alpha \in Act_i$, it waits in its current state to receive $sync_{i'}(I)$ from each $i' \in I$. Assuming lossless communication and instant synchronization upon receiving the needed synchronization requests, the agents can this way enforce waiting for each other and executing actions simultaneously. We denote by $Sync_i = \{sync_i(I) \mid \{i\} \subseteq I \subseteq \mathscr{N}\}$ the set of all synchronization requests of agent i.

Each agent $i \in \mathscr{N}$ is given a specification that consists of

- a *motion specification* ϕ_i, which is an $\text{LTL}_{\setminus X}$ formula over Π_i that captures requirements on the states the agent passes through, such as safety, reachability, persistent surveillance, and their combination. The motion specification is interpreted over the word $w(\tau_i)$; and
- a *task* specification ψ_i, which is an LTL formula over $\Sigma = \bigcup_{i' \in \mathscr{N}} \Sigma_{i'}$ that captures requirements on services provided along the system execution. In contrast to the motion specification, the task specification is collaborative and yields dependencies between the agents. Each task specification is interpreted over the set of all words $\omega(\tau_i')$, $i' \in \mathscr{N}$. In particular, agent i decides whether ψ_i is satisfied from its local point of view by looking at the subsequence of non-silent services, i.e., services of interest of $\omega(\tau_i)$ and the services provided by the remainder of the team at the corresponding times.

Problem 12.1 *Consider* a set of agents $\mathscr{N} = \{1, \ldots, N\}$, each of which is modeled as a transition system $T_i = (S_i, s_{init,i}, A_i, \to_i, \Pi_i, L_i, \Sigma_i, \mathscr{L}_i, Sync_i)$, and assigned a task in the form of an $\text{LTL}_{\setminus X}$ formula ϕ_i over Π_i and ψ_i over $\Sigma = \bigcup_{i' \in \mathscr{N}} \Sigma_{i'}$. *For each $i \in \mathscr{N}$ find a plan, i.e.,* (i) a trace $\tau_i = s_{i,1}\alpha_{i,1}s_{i,2}\alpha_{i,2}\ldots$ of T_i and (ii) a synchronization sequence $\gamma_i = r_{i,1}r_{i,2}\ldots$ over $Sync_i$ with the property that the set of induced behaviors is nonempty, and both ϕ_i and ψ_i are satisfied from the agent i's viewpoint.

As each LTL formula can be translated into a BA, from now on, we pose the problem equivalently with the motion specification of each agent i given as a BA $B_i^\phi = (Q_i^\phi, q_{init,i}^\phi, \delta_i^\phi, 2^{\Pi_i}, F_i^\phi)$, and the task one as a BA $B_i^\psi = (Q_i^\psi, q_{init,i}^\psi, \delta_i^\psi, 2^\Sigma, F_i^\psi)$.

12.3.2 Problem Solution

Even though the agents' motion specifications are mutually independent, each of them is dependent on the respective agent's task specification, which is dependent on the task specifications of the other agents. As a result, the procedure of synthesizing the desired N strategies cannot be decentralized in an obvious way. However, one can quite easily obtain a centralized solution when viewing the problem as a synthesis of a single team plan. A major drawback of the centralized solution is the state-space explosion, which makes it practically intractable. We aim to decentralize the solution as much as possible. Namely, we aim to separate the synthesis of service plans yielding the local satisfaction of the task specifications from the syntheses of traces that guarantee the motion specifications. Our approach is to precompute possible traces and represent them efficiently, while abstracting away the features that are not significant for the synthesis of action plans. This abstraction serves as a guidance for the action and synchronization planning, which, by construction, allows for finding a trace complying with both the synthesized action and synchronization plans and the motion specification.

12.3.2.1 Preprocessing the Motion Specifications

Consider for now a single agent $i \in \mathcal{N}$, and its motion specification BA B_i^ϕ. We slightly modify the classical construction of a product automaton of T_i and B_i^ϕ to obtain a BA that represents the traces of T_i accepted by B_i^ϕ, and furthermore explicitly captures the services provided along the trace.

Definition 12.1 *(Motion product)* The *motion product* of a TS T_i, and a BA B_i is a BA $P_i = (Q_i, q_{init,i}, \delta_i, 2^{\Sigma_i} \cup 2^{\mathcal{E}_i}, F_i)$, where $Q_i = S_i \times Q_i^\phi$; $q_{init,i} = (s_{init,i}, q_{init,i}^\phi)$; $((s,q), \mathcal{L}_i(\alpha), (s',q')) \in \delta_i$ if and only if $s, \xrightarrow{\alpha}_i s'$, and $(q, L_i(s), q') \in \delta_i^\phi$; and $F_i = \{(s,q) \mid q \in F_i^\phi\}$.

We introduce a way to reduce the size of the motion product by removing all states and transitions that are insignificant with respect to the local satisfaction of the task specification, and hence with respect to the collaboration with others. Specifically, the significant states are only the ones that have an outgoing transition labeled with $\mathcal{L}_i(\alpha) \neq \mathcal{E}_i$.

First, we remove all insignificant non-accepting states and their incoming and outgoing transitions and we replace each state with a set of transitions leading directly from the state's predecessors to its successors, i.e., we concatenate the incoming and the outgoing transitions. The labels of the new transitions differ: if both labels of the concatenated incoming and outgoing transition are $\mathcal{E}_i$, then the new label will stay $\mathcal{E}_i$ to indicate that the transition represents a sequence of actions that are not interesting with respect to the local satisfaction of task specifications. On the other hand, if the label σ of the incoming transition belongs to 2^{Σ_i}, we use the action σ as the label for the new transition. Each path between two significant states in P_i then maps onto a path between the same states in the reduced motion product and the sequences of non-silent services read on the labels of the transitions of the two paths are equal; and vice versa. Second, we handle the insignificant accepting states similarly to the non-accepting ones, however, we do not remove the states whose predecessors include a significant state in order to preserve the accepting condition. Moreover, we remove all states from which none of the accepting states is reachable,

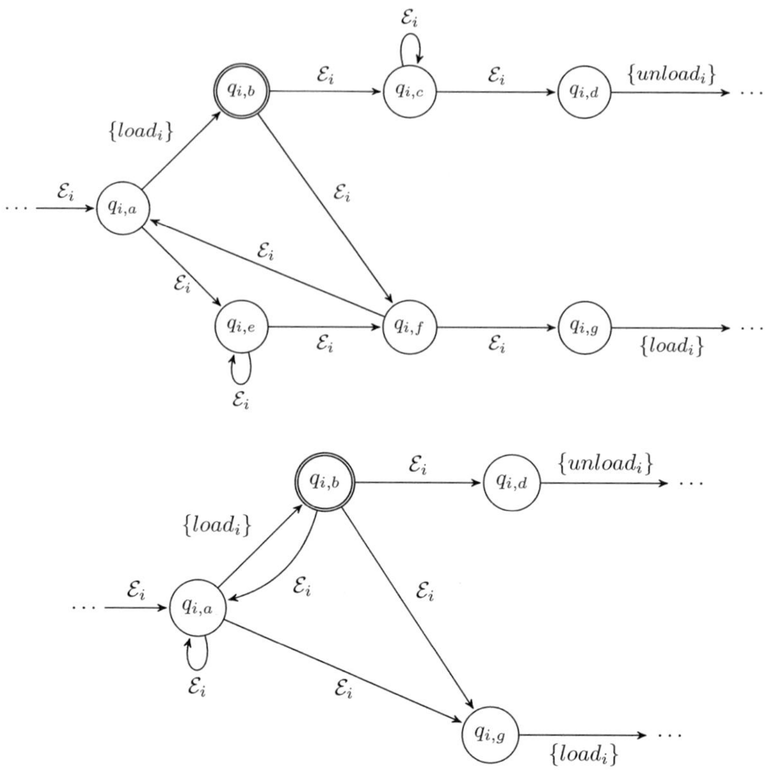

Fig. 12.4 An example of a part of a product automaton P_i (top), and the corresponding part of the reduced product automaton $\ddot{P}_i$ (bottom)

and we can keep only one copy of duplicate states that have analogous incoming and outgoing edges. An example of the reduction is given in Fig. 12.4.

There is a correspondence between the infinite runs of P_i and the infinite runs of the reduced motion product, which we denote by $\ddot{P}_i$: for each run of P_i there exists a run of $\ddot{P}_i$, such that the states of the latter one are a subsequence of the states of the former one, the sequences of non-silent services read on the labels of the transitions of the two runs are equal, and that the latter one is acceptable if and only if the former one is accepting; and vice versa. This correspondence will allow us to reconstruct a desired run of P_i from a run of $\ddot{P}_i$, as we will discuss in Sect. 12.3.2.3.

12.3.2.2 Preprocessing the Task Specifications

The next two steps of the solution follow similar ideas as in Sect. 12.3.2.1: We build a local task and motion product $\bar{P}_i$ of the reduced motion product $\ddot{P}_i$ and the task specification BA B_i^{ψ} for each agent i separately, to capture the admissible traces of i that comply both with its motion and task specification. At this stage, the other agents' collaboration capabilities are not included, yet. We again remove insignificant states, which are now ones that do not have an outgoing dependent transition, i.e., a transition labeled with a non-silent service $\sigma \in \Sigma \setminus \Sigma_i$. We thus reduce $\bar{P}_i$ to $\widehat{P}_i$. Similarly as before, there is a correspondence between the infinite runs of $\bar{P}_i$ and the infinite runs of $\widehat{P}_i$: for each run of $\bar{P}_i$ there exists a run of $\widehat{P}_i$, such that the states of the latter one are a subsequence of the states of the former one, the sequences of services read on the labels of the transitions leading from the significant states of the two runs are equal, and that the latter one is acceptable if and only if the former one is accepting; and vice versa.

Finally, we build the *global product* P of the reduced task and motion product automata $\widehat{P}_1, \ldots \widehat{P}_N$. Each accepting run of the global product P maps directly on the accepting runs of the reduced task and motion product automata and vice versa, for each collection of accepting runs of the reduced task and motion product automata, there exists an accepting run of the global product P.

12.3.2.3 Plan Synthesis

The final step of our solution is the generation of the plan in P and its mapping onto a trace τ_i of T_i and a synchronization sequence γ_i over $Sync_i$, for all $i \in \mathcal{N}$. Using standard graph algorithms (see, e.g., [2]), we find an accepting run $q_1 q_2 \ldots$ over a word $\sigma_1 \sigma_2 \ldots$ in P, where $q_j = (\widehat{q}_{1,j}, \ldots, \widehat{q}_{N,j}, k)$, for all $j \geqslant 1$. For each agent $i \in \mathcal{N}$, we can project this accepting run onto the states of $\widehat{P}_i$, and then, due to the above discussed correspondences between runs of the product automata, we can also find sequences of the states in $\bar{P}_i$, $\ddot{P}_i$, P_i, such that the projection from an accepting run of P_i onto the states of T_i yield the desired trace τ_i (and also the desired sequence of services $\sigma_{i,1}\sigma_{i,2}\ldots$). The synchronization sequence γ_i is constructed by setting $r_{i,j}$ to be the set of agents that need to collaborate on executing the transition from $s_{i,j}$ to $s_{i,j+1}$ in order to provide the service $\sigma_{i,j}$.

12.3.2.4 Receding Horizon Approach

Although we have reduced the size of each local product automaton before constructing the global product P, an additional improvement can be achieved by decomposing the infinite-horizon planning into an infinite sequence of finite-horizon planning problem that can further significantly reduce the size of the global product P.

In particular, following the ideas from [22], we propose to (1) partition the agents into classes based on their dependency observed in $\widehat{P}_1, \ldots, \widehat{P}_N$ within a horizon H; and then for each of the classes separately: (2) build a product automaton up to a predefined horizon h and synthesize a plan that leads to progress in satisfaction of the task specifications; (3) execute part of the plan till the first non-silent service is provided and repeat steps (1), (2), (3).

The benefit of the receding horizon approach reaches beyond tackling the tractability of multi-agent plan synthesis. It builds on Event-Triggered synchronization, and hence, it is especially useful in cases where the agents travel at different speeds than originally assumed.

12.3.2.5 Complexity

In the worst case, our solution meets the complexity of the centralized solution. However, this is often not the case. Since the size of the global product is highly dependent on the number of dependent services available in the agents' workspace, our solution is particularly suitable for systems with complex motion capabilities, sparsely distributed services of interest, and occasional needs for collaboration.

12.4 Decentralized Control Under Local Tasks and Coupled Constraints

In this section, we tackle the multi-agent control problem under local LTL tasks from the bottom-up perspective. We aim for a decentralized solution while taking into account the constraints that the agents can exchange messages only if they are close enough. Following the hierarchical approach to LTL planning, we first generate for each agent a sequence of actions as a high-level plan that, if followed, guarantees the accomplishment of the respective agent's LTL task. Second, we merge and implement the synthesized plans in real time, upon the run of the system. Namely, we introduce a distributed continuous controller for the leader–follower scheme, where the current leader guides itself and the followers toward the satisfaction of the leader's task. At the same time, the connectivity of the multi-agent system is maintained. By a systematic leader reelection, we ensure that each agent's task will be met in long term. This section is a brief summary of the results from the conference publication [11] and an extended study of related problems can be found in [12].

12.4.1 Related Work

The consideration of relative-distance constraints is closely related to the connectivity of the multi-agent network in robotic tasks [18]. As pointed out in [13, 24], maintaining this connectivity is of great importance for the stability, safety, and integrity of the overall team, for global objectives like rendezvous, formation, and flocking. Very often the connectivity of underlying interaction graphs is imposed by assumption rather than treated as an extra control objective. Here, the proposed distributed motion controller guarantees global convergence and the satisfaction of relative-distance constraints for all time. Moreover, different from [8] where a satisfying discrete plan is enough, the proposed initial plan synthesis algorithm here minimizes a cost of a satisfying plan, along with the communication constraints. Lastly, the same bottom-up planning problem from LTL specifications are considered in [23], where it is assumed that the agents are synchronized in their discrete abstractions and the proposed solutions rely on construction of the synchronized product system between the agents, or at least of its part. In contrast, in this work, we avoid the product construction completely. Compared with [15], these coordination policies are fully distributed and can be applied to agents with limited communication capabilities.

12.4.2 Problem Formulation

In mathematical terms, we consider a team of N autonomous agents with unique identities (IDs) $i \in \mathscr{N} = \{1, \ldots, N\}$. They all satisfy the single-integrator dynamics $\dot{x}_i(t) = u_i(t)$, where $x_i(t),\ u_i(t) \in \mathbf{R}^2$ are the respective state and the control input of agent i at time $t > 0$. The agents are modeled as point masses without volume. Each agent has a limited communication radius of $r > 0$. Namely, agent i can communicate *directly* with agent j if $\|x_i(t) - x_j(t)\| \leqslant r$ or *indirectly* via a chain of connected robots. We assume that initially all agents are connected.

Each robot $i \in \mathscr{N}$ has a local task φ_i specified over $\Sigma_i = \{\sigma_{ih}, h \in \{1, \cdots, M_i\}\}$, which is a set of services that robot i can provide at different regions $\mathscr{R}_i = \{R_{ig}, g \in \{1, \cdots, K_i\}\}$. Note that $R_{ig} = \{y \in \mathbf{R}^2 | \|y - c_{ig}\| \leqslant r_{ig}\}$ is a circular area with the center c_{ig} and radius r_{ig}. Furthermore, some of the services in Σ_i can be provided solely by the agent i, while others require cooperation with some other agents. A service σ_{ih} is provided if the agent's relevant service-providing action π_{ih} and the corresponding cooperating agents' actions $\bigwedge_{i' \in C_{ih}} \varpi_{i'ih}$ are executed at the same time, i.e., $\sigma_{ih} = \pi_{ih} \wedge \bigwedge_{i' \in C_{ih}} \varpi_{i'ih}$. Lastly, a LTL task φ_i is fulfilled if the sequence of services provided by robot i satisfies φ_i. Thus, the problem is to synthesize the control input u_i, time sequence of executed actions T_i^A and the associated sequence of actions A_i for each robot $i \in \mathscr{N}$.

12.4.3 Solution Outline

Our approach to the problem involves an offline and an online step. In the offline step, we synthesize a high-level plan in the form of a sequence of services for each of the agents. In the online step, we dynamically switch between the high-level plans through leader election and choose the associated continuous controllers. The whole team then follows the leader towards until its next service is provided and then a new leader is selected.

12.4.3.1 Offline Discrete Plan Synthesis

Given an agent $i \in \mathcal{N}$, a set of services Σ_i, and an LTL formula φ_i over Σ_i, a high-level plan for i can be computed via standard model-checking methods [2, 10]. Roughly, by translating φ_i into a equivalent Büchi automaton and by consecutive analysis of the automaton, a sequence of services with the prefix–suffix format $\Omega_i = \sigma_{i1} \ldots \sigma_{ip_i}(\sigma_{ip_i+1} \ldots \sigma_{is_i})^{\omega}$, such that $\Omega_i \models \varphi_i$ can be found, where σ_{i1} can be independent or dependent services for robot $i \in \mathcal{N}$.

12.4.3.2 Dynamic Leader Selection

In this part, we describe how to elect a leader from the team in a repetitive online procedure, such that each of the agents is elected as a leader infinitely often. Intuitively, each agent $i \in \mathcal{N}$ is assigned a value that represents the agent's urge to provide the next service in its high-level plan. Using ideas from bully leader election algorithm [9], an agent with the strongest urge is always elected as a leader within the connectivity graph.

Particularly, let i be a fixed agent, t the current time and $\sigma_{i1} \ldots \sigma_{ik}$ a prefix of services of the high-level plan Ω_i that have been provided till t. Moreover, let $\tau_{i\lambda}$ denote the time, when the latest service, i.e., $\sigma_{i\lambda} = \sigma_{ik}$ was provided, or $\tau_{i\lambda} = 0$ in case no service prefix of Ω_i has been provided, yet. Using $\tau_{i\lambda}$, we could define agent i's *urge* at time t as a tuple $\Upsilon_i(t) = (t - \tau_{i\lambda},\ i)$. Furthermore, to compare the agents' urges at time t, we use lexicographical ordering: $\Upsilon_i(t) > \Upsilon_j(t)$ if and only if (1) $t - \tau_{i\lambda} > t - \tau_{j\lambda}$, or (2) $t - \tau_{i\lambda} = t - \tau_{j\lambda}$, and $i > j$. Note that $i \neq j$ implies that $\Upsilon_i(t) \neq \Upsilon_j(t)$, for all $t \geqslant 0$. As a result, the defined ordering is a linear ordering and at any time t, there exists exactly one agent i maximizing its urge $\Upsilon_i(t)$. As a result, there is always a single agent that has the highest urge within $\mathcal{N}$ for any given time t. The robot with the highest urge is selected as the leader, which has the opportunity to execute its local plan Ω_i. However, due to the relative-distance constraints and the depended services, it can not simply move there without adopting a collaborative motion controller described below.

12.4.3.3 Collaborative Controller Design

Let us first introduce the notion of agents' connectivity graph that will allow us to handle the constraints imposed on communication between the agents. Recall that each agent has a limited communication radius $r > 0$. Moreover, let $\varepsilon \in (0,\ r)$ be a given constant, which plays an important role for the edge definition. In particular, let $G(t) = (\mathcal{N},\ E(t))$ denote the undirected time-varying connectivity graph formed by the agents, where $E(t) \subseteq \mathcal{N} \times \mathcal{N}$ is the edge set for $t \geqslant 0$. At time $t = 0$, we set $E(0) = \{(i,\ j) | \|x_i(0) - x_j(0)\| < r\}$.At time $t > 0$, $(i,\ j) \in E(t)$ if and only if one of the following conditions hold: (i) $\|x_i(t) - x_j(t)\| \leqslant r - \varepsilon$, or (ii) $r - \varepsilon < \|x_i(t) - x_j(t)\| \leqslant r$ and $(i,\ j) \in E(t^-)$, where $t^- < t$ and $|t - t^-| \to 0$. Note that the condition (ii) in the above definition guarantees that a new edge will only be added when the distance between two unconnected agents decreases below $r - \varepsilon$.

Now consider the following problem: given a leader $\ell \in \mathcal{N}$ at time t and a goal region $R_{\ell g} \in \mathcal{R}_\ell$, propose a decentralized continuous controller that (1) $G(t')$ remains connected for all $t' \in [t, \bar{t}]$; (2) guarantees that all agents $i \in \mathcal{N}$ reach $R_{\ell g}$ at a finite time $\bar{t} < \infty$. Both objectives are critical to ensure sequential satisfaction of φ_i for each $i \in \mathcal{N}$.

Denote by $x_{ij}(t) = x_i(t) - x_j(t)$ the pairwise relative position between neighboring agents, $\forall (i, j) \in E(t)$. Thus $\|x_{ij}(t)\|^2 = \big(x_i(t) - x_j(t)\big)^T \big(x_i(t) - x_j(t)\big)$ denotes the corresponding distance. We propose the following continuous controller:

$$u_i(t) = -b_i\big(x_i - c_{ig}\big) - \sum_{j \in \mathcal{N}_i(t)} \frac{2r^2}{(r^2 - \|x_{ij}\|^2)^2}(x_i - x_j), \tag{12.11}$$

where $b_i \in \{0, 1\}$ indicates if agent i is the leader; $c_{ig} \in \mathbf{R}^2$ is the center of the next goal region for agent i; b_i and c_{ig} are derived from the leader selection scheme earlier. It can be seen that the above controller is fully distributed as it only depends x_i and $x_j, \forall j \in \mathcal{N}_i(t)$.

Given the controller (12.11), we can prove the following two important properties of the complete system: Assume that $G(t)$ is connected at $t = T_1$ and agent $\ell \in \mathcal{N}$ is the fixed leader for all $t \geqslant T_1$. By applying the controller in (12.11), the following two statements hold:

- The graph $G(t)$ remains connected and $E(T_1) \subseteq E(t)$ for $t \geqslant T_1$.
- There exist a finite time $T_1 \leqslant \bar{t} < +\infty$, $x_i(\bar{t}) \in R_{\ell g}$, $\forall i \in \mathcal{N}$.

To briefly prove the above two statements, we consider the following potential function of the complete system:

$$V(t) = \frac{1}{2}\sum_{i=1}^{N} \sum_{j \in \mathcal{N}_i(t)} \phi(\|x_{ij}\|) + \frac{1}{2}\sum_{i=1}^{N} b_i(x_i - c_{ig})^T(x_i - c_{ig}), \tag{12.12}$$

where the potential function $\phi\big(\|x_{ij}\|\big) = \frac{\|x_{ij}\|^2}{r^2 - \|x_{ij}\|^2}$ for $\|x_{ij}\| \in [0, r)$, and thus $V(t)$ is positive semidefinite. Assume that $G(t)$ remains *invariant* during $[t_1, t_2) \subseteq [T_1, \infty)$. The time derivative of (12.12) during $[t_1, t_2)$ is given by

$$\begin{aligned} \dot{V}(t) = &- \sum_{i=1,\, i \neq \ell}^{N} \| \sum_{j \in \mathcal{N}_i(t)} \nabla_{x_i}\phi(\|x_{ij}\|)\|^2 \\ &- \|(x_\ell - c_{\ell g}) + \sum_{j \in \mathcal{N}_\ell(t)} \nabla_{x_\ell}\phi(\|x_{\ell j}\|)\ \|^2 \leqslant 0. \end{aligned} \tag{12.13}$$

Thus $V(t) \leqslant V(0) < +\infty$ for $t \in [t_1, t_2)$. It means that during $[t_1, t_2)$, no existing edge can have a length close to r, i.e., no existing edge will be *lost* by the definition of an edge. On the other hand, assume a *new* edge (p, q) is added to the graph $G(t)$ at $t = t_2$, where $p, q \in \mathcal{N}$. It holds that $\|x_{pq}(t_2)\| \leqslant r - \varepsilon$ and $\phi(\|x_{pq}(t_2)\|) = \frac{r-\varepsilon}{\varepsilon(2r-\varepsilon)} < +\infty$ since $0 < \varepsilon < r$. Denote the set of newly added edges at $t = t_2$ as $\widehat{E} \subset \mathcal{N} \times \mathcal{N}$. Let $V(t_2^+)$ and $V(t_2^-)$ be the value of function from (12.12) before and after adding the set of new edges to $G(t)$ at $t = t_2$. We get $V(t_2^+) \leqslant V(t_2^-) + |\widehat{E}| \frac{r-\varepsilon}{\varepsilon(2r-\varepsilon)} < +\infty$. As a result, $V(t) < +\infty$ for $t \in [T_1, \infty)$. Since one existing edge $(i, j) \in E(t)$ will be lost only if $x_{ij}(t) = r$, it implies that $\phi(\|x_{ij}\|) \to +\infty$, i.e., $V(t) \to +\infty$ by (12.12). By contradiction, we can conclude that new edges will be added but no existing edges will be lost. This proves the first statement above.

Regarding the second statement, we need to show that all agents converge to the goal region of the leader in finite time. By (12.13), $\dot{V}(t) \leqslant 0$ for $t \geqslant T_1$ and $\dot{V}(t) = 0$ when the following conditions hold: (i) for $i \neq \ell$ and $i \in \mathcal{N}$, it holds that $\sum_{j \in \mathcal{N}_i(t)} h_{ij}(x_i - x_j) = 0$; and (ii) for the leader $\ell \in \mathcal{N}$, it holds that $(x_\ell - c_{\ell g}) + \sum_{j \in \mathcal{N}_\ell(t)} h_{ij}(x_\ell - x_j) = 0$, where $h_{ij} = \frac{2r^2}{(r^2 - \|x_{ij}\|^2)^2}$, $\forall (i, j) \in E(t)$. Then we can combine the above two conditions into $H \otimes I_2 \cdot \mathbf{x} + (\mathbf{x} - \mathbf{c}) = 0$, where H is a $N \times N$ matrix satisfying $H(i, i) = \sum_{j \in \mathcal{N}_i} h_{ij}$ and $H(i, j) = -h_{ij}$, where $i \neq j \in \mathcal{N}$. Note that H is positive-semidefinite with a single eigenvalue at the origin, of which the corresponding eigenvector is the unit column vector of length N. Thus, the only equilibrium is $\mathbf{x} = \mathbf{c}$, i.e., $x_i = c_{\ell g}, \forall i \in \mathcal{N}$. By LaSalle's Invariance principle [14], there exists $\bar{t} < +\infty$ that $x_i(\bar{t}) \in R_{\ell g}, \forall i \in \mathcal{N}$.

12.4.3.4 Integrated System

The integrated system combines the leader selection scheme from Sect. 12.4.3.2 and the continuous control scheme from Sect. 12.4.3.3, such that the discrete plan Ω_i synthesized in Sect. 12.4.3.1 can be executed. Particularly, via communicating and comparing the urge function Υ_i among all robots, one robot with the highest urge is selected as the leader, denoted by $\ell \in \mathcal{N}$. Then robot ℓ finds its next goal region according to its plan Ω_i as $R_{\ell g}$. After that, all robots applies the control input from (12.11) where the leader ℓ sets $b_\ell = 1$ while the rest sets $b_i = 0, \forall i \in \mathcal{N}$ and $i \neq \ell$. Consequently, as proven in Sect. 12.4.3.3, there exists a finite time that all robots are within the region $R_{\ell g}$, where robot ℓ can provide action $\pi_{\ell h}$and its collaborating robots can provide the action $\varpi_{\ell' \ell h}, \forall \ell' \in C_{\ell h}$. As a result, the service $\sigma_{\ell h}$ is provided at region $R_{\ell g}$. Afterward, the urge function of robot ℓ is updated and a new leader is selected for the team. This process repeats itself indefinitely such that all robots can fulfill its local task. Detailed algorithms can be found in [11].

12.4.3.5 Simulation

We simulate a system of 4 robots (R_1, R_2, R_3, R_4) with regions of interested in a 4×4 m workspace as shown in Fig. 12.5. They initially start from positions $(0.0, 0.0)$, $(1.0, 0.0)$, $(2.0, 0.0)$, $(3.0, 0.0)$ and they all have the communication

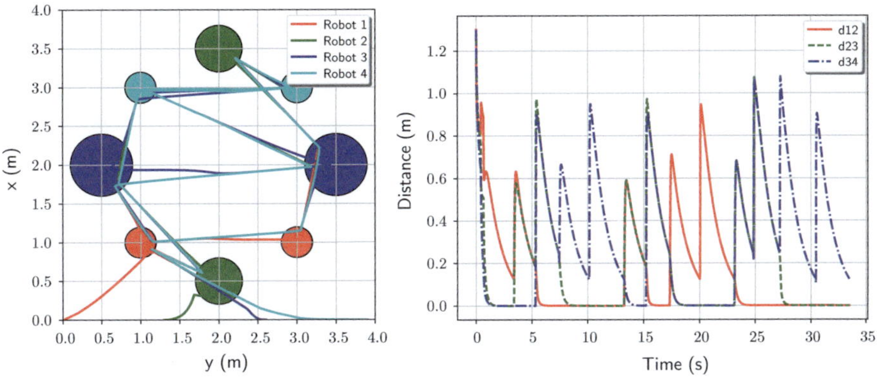

Fig. 12.5 Left: trajectories of 4 robots that satisfy their local service tasks. Right: the relative distances of initially connected neighboring robots

radius 1.5 m. Furthermore, each robot is assigned a local service task. For instance, the local task for robot 1 is to provide services sequentially in the circular region (1.0, 1.0, 0.2) and the circular region (3.0, 1.0, 0.2), where the service at region (3.0, 1.0, 0.2) requires collaboration from other robots. The tasks of other robots are defined similarly. We apply the proposed control and coordination framework as described above. The resulting trajectories of all robots are shown in Fig. 12.5, which verify that all local tasks are satisfied. Moreover, the relative distances between initially connected neighboring robots, i.e., (R_1, R_2), (R_2, R_3), (R_3, R_4), are also shown in Fig. 12.5, all of which stay below the communication radius 1.5 m at all time. More numerical examples are given in [11].

12.4.4 Conclusion and Future Work

To summarize, in this section we present the decentralized control scheme of a team of agents that are assigned local tasks expressed as LTL formulas. The solution follows the automata-theoretic approach to LTL model checking, however, it avoids the computationally demanding construction of synchronized product system between the agents. The decentralized coordination among the agents relies on a dynamic leader–follower scheme, to guarantee the low-level connectivity maintenance at all times and a progress toward the satisfaction of the leader's task. By a systematic leader switching, we ensure that each agent's task will be accomplished.

Acknowledgements This work was supported by the Swedish Research Council (VR), the Knut och Alice Wallenberg Foundation (KAW), the H2020 Co4Robots project, and the H2020 ERC Starting Grant BUCOPHSYS.

References

1. Andreasson, M., Dimarogonas, D.V., Sandberg, H., Johansson, K.H.: Distributed control of networked dynamical systems: static feedback, integral action and consensus. IEEE Trans. Autom. Control **59**, 1750–1764 (2014)
2. Baier, C., Katoen, J.-P.: Principles of Model Checking. MIT press, Cambridge (2008)
3. Boskos, D., Dimarogonas, D.V.: Decentralized abstractions for feedback interconnected multi-agent systems. In: IEEE Conference on Decision and Control (CDC), pp. 282–287 (2015)
4. Boskos, D., Dimarogonas, D.V.: Abstractions of varying decentralization degree for coupled multi-agent systems. In: IEEE Conference on Decision and Control (CDC), pp. 81–86 (2016)
5. Boskos, D., Dimarogonas, D.V.: Robustness and invariance of connectivity maintenance control for multi-agent systems. SIAM J. Control Optim. **55**(3), 1887–1914 (2017)
6. Boskos, D., Dimarogonas, D.V.: Online abstractions for interconnected multi-agent control systems. In: Proceedings of the 20th IFAC World Congress, Toulouse, France. IFAC-PapersOnLine, vol. 50, Iss. 1, pp. 15810–15815 (2017)
7. Dallal, E., Tabuada, P.: On compositional symbolic controller synthesis inspired by small-gain theorems. In: IEEE Conference on Decision and Control (CDC), pp. 6133–6138 (2015)
8. Filippidis, I., Dimarogonas, D.V., Kyriakopoulos, K.J.: Decentralized multi-agent control from local LTL specifications. In: IEEE Conference on Decision and Control (CDC), pp. 6235–6240 (2012)
9. Garcia-Molina, H.: Elections in a distributed computing system. IEEE Trans. Comput. **C-31**(1), 48–59 (1982)
10. Guo, M., Dimarogonas, D.V.: Multi-agent plan reconfiguration under local LTL specifications. Int. J. Robot. Res. **34**(2), 218–235 (2015)
11. Guo, M., Tumova, J., Dimarogonas, D.V.: Cooperative decentralized multi-agent control under local LTL tasks and connectivity constraints. In: IEEE Conference on Decision and Control (CDC), pp. 75–80 (2014)
12. Guo, M., Tumova, J., Dimarogonas, D.V.: Communication-free multi-agent control under local tasks and relative-distance constraints. IEEE Trans. Autom. Control **61**(12), 3948–3962 (2016)
13. Guo, M., Zavlanos, M.M., Dimarogonas, D.V.: Controlling the relative agent motion in multi-agent formation stabilization. IEEE Trans. Autom. Control **59**(3), 820–826 (2014)
14. Khalil, H.K.: Nonlinear Systems. Prentice Hall, Upper Saddle River (2002)
15. Kloetzer, M., Ding, X.C., Belta, C.: Multi-robot deployment from LTL specifications with reduced communication. In: IEEE Conference on Decision and Control and European Control Conference (CDC-ECC), pp. 4867–4872 (2011)
16. Mesbahi, M., Egerstedt, M.: Graph Theoretic Methods for Multiagent Networks. Princeton University Press, Princeton (2010)
17. Meyer, P-J., Girard, A., Witrant, E.: Safety control with performance guarantees of cooperative systems using compositional abstractions. In: Proceedings of the 5th IFAC Conference on Analysis and Design of Hybrid Systems, pp. 317–322 (2015)
18. Ögren, P., Egerstedt, M., Hu, X.: A control lyapunov function approach to multi-agent coordination. In: IEEE Conference on Decision and Control (CDC), pp. 1150–1155 (2001)
19. Pola, G., Pepe, P., Di Benedetto, M.D.: Symbolic models for networks of control systems. IEEE Trans. Autom. Control **61**, 3663–3668 (2016)
20. Rungger, M., Zamani, M.: Compositional construction of approximate abstractions. In: Proceedings of the 18th International Conference on Hybrid Systems: Computation and Control, pp. 68–77 (2015)
21. Tazaki, Y., Imura, J.: Bisimilar finite abstractions of interconnected systems. In: International Workshop on Hybrid Systems: Computation and Control, pp. 514–527. Springer, Berlin (2008)
22. Tumova, J., Dimarogonas, D.V.: Multi-agent planning under local LTL specifications and event-based synchronization. Automatica **70**(C), 239–248 (2016)
23. Tumova, J., Dimarogonas, D.V.: Decomposition of multi-agent planning under distributed motion and task LTL specifications. In: IEEE Conference on Decision and Control (CDC), pp. 7448–7453 (2015)
24. Zavlanos, M.M., Egerstedt, M.B.: Graph-theoretic connectivity control of mobile robot networks. Proc. IEEE **99**(9), 1525–1540 (2011)

Chapter 13
Modeling and Co-Design of Control Tasks over Wireless Networking Protocols

A. D'Innocenzo

Abstract In this chapter, we provide a brief overview of the state of the art on control over wireless communication protocols and present some recent advances in the co-design of controller and communication protocol configuration (i.e., scheduling and routing) subject to stochastic packet drops.

13.1 Introduction

Wireless networked control systems (WNCS) are distributed control systems where the communication between sensors, actuators, and computational units is supported by a wireless communication network. WNCSs have a wide spectrum of applications, ranging from smart grids to remote surgery, passing through industrial automation, environment monitoring, intelligent transportation, and unmanned aerial vehicles, to name few.

The use of WNCS in industrial automation results in flexible architectures and generally reduces installation, debugging, diagnostic, and maintenance costs with respect to wired networks (see e.g., [3, 33] and references therein). However modeling, analysis, and co-design of WNCS are challenging open research problems since they require to take into account the joint dynamics of physical systems, communication protocols, and network infrastructures. Recently, a huge effort has been made in scientific research on WNCSs, see e.g., [5, 8, 10, 19, 24, 28, 31, 40, 43, 68, 74, 78, 81, 84] and references therein for a general overview.

The challenges in analysis and co-design of WNCSs are best explained by considering wireless industrial control protocols. In this chapter, we focus on a networking protocol specifically developed for wireless industrial automation, i.e., WirelessHART, [35–37]. Indeed WirelessHART is not a niche technology, as many high-impact technological companies, such as Siemens, ABB, Emerson, sent to the

A. D'Innocenzo (✉)
Department of Information Engineering, Computer Science and Mathematics, Center of Excellence DEWS, University of L'Aquila, Via Vetoio, L'Aquila, Italy
e-mail: alessandro.dinnocenzo@univaq.it

S. Tarbouriech et al. (eds.), *Control Subject to Computational and Communication Constraints*, Lecture Notes in Control and Information Sciences 475, https://doi.org/10.1007/978-3-319-78449-6_13

market devices and industrial automation solutions based on the WirelessHART protocol. Due to its novelty, plenty of research activity on WirelessHART is still in progress in order to analyze the real capabilities of the standard and the application limits. In particular, the WirelessHART specification creates the opportunity for designers to implement ad hoc algorithms to configure the network configuration: we fill this gap by providing novel algorithms for co-designing controller and network configuration from a control performance point of view, and in particular, we investigate the design of redundancy when routing actuation data to a LTI system connected to the controller via a wireless network. We show with an example that the optimal co-design of controller gain and routing can strongly improve the control performance [21].

Pursuing the objective described above, we first consider the modeling, stability analysis, and controller design problems in a purely nondeterministic setting, when the actuation signal is subject to switching propagation delays due to dynamic routing [50]. We show how to model these systems as pure switching linear systems and provide an algorithm for robust stability analysis. We show that the stability analysis problem is NP-hard in general and provide an algorithm that computes in a finite number of steps the look-ahead knowledge of the routing policy necessary to achieve controllability and stabilizability.

We then consider, in a stochastic setting, the case when actuation packets can be delivered from the controller to the actuator via multiple paths, each associated with a delay and with time-varying packet loss probability. The *packet dropouts* have been modeled in the WNCS literature either as stochastic or nondeterministic phenomena [43]. The proposed nondeterministic models specify packet losses in terms of time averages or in terms of worst-case bounds on the number of consecutive dropouts (see e.g., [40]). For what concerns stochastic models, a vast amount of research assumes memoryless packet drops, so that dropouts are realizations of a Bernoulli process [31, 68, 74]. Other works consider more general correlated (bursty) packet losses and use a transition probability matrix of a finite-state (time-homogeneous) Markov chain (see e.g., the finite-state Markov modeling of Rayleigh, Rician, and Nakagami fading channels in [72] and references therein) to describe the stochastic process that rules packet dropouts (see [30, 74]). In these works, WNCS with missing packets are modeled as time-homogeneous Markov jump linear systems, which are an important family of stochastic hybrid systems that we use to model packet losses. In particular, it has been shown (e.g., in [21, 30, 74, 77]) that discrete-time Markov jump linear systems (MJLS, [17]) represent a promising mathematical model to jointly take into account the dynamics of a physical plant and nonidealities of wireless communication such as packet losses. A MJLS is, basically, a switching linear system where the switching signal is a Markov chain. The transition probability matrix of the Markov chain can be used to model the stochastic process that rules packet losses due to wireless communication. However, in most real cases, such probabilities cannot be computed exactly and are time-varying. We can take into account this aspect by assuming that the Markov chain of a MJLS is time-inhomogeneous, i.e., a Markov chain having its transition probability matrix varying over time, with variations that are arbitrary within a polytopic set of stochastic matrices.

Given such mathematical model, the first problem we address is providing necessary and sufficient conditions for the stochastic notion of mean square stability (MSS). Some recent works addressed the above problem: in [2], a sufficient condition for stochastic stability in terms of linear matrix inequality feasibility problem is provided, while in [16] a sufficient condition for MSS of system with interval transition probability matrix, which in turn can be represented as a convex polytope [38], is presented in relation to spectral radius; in general, only sufficient stability conditions have been derived for MJLS with time-inhomogeneous Markov chains having transition probability matrix arbitrarily varying within a polytopic set of stochastic matrices. We derive necessary and sufficient conditions [60] for MSS of discrete-time MJLS with time-inhomogeneous Markov chains. Having solved the stability problem, we extend the framework of MJLSs replacing the time-inhomogeneous Markov chain with a time-inhomogeneous Markov Decision Process and provide the optimal solution of the finite time horizon LQR problem considering the issue of joint minimization of costs of continuous and discrete control inputs for the worst possible disturbance in transition probabilities [61].

In addition to the investigations described above, we also addressed the problem of stabilizing a WNCS in presence of long- term link failures and malicious attacks. More precisely, we addressed the co-design problem of controller and communication protocol, and in particular routing and scheduling, when the physical plant is a MIMO LTI system and the communication nodes are subject to failures and/or malicious attacks. We first characterize by means of necessary and sufficient conditions the set of network configurations that invalidate controllability and observability of the plant. Then, we investigate the problem of detecting and isolating communication nodes affected by failures and/or malicious attacks and provide necessary and sufficient conditions for the solvability of this problem. This latter line of research is not illustrated in this chapter for space limitations, and we refer the interested reader to the papers [24–26].

This chapter is organized as follows. In Sect. 13.2, we provide a high-level description of the WirelessHART communication protocol. In Sect. 13.3, we first define our mathematical framework (i.e., time-inhomogeneous MJSLSs), which takes into account accurate models of packet dropouts; then we show with a motivating example that co-designing the controller and the routing strategy can lead to a strong improvement of the control performance. In Sect. 13.4, we summarize and discuss our technical results on co-design of controller and network configuration. In Sect. 13.5 we draw conclusions and directions for future work.

13.2 The WirelessHART Protocol

In this section, we introduce WirelessHART, one of the most relevant protocols currently used in industrial environments, emphasizing the features that will be analyzed and addressed in our mathematical models and co-design algorithms for WNCS. WirelessHART, [35–37] is one of the first wireless communication standards specifically designed for process automation applications. The standard has been finalized in 2007, at the beginning of 2010 it has been ratified as an IEC standard, and is based upon the physical layer of IEEE 802.15.4. When WirelessHART was developed, many requirements deemed critical for the industrial environment were not defined in IEEE 802.15.4, thus further specifications have been added in the data link layer (DLL). This new MAC protocol combines frequency hopping with a TDMA scheme utilizing a centralized a priori slot allocation mechanism. Indeed, it is commonly thought that TDMA-based protocols offer good opportunities for energy-efficient operation of sensor nodes, as they allow them to enter sleep mode when they are not involved in any communications. Such allocation of the shared channel regularizes the dynamical behavior introduced by multi-hop transmission: indeed, TDMA schemes avoid collisions and thus induce a periodic time-varying behavior, where delays and transmission times are well predictable, which can be nicely analyzed by considering sophisticated mathematical models like Markov Jump Linear Systems or Markov Decision Processes. In the results, illustrated in this chapter, we concentrate on modeling the joint dynamics of a closed-loop dynamical system and the WirelessHART data link and network layers.

About **data link layer**, the timing hierarchy of WirelessHART can be split in three timescale layers, as depicted in Fig. 13.1. The lowest layer consists of individual time slots: within each time slot, one data packet and the corresponding immediate acknowledgment packet are exchanged. A time slot in WirelessHART has a fixed length of 10 ms., and two types of time slots are available: *dedicated time slots* (slot is allocated to one specific sender–receiver pair) and *shared time slots* (more than one device may try to transmit a message). Within a dedicated time slot, transmission

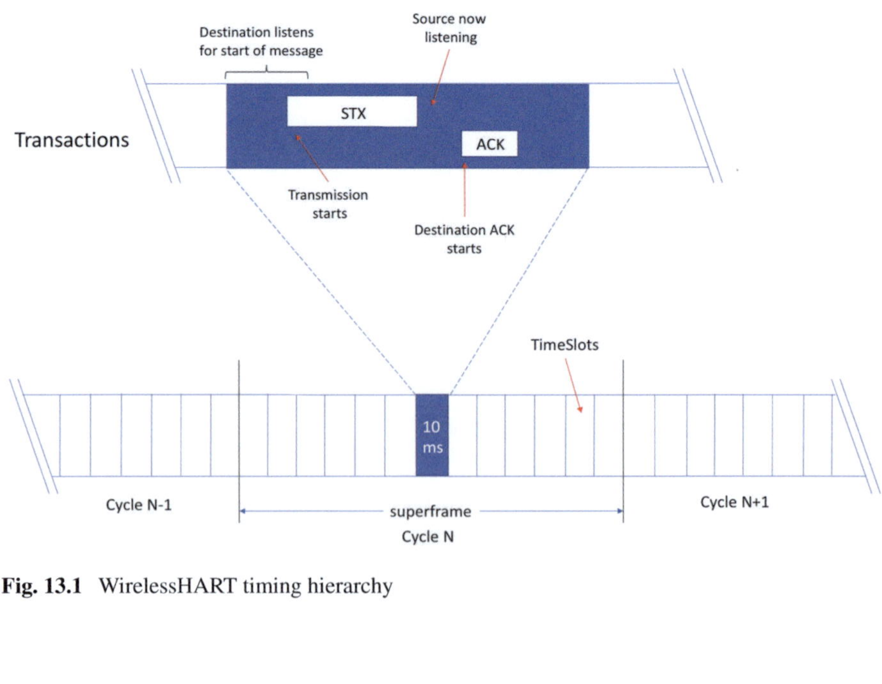

Fig. 13.1 WirelessHART timing hierarchy

of the source message starts at a specified time after the beginning of a slot. This short time delay allows the source and destination to set their frequency channel and allows the receiver to begin listening on the specified channel. Since there is a tolerance on clocks, the receiver must start to listen before the ideal transmission start time and continue listening after that ideal time. Once the transmission is complete, the destination device indicates by transmitting an ACK whether it received the source device data link packet successfully or not, indicating in this latter case a specific class of detected errors. At the second layer, a contiguous group of time slots of fixed length forms a superframe. At the third layer, a contiguous group of superframes forms a network cycle. Within each cycle, each field device obtains at least one time slot for data transmission, but certain devices may have more time slots than others because they provide data with more important requirements or have additional forwarding duties.

About **network layer**, in advanced applications point-to-point communications are unreliable also because it is very difficult to place devices to maintain line of sight all the time. The best architecture solution for wireless communication is a mesh topology network, which can provide multipath redundancy. Nodes, in fact, can multicast the same information to one or more of its neighbors with the goal of mitigating the risk of unplanned outages and ensure continuity of operation by instantly responding to and reducing the effects of a point of failure anywhere along the critical data path. In order to guarantee timely and reliable data delivery, routing topology and transmission schedule are centrally computed by a network manager device (which has global knowledge of the network state) and then disseminated to all devices in the network. WirelessHART allows two strategies to route packets: graph routing and source routing.

In *source routing*, a single fixed route of devices is decided by the source node and written in the header of the packet. Then, each device in the route forwards the packet to the next specified device until the destination is reached. There are no alternate routes in this mode, so if any device fails on a route, the whole route fails. Source routing is mostly used for network diagnostics.

In *graph routing*, the network manager defines a set of *routing graphs*, consisting of a set of acyclic directed graphs each connecting a source node to a destination node through some relay nodes on the network, and communicates them to each device. When a source node needs to send a packet, it writes a routing graph ID in the header of the packet to be sent. As the packet arrives at each node, the node forwards (or consumes, if it is the destination node) the packet according to the corresponding routing graph. Each relay node can be configured with multiple neighbors to create redundancy in the packet's forwarding. Thanks to such redundancy, graph routing is mostly used for sensing and actuation data communication.

As discussed above, the WirelessHART standard specifies the communication stack as well as the interfaces and tasks for the devices comprising a WirelessHART network. However it does not specify how these tasks should be accomplished, which provides interesting opportunities to develop improved and optimized solutions. An example is the exploitation of routing redundancy by jointly configuring the scheduling at the data link layer and the routing graph at the network layer. WirelessHART does not specify the algorithms and performance metrics to be used for scheduling and routing, and a designer must implement the best policies according to the specific application: we will provide in the next sections methods for co-designing controller and network from the control performance point of view.

13.3 Mathematical Framework and Motivating Example

In this section, we first introduce our reference mathematical framework and then provide an example showing that the joint design of network communication policies and control has a relevant impact on the control performance of a closed-loop system.

13.3.1 Time-Inhomogeneous Discrete-Time Markov Jump (switched) Linear Systems

Linear systems subject to abrupt parameter changes due, for instance, to environmental disturbances, component failures, changes in subsystems interconnections, changes in the operation point for a nonlinear plant, etc., can be modeled by a set of discrete-time linear systems with modal transition given by a discrete-time finite-state Markov chain. This family of systems is known as discrete-time Markov(ian) jump linear systems, often abbreviated as MJLSs.

The transition probabilities of a Markov chain are frequently time-varying and unavailable to the modeler, and a large body of research has been devoted to deal with these uncertainties and also to the identification of the Markov chain using available observations (see [13] and references therein for an introduction to the topic of estimation of such transition probabilities, which always introduces estimation errors). In order to account for uncertainties and time-variance inherent to real-world scenarios, the time-inhomogeneous polytopic model of transition probabilities is very general and widely used in the literature.

In this section, we present a rigorous mathematical model of MJLSs with polytopic uncertainties on transition probabilities and also the model of their natural extention, i.e., Markov jump switched linear systems (MJSLSs).

A discrete-time Markov jump linear systems can be defined as follows:

$$\begin{cases} \mathrm{x}_{k+1} = A_{\theta_k}\mathrm{x}_k + B_{\theta_k}\mathrm{u}_k + H_{\theta_k}\mathrm{v}_k, \\ \mathrm{y}_k = F_{\theta_k}\mathrm{x}_k + G_{\theta_k}\mathrm{w}_k, \\ \mathrm{z}_k = C_{\theta_k}\mathrm{x}_k + D_{\theta_k}\mathrm{u}_k, \\ \mathrm{x}_0 = \boldsymbol{x}_0,\ \theta_0 = \vartheta_0,\ \mathrm{p}_0 = \boldsymbol{p}_0, \end{cases} \tag{13.1}$$

where $k \in \mathbb{T}$ is a discrete-time instant, $\mathbb{T}$ is a discrete-time set, $\mathbb{T} = \mathbb{Z}_0$, with $\mathbb{Z}_0$ indicating the set of all nonnegative integers and $\mathbb{Z}$ the set of integers. Then, x_k is a vector of n_x either real or complex *state* variables of the Markov jump linear system, where $n_x \in \mathbb{Z}_+$ with $\mathbb{Z}_+$ the set of positive integers, and $x_k \in \mathbb{F}^{n_x}$ with $\mathbb{F}^{n_x}$ an n_x-dimensional linear space with entries in $\mathbb{F}$. Note that $\mathbb{F}$ indicates the set of either real numbers $\mathbb{R}$ or complex numbers $\mathbb{C}$: in general, when studying MJLSs, it is a standard practice to work with complex fields [17], but one can consider complex operators acting on $\mathbb{C}^{m,n}$ as real (block) matrices acting on $\mathbb{R}^{2m,2n}$ [49].

For what concerns other system variables in the aforementioned state-space representation of an MJLS, u_k stands for a vector of n_u *control input* variables, $u_k \in \mathbb{F}^{n_u}$; then, $v_k \in \mathbb{F}^{n_v}$ and $w_k \in \mathbb{F}^{n_w}$ are vectors of exogenous input variables, known as *process noise* and *observation noise*, respectively; $y_k \in \mathbb{F}^{n_y}$ represents a vector of *measured state* variables available to the controller; $z_k \in \mathbb{F}^{n_z}$ denotes a vector of measured *system output* variables. Clearly, $n_u, n_v, n_w, n_y, n_z \in \mathbb{Z}_+$.

In the following, we write $\mathbb{F}^{m,n}$ to denote a set of matrices with m rows, n columns, and entries in $\mathbb{F}$. Consequently, the elements of the system matrices A_{θ_k}, B_{θ_k}, etc., are also defined on a field of either real or complex numbers $\mathbb{F}$. The subscript θ_k indicates that system matrices active in a time instant k are determined by the value of the *jump* variable θ_k, which is a random variable having the set $\mathbb{M} \triangleq \{i \in \mathbb{Z}_+ : i \leq N\}$ as its *state space*, where $N \in \mathbb{Z}_+$ is the cardinality of the set, formally $|\mathbb{M}| = N$. The set $\mathbb{M}$ is generally referred to as the (index) set of *operational modes* of the Markov jump linear system. We denote by θ_k as the identity function of the set of operational modes, i.e., $\theta_k : \mathbb{M} \rightarrow \mathbb{M}$, and $\forall i \in \mathbb{M}$, we have that $\theta_k(i) = i$.

For every operational mode, there is a correspondent system matrix, and the collection of the system matrices of each type is generally represented by a sequence of N matrices, which are not necessarily all distinct. Specifically, $\boldsymbol{A} \triangleq (A_i)_{i=1}^N \in {}_{\mathrm{N}}\mathbb{F}^{n_x,n_x}$ is a sequence of the so-called *state matrices*, each of which is associated to an operational mode of the (switching) system. Noticeably, ${}_{\mathrm{N}}\mathbb{F}^{m,n}$ indicates a *linear space* made up of all *N-sequences of* $m \times n$ *matrices* with entries in $\mathbb{F}$. Similarly, $\boldsymbol{B} \triangleq (B_i)_{i=1}^N \in {}_{\mathrm{N}}\mathbb{F}^{n_x,n_u}$ is an N-sequence of *input matrices*; $\boldsymbol{C} \triangleq (C_i)_{i=1}^N \in {}_{\mathrm{N}}\mathbb{F}^{n_z,n_x}$ is a sequence of *output matrices*; $\boldsymbol{D} \triangleq (D_i)_{i=1}^N \in {}_{\mathrm{N}}\mathbb{F}^{n_z,n_u}$ is a sequence of *direct transition* (also known as feed-forward or feedthrough) *matrices*; $\boldsymbol{F} \triangleq (F_i)_{i=1}^N \in {}_{\mathrm{N}}\mathbb{F}^{n_y,n_x}$ is a sequence of *observation matrices*; $\boldsymbol{G} \triangleq (G_i)_{i=1}^N \in {}_{\mathrm{N}}\mathbb{F}^{n_y,n_w}$ is a sequence of *observation noise matrices*; and $\boldsymbol{H} \triangleq (H_i)_{i=1}^N \in {}_{\mathrm{N}}\mathbb{F}^{n_x,n_v}$ is a sequence of *process noise matrices*.

The transitions, or jumps, between operational modes of an MJLS are governed by a discrete-time Markov chain θ, which is a collection of random variables θ_t all taking values in the same state space, i.e., $\{\theta_t : t \in \mathbb{T}\}$, and satisfying the Markov property. The initial probability distribution of the Markov chain is defined $\forall i \in \mathbb{M}$ by $p_i(0) \triangleq \Pr(\theta_0 = i)$, and the initial probability distribution of all the operational modes is defined as the vector $p_0 \triangleq [p_1(0) \dots p_N(0)]' \in \mathbb{R}^{N,1}$. The transition probability between the operational modes $i, j \in \mathbb{M}$ of a Markov jump linear system is formally defined as

$$p_{ij}(k) \triangleq \Pr(\theta_{k+1} = j \mid \theta_k = i)\,, \tag{13.2}$$

where $\forall i \in \mathbb{M}$ and $\forall k \in \mathbb{T}$, $\sum_{j=1}^N p_{ij}(k) = 1$. The corresponding transition probability matrix is defined as

$$P(k) \triangleq \left[p_{ij}(k)\right] \in \mathbb{R}^{N,N}. \tag{13.3}$$

The initial conditions for a Markov jump linear system consist of the initial state of the dynamical system $\mathrm{x}_0 = \boldsymbol{x}_0 \in \mathbb{F}^{n_\mathrm{x}}$, the initial state of the Markov chain $\theta_0 = \vartheta_0 \in \mathbb{M}$ and the initial probability distribution of the states of the Markov Chain denoted by $\mathrm{p}_0 = \boldsymbol{p}_0 \in \mathbb{R}_0^N$ s.t. $\left\| \boldsymbol{p}_0 \right\|_1 = 1$, with $\| \cdot \|_1$ the standard 1−norm of a vector and $\mathbb{R}_0^N$ the N-dimensional linear space with entries in the set of nonnegative real numbers $\mathbb{R}_0$.

Although in engineering problems the operation modes are not often available, there are enough cases where the knowledge of random changes in system structure is directly available to make these applications of great interest [9, 17]. The typical examples include a ship steering autopilot, control of pH in a chemical reactor, combustion control of a solar-powered boiler, fuel–air control in a car engine, and flight control systems [17]. In this chapter, we focus on the fact that MJLSs' model is well suited also for the WNCS scenario, when a channel estimation is performed (see e.g., [57] and the references therein), so the channel state information is known at each time step. In fact, the knowledge of θ_k at each time instant k is a standard assumption in the setting of MJLSs, and in this chapter we do the following assumption.

Assumption 13.1 At every time step $k \in \mathbb{T}$, the jump variable θ_k is measurable and available to a controller.

Depending on the considered problem, (some of) the system's vector variables x_k, u_k, y_k, and z_k may also be viewed as measurable.

We have previously discussed that in most real cases the transition probability matrix $P(k)$ introduced in (13.3) cannot be computed exactly and is time-varying, and that there exists a considerable number of works on discrete-time Markov jump systems (both linear and nonlinear) with polytopic uncertainties, which can be either time-varying or time-invariant. From now on, we assume that $P(k)$ is varying over time, with variations that are arbitrary within a polytopic set of stochastic matrices. In order to express this statement formally, let $V \in \mathbb{Z}_+$ be a number of vertices of a convex polytope, and $\mathbb{V}$ be an index set of vertices of a convex polytope, i.e., $\mathbb{V} \triangleq \{i \in \mathbb{Z}_+ : i \leq V\}$. Then, the set of vertices of a convex polytope of transition probability matrices is formally defined as

$${}_{\mathbb{V}}\mathbb{P} \triangleq \left\{P_l \in \mathbb{R}^{N,N} : l \in \mathbb{V}\right\}. \tag{13.4}$$

Clearly, being a transition probability matrix, each vertex P_l satisfies (13.2) and (13.3). These vertices are obtained from measurement on the real system or via numerical reasoning, taking into account accuracy and precision of the measuring instruments and/or numerical algorithms. They bound the possible values each transition probability can assume. Then, the polytopic time-inhomogeneous assumption is stated as follows.

Assumption 13.2 The time-varying transition probability matrix $P(k)$ is **polytopic**, that is, for all $k \in \mathbb{T}$, one has that

$$P(k) = \sum_{l=1}^{V} \lambda_l(k) P_l, \quad \lambda_l(k) \geq 0, \quad \sum_{l=1}^{V} \lambda_l(k) = 1, \tag{13.5}$$

where for each $l \in \mathbb{V}$, $P_l \in {}_{\mathbb{V}}\mathbb{P} \subset {}_{\mathbb{V}}\mathbb{R}^{N,N}$, i.e., P_l are elements of a given finite set of transition probability matrices, which are the vertices of a convex polytope; moreover, $\lambda_l(k)$ are unmeasurable.

Assumption 13.2 plays an important role also in our model of Markov jump switched linear system, which is a dynamical system having the same form as (13.1), with the only difference being that the operational modes of the system are determined by the stochastic variable s_k that represents a standard Markov Decision Process. A Markov Decision Process is a quintuple $(\mathbb{M}, \mathbb{A}, \Pr, g, \gamma)$, where

- $\mathbb{M}$ is a finite set of states of a process, with $|\mathbb{M}| = N$.
- $\mathbb{A}$ is a finite (index) set of actions among which a decision maker (a.k.a. a discrete controller or a supervisor) is able to chose, i.e., $\mathbb{A} \triangleq \{i \in \mathbb{Z}_+ : i \leq M\}$. Typically, only a subset of $\mathbb{A}$ is available in any given state of an MDP.[1] We take this into account by defining for each state $i \in \mathbb{M}$ the related set $\mathbb{A}_i$ of actions α available in that state. We write this statement symbolically as $\mathbb{A}_i \subseteq \mathbb{A}$, $\alpha \in \mathbb{A}_i$.
- Pr is state- and action-dependent transition probability distribution. For any $k \in \mathbb{T}$, $i, j \in \mathbb{M}$, $\alpha \in \mathbb{A}_i$, the future transition probability distribution, conditioned on the present state s_k of the MDP and the action α_k to be taken from that state, is denoted by
$$p_{ij}^{\alpha}(k) \triangleq \Pr\{s_{k+1} = j \mid s_k = i, \ \alpha_k = \alpha\}. \tag{13.6}$$
Being a probability distribution, $p_{ij}^{\alpha}(k) \in \mathbb{R}_0$ and satisfies $\forall k \in \mathbb{T}$, $i, j \in \mathbb{M}$, and $\alpha \in \mathbb{A}_i$
$$\sum_{j=1}^{N} p_{ij}^{\alpha}(k) = 1. \tag{13.7}$$
For any $\alpha \notin \mathbb{A}_i$, the action is not available in a given state of the MDP. Hence, $\forall j \in \mathbb{M}$
$$p_{ij}^{\alpha}(k) \triangleq 0. \tag{13.8}$$
In the following, we assume that the transition probability matrices constituting Pr are varying over time, with variations that are arbitrary within a polytopic set of stochastic matrices according to Assumption 13.2.
- Selecting an (available) action in any given state of a Markov decision process entails a (nonnegative) cost, which is seen as a function $g : \mathbb{M} \times \mathbb{A} \to \mathbb{G}$, where $\mathbb{G} \subseteq \mathbb{R}_0$ is a set of immediate costs.
- γ is a discounting factor, which represents the difference in importance between future costs and present costs; $\gamma \in \mathbb{R}_0$, $\gamma \leq 1$. Since taking the discount factor into

---

[1]For instance, in a decision problem of the optimal transmission power management in a wireless communication, the possible actions available to a controller may be those of increasing or decreasing of a transmission power: in a finite set of transmission power levels, it is impossible to increase a power from a maximum level or decrease it from a minimum level.

account does not affect any theoretical results or algorithms in the finite-horizon case (but might affect the decision maker's preference for policies) [71, p. 79], we do not consider a discounting factor here.

In the same way as done before for a Markov jump variable, we make the following assumption.

Assumption 13.3 The state s_k of the Markov decision process is measurable and available for the discrete controller at each time step $k \in \mathbb{T}$.

We are now ready to define a MJSLS as the following system of recursive equations, where the system's variables and matrices are the same as above:

$$\begin{cases} \mathrm{x}_{k+1} &= A_{s_k}\mathrm{x}_k + B_{s_k}\mathrm{u}_k \\ \mathrm{z}_k &= C_{s_k}\mathrm{x}_k + D_{s_k}\mathrm{u}_k, \\ \mathrm{x}_0 &= \boldsymbol{x}_0,\ s_0 = \boldsymbol{s}_0,\ \mathrm{p}_0 = \boldsymbol{p}_0. \end{cases} \tag{13.9}$$

13.3.2 Co-design of Controller and Routing Redundancy

As discussed in Sect. 13.2, routing and scheduling schemes in a WNCS where the network is based on a WirelessHART-like protocol have a direct and relevant impact on closed-loop performance and the aforementioned standards leave the possibility, for the designers, to implement complex and time-varying routing and scheduling ad hoc optimal policies. Indeed, to make a WirelessHART WNCS robust to transmission nonidealities routing redundancy can be exploited by relaying data via multiple paths and then appropriately recombining them, which is reminiscent of network coding. Basically, if a communication path fails, another one can be available to maintain the communication flow. As paths are characterized by different communication properties, such as delays and packet losses, the routing design must take into account the effect on the control performance of the closed-loop system.

In this section, we illustrate a motivating example from [21], where we model a WNCS implementing WirelessHART as a MJSLS and address the problem of optimally co-designing controller and routing with respect to a control performance index, e.g., the classical quadratic cost used in LQR.

Consider a state-feedback WNCS as in Fig. 13.2, where the communication between the controller and the actuator can be performed via a set of r routing paths $\{\rho_i\}_{i=1}^r$ in a wireless multi-hop communication network. Each path ρ_i is characterized by a delay $d_i \in \mathbb{N}^+$ and a packet loss probability $p_i \in [0, 1]$ that represents the probability that the packet transmitted on that path will not reach the actuator due to communication failure. Therefore let us define, for each path ρ_i, the stochastic process $\sigma_i(k) \in \{0, 1\}$, with $\sigma_i(k) = 0$ if the packet expected to arrive via the routing path ρ_i at time k suffered a packet drop and $\sigma_i(k) = 1$ if the packet is successfully received at time k. For simplicity we assume that $\sigma_i(k)$ is a sequence of

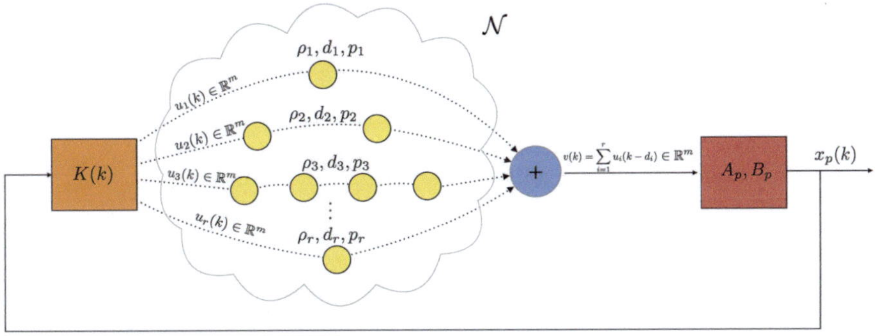

Fig. 13.2 State-feedback control scheme

i.i.d. random variables, each characterized by a Bernoulli distribution with probability measure $\mathbb{P}[\sigma_i(k) = 0] = p_i$. It is also assumed here that the events of occurrence of packet losses in the different paths are i.i.d.: as a consequence the stochastic process $\sigma(k) \doteq [\sigma_1(k), \ldots, \sigma_r(k)]'$ is a vector of i.i.d. random variables, where $\sigma(k)$ can assume 2^r values. The controller cannot measure the signal $\sigma(k)$, i.e., it is not possible to measure the occurrence of packet losses. It is assumed that, in general, the controller can decide for each time instant k the set of paths where data will be sent: i.e., the controller can decide to send data at time k on all paths, on a subset of paths, on one path, or even not to send any data. To this aim let us define for each path i the discrete control signal $a_i(k) \in \{0, 1\}$, with $a_i(k) = 1$ if the controller decides to send a packet via the routing path i at time k, and $a_i(k) = 0$ if no packet is sent via path i at time k. Consequently, the discrete control signal $a(k) \doteq [a_1(k), \ldots, a_r(k)]'$, where $a(k)$ can be chosen among 2^r different values.

Let the plant be a discrete-time LTI system described by the matrices $A_P \in \mathbb{R}^{\ell \times \ell}$, $B_P \in \mathbb{R}^{\ell \times m}$ and assume that we can measure the full system's state, then the dynamics of the networked system are as follows:

$$\begin{cases} x(k+1) = A_{\sigma(k)}x(k) + B_{a(k)}u(k) \\ y(k) = x(k) \end{cases} \tag{13.10}$$

with

$$A_{\sigma(k)} = \begin{bmatrix} A_P & \Lambda_1(\sigma(k)) & \Lambda_2(\sigma(k)) & \cdots & \Lambda_r(\sigma(k)) \\ 0 & \Gamma_1 & 0 & \cdots & 0 \\ 0 & 0 & \Gamma_2 & \cdots & 0 \\ \vdots & \vdots & \vdots & \ddots & \vdots \\ 0 & 0 & 0 & \cdots & \Gamma_r \end{bmatrix} \in \mathbb{R}^{(\ell+\nu(r)\times\ell+\nu(r))},$$

$$B_{a(k)} = \begin{bmatrix} 0 & 0 & \cdots & 0 \\ a_1(k) I_m \otimes \mathbf{e}_{d_1} & 0 & \cdots & 0 \\ 0 & a_2(k) I_m \otimes \mathbf{e}_{d_2} & \cdots & 0 \\ \vdots & \vdots & \ddots & \vdots \\ 0 & 0 & \cdots & a_r(k) I_m \otimes \mathbf{e}_{d_r} \end{bmatrix} \in \mathbb{R}^{(\ell+\nu(r) \times mr)},$$

with

$$\Lambda_i(\sigma(k)) \doteq \sigma_i(k) \begin{bmatrix} B_P & 0 & \cdots & 0 \end{bmatrix} \in \mathbb{R}^{\ell \times md_i},$$

$$\Gamma_i \doteq \begin{bmatrix} 0 & I_m & \cdots & 0 & 0 \\ \vdots & \vdots & \ddots & \vdots & \vdots \\ 0 & 0 & \cdots & I_m & 0 \\ 0 & 0 & \cdots & 0 & I_m \\ 0 & 0 & \cdots & 0 & 0 \end{bmatrix} \in \mathbb{R}^{md_i \times md_i},$$

and with $\nu(i) \doteq m \sum_{j=1}^{i} d_j$, I_m the m-dimensional identity matrix, $\mathbf{e}_i$ a column vector of appropriate dimension with all zero entries except the $i-th$ entry equal to 1, and $\otimes$ the Kronecker product. Note that in the feedback scheme presented, it is assumed that the controller can measure the whole state $x(k) = [x_P(k)' x_N(k)']'$ of (13.10), where $x_P(k) \in \mathbb{R}^{\ell}$ is the state of the plant and $x_N(k) \in \mathbb{R}^{\nu(r)}$ are state variables modeling the delay induced by each path. It is assumed that the controller can measure the state $x_P(k)$ of the plant via sensors. Also, the controller is aware of the current and past actuation signals $u(k)$ that have been sent to the actuator, as well as of the current and past signals $a(k)$: as a consequence the controller has direct access to the state of $x_N(k)$, which models the actuation commands that are expected to arrive at the actuator, but is not aware of their actual arrival to the actuator since $\sigma(k)$, which models packet drops, is not measurable.

We consider in this example the simpler case when routing is designed a priori, i.e., $\forall k \geq 0$, $a(k) = a_k$: note that with this assumption system (13.10) is a MJLS as defined in (13.2). Let us now consider an instance of system (13.10) characterized by a four-dimensional unstable randomly generated plant

$$A_P = \begin{bmatrix} 1.1062 & -1.0535 & 0.7944 & -0.4543 \\ 0.0202 & -0.0654 & 0.9697 & -0.6888 \\ 0.1131 & -0.5755 & 1.7434 & -0.7174 \\ 0.0745 & -0.2565 & 0.2999 & 0.7252 \end{bmatrix}, \quad B_P = \begin{bmatrix} -0.1880 \\ 0.0182 \\ 0.1223 \\ 0.2066 \end{bmatrix},$$

and by a wireless network characterized by two paths: ρ_1 with packet loss probability $p_1 = 0.25$ and delay $d_1 = 1$ and ρ_2 with packet loss probability $p_2 = 0$ and delay $d_2 = 5$. We setup the following standard LQR optimization problem:

Problem 13.1 Given System (13.10) and a routing sequence a_k, $k = 0, 1, \ldots, N-1$, design for any $k \in \{0, \ldots, N-1\}$ an optimal state-feedback control policy $u^*(k) = K^*(\theta(k), k)x(k)$ minimizing the following objective function:

$$\mathscr{J}(\theta_0, \mathrm{x}_0) \triangleq \min_{\mathrm{u}} \sum\nolimits_{k=0}^{T-1} \mathbb{E}\left(\|\mathrm{z}_k\|_2^2\right) + \mathbb{E}\left(\mathrm{x}_T^* Z_{\theta_T} \mathrm{x}_T\right) \tag{13.11}$$

with θ and z_k as in (13.1) and $\mathbf{Z} \triangleq (Z_i)_{i=1}^N \in {}_{\mathrm{N}}\mathbb{F}_0^{n_x, n_x}$ a sequence of the terminal cost weighting matrices.

We solve Problem 13.1 using the optimal LQR solution for MJLS for comparing the performance of three simple routing strategies: (1) using for all time instants only path ρ_1; (2) using for all time instants only path ρ_2; using for all time instants both paths simultaneously. Solution is computed for a time horizon $T = 300$. For a detailed description of the weight matrices of the cost function (13.11) and the initial conditions we refer the reader to [21]. For each routing strategy, $5K$ MC simulations of the state trajectories are performed. Figure 13.3 shows the trajectories of the first component of the extended state vector when only path ρ_1 is used. The system can be stabilized, but clearly the variance of the trajectories is large. This routing policy is clearly a bad choice. Figure 13.4 shows the trajectories when only path ρ_2 is used (red) and when both paths ρ_1 and ρ_2 are used (blue and green). Routing data only to path ρ_2 clearly generates always the same trajectory since $p_2 = 0$. The system trajectories are stable but the associated cost is quite large because of the delay, as evidenced by the overshoot and the settling time performances. Figure 13.4 evidences that routing data via both paths ρ_1 and ρ_2 the control performance strongly

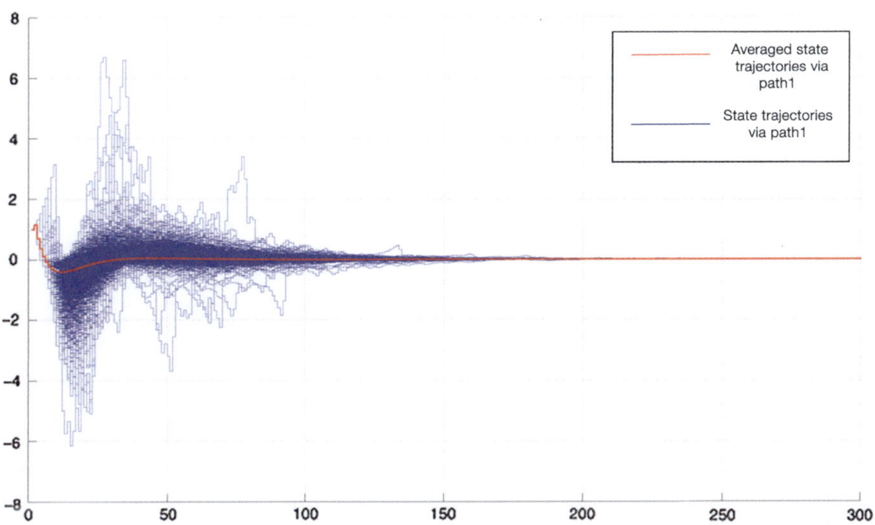

Fig. 13.3 State trajectories routing only via path ρ_1 (blue) and their average (red)

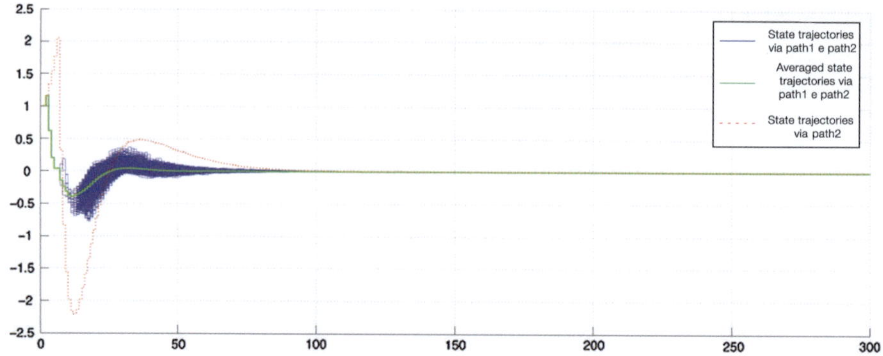

Fig. 13.4 State trajectory routing only via path ρ_2 (red dashed); state trajectories routing via both paths ρ_1 and ρ_2 (blue) and their average (green)

Table 13.1 Cost averaged over 5K MC simulations

	Averaged cost
Via path ρ_1	~ 900
Via path ρ_2	~ 250
Via paths ρ_1, ρ_2	~ 100

improves: in particular, the trajectory of the system computed by averaging over all MC simulations is characterized by much smaller overshoot and faster settling time. The single trajectories generated routing data via both paths ρ_1 and ρ_2 clearly have some variance due to the high packet loss probability p_1: however, in the $5K$ MC simulations, the performance of any of the single trajectories is much better than the case when only path ρ_2 is used. Table 13.1 shows the tremendous improvement of the controller performance obtained by exploiting both paths and co-designing optimal control and routing redundancy.

13.4 Main Results

In the previous section, we have illustrated that time-inhomogeneous MJSLSs represent a mathematical model to jointly take into account the dynamics of a physical plant and nonidealities of wireless communication such as packet losses, and that their exploitation for optimal design of routing redundancy can strongly improve the closed-loop control performance. In this section, we illustrate recent advances related to the co-design of controller and communication protocol configuration subject to stochastic packet drops.

13.4.1 Stability Analysis of Linear Systems with Switching Delays

In our research line, as a first approach to address the above problem, we considered in [50] the case when routing is purely nondeterministic and packet losses are not present. We will exploit the mathematical framework of switching linear systems, which can be considered a special case of a time-inhomogeneous MJLS, where the polytopic uncertainty set contains all stochastic transition probability matrices. Because of the TDMA scheduling each path is characterized by a fixed delay in forwarding the data (see [24] for details), as a consequence each actuation data is delayed of a finite number of time steps according to the chosen routing and our system is characterized by switching time-varying delays of the input signal.

Systems with time-varying delays have attracted increasing attention in recent years (see e.g., [41, 44, 75] and references therein). In [46], it is assumed that the time-varying delay is approximatively known and numerical methods are proposed to exploit this partial information for adapting the control law in real time. The LMI-based design procedures that have been developed for switching systems with time-varying delays (see e.g., [45, 86]) do not take into account the specific structure of the systems induced by the fact that the switching is restricted on the delay-part of the dynamics. Our goal is to leverage this particular structure in order to improve our theoretical understanding of the dynamics at stake in these systems. This enables us to design tailored controllers, whose performance or guarantees are better than for classical switching systems. Our modeling choice is close to the framework in [44]. However, our setting is more general and realistic in that, differently from [44], it allows for several critical phenomena to happen: in our model, control commands generated at different times can reach the actuator simultaneously, their arrival time can be inverted, and it is even possible that at certain times no control commands arrive at the actuator. An investigation similar to ours, applied to the different setting of Lyapunov exponents of randomly switching systems, has been pursued in [76].

In our setting assume that, at each time t, the controller is aware of the propagation delays of the actuation signals sent at times $t, t+1, \ldots, t+N-1$. We assume that N can be larger than 1, i.e., that the controller is aware of the current and $N-1$ next future routing path choices and keeps memory of the past delays: we define N the *look-ahead* parameter and we call this situation the *delay-dependent case*.

The practically admissible values for N depend on the protocol used to route data: indeed, note that in several practical situations, the networking protocol can be designed to choose at any time t the future routing paths up to $t+N-1$.

As a first contribution, we show that our particular networked systems can be modeled by pure switching systems, where the switching matrices assume a particular form. As a direct consequence, the well-known LMI stability conditions for switching systems (see e.g., [70]) can be directly used to compute the worst rate of growth with fixed and arbitrarily small conservativeness. Also, while it is well known that the stability analysis problem is NP-hard for general switching systems [14], we prove that it is NP-hard even in our particular case of switching delays.

As a second contribution, we address the controller design problem. We first consider the case when one can design the communication system such that we have an arbitrarily large but finite look-ahead N. Of course, we are interested in requiring the smallest N: to this aim, we first provide an algorithm for efficiently constructing this controller, or deciding it does not exist. In case it exists, we prove a general upper bound N^* on the needed look-ahead, depending only on the dimension of the plant and the set of delays. This result has strong practical implications, since it implies (if a system is controllable) that it is never necessary to have infinite look-ahead and moreover a look-ahead equal to N^* is always sufficient. If $N \geq N^*$, stabilizability is equivalent to controllability of a projection of the initial system (as is customary for linear time-invariant systems). This implies that our techniques are also valid for the stabilizability problem.

The results described above, presented in [50], provide necessary stability and controllability conditions when extending our modeling framework to the stochastic setting in order to address packet loss models, which is the main topic of the next section.

13.4.2 Analysis and Design of Time-Inhomogeneous Discrete-Time MJSLS

In this section, we first provide necessary and sufficient stability conditions for time-inhomogeneous discrete-time MJLS ([60, 62]). Then we illustrate optimal solutions for the LQR problem for time-inhomogeneous discrete-time MJSLS ([61]). In [59] we recently addressed the optimal filtering problem and proved a separation principle.

The robust stability problem: Let us consider an autonomous discrete-time Markov jump linear system described by the following state-space model:

$$\begin{cases} \mathrm{x}_{k+1} = A_{\theta_k}\mathrm{x}_k + H_{\theta_k}\mathrm{v}_k, \\ \mathrm{x}_0 = \boldsymbol{x}_0, \ \theta_0 = \vartheta_0. \end{cases} \tag{13.12}$$

Let us denote by $\mathbb{E}(\cdot)$ the expected value of a random variable, and by $\|\cdot\|$ either any vector norm or any matrix norm. Then, the *mean square stability* of a system (13.12) is defined as follows.

Definition 13.1 [17, p. 36–37] A Markov jump linear system (13.12) is **mean square stable** if for any initial condition $\mathrm{x}_0 \in \mathbb{F}^{n_\mathrm{x}}$ and $\theta_0 \in \Theta_0$ there exist $\boldsymbol{x}_e \in \mathbb{F}^{n_\mathrm{x}}$ and $Q_e \in \mathbb{F}_+^{n_\mathrm{x}, n_\mathrm{x}}$ (independent from initial conditions x_0 and θ_0), such that

$$\lim_{k \to \infty} \|\mathbb{E}(\mathrm{x}_k) - \boldsymbol{x}_e\| = 0, \tag{13.13a}$$

$$\lim_{k \to \infty} \left\|\mathbb{E}\left(\mathrm{x}_k \mathrm{x}_k^*\right) - Q_e\right\| = 0. \tag{13.13b}$$

Remark 13.1 It is worth mentioning [17, p. 37, Remark 3.10] that in noiseless case, i.e., when $\mathrm{v}_k = 0$ in (13.12), the conditions (13.13) defining mean square stability become

$$\lim_{k\to\infty} \mathbb{E}(\mathrm{x}_k) = 0, \qquad \lim_{k\to\infty} \mathbb{E}\left(\mathrm{x}_k \mathrm{x}_k^*\right) = 0. \tag{13.14}$$

There exist also other forms of stability for Markov jump linear systems without process noise, notably *exponential mean square stability* (EMSS) and *stochastic stability* (SS), that we define as follows.

Definition 13.2 [17] An MJLS (13.12) is **exponentially mean square stable** if for some reals $\beta \geq 1, 0 < \zeta < 1$, we have for all initial conditions $\mathrm{x}_0 \in \mathbb{F}^{n_\mathrm{x}}$ and $\theta_0 \in \Theta_0$ that, for every $k \in \mathbb{T}$, if $\mathrm{v}_k = 0$, then

$$\mathbb{E}\left(\|\mathrm{x}_k\|^2\right) \leq \beta\zeta^k \|\mathrm{x}_0\|_2^2 \tag{13.15}$$

We observe that $\|\cdot\|_2$ denotes the Euclidean norm, also known as $\mathbb{L}^2$ norm or simply 2-norm.

Definition 13.3 [17] A Markov jump linear system (13.12) is **stochastically stable** if for all initial conditions $\mathrm{x}_0 \in \mathbb{F}^{n_\mathrm{x}}$ and $\theta_0 \in \Theta_0$, we have that, if $\mathrm{v}_k = 0$ for every $k \in \mathbb{T}$, then

$$\sum_{k=0}^{\infty} \mathbb{E}\left(\|\mathrm{x}_k\|^2\right) \leq \infty. \tag{13.16}$$

In the *time-homogeneous* case, i.e., when the transition probability matrix defined by (13.2) and (13.3), is such that $P(k) = P$ for all $k \in \mathbb{T}$, there is a condition based on a value of a spectral radius of a matrix associated to the second moment of x_k that is necessary and sufficient for the mean square stability of system (13.12); furthermore, in the noiseless setting, MSS, EMSS, and SS are equivalent [17, pp. 36–44]. Specifically, the matrix related to the second moment of x_k that we have mentioned above is

$$\Lambda \triangleq \left(P^T \otimes I_{n_\mathrm{x}^2}\right) \left(\bigoplus_{i=1}^{N} \left(\bar{A}_i \otimes A_i\right),\right) \tag{13.17}$$

where $\otimes$ denotes the Kronecker product, $I_{n_\mathrm{x}^2}$ is the identity matrix of size n_x^2, and the direct sum $\oplus$ of the manipulated elements of a sequence of state matrices $\boldsymbol{A}$ produces a block diagonal matrix, having the matrices $\left(\bar{A}_i \otimes A_i\right)$ on the main diagonal blocks.

The necessary and sufficient condition for the mean square stability of time-homogeneous Markov jump linear systems we have hinted at before is

$$\rho(\Lambda) < 1, \tag{13.18}$$

where $\rho(\cdot)$ denotes the spectral radius of a matrix. This condition for mean square stability does not hold in time-inhomogeneous case. The results of this section are based on a noiseless version of (13.12), i.e., when $\mathrm{v}_k = 0$ for every $k \in \mathbb{T}$. They are based on our first work on MJLSs [60].

Let us consider a noiseless autonomous discrete-time Markov jump linear system described by the following system of difference equations

$$\begin{cases} \mathrm{x}_{k+1} = A_{\theta_k}\mathrm{x}_k, \\ \mathrm{x}_0 = \boldsymbol{x}_0, \ \theta_0 = \vartheta_0 \end{cases} \tag{13.19}$$

where, as before, $\mathrm{x}_k \in \mathbb{F}^{n_\mathrm{x}}$ is a system's state vector, $\boldsymbol{A} \triangleq (A_i)_{i=1}^N \in_{\mathrm{N}} \mathbb{F}^{n_\mathrm{x}, n_\mathrm{x}}$ is a sequence of *state matrices*, each of which is associated to an operational mode; while $\mathrm{x}_0 \in \mathbb{F}^{n_\mathrm{x}}$ and $\theta_0 \in \Theta_0$ are initial conditions. Let the transition probability matrix $P(k) = [p_{ij}(k)]$ of the system (13.19) be polytopic time-inhomogeneous, i.e., satisfying Assumption 13.2.

Theorem 13.4 [60] *The discrete-time Markov jump linear system (13.19) with unknown and time-varying transition probability matrix $P(k) \in \mathrm{conv}_{\mathbb{V}}\mathbb{P}$ is mean square stable if and only if $\hat{\rho}(_{\mathbb{V}}\boldsymbol{A}) < 1$.*

In Theorem 13.4 $\mathrm{conv}_{\mathbb{V}}\mathbb{P}$ denotes the convex hull of the set of transition probability matrix vertices as defined in (13.4) and $\hat{\rho}(\cdot)$ denotes the joint spectral radius (JSR)[2] of the set of matrix vertices $_{\mathbb{V}}\boldsymbol{A}$ obtained by replacing in (13.17) the transition probability matrix P with the vertices $_{\mathbb{V}}\mathbb{P}$. While it is well known that the stability analysis problem for general switching systems (that is, deciding whether the joint spectral radius is smaller than 1) is NP-hard [14], we proved that it is NP-hard even in our particular model.

Theorem 13.5 [60] *Given a discrete-time Markov jump linear system (13.19) with unknown time-varying transition probability matrix $P(k) \in \mathrm{conv}_{\mathbb{V}}\mathbb{P}$, unless $P = NP$, there is no polynomial-time algorithm that decides whether it is mean square stable.*

Our last but not least important result on stability of autonomous noiseless Markov jump linear systems as in (13.19) having polytopic time-inhomogeneous transition probabilities has been presented in the following theorem.

Theorem 13.6 [60] *The following assertions are equivalent.*

1. *The system* (13.19) *is mean square stable (MSS);*
2. *The system* (13.19) *is exponentially mean square stable (EMSS);*
3. *The system* (13.19) *is stochastically stable (SS).*

We developed an extension of the above results in presence of bounded-energy disturbance in [62].

The switched LQR problem. Using the approach in Sect. 13.3.2 it is possible to compute, for a finite set of predefined routing policies, the associated expected quadratic cost and choose the optimal policy. To further improve the performance one can

[2]It is well known that the maximal rate of growth among all products of matrices from a bounded set is given by its JSR $\hat{\rho}(\cdot)$, which is the generalization of the notion of spectral radius to sets of matrices. See [49] and references therein for a detailed treatment of the JSR theory.

dynamically choose, for each time step and according to the plant state measurement, the routing choice: we address this problem by considering the mathematical framework of time-inhomogeneous MJSLS. In particular, we consider the problem of joint cost minimization of continuous and discrete control inputs for the worst possible disturbance of the transition probabilities. The provided solution has been derived in [61] and consists of a finite set of recursive-coupled Riccati difference equations. This result is an extension of state of the art which is nontrivial from the technical point of view, since in the proof we needed to show that due to the time-varying nature of perturbations, at generic time step k the vertex that attains the maximum is unknown and state dependent. With respect to previous works on MJLSs having exactly known transition probabilities, we also needed to define and address the issue of explosion of the number of coupled Riccati difference equations.

Let us consider the discrete-time Markov jump switched linear system (13.9) with the switching between operational modes of the system being governed by a Markov decision process $(\mathbb{M}, \mathbb{A}, \Pr, g, \gamma)$. Its transition probabilities associated to each action available in an operational mode are polytopic time-inhomogeneous, as by Assumption 13.2. Also, all the operational modes of the system are considered to be measurable (Assumptions 13.3). We recall that the state-space representation of the system (13.9) under consideration is

$$\begin{cases} \mathrm{x}_{k+1} = A_{s_k}\mathrm{x}_k + B_{s_k}\mathrm{u}_k \\ \mathrm{z}_k = C_{s_k}\mathrm{x}_k + D_{s_k}\mathrm{u}_k, \\ \mathrm{x}_0 = \boldsymbol{x}_0,\ s_0 = \boldsymbol{s}_0,\ \mathrm{p}_0 = \boldsymbol{p}_0 \end{cases}$$

where the system variables and matrices are those of Sect. 13.3.1. Without loss of generality [17, Remark 4.1, p. 74] we assume that for each $i \in \mathbb{M}$ $C_i^* D_i = 0$ and $D_i^* D_i \succ 0$.

For each $k \in \mathbb{T}$, we denote by π_k the *hybrid control pair* (α_k, u_k), where $\alpha_k \in \mathbb{A}_i$ and u_k are respectively a discrete and a continuous action at time instant k. The sequence π of hybrid control pairs $(\pi_k)_{k=0}^{T-1}$ is called *hybrid control sequence*. At each time step (or decision epoch, in MDP terminology) k, a *particular choice* $\boldsymbol{u}_k$ of u_k is called the continuous control law; similarly, $\boldsymbol{\alpha}_k$ is denominated discrete switching control law. The pair $(\boldsymbol{\alpha}_k, \boldsymbol{u}_k)$ forms the hybrid control *law* $\boldsymbol{\pi}_k$, and the sequence of hybrid control laws over the horizon T constitutes a finite horizon *feedback policy*, $\boldsymbol{\pi} \triangleq (\boldsymbol{\pi}_k)_{k=0}^{T-1} \triangleq (\boldsymbol{\alpha}_k, \boldsymbol{u}_k)_{k=0}^{T-1}$. We also indicate by $\mathrm{p}^{\alpha}_{s\bullet} \triangleq \left(\mathrm{p}^{\alpha}_{s_k\bullet}(k)\right)_{k=0}^{T-1}$ the sequence of length $T \in \mathbb{T}$ of the transition probability row vectors $\mathrm{p}^{\alpha}_{i\bullet}(k)$, with $k \in \mathbb{T}_{T-1}$. Note that the transition probability row vectors $\mathrm{p}^{\alpha}_{i\bullet}(k)$ belong to a polytopic set of transition probability row vectors induced by the tranisition probability matrix vertices ${}_{\mathbb{V}_\alpha}\mathbb{P}$ similarly to Assumption 13.2. For more details the reader is referred to [61].

We cast an optimal linear quadratic state-feedback control problem for Markov jump switched linear systems with bounded perturbations of the transition probabilities as a min-max *problem of optimizing robust performance*, i.e., *finding the minimum over the finite-horizon feedback policy of the maximum over the transition probability disturbance obtained in correspondence of the chosen feedback policy*.

This problem can be cast from the game-theoretic point of view, where at each time step $k \in \mathbb{T}$ the perturbation-player (environment and/or malicious adversary) tries to maximize the cost while the controller tries to minimize the cost. The game-theoretic formulation of the optimal robust control problem requires to make explicit the following assumption on the information structure for the controller and the adversary.

Assumption 13.7 The perturbation-player has no information on the choice of the controller and vice versa.

The problem of designing the optimal mode-dependent state-feedback Markov jump controller, which is robust to all possible polytopic perturbations in transition probabilities, is formally defined as follows.

Problem 13.2 Given a discrete-time Markov jump switched linear system (13.9) with unknown and time-varying transition probability row vectors $\mathrm{p}^{\alpha}_{i\bullet}(k) \in {}_{\mathbb{V}_\alpha}\mathbb{P}$ and satisfying Assumption 13.2, find the mode-dependent state feedback policy $\boldsymbol{\pi}$ that achieves the following optimal cost of robust control.

$$\mathscr{J}(\mathrm{s}_0, \mathrm{x}_0) \triangleq \min_{\pi} \max_{\mathrm{p}^{\alpha}_{s\bullet}} \sum_{k=0}^{T-1} \mathbb{E}\big(\|\mathrm{z}_k\|_2^2 + g(s_k, \alpha_k)\big) + \mathbb{E}\big(\mathrm{x}_T^* Z_{s_T} \mathrm{x}_T\big) \qquad (13.20)$$

with $\mathbf{Z} \triangleq (Z_i)_{i=1}^N \in {}_{\mathrm{N}}\mathbb{F}_0^{n_x, n_x}$ being a sequence of the terminal cost weighting matrices.

Our solution to Problem 13.2 has been derived in [61] based on the *dynamic programming* approach in *Bellman's optimization formulation* [12], by backward induction. Note that even if the cost $g(s_k, \alpha_k)$ of performing a discrete action α_k in an operational mode s_k here is treated as time-invariant, the result will obviously remain the same in the case of the time-varying cost $g(s_k, \alpha_k, k)$, as long as the current value of the cost is known by the decision maker.

Exploiting the optimal solution defined in Problem 13.2 in the example of Sect. 13.3.2 the dynamic routing choice results in an event-driven policy that depends at each time step on the current state measurement. Note that the controller may also decide not to send control data over the network. This approach is closely related to the Event-Triggered control paradigm (see ([7, 39, 65] and references therein), where a triggering condition based on current state measurements is continuously monitored and control actuations are generated and applied when the plant state deviates more than a certain threshold from a desired value.

13.5 Conclusions and Future Work

This chapter presents an overview of some recent results on co-design of controller and network parameters of WNCS implementing communication protocols similar to the WirelessHART standard.

We leverage the class of discrete-time Markov jump linear systems, putting a specific focus on dealing with abrupt and unpredictable dynamic perturbations of transition probabilities between the operational modes of such systems and adding to the model the possibility to make discrete decisions, i.e., defining the class of time-inhomogeneous discrete-time Markov jump switching linear systems. In order to account for uncertainties and time-variance inherent to real world scenarios, we use the time-inhomogeneous polytopic model of transition probabilities, which is very general and widely used. We illustrate that time-inhomogeneous MJSLS represent a mathematical model to jointly take into account the dynamics of a physical plant and non-idealities of wireless communication such as packet losses, and that their exploitation for optimal design of routing redundancy can strongly improve the closed-loop control performance. We provide novel results in this setting addressing the robust stability and the switched LQR problems.

Our interest in this particular class of systems is inspired by their application as possible models for WNCS implementing communication protocols specifically developed for automation applications: we believe that this topic is timely, especially in view of the ongoing efforts made by academia and industry in developing a fifth generation of mobile technology (5G), which also uses models based on Markov chains and is expected to meet the requirements of ultra-reliable, low-latency communications for factory automation and safety-critical internet of things. Based on the research illustrated in this chapter we will attempt to improve our models of the communication protocols and wireless communication non-idealities and our analysis and design algorithms, with the aim of bringing substantial improvements in wireless closed-loop automation systems of the next generation by optimally co-designing the controller as well as the different layers of the communication protocol stack.

Acknowledgements The author would like to acknowledge all co-authors of the papers constituting the line of research illustrated in this chapter: Marika Di Benedetto, Gianni Di Girolamo, Raphaël Jungers, Emmanuele Serra, Francesco Smarra, and Yuriy Zacchia Lun. This line of research has been partially supported by EU FP6 NoE HYCON, EU FP7 NoE HYCON2, Cipe resolution n.135 INCIPICT, H2020-ECSEL-2015 SAFECOP and H2020-ECSEL-2016-1 AQUAS.

References

1. Abate, A., D'Innocenzo, A., Di Benedetto, M.D.: Approximate abstractions of stochastic hybrid systems. IEEE Trans. Autom. Control **56**(11), 2688–2694 (2011)
2. Aberkane, S.: Stochastic stabilization of a class of nonhomogeneous Markovian jump linear systems. Syst. Control Lett. **60**(3), 156–160 (2011)
3. Akyildiz, I.F., Kasimoglu, I.H.: Wireless sensor and actor networks: research challenges. Ad Hoc Netw. **2**(4), 351–367 (2004)
4. Alderisi, G., Girs, S., Lo Bello, L., Uhlemann, E., Bjorkman, M.: Probabilistic scheduling and adaptive relaying for wirelesshart networks. In: 2015 IEEE 20th Conference on Emerging Technologies Factory Automation (ETFA), pp. 1–4 (2015)

5. Alur, R., D'Innocenzo, A., Johansson, K.H., Pappas, G.J., Weiss, G.: Compositional modeling and analysis of multi-hop control networks. IEEE Trans. Autom. Control., Special Issue on Wirel. Sens. Actuator Netw. **56**(10), 2345–2357 (2011)
6. Alur, R., D'Innocenzo, A., Johansson, K.H., Pappas, G.J., Weiss, G.: Modeling and Analysis of Multi-Hop Control Networks. In: Proceedings of the 15th IEEE Real-Time and Embedded Technology and Applications Symposium, San Francisco, CA, United States, pp. 223–232. 13–16 April 2009
7. Antunes, D., Heemels, W.P.M.H.: Rollout event-triggered control: beyond periodic control performance. IEEE Trans. Autom. Control **59**(12), 3296–3311 (2014)
8. Årzén, K.-E., Bicchi, A., Hailes, S., Johansson, K.H., Lygeros, J.: On the design and control of wireless networked embedded systems. In: Proceedings of the 2006 IEEE Conference on Computer Aided Control Systems Design, Munich, Germany (2006)
9. Åström, K.J., Wittenmark, B.: Adaptive control. Dover Books on Electrical Engineering, 2nd edn. Dover Publications (2008)
10. Aström, K., Wittenmark, B.: Computer-Controlled Systems: Theory and Design. Prentice Hall (1997)
11. Beckert, B., Hähnle, R.: Reasoning and verification: state of the art and current trends. IEEE Intell. Syst. **29**(1), 20–29 (2014)
12. Bertsekas, D.P.: Dynamic Programming and Optimal Control, vols. I and II. Athena Scientific, Belmont, MA (1995)
13. Bertuccelli, L.F., How, J.P.: Estimation of non-stationary Markov chain transition models. In: 2008 47th IEEE Conference on Decision and Control (CDC), IEEE, New York, pp. 55–60 (2008)
14. Blondel, V.D., Tsitsiklis, J.N.: The Lyapunov exponent and joint spectral radius of pairs of matrices are hard - when not impossible - to compute and to approximate. Math. Control Signals Syst. **10**, 31–40 (1997)
15. Borrelli, F., Baotic, M., Bemporad, A., Morari, M.: Dynamic programming for constrained optimal control of discrete-time linear hybrid systems. Automatica **41**, 1709–1721 (2005)
16. Chitraganti, S., Aberkane, S., Aubrun, C.: Mean square stability of non-homogeneous Markov jump linear systems using interval analysis. In: Proceedings of the European Control Conference, pp. 3724–3729 (2013)
17. Costa, O.L.V., Fragoso, M.D., Marques, R.P.: Discrete-Time Markov Jump Linear Systems. Springer, London (2005)
18. De Guglielmo, D., Anastasi, G., Seghetti, A.: From IEEE 802.15.4 to IEEE 802.15.4e: A Step Towards the Internet of Things. Springer International Publishing, Berlin (2014)
19. Di Benedetto, M.D., Bicchi, A., D'Innocenzo, A., Johansson, K.H., Robertsson, A., Santucci, F., Tiberi, U., Tzes, A.: Networked control. In: Lunze, J., Lamnabhi, F. (eds.) Handbook of Hybrid Systems Control: Theory, Tools, Applications, pp. 106–112, Cambridge University Press (2009). ISBN:978-0-521-76505-3
20. Di Girolamo, G.D., D'Innocenzo, A., Di Benedetto, M.D.: Co-design of control, scheduling and routing in a wirelesshart network. Submitted to Conference on decision and control CDC17 (2017)
21. Di Girolamo, G.D., D'Innocenzo, A., Di Benedetto, M.D.: Co-design of controller and routing redundancy over a wireless network. In: Proceedings of the 5th IFAC Workshop on Estimation and Control of Networked Systems, 10–11 September 2015
22. Di Girolamo, G.D., D'Innocenzo, A., Di Benedetto, M.D.: Data-rate and network coding co-design with stability and capacity constraints. In: 20th IFAC World Congress, Toulouse, France, 9–14 July 2017
23. Di Marco, P., Fischione, C., Santucci, F., Johansson, K.H.: Modeling ieee 802.15.4 networks over fading channels. IEEE Trans. Wirel. Commun. **13**(10), 5366–5381 (2014)
24. D'Innocenzo, A., Di Benedetto, M.D., Serra, E.: Fault tolerant control of multi-hop control networks. IEEE Trans. Autom. Control **58**(6), 1377–1389 (2013)
25. D'Innocenzo, A., Smarra, F., Di Benedetto, M.D.: Further results on fault detection and isolation of malicious nodes in multi-hop control networks. In: Proceedings of the 14th European Control Conference (ECC'15), Linz, Austria, 15–17 July 2015

26. D'Innocenzo, A., Smarra, F., Di Benedetto, M.D.: Resilient stabilization of multi-hop control networks subject to malicious attacks. Automatica **71**, 1–9 (2016)
27. D'Innocenzo, A., Weiss, G., Alur, R., Isaksson, A.J., Johansson, K.H., Pappas, G.J.: Scalable scheduling algorithms for wireless networked control systems. In: Proceedings of the 5th IEEE International Conference on Automation Science and Engineering, Bangalore, India, pp. 409-414. 22–25 August 2009
28. Donkers, M.C.F., Heemels, W.P.M.H., Hetel, L., van de Wouw, N.: Stability analysis of networked control systems using a switched linear systems approach. IEEE Trans. Autom. Control **56**(9), 2101–2115 (2011)
29. Gatsis, K., Ribeiro, A., Pappas, G.J.: Optimal power management in wireless control systems. IEEE Trans. Autom. Control **59**(6), 1495–1510 (2014)
30. Gonçalves, A.P.C., Fioravanti, A.R., Geromel, J.C.: Markov jump linear systems and filtering through network transmitted measurements. Signal Process. **90**(10), 2842–2850 (2010)
31. Gupta, V., Dana, A.F., Hespanha, J.P., Murray, R.M., Hassibi, B.: Data transmission over networks for estimation and control. IEEE Trans. Autom. Control **54**(8), 1807–1819 (2009)
32. Haesaert, S., Abate, A., Van den Hof, P.M.J.: Correct-by-design output feedback of LTI systems. In: 2015 54th IEEE Conference on Decision and Control (CDC), pp. 6159–6164 (2015)
33. Han, S., Zhu, X., Mok, A.K., Nixon, M., Blevins, T., Chen, D.: Control over WirelessHART network. In: Proceedings of the 36th Annual Conference of IEEE Industrial Electronics Society (IECON), pp. 2114–2119 (2010)
34. Hansson, H., Jonsson, B.: A logic for reasoning about time and reliability. Formal Asp. Comput. **6**(5), 512–535 (1994)
35. HART Communication Foundation. Network management specification. In: HART Communication Protocol Specification (2008)
36. HART Communication Foundation. Tdma data link layer specification. HART Communication Protocol Specification (2008)
37. HART Communication Foundation. Wirelesshart device specification. HART Communication Protocol Specification (2008)
38. Hartfiel, D.J.: Markov Set-Chains. Lecture Notes in Mathematics, vol. 1695. Springer, Berlin (1998)
39. Heemels, W.P.M.H., Johansson, K.H., Tabuada, P.: An introduction to event-triggered and self-triggered control. In: 2012 IEEE 51st IEEE Conference on Decision and Control (CDC), pp. 3270–3285 (2012)
40. Heemels, W.P.M.H., Teel, A.R., van de Wouw, N., Nesic, D.: Networked control systems with communication constraints: tradeoffs between transmission intervals, delays and performance. IEEE Trans. Autom. Control **55**(8), 1781–1796 (2010)
41. Heemels, W.P.M.H., Teel, A.R., van de Wouw, N., Nesic, D.: Networked control systems with communication constraints: tradeoffs between transmission intervals, delays and performance. IEEE Trans. Autom. Control **55**(8), 1781–1796 (2010)
42. Heemels, W.P.M.H., van De Wouw, N.: Stability and stabilization of networked control systems. In: Bemporad, A., Maurice, W.P., Heemels, H., Johansson, M. (eds), Networked Control Systems. Lecture Notes in Control and Information Sciences, vol. 406, chapter 7, pp. 203–253. Springer, London (2010)
43. Hespanha, J.P., Naghshtabrizi, P., Xu, Y.: A survey of recent results in networked control systems. Proc. IEEE **95**(1), 138–162 (2007)
44. Hetel, L., Daafouz, J., Iung, C.: Stability analysis for discrete time switched systems with temporary uncertain switching signal. In: Proceedings of the 46th IEEE Conference on Decision and Control (CDC2007), New Orleans, LA, USA, pp. 5623–5628. 12–14 December (2007)
45. Hetel, L., Daafouz, J., Iung, C.: Stabilization of arbitrary switched linear systems with unknown time-varying delays. IEEE Trans. Autom. Control **51**(10), 1668–1674 (2006)
46. Hetel, L., Daafouz, J., Richard, J.-P., Jungers, M.: Delay-dependent sampled-data control based on delay estimates. Syst. Control Lett. **60**(2), 146–150 (2011)
47. Hwang, I., Kim, S., Kim, Y., Seah, C.E.: A survey of fault detection, isolation, and reconfiguration methods. IEEE Trans. Control Syst. Technol. **18**(3), 636–653 (2010)

48. IEEE. 802.15 wpan task group 4 (2017)
49. Jungers, Raphaël: The joint spectral radius: theory and applications. Lecture Notes in Control and Information Sciences, vol. 385. Springer, Berlin (2009)
50. Jungers, R.M., D'Innocenzo, A., Di Benedetto, M.D.: Controllability of linear systems with switching delays. IEEE Trans. Autom. Control **61**(4), 1117–1122 (2016)
51. Jungers, R.M., D'Innocenzo, A., Di Benedetto, M.D.: Feedback stabilization of dynamical systems with switched delays. In: 51st IEEE Conference on Decision and Control, Maui, Hawaii, 10–13 December 2012
52. Jungers, R.M., D'Innocenzo, A., Di Benedetto, M.D.: Further results on controllability of linear systems with switching delays. In: 9th IFAC World Congress, Cape Town, South Africa, 24–29 August 2014
53. Jungers, R.M., D'Innocenzo, A., Di Benedetto, M.D.: How to control Linear Systems with switching delays. In: 13th European Control Conference (ECC14), Strasbourg, France, 24–27 June 2014
54. Kalman, R.E.: Contributions to the theory of optimal control. Boletin de la Sociedad Matematica Mexicana **5**, 102–119 (1960)
55. Khader, O., Willig, A., Wolisz, A.: Wirelesshart tdma protocol performance evaluation using response surface methodology. In: 2011 International Conference on Broadband and Wireless Computing, Communication and Applications, pp. 197–206 (2011)
56. Khader, O., Willig, A.: An energy consumption analysis of the wireless hart tdma protocol. Comput. Commun. **36**(7), 804–816 (2013)
57. Komninakis, C., Wesel, R.D.: Joint iterative channel estimation and decoding in flat correlated Rayleigh fading. IEEE J. Sel. Areas Commun. **19**(9), 1706–1717 (2001)
58. Lahijanian, M., Andersson, S.B.: Formal verification and synthesis for discrete-time stochastic systems. IEEE Trans. Autom. Control **60**(8), 2031–2045 (2015)
59. Lun, Y.Z., D'Innocenzo, A., Abate, A., Di Benedetto, M.D.: Optimal robust control and a separation principle for polytopic time-inhomogeneous Markov jump linear systems. In: 2017 IEEE 56th Conference on Decision and Control (CDC) (2017)
60. Lun, Y.Z., D'Innocenzo, A., Di Benedetto, M.D.: On stability of time-inhomogeneous Markov jump linear systems. In: 2016 IEEE 55th Conference on Decision and Control (CDC), pp. 5527–5532 (2016)
61. Lun, Y.Z., D'Innocenzo, A., Di Benedetto, M.D.: Robust LQR for time-inhomogeneous Markov jump switched linear systems. In: The 20th World Congress of the International Federation of Automatic Control (2017)
62. Lun, Y.Z., D'Innocenzo, A., Di Benedetto, M.D.: Robust stability of time-inhomogeneous Markov jump linear systems. In: The 20th World Congress of the International Federation of Automatic Control (2017)
63. Lun, Y.Z., D'Innocenzo, A., Malavolta, I., Di Benedetto, M.D.: Cyber-physical systems security: a systematic mapping study. arXiv:1605.09641 (2016)
64. Matei, I., Martins, N.C., Baras, J.S.: Optimal linear quadratic regulator for markovian jump linear systems, in the presence of one time-step delayed mode observations. IFAC Proc. **41**(2), 8056–8061 (2008). (17th IFAC World Congress)
65. Mazo, M., Tabuada, P.: Decentralized event-triggered control over wireless sensor/actuator networks. IEEE Trans. Autom. Control **56**(10), 2456–2461 (2011)
66. Mesquita, A.R., Hespanha, J., Nair, G.N.: Redundant data transmission in control/estimation over lossy networks. Automatica **48**, 1020–1027 (2012)
67. Nobre, M., Silva, I., Guedes, L.A.: Routing and scheduling algorithms for wirelesshart networks: a survey. In: Sensors, pp. 9703–9740 (2015)
68. Pajic, M., Sundaram, S., Pappas, G.J., Mangharam, R.: The wireless control network: a new approach for control over networks. IEEE Trans. Autom. Control **56**(10), 2305–2318 (2011)
69. Peters, E.G.W., Quevedo, D.E., Fu, M.: In: 54th IEEE Conference on Decision and Control (CDC), pp. 2459–2464 (2015)
70. Protasov, V.Y., Jungers, R.M., Blondel, V.D.: Joint spectral characteristics of matrices: a conic programming approach. SIAM J. Matrix Anal. Appl. **31**(4), 2146–2162 (2010)

71. Puterman, M.L.: Markov Decision Processes: Discrete Stochastic Dynamic Programming. Wiley Series in Probability and Statistics, vol. 594. Wiley-Interscience, New Jersey (2005)
72. Sadeghi, P., Kennedy, R.A., Rapajic, P.B., Shams, R.: Finite-state Markov modeling of fading channels - a survey of principles and applications. IEEE Signal Process. Mag. **25**(5), 57–80 (2008)
73. Saifullah, A., Xu, Y., Lu, C., Chen, Y.: Real-time scheduling for wirelesshart networks. In: 31st IEEE Real-Time Systems Symposium, pp. 150–159 (2010)
74. Schenato, L., Sinopoli, B., Franceschetti, M., Poolla, K., Sastry, S.S.: Foundations of control and estimation over lossy networks. Proc. IEEE **95**(1), 163–187 (2007)
75. Shao, H., Han, Q.-L.: New stability criteria for linear discrete-time systems with interval-like time-varying delays. IEEE Trans. Autom. Control **56**(3), 619–625 (2011)
76. Sinopoli, B., Schenato, L., Franceschetti, M., Poolla, K., Jordan, M., Sastry, S.: Kalman filtering with intermittent observations. IEEE Trans. Autom. Control **49**(9), 1453–1464 (2004)
77. Smarra, F., D'Innocenzo, A., Di Benedetto, M.D.: Approximation methods for optimal network coding in a multi-hop control network with packet losses. Control Conference (ECC). 2015 European, pp. 1962–1967. IEEE, Linz (2015)
78. Tabbara, M., Nešić, D., Teel, A.R.: Stability of wireless and wireline networked control systems. IEEE Trans. Autom. Control **52**(7), 1615–1630 (2007)
79. Tabuada, P., Pappas, G.J.: Linear time logic control of discrete time linear systems. IEEE Trans. Autom. Control **51**(12), 1862–1877 (2006)
80. Vargas, A.N., Ishihara, J.Y., do Val, J.B.R.: Linear Quadratic Regulator for a Class of Markovian Jump Systems with Control in Jumps, pp. 2282–2285 (2010)
81. Walsh, G.C., Ye, H., Bushnell, L.G.: Stability analysis of networked control systems. IEEE Trans. Control Syst. Technol. **10**(3), 438–446 (2002)
82. Weiss, G., D'Innocenzo, A., Alur, R., Johansson, K.H., Pappas, G.J.: Robust Stability of Multi-Hop Control Networks. In: Proceedings of the 48th IEEE Conference on Decision and Control and the 28th Chinese Control Conference, Shangai, China, pp. 2210-2215. 15–18 December 2010
83. Zhang, W., Abate, A., Hu, J.: Efficient Suboptimal Solutions of Switched LQR Problems, pp. 1084–1091 (2009)
84. Zhang, W., Branicky, M.S., Phillips, S.M.: Stability of networked control systems. IEEE Control Syst. Mag. **21**(1), 84–99 (2001)
85. Zhang, W., Hu, J., Abate, A.: Infinite-horizon switched LQR problems in discrete time: a suboptimal algorithm with performance analysis. IEEE Trans. Autom. Control **57**(7), 1815–1821 (2012)
86. Zhang, L., Shi, P., Basin, M.: Robust stability and stabilisation of uncertain switched linear discrete time-delay systems. IET Control Theory Appl. **2**(7), 606–614 (2008)

Chapter 14
Discontinuities, Generalized Solutions, and (Dis)agreement in Opinion Dynamics

F. Ceragioli and P. Frasca

Abstract This chapter is devoted to the mathematical analysis of some continuous-time dynamical systems defined by ordinary differential equations with discontinuous right-hand side, which arise as models of opinion dynamics in social networks. Discontinuities originate because of specific communication constraints, namely, *quantization* or *bounded confidence*. Solutions of these systems may or may not converge to a state of agreement, where all components of the state space are equal. After presenting three models of interest, we elaborate on the properties of their solutions in terms of existence, completeness, and convergence.

14.1 Discontinuous Consensus-Seeking Systems

This chapter studies some continuous-time dynamical systems defined by ordinary differential equations with discontinuous right-hand side. The dynamics under consideration have been proposed in the last 15 years in the context of "consensus-seeking" systems, which describe coordination phenomena in engineering, biology, and social sciences. Given this range of applications, the reader will not be surprised that we are dealing with rather abstract representations of reality.

The most basic consensus-seeking system takes the following form. Let x be an N-dimensional vector, where each component x_i is associated to an individual $i \in \mathscr{I} = \{1, \dots, N\}$ and evolves in time according to the ordinary differential equation

$$\dot{x}_i(t) = \sum_{j=1}^{N} a_{ij}\big(x_j(t) - x_i(t)\big) \qquad i \in \mathscr{I}. \tag{14.1}$$

F. Ceragioli
Politecnico di Torino, c.so Duca degli Abruzzi 24, 10129 Torino, Italy
e-mail: francesca.ceragioli@polito.it

P. Frasca (✉)
Univ. Grenoble Alpes, CNRS, Inria, Grenoble INP, GIPSA- Lab, 38000 Grenoble, France
e-mail: paolo.frasca@gipsa-lab.fr

S. Tarbouriech et al. (eds.), *Control Subject to Computational and Communication Constraints*, Lecture Notes in Control and Information Sciences 475, https://doi.org/10.1007/978-3-319-78449-6_14

Since we assume that the interaction weights a_{ij} are nonnegative, this dynamics postulates that each individual is attracted by the other individuals with whom it interacts. Under very mild assumptions on the interaction pattern, this dynamics converges to a state of *agreement* where all components x_i are equal.

Variations to this dynamics have been proposed in order to accommodate a host of phenomena, including time- and state-dependent interactions $a_{ij}(t, x)$. An interesting case of state-dependent interactions is the following, which is termed *bounded confidence* in the literature. Two individuals are assumed to influence each other if their states are closer than a certain threshold (that we choose to be 1 for simplicity):

$$\dot{x}_i(t) = \sum_{j=1}^{N} a(x_i(t), x_j(t))\big(x_j(t) - x_i(t)\big) \qquad i \in \mathscr{I} \tag{BC}$$

$$\text{where } a(y, z) = \begin{cases} 1 & \text{if } |y - z| < 1 \\ 0 & \text{if } |y - z| \geqslant 1. \end{cases}$$

This model, which is a continuous-time counterpart of the opinion dynamics studied by Hegselmann and Krause [34], has been proposed by [8] and further considered in [14]. Very similar models have been considered in [19, 35, 40, 49, 51]. We will see that (BC) does not produce agreement, but clustering of individuals into groups characterized by agreement within each group and disagreement between groups.

Another relevant phenomenon is *quantization*, which occurs both in engineering and in social systems. In engineering, it can represent communication constraints, where the state variable is communicated between individuals via a digital channel with finite data rate, and thus constrained to take on discrete values. For the sake of this analysis, we shall define the quantization of a real number simply by rounding it to the closest integer: $q(s) = \lfloor s + \frac{1}{2} \rfloor$. In this context, an effective consensus-seeking system is the following "quantized states" system studied in [17]:

$$\dot{x}_i(t) = \sum_{j=1}^{N} a_{ij}\big(q(x_j(t)) - q(x_i(t))\big) \qquad i \in \mathscr{I}. \tag{QS}$$

Note that the right-hand side features the quantized values of both states x_j and x_i: the presence of the quantized state $q(x_i)$ is crucial to ensure the "good" properties of this dynamics, which will be discussed below.

In social systems, quantization may originate because the state variable is "communicated" as the display of an action or behavior, which can take on discrete values only: for instance, the purchase of certain products. In this context, we have recently proposed [15, 16] to investigate the following "quantized behaviors" model:

$$\dot{x}_i(t) = \sum_{j=1}^{N} a_{ij}[q(x_j(t)) - x_i(t)] \qquad i \in \mathscr{I} = \{1, \dots, N\}. \tag{QB}$$

Note that, in contrast with (QS), the right-hand side features the quantized value of x_j, but not of x_i, which leads to more complex dynamics than (QS).

The reader can notice that the right-hand sides of equations (BC), (QS), and (QB) are discontinuous in the state variable. Since the study of these non-smooth systems relies on relatively sophisticated instruments that might not always be accessible to the nonspecialists, we have included in this chapter an extended review of the necessary mathematical machinery, which we hope can be of independent interest. Nevertheless, readers are advised to consult the literature specific to the topic, for instance, the tutorial [21] and the books [4, 24], as well as more specific works about stability [6] or generalized solutions [12]. Additional references are provided in the following sections.

The rest of this chapter is organized as follows. Section 14.2 defines some useful notation and summarizes well-known results from graph theory together with their consequences for the consensus dynamics (14.1). Section 14.3 presents some notions of solutions that are relevant in this context, namely, those of Carathéodory and Krasovskii. The section also contains results on existence and completeness of these solutions for a general class of piecewise affine systems and specifically for the three dynamics at hand. Section 14.4 deals with equilibria (again, declined according to the relevant notions of solutions) and describes the sets of equilibria for our three dynamics. Section 14.5 deals with convergence of their trajectories. We provide two kinds of results: on the one hand, sufficient conditions to reach agreement, and on the other hand, general statements about convergence to the equilibria (or to their proximity).

Owing to the survey purpose that we set us for this chapter, we have avoided reporting the details of some proofs that can be easily found in the literature. Furthermore, we restrict our presentation to assume symmetric interactions, namely, such that $a_{ij} = a_{ji}$ for all i, j in $\mathscr{I}$. We made this choice for simplicity of exposition, even though most results can be extended to nonsymmetric interactions.

14.2 Preliminaries

Notation. Given a subset S of $\mathbb{R}^N$, we denote by $\overline{S}$ its topological closure, by ∂S its border, and by $\overline{co}S$ its closed convex hull. We let $\mathbf{0} = (0, \dots, 0)^\top$, $\mathbf{1} = (1, \dots, 1)^\top$ and e_i, $i = \{1, \dots, N\}$, the vectors of the canonical basis of $\mathbb{R}^N$. We call *consensus point* a point of the form $\alpha\mathbf{1}$ with $\alpha \in \mathbb{R}$. The N-dimensional identity matrix is denoted by I, and $\|\cdot\|$ denotes the Euclidean norm both for vectors and matrices. Given the vector $x \in \mathbb{R}^N$, we denote its average by $x_{ave} = \frac{1}{N}\mathbf{1}^\top x = \frac{1}{N}\sum_{i=1}^N x_i$. When $x = x(t)$ we shall write $x_{ave}(t) = \frac{1}{N}\sum_{i=1}^N x_i(t)$. The notation $q(x)$ with $x \in \mathbb{R}^N$ will denote the vector whose ith component is $q(x_i)$.

Graph theory. A *weighted (undirected) graph* $\mathcal{G} = (\mathcal{I}, \mathcal{E}, A)$ consists of a node set $\mathcal{I} = \{1, \ldots, N\}$, an edge set $\mathcal{E} \subset \mathcal{I} \times \mathcal{I}$, and a *symmetric adjacency matrix* $A \in \mathbb{R}_+^{N\times N}$ such that $a_{ij} > 0$ if $(i, j) \in \mathcal{E}$, and $a_{ij} = 0$ if $(j, i) \notin \mathcal{E}$. We assume no self-loops in the graph, that is $a_{ii} = 0$ for all $i \in \mathcal{I}$. Nodes (vertices) are referred to as agents or individuals, edges as links. Let $d_i := \sum_{j=1}^{N} a_{ij}$ be the degree of node $i \in \mathcal{I}$. Let $D = diag(A\mathbf{1})$ be the diagonal matrix whose diagonal entries are the degrees of each node. Let $L = D - A$ be the *Laplacian matrix* of the graph $\mathcal{G}$. Note that by construction $L\mathbf{1} = \mathbf{0}$ and by symmetry $\mathbf{1}^\top L = \mathbf{0}^\top$. In case the graph is state-dependent, we write $\mathcal{G}(x)$, $\mathcal{E}(x)$, $A(x)$, $D(x)$, $L(x)$.

Given an edge (i, j), we shall refer to i and to j as the tail and the head of the edge, respectively. A path is an ordered list of edges such that the head of each edge is equal to the tail of the following one. The graph $\mathcal{G}$ is said to be *connected* if for any $i, j \in \mathcal{I}$ there is a path from i to j in $\mathcal{G}$. If the graph is connected, then the eigenvalue 0 of the Laplacian matrix L has algebraic multiplicity 1. The vector $x - x_{ave}\mathbf{1}$ is the projection of x on the subspace orthogonal to $\mathbf{1}$: consequently, if we denote by λ_* the smallest nonzero eigenvalue of L, one has

$$(x - x_{ave}\mathbf{1})^\top L(x - x_{ave}\mathbf{1}) \geqslant \lambda_* \|x - x_{ave}\mathbf{1}\|^2 \ \forall x \in \mathbb{R}^N.$$

Convergence to agreement. Using the Laplacian matrix, dynamics (14.1) can be compactly rewritten as

$$\dot{x} = -Lx. \tag{14.2}$$

Its key properties, descending from the properties of the Laplacian that we recalled above, are summarized by the following well-known result and illustrated in Fig. 14.1.

Theorem 14.1 (Real consensus) *If the graph underlying* (14.1) *is connected and the adjacency matrix A is symmetric, then for any solution $x(t)$ of* (14.1)*, the following properties hold true:*

1. *(contractivity and boundedness)* $\overline{co}\{x_i(t), i \in \mathcal{I}\} \subseteq \overline{co}\{x_i(0), i \in \mathcal{I}\}$;
2. *(average preservation)* $x_{ave}(0) = x_{ave}(t)$;
3. *(equilibria) x^* is an equilibrium point of* (14.1) *if and only if x^* is a consensus point;*
4. *(average consensus)* $\lim_{t\to+\infty} x(t) = x_{ave}(0)\mathbf{1}$.

14.3 Generalized Solutions and Basic Properties of the Dynamics

In this section, we summarize some notions which are essential in order to deal with systems whose right-hand side is discontinuous with respect to the state variable.

Let us consider the Cauchy problem

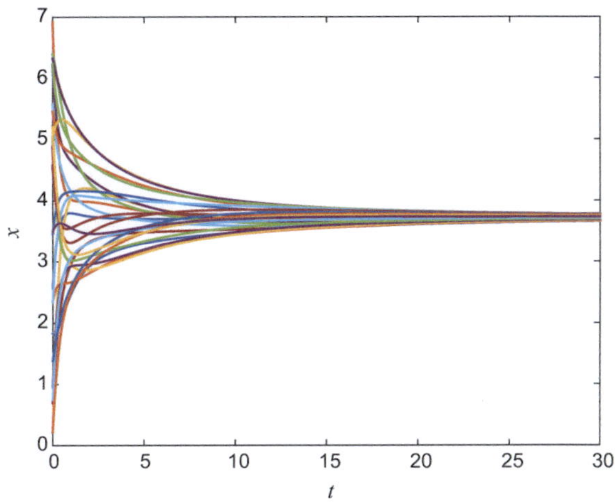

Fig. 14.1 Evolution of a solution of (14.1) from random initial conditions on a cycle graph on 25 nodes

$$\dot{x} = f(x) \qquad x(0) = x_0, \tag{14.3}$$

where $x_0 \in \mathbb{R}^N$ and $f : \mathbb{R}^N \to \mathbb{R}^N$ is measurable and locally bounded. We will denote by Δ_f the subset of $\mathbb{R}^N$ where f is discontinuous. When facing system (14.3), one should first of all choose which type of generalized solution is the most suitable for the system of interest. We shall consider Carathéodory solutions and Krasovskii solutions.

14.3.1 Carathéodory Solutions

The notion of solution nearest to the classical one is that of Carathéodory solution.

Definition 14.1 (*Carathéodory solution*) Let $I \subset \mathbb{R}$ be an interval with $0 \in I$ and let $x_0 \in \mathbb{R}^N$. An absolutely continuous function $\varphi : I \to \mathbb{R}^N$ is a *Carathéodory solution* of equation (14.3) on I with initial condition x_0 if $\varphi(0) = x_0$ and if it satisfies (14.3) for almost all $t \in I$ or, equivalently, if it is a solution of the integral equation

$$\varphi(t) = x_0 + \int_0^t f(\varphi(s))ds.$$

We say that a *local Carathéodory solution* corresponding to the initial condition $x_0 \in \mathbb{R}$ exists if there exist a neighborhood $I(x_0)$ of x_0, an interval of the form $[0, T)$, and an absolutely continuous function $\varphi : [0, T) \to I(x_0)$ such that $\varphi(0) = x_0$ and $\varphi(t)$ is a Carathéodory solution of (14.3) on $[0, T)$.

Note that in the models we are considering the set Δ_f of discontinuity points of the vector field $f(x)$ has a particularly simple structure, as it can be locally represented as the union of a finite number of hyperplanes. This observation is made rigorous by the following assumption.

Assumption 14.1 (*On the discontinuity set*) For any $x_0 \in \Delta_f$, there exists a neighborhood $I(x_0)$ of x_0 and m affine functions $s_1, \dots, s_m : I(x_0) \to \mathbb{R}^N$ defined by

$$s_\ell(x) = p_\ell^\top x - c_\ell \qquad \ell \in \{1, \dots, m\}$$

with $p_\ell \in \mathbb{R}^N$ and $c_\ell \in \mathbb{R}$, such that $s_\ell(x_0) = 0$ for all $\ell \in \{1, \dots, m\}$ and

$$\Delta_f \cap I(x_0) = \{x \in I(x_0) : s_1(x) = 0\} \cup \ldots \cup \{x \in I(x_0) : s_m(x) = 0\}.$$

Under Assumption 14.1, the neighborhood $I(x_0)$ of x_0 is partitioned in 2^m sectors $S_{\mathbf{b}}(x_0)$ defined by the signs of the functions $s_1, \dots, s_m$, indexed by means of $\mathbf{b} \in \{-1, 1\}^m$, and defined in the following way[1]:

$$S_{\mathbf{b}}(x_0) = \{x \in I(x_0) : s_\ell(x) < 0 \text{ if } b_\ell = -1 \text{ and } s_\ell(x) \geqslant 0 \text{ if } b_\ell = 1,\ \ell = 1, \dots, m\}.$$

Assumption 14.2 (*On the discontinuous vector field*) The parts $S_{\mathbf{b}}(x_0)$ are defined so that the vector field $f(x)$ is continuous on $S_{\mathbf{b}}(x_0)$ for all $\mathbf{b} \in \{-1, 1\}^m$.

Without Assumption 14.2, the choice of the representation of the discontinuity hyperplanes by means of p_ℓ and c_ℓ would not be unique, since the orientation of the normal vector is arbitrary. Assumption 14.2 makes sure that the choice of the representation is consistent with the functions $a(\cdot, \cdot)$ and $q(\cdot)$ in Eqs. (BC), (QS), and (QB).

Under Assumptions 14.1 and 14.2, the vector field $f(x)$ has 2^m limit values as $x \to x_0$, namely,

$$f^{\mathbf{b}}(x_0) = \lim_{x \in S_{\mathbf{b}}, x \to x_0} f(x).$$

Example 14.1 (*BC dynamics with three individuals*) Consider dynamics (BC) with $N = 3$ and $x_0 = (0, 1, 2)^\top$. Clearly, point x_0 lies at the intersection of the two planes of discontinuity $x_2 - x_1 - 1 = 0$ and $x_3 - x_2 - 1 = 0$, namely, defined by the normal vectors $p_1 = (-1, 1, 0)^\top$ and $p_2 = (0, -1, 1)^\top$. In the sectors $S_{(1,1)}$, $S_{(-1,1)}$, $S_{(1,-1)}$, $S_{(-1,-1)}$, we, respectively, identify the four limit values of the vector field

[1] It would be more precise to write $S_{\mathbf{b}}(x_0, I(x_0))$ instead of $S_{\mathbf{b}}(x_0)$, as it depends on $I(x_0)$. Note however that if $I(x_0)$ and $I'(x_0)$ are two distinct neighborhoods of x_0, then the sets $S_{\mathbf{b}}(x_0, I(x_0))$ and $S_{\mathbf{b}}(x_0, I'(x_0))$ coincide on $I(x_0) \cap I'(x_0)$. Hence, neglecting $I(x_0)$ from the notation brings no ambiguity.

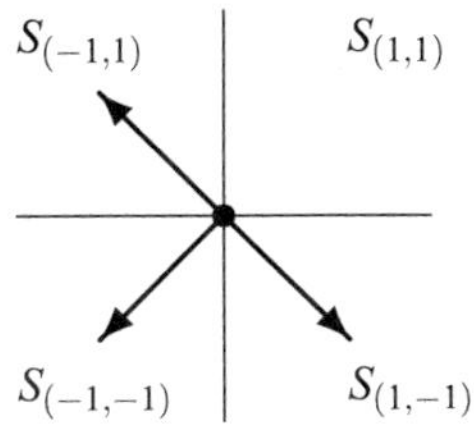

Fig. 14.2 Representation of the limit values of the vector field at a discontinuity point where two hyperplanes intersect, namely, dynamics (BC) at point $(0, 1, 2)^\top$

$$f^{(1,1)}(x_0) = \begin{pmatrix} 0 \\ 0 \\ 0 \end{pmatrix} \quad f^{(-1,1)}(x_0) = \begin{pmatrix} 1 \\ -1 \\ 0 \end{pmatrix}$$

$$f^{(1,-1)}(x_0) = \begin{pmatrix} 0 \\ 1 \\ -1 \end{pmatrix} \quad f^{(-1,-1)}(x_0) = \begin{pmatrix} 1 \\ 0 \\ -1 \end{pmatrix}.$$

This situation is represented in Fig. 14.2.

Under Assumptions 14.1 and 14.2, the study of existence and completeness of Carathéodory solutions can be relatively simple. Nevertheless, one cannot expect to have local existence in general: the negative example is given by the following proposition [13, 33], illustrated in the fourth diagram of Fig. 14.3.

Proposition 14.1 (Nonexistence of Carathéodory solutions) *Let $x_0 \in \Delta_f$ and assume $m = 1$ in Assumption 14.1. If $p_1^\top f^{(-1)}(x_0) > 0$ and $p_1^\top f^{(1)}(x_0) < 0$, then there exists no Carathéodory solution of* (14.3) *with initial condition x_0.*

An example of this situation is given by dynamics (QS).

Example 14.2 (*Nonexistence in QS dynamics*) Consider dynamics (QS) over an undirected path graph with $N = 3$ whose adjacency matrix A has all non-null entries equal to 1 and the initial condition $\bar{x}_0 = (1, 3/2, 2)^\top$. The right-hand side of the system is clearly discontinuous at $\bar{x}_0$. There exists a neighborhood $I(\bar{x}_0)$ of $\bar{x}_0$ such that $\Delta_f \cap I(\bar{x}_0) = \{x \in \mathbb{R}^N : x_2 - 3/2 = 0\}$ and we thus define $s_1(x) = x_2 - 3/2 = (0, 1, 0)x - 3/2$. We get that $f^{(-1)}(\bar{x}_0) = (0, 1, -1)^\top$ and $f^{(1)}(\bar{x}_0) = (1, -1, 0)^\top$, then $(0, 1, 0) f^{(1)}(\bar{x}_0) = -1 < 0$ and $(0, 1, 0) f^{(-1)}(\bar{x}_0) > 0$. By applying Proposition 14.1 we conclude that there are no Carathéodory solutions issuing from $\bar{x}_0$.

Instead, the following result provides a sufficient condition for the existence of local Carathéodory solutions. It is inspired by the concept of directional continuity in [44] but it allows also for solutions lying on the discontinuity set: this case, which can be particularly subtle to be treated, is simplified here by the discontinuity being a union of hyperplanes. Informally, the sufficient condition requires that, for each discontinuity point, at least one among the "pieces" of the vector field either pulls away from the discontinuity surface in its own sector, or is parallel to a discontinuity hyperplane.

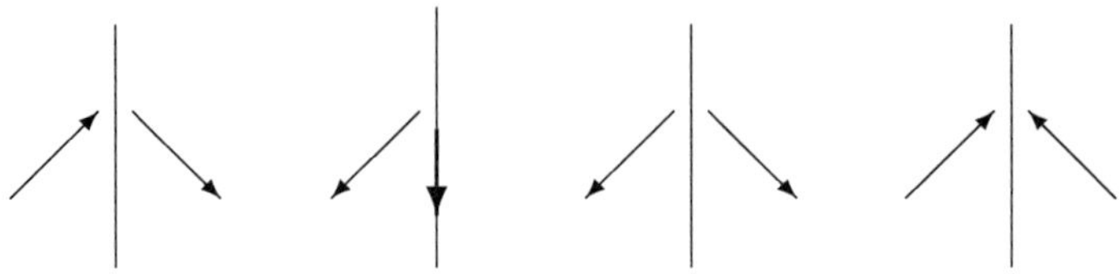

Fig. 14.3 Representations of the possible orientations of the vector fields in the neighborhood of a discontinuity hyperplane

Theorem 14.2 (Sufficient condition for Carathéodory solutions) *Assume that Assumptions 14.1 and 14.2 hold. Assume that for any* $x_0 \in \Delta_f$*, there exists* $\tilde{\mathbf{b}} \in \{-1, 1\}^m$ *such that*

1. $[p_\ell^\top f^{\tilde{\mathbf{b}}}(x_0)]\,\tilde{b}_\ell \geqslant 0$ *for all* $\ell \in \{1, \dots, m\}$*;*
2. *if there exists* $\bar{\ell} \in \{1, \dots, m\}$ *such that* $[p_{\bar{\ell}}^\top f^{\tilde{\mathbf{b}}}(x_0)]\,\tilde{b}_{\bar{\ell}} = 0$*, then there exists a neighborhood* $J(x_0)$ *such that, for all* $x \in J(x_0) \cap \partial S_{\tilde{\mathbf{b}}}$*, both* $[p_{\bar{\ell}}^\top f(x)]\,\tilde{b}_{\bar{\ell}} = 0$ *and the restriction* $f|_{(J(x_0) \cap \partial S_{\tilde{\mathbf{b}}}) \setminus \{x_0\}}$ *is continuous.*

Then, there exists a local Carathéodory solution issuing from x_0.

Proof If $p_\ell^\top f^{\tilde{\mathbf{b}}}(x_0)\,\tilde{b}_\ell > 0$ for all $\ell \in \{1, \dots, m\}$, then the vector field points into the interior of $S_{\tilde{\mathbf{b}}}$ and a local solution can be easily constructed as in [44]. In the case of condition 2, i.e., when the vector field $f(x)$ is parallel to one of the discontinuity hyperplanes in a neighborhood of x_0 except, possibly, in x_0, one can still construct a sequence of Euler polygonal chains that lie on the hyperplane and converge to a Carathéodory solution. □

Condition 2 in Theorem 14.2 is a relaxation of [13, Assumption (H3)], which is made possible by the fact that discontinuities are (locally) hyperplanes. A simple application of Theorem 14.2 is the following.

Example 14.3 (BC dynamics with three individuals—continued) Consider again dynamics (BC) with $N = 3$ and $x_0 = (0, 1, 2)^\top$, as illustrated in Fig. 14.2. We observe that in the four sectors around x_0,
$p_1^\top f^{(1,1)}(x_0)\,(1) = 0$ and $p_2^\top f^{(1,1)}(x_0)\,(1) = 0$.
$p_1^\top f^{(-1,1)}(x_0)\,(-1) = (-2)(-1) > 0$ and $p_2^\top f^{(-1,1)}(x_0)\,(1) = (1)(1) > 0$.
$p_1^\top f^{(1,-1)}(x_0)\,(1) = (1)(1) > 0$ and $p_2^\top f^{(1,-1)}(x_0)\,(-1) = (-2)(-1) > 0$.
$p_1^\top f^{(-1,-1)}(x_0)\,(-1) = (-1)(-1) > 0$ and $p_2^\top f^{(-1,-1)}(x_0)\,(-1) = (-1)(-1) > 0$.
Then, Theorem 14.2 implies local existence. Actually, one can see that in this case, four solutions originate from x_0, one for each sector: they are shown in Fig. 14.4 as functions of time.

More in general, we can prove the existence of Carathéodory solutions of (BC) for any initial condition. Carathéodory solutions of (BC) were studied in [7] where existence and uniqueness of solutions were proved for almost all initial conditions.

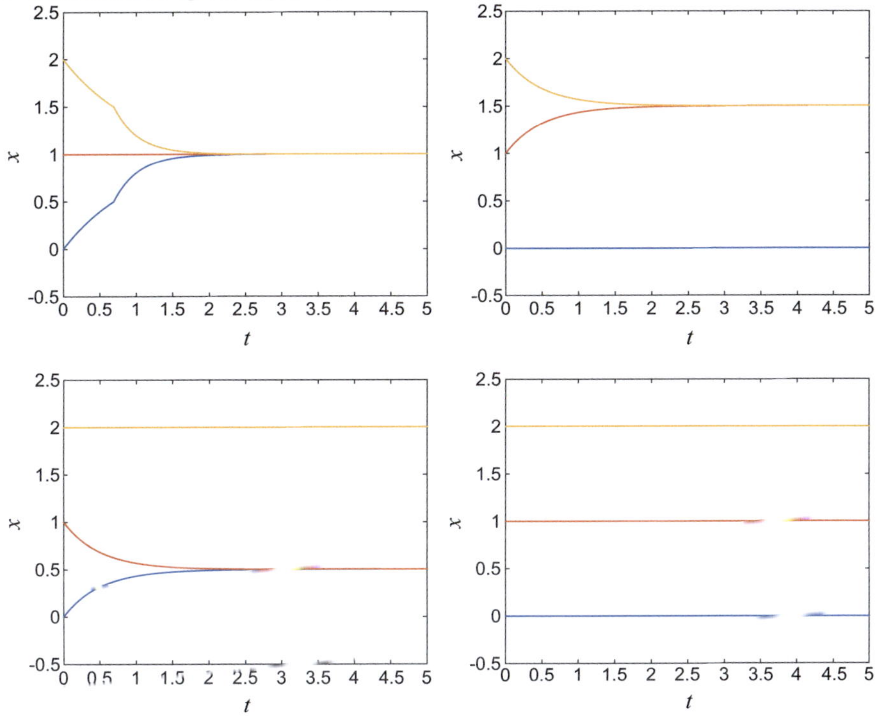

Fig. 14.4 Evolutions of the four solutions of (BC) that originate from $x_0 = (0, 1, 2)^\top$

Here, we prove[2] the existence for all initial conditions and we remark that in general, we do not have uniqueness. The proof is a verification of the assumptions of Theorem 14.2, where we see that the sole strictly positive case suffices.

Corollary 14.1 (Existence for BC) *For any initial condition, there exists a local Carathéodory solution of* (BC).

Proof We denote by $f(x)$ the right-hand side of (BC) and observe that $\Delta_f = \{x \in \mathbb{R} : \exists\, i, j \in \mathscr{I}$ such that $x_i - x_j = 1\}$. We first consider $x_0 \in \Delta_f$ in the case $x_{0i} - x_{0j} = 1$ for only one pair of indices $i, j \in \mathscr{I}$. In this case $m = 1$ in Assumption 14.1, $s_1(x) = x_i - x_j - 1 = (e_i - e_j)^\top x - 1$, and as b is either -1 or 1,

$S_{(-1)}(x_0) = \{x \in \mathbb{R}^N : x_i - x_j - 1 < 0\}$,
$(f^{(-1)}(x_0))_i = \sum_{h \neq j : |x_{0h} - x_{0i}| < 1} (x_h - x_i) - 1$,
$(f^{(-1)}(x_0))_j = \sum_{h \neq i : |x_{0i} - x_{0j}| < 1} (x_h - x_j) + 1$,

as well as

$S_{(1)}(x_0) = \{x \in \mathbb{R}^N : x_i - x_j - 1 \geqslant 0\}$,
$(f^{(1)}(x_0))_i = \sum_{h \neq j : |x_{0i} - x_{0i}| < 1} (x_h - x_i)$,

[2]Even though the corollary is new, it could have been deduced by inspecting the proofs in [7].

$(f^{(1)}(x_0))_j = \sum_{h \neq i:|x_{0i}-x_{0j}|<1}(x_h - x_j)$.

We then get $(e_i - e_j)^\top f^{(-1)}(x_0) = \sum_{h\neq j:|x_h-x_i|<1}(x_h - x_i) - \sum_{h\neq i:|x_h-x_i|<1}(x_h - x_i) - 2$ and $(e_i - e_j)^\top f^{(1)}(x_0) = \sum_{h\neq j:|x_h-x_i|<1}(x_h - x_i) - \sum_{h\neq i:|x_h-x_i|<1}(x_h - x_i)$. If $\left[\sum_{h\neq j:|x_h-x_i|<1}(x_h - x_i) - \sum_{h\neq i:|x_h-x_i|<1}(x_h - x_i)\right] > 0$, then $[(e_i - e_j)^\top f^{(1)}(x_0)](1) > 0$ and condition 1 of Theorem 14.2 is verified and a Carathéodory solution entering $S_{(1)}$ exists. If $\left[\sum_{h\neq j:|x_h-x_i|<1}(x_h - x_i) - \sum_{h\neq i:|x_h-x_i|<1}(x_h - x_i)\right] \leqslant 0$ then $[(e_i - e_j)^\top f^{(-1)}(x_0)] = [(e_i - e_j)^\top f^{(1)}(x_0)] - 2 < 0$ and $[(e_i - e_j)^\top f^{(-1)}(x_0)](-1) > 0$, so that a Carathéodory solution issuing from x_0 and entering $S_{(-1)}$ exists.

In case $m > 1$ in Assumption 14.1, i.e., $x_{0i} - x_{0j} = 1$ for more than one pair (i, j), one starts by considering the set $S_{(1,1,\dots,1)}$ and the vector $f^{(1,1,\dots,1)}(x_0)$. If $[p_\ell^\top f^{(1,1,\dots,1)}(x_0)](1) > 0$ for all $\ell \in \{1, \dots, m\}$ then condition 1 of Theorem 14.2 is verified. Otherwise, there exists $\bar\ell \in \{1, \dots, m\}$ such that $[p_{\bar\ell}^\top f^{(1,1,\dots,1)}(x_0)](1) \leqslant 0$. In this case, we consider the set $S_{(1,\dots,-1,\dots,1)}$ and the vector $f^{(1,\dots,-1,\dots,1)}(x_0)$. From the previous step, we know that for $\ell = 1, \dots, \bar\ell - 1$ one has $[p_\ell^\top f^{(1,\dots,-1,\dots,1)}(x_0)](1) > 0$ and $[p_{\bar\ell}^\top f^{(1,\dots,-1,\dots,1)}(x_0)](-1) = [p_{\bar\ell}^\top f^{(1,1,\dots,1)}(x_0) - 2](-1) > 0$. Now, if for $\ell = \bar\ell + 1, \dots, m$ one has $[p_\ell^\top f^{(1,\dots,-1,\dots,1)}(x_0)](1) > 0$, then condition 1 is satisfied; otherwise, one goes on with the same procedure. We remark that if $[p_{\bar\ell}^\top f^{(1,1,\dots 1)}(x_0)] \leqslant 0$, then $[p_{\bar\ell}^\top f^{\mathbf{b}}(x_0)] \leqslant 0$ for all $\mathbf{b} \in \{-1, 1\}^m$. The procedure stops after at most m steps, having checked all the sectors, returning a certificate for condition 1. □

Regarding (QB), the following corollary of Theorem 14.2 is proved in [16]: in this case, condition 2 of Theorem 14.2 needs to be applied.

Corollary 14.2 (Existence for QB) *For any initial condition, there exists a local Carathéodory solution of* (QB).

We emphasize that Carathéodory solutions may not be unique, as shown in Example 14.3. Another example of nonuniqueness is given by (QB).

Example 14.4 Consider the discrete behavior dynamics (QB) over the undirected path graph with $N = 2$ whose adjacency matrix A has non-null entries equal to 1 and the initial condition $\bar x_0 = (1/2, 1/2)^\top$. The right-hand side of the system is clearly discontinuous at $\bar x_0$. There are two solutions issuing from this point which correspond to the limit values of $f(x)$ when restricted to the two sets $S_{(-1,1)} = \{x \in \mathbb{R}^2 : x_i - \frac{1}{2} < 0, i = 1, 2\}$ and $S_{(1,1)} = \{x \in \mathbb{R}^2 : x_i - \frac{1}{2} \geqslant 0, i = 1, 2\}$. These solutions converge to $(0, 0)^\top$ and $(1, 1)^\top$, respectively. Their trajectories are the line segments joining the initial condition with the points $(0, 0)^\top$ and $(1, 1)^\top$, see Fig. 14.5.

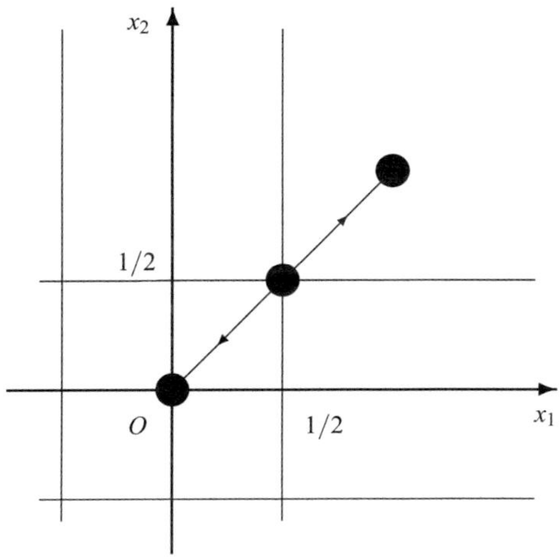

Fig. 14.5 Lack of unicity in dimension 2 for dynamics (QB), see Example 14.4

14.3.2 Krasovskii Solutions

In order to cope with nonexistence of solutions, other generalized solutions have been introduced in the literature. In the context described here, Krasovskii solutions can be easily and successfully used.

Definition 14.2 (*Krasovskii solutions*) Let $I \subset \mathbb{R}$ be an interval with $0 \in I$ and let $x_0 \in \mathbb{R}^N$. An absolutely continuous function $\varphi : I \to \mathbb{R}^N$ is a *Krasovskii solution* of (14.3) with initial condition x_0 if $\varphi(0) = x_0$ and if for almost all $t \in I$ it satisfies the differential inclusion

$$\dot{\varphi}(t) \in \mathscr{K} f(\varphi(t)), \tag{14.4}$$

where

$$\mathscr{K} f(x) = \bigcap_{\delta>0} \overline{co}\{f(y) : \ y \ \text{ such that } \|x - y\| < \delta\}.$$

We say that a *local Krasovskii solution* corresponding to the initial condition $x_0 \in \mathbb{R}^N$ exists if there exists a neighborhood $I(x_0)$ of x_0, an interval of the form $[0, T)$ and an absolutely continuous function $\varphi : [0, T) \to I(x_0)$ such that $\varphi(0) = x_0$ and $\varphi(t)$ is a Krasovskii solution of (14.3) on $[0, T)$.

The following existence theorem is an immediate consequence of [4, Theorem 3, page 98], as the vector field $f(x)$ is measurable and locally bounded.

Theorem 14.3 *For any initial condition, $x_0 \in \mathbb{R}^N$ there exists a local Krasovskii solution of* (14.3).

We underline that *any Carathéodory solution is also a Krasovskii solution.* Another type of generalized solutions often adopted for discontinuous systems is

Filippov solutions [24]. Under Assumptions 14.1 and 14.2, Krasovskii solutions coincide with Filippov solutions (see [33]).

14.3.3 Completeness of Solutions

Besides local existence, we are interested in solutions that are defined on unbounded intervals. The following condition is well known: its proof is included for completeness and tutorial purposes.

Proposition 14.2 (Prolongation by boundedness) *Let* (14.3) *be a system that admits local Carathéodory (Krasovskii) solutions for every initial condition in* $\mathbb{R}^N$. *Let* $\varphi : [0, T) \to \mathbb{R}^N$ *be a Carathéodory (Krasovskii) solution of* (14.3). *If* $\varphi(t)$ *is bounded on* $[0, T)$, *then it can be continued over* $[0, T')$ *with* $T' > T$.

Proof Assume by contradiction that there exists an initial condition x_0 whose corresponding maximal Carathéodory (Krasovskii) right solution has domain $[0, T)$ with $T < +\infty$ and let $\{t_n\}$ be a sequence such that $t_n > 0$ and $\lim t_n = T$. Since $x(t)$ is bounded also the sequence $\{x(t_n)\}$ is bounded, and thus, there exists a subsequence $x(t_{n_k})$ converging to a point x^*. Since $x(t)$ is continuous this implies that there exists $\lim_{t\to T} x(t) = x^* \in \mathbb{R}^N$. One can then pose a new Cauchy problem with initial condition $x(T) = x^*$ and then continue the Carathéodory (Krasovskii) solution on an interval $[T, T')$. We then get to a contradiction as the interval $[0, T)$ is not maximal for the considered Carathéodory (Krasovskii) solution with initial condition $x(0) = x_0$. □

This fact is useful because boundedness is easily established in our examples.

Proposition 14.3 (Boundedness) *Any Krasovskii solution of* (BC), (QS), *and* (QB) *defined on an interval of the form* $[0, T)$ *is bounded.*

Proof Let m be any index in $\mathcal{I}$ such that $x_m(t) = \min\{x_i(t), i \in \mathcal{I}\}$ and M any index such that $x_M(t) = \max\{x_i(t), i \in \mathcal{I}\}$. In the cases of (BC) and (QS), it is straightforward to verify that $x_m(t)$ is a nondecreasing function of time and, similarly, that $x_M(t)$ is nonincreasing. More delicate is the case of (QB), which we verify in detail. Let $q_m(t) = q(x_m(t))$. We have to distinguish three cases. If $x_m(t) \in \left(q_m(t) - \frac{1}{2}, q_m(t)\right]$, then $\dot{x}_m(t) = \sum_j a_{mj}[q(x_j(t)) - x_m(t)] \geqslant 0$, because by definition $x_i(t) \geqslant x_m(t)$ for $i \in \mathcal{I}$. If $x_m(t) \in \left(q_m(t), q_m(t) + \frac{1}{2}\right)$, then $x_m(t)$ may be decreasing as there may be other indices i such that $q(x_i(t)) = q_m(t)$. Nevertheless, $\dot{x}_m(t) \geqslant 0$ when $x_m(t) = q_m(t)$ and then $x_m(t)$ remains lower bounded by $\min\{x_m(0), q_m(0)\}$ The remaining case when $x_m(t) = q_m(t) - \frac{1}{2}$ is more delicate and specific to Krasovskii solutions. Indeed, there can exist an index ℓ such that $x_\ell(t) = x_m(t)$ but $q(y) = q_m(t) - 1$ for some points y in the neighborhood of x_ℓ, which makes the (set valued) right-hand side include negative values. In such a case, $x_m(t)$ would be allowed to decrease, but this fact would in turn lead to the situation of the second case. In conclusion, $x_m(t)$ is lower bounded by $q_m(0) - 1$ and similarly $x_M(t)$ is upper bounded by $q_M(0) + 1$. □

Combining the previous two results, we readily obtain the following.

Corollary 14.3 (Completeness) *Any Carathéodory solution of* (QB) *or of* (BC), *as well as any Krasovskii solution of* (QB), (QS), *or* (BC) *is defined on* $[0, +\infty)$.

This completeness result justifies the analysis of the limit behavior for the dynamics of interest, which we shall undergo in Sect. 14.5. Before that, however, we turn our attention to the study of equilibria.

14.4 Equilibria: Agreement and Beyond

In this section, we recall the definitions of equilibria that are natural in our context, we briefly discuss some counterintuitive facts about generalized equilibria, and then we study the equilibria of systems (BC), (QB), and (QS).

Equilibria are points where a solution can remain indefinitely.[3] In the context of generalized solutions, this general definition leads to distinguish between Carathéodory equilibria and Krasovskii equilibria.

Definition 14.3 (*Equilibria*) A point x^* is a Carathéodory (Krasovskii) equilibrium of (14.3) if the function $\varphi(t) \equiv x^*, t \geqslant 0$ is a Carathéodory (Krasovskii) solution of (14.3).

Carathéodory equilibria are characterized by the equation $f(x) = 0$ while Krasovskii equilibria are characterized by the inclusion $0 \in \mathscr{K} f(x)$. Thanks to the multiplicity of solutions, there are examples of non-constant solutions issuing from an equilibrium point.

Example 14.5 (*Escaping from equilibria*) Consider the bounded confidence system (BC) with $N = 2$ and the initial condition $x_0 = (-1/2, 1/2)^\top$. Let us denote by $f(x)$ the vector field defined by the right-hand side of (BC). Clearly, x_0 is a discontinuity point as $x_{02} - x_{01} - 1 = 0$ and $m = 1$ in Assumption 14.1. Let $s_1(x) = x_2 - x_1 - 1 = (-1, 1)x - 1$. Note that $f(x_0) = 0$, then x_0 is a Carathéodory equilibrium point for the system. On the other hand, $f^{(-1)}(x_0) = (1, -1)$ and $[(-1, 1)f^{(-1)}(x_0)](-1) = (-2)(-1) = 2 > 0$. A Carathéodory solution starts from x_0, enters $S_{(-1)}$, and converges to $(0, 0)$.

Being $f(x)$ in (14.3) allowed to be discontinuous, there may be points which are attractive for Carathéodory solutions without being Carathéodory equilibria: actually, these pathological points are Krasovskii equilibria.

[3]Note that this is a "weak" notion of equilibrium: in case of multiple solutions, we do not require that all solutions remain at the equilibrium.

Example 14.6 (*Attractive non-Carathéodory equilibria*) Let us consider the quantized behavior system (QB) over an undirected 4-node path graph with adjacency matrix

$$A = \begin{pmatrix} 0 & 1 & 0 & 0 \\ 1 & 0 & 1 & 0 \\ 0 & 1 & 0 & 1 \\ 0 & 0 & 1 & 0 \end{pmatrix}.$$

The point $x_0 = (0, \frac{1}{2}, \frac{1}{2}, 1)^\top$ is attractive for Carathéodory solutions issuing from points in the set $\{x \in \mathbb{R}^4 : -1/2 \leqslant x_1, x_2 < 1/2,\ \ 1/2 \leqslant x_3, x_4 \leqslant 3/2)\}$. Point x_0 cannot be a Carathéodory equilibrium, because $q(x_{02}) = 1 \neq x_{01} = 0$, even though it is a Krasovskii equilibrium. A Carathéodory solution originating from x_0 converges to $(1, 1, 1, 1)^\top$.

The following propositions concern equilibria of the systems under consideration. For the dynamics (BC) the sets of Carathéodory and Krasovskii equilibria coincide. The equilibria are precisely those states where individuals either agree or are enough apart not to influence each other, as specified in the following simple result already available in [8].

Proposition 14.4 (Equilibria of BC) *The set of Krasovskii equilibria of* (BC) *is*

$$\mathscr{F} = \{x \in \mathbb{R}^N : \textit{for every } (i, j) \in \mathscr{I} \times \mathscr{I},\ \textit{either } x_i = x_j \textit{ or } |x_i - x_j| \geqslant 1\}.$$

In the case of the quantized states dynamics (QS), Carathéodory equilibria and Krasovskii equilibria differ. Carathéodory equilibria are not necessarily consensus points, but the quantizations of their states must agree. The following proposition was proved in [17].

Proposition 14.5 (Equilibria of QS) *The set of Carathéodory equilibria of* (QS) *is*

$$\mathscr{D} = \{x \in \mathbb{R}^N : \exists h \in \mathbb{Z} \textit{ such that } h - \frac{1}{2} \leqslant x_i < h + \frac{1}{2},\ \forall i \in \mathscr{I}\}.$$

The set of Krasovskii equilibria of (QS) *is* $\overline{\mathscr{D}}$.

In case of the quantized behaviors Eq. (QB), we do not have a characterization of the set of equilibria. On the one hand, we observe that consensus points of the form $h\mathbf{1}$ with $h \in \mathbb{Z}$ are Carathéodory equilibria. On the other hand, there exist equilibria that are far from consensus and are attractive for some Carathéodory solutions. An example is provided in the next result.

Proposition 14.6 (Far-from-consensus equilibrium of QB) *Consider* (QB) *with an N-node path as underlying graph and all nonzero entries of the adjacency matrix A equal to 1. Then, there exists a Krasovskii equilibrium x^* such that*

$$x_N^* - x_1^* = \begin{cases} \frac{(N-2)^2}{4} & \textit{if } N \textit{ is even} \\ \frac{(N-1)(N-3)}{4} & \textit{if } N \textit{ is odd.} \end{cases}$$

Proof The equilibrium can be constructed as follows. We select $\mathbf{k} \in \mathbb{Z}^N$ such that $k_1 = 0$ and

$$k_i - k_{i-1} = \begin{cases} i-2, & 2 \leqslant i \leqslant \frac{N+2}{2} \\ N-i, & \frac{N+2}{2} < i \leqslant N \end{cases}$$

and then we set

$$\begin{aligned} x_1^* &= k_2 \\ x_i^* &= \tfrac{k_{i-1}+k_{i+1}}{2}, \quad i = 2, \ldots, N-1 \\ x_N^* &= k_{N-1}. \end{aligned}$$

It can be easily verified (details in [16]) that x^* is a Krasovskii equilibrium. □

14.5 Disagreement and Distance from Consensus

In the models (BC), (QS), and (QB), we cannot expect to have the same converge properties as (14.1): in fact, they are interesting as they attempt to explain agreement and disagreement at the same time. For these models, we point out the occurrence of disagreement, we give estimates of distance from consensus, and, when possible, we give sufficient conditions for convergence to consensus.

14.5.1 Bounded Confidence Dynamics

In general, the following is the strongest convergence result that has been given about (BC). An example of evolution is in Fig. 14.6.

Theorem 14.4 (Asymptotic behavior [14]) *Any Krasovskii solution of* (BC) *converges to a point in* $\mathscr{F}$.

In the wake of this fact, much research (from [8] to [49]) has been devoted to understand to *which* point in $\mathscr{F}$ a solution converges. Since the interaction topology is encoded in the state x by the definition of function $a(\cdot, \cdot)$, conditions should be given in terms of the initial condition $x(0)$. For instance, one can immediately observe that if $\mathscr{G}(x_0)$ is a complete graph, then the dynamics converges to a consensus. More general, though not necessary, conditions for consensus are stated in [51].

Theorem 14.5 (Sufficient condition for consensus) *If* $x_0 \in \mathbb{R}^N$ *is such that*

1. $\mathscr{G}(x_0)$ *is connected and*
2. *for any edge* $(i, j) \in \mathscr{E}(x_0)$*, the set* $\{k \in \mathscr{I} : (i, k) \in \mathscr{E}(x_0) \text{ and } (j, k) \in \mathscr{E}(x_0)\}$ *has cardinality not smaller than* $\frac{N}{2} - 2$,

then Krasovskii solutions issuing from x_0 *converge to the consensus point* $x_{ave}(0)\mathbf{1}$.

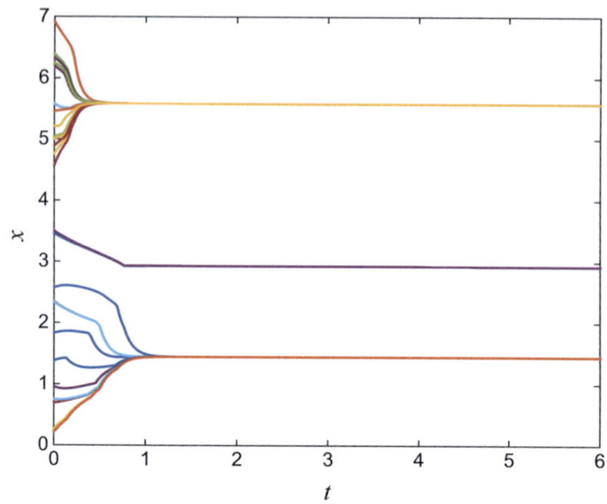

Fig. 14.6 Evolution of a solution of (BC) from a random initial condition on 25 nodes

The previous conditions imply that initial values cannot be too much spread. For example, in the case of 10 agents, the distance among agents for $\mathscr{G}(x_0)$ to be connected can be as large as 9, but in order to satisfy the second condition of Theorem 14.5, it can be at most 3. Other sufficient conditions for consensus may be found by applying the methods in [47].

14.5.2 Quantized States Dynamics

Convergence to the set of equilibria can also be proved for dynamics (QS), an example of which is given in Fig. 14.7.

Theorem 14.6 (Sufficient conditions for discrete consensus [17]) *Any Krasovskii solution* $\varphi(t)$ *of* (QS) *is such that* $dist(\varphi(t), \mathscr{D}) \to 0$ *as* $t \to +\infty$.

We remark that the set $\mathscr{D}$ is not formed by consensus points, but points in $\mathscr{D}$ are such that $q(x_i) = q(x_j)$ for all $i, j \in \mathscr{I}$. Thus, the 2-norm distance of Krasovskii solutions from consensus is, asymptotically, at most $\sqrt{N/2}$.

The assumptions that L be symmetric and the interaction graph be undirected can be lifted: recently, [50] has proved the same convergence property for more general functions q and weaker connectivity. Namely, q only needs to be nondecreasing and the graph can be directed and only needs to have a globally reachable node.[4]

[4]We refer the reader to [[23], Chap. 1] for the relevant definitions about directed graphs.

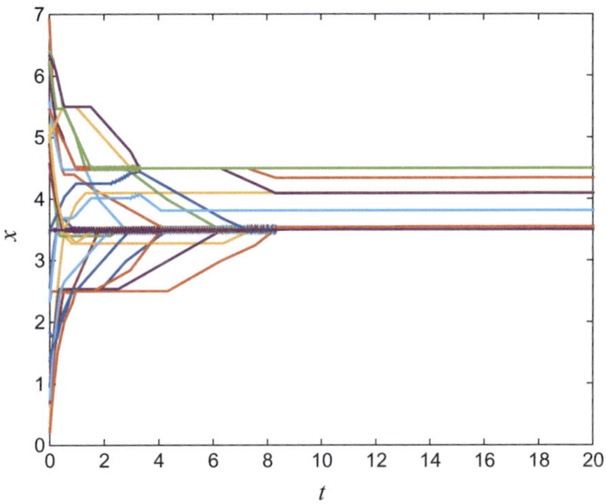

Fig. 14.7 Evolution of a solution of (QS) on a cycle graph on 25 nodes from the same initial conditions as in Fig. 14.6

14.5.3 Quantized Behavior Dynamics

For this system, a general proof of convergence to the equilibria is missing. However, some properties of solutions for large times can be established and are confirmed by simulations, see Fig. 14.8.

Theorem 14.7 (Distance from consensus) *If $\varphi(t)$ is any Krasovskii solution of* (QB) *and*

$$M = \left\{ x \in \mathbb{R}^N : \inf_{\alpha \in \mathbb{R}} \|x - \alpha \mathbf{1}\| \leqslant \frac{||A||}{\lambda_*} \frac{\sqrt{N}}{2} \right\},$$

then $dist(\varphi(t), M) \to 0$ as $t \to +\infty$.

Proof First of all, we observe that system (QB) can be written

$$\dot{x} = -Lx + A(q(x) - x). \tag{14.5}$$

Let $y(t) = x(t) - x_{ave}(t)\mathbf{1}$. Then $\dot{y}(t) = \dot{x}(t) - \dot{x}_{ave}(t)\mathbf{1}$. Consider the function $V(y) = \frac{1}{2} y^\top y$. We have that

$$\begin{aligned}
\nabla V(y)^\top \dot{y} =& y^\top \dot{y} \\
=& (x - x_{ave}\mathbf{1})^\top [\dot{x} - \dot{x}_{ave}\mathbf{1}] \\
=& (x - x_{ave}\mathbf{1})^\top \dot{x} - x^\top \dot{x}_{ave}\mathbf{1} + x_{ave}\mathbf{1}^\top \dot{x}_{ave}\mathbf{1} \\
=& (x - x_{ave}\mathbf{1})^\top \dot{x} - \dot{x}_{ave} x^\top \mathbf{1} + \dot{x}_{ave} N x_{ave} \\
=& (x - x_{ave}\mathbf{1})^\top \dot{x} - \dot{x}_{ave} N x_{ave} + \dot{x}_{ave} N x_{ave} \\
=& (x - x_{ave})^\top \dot{x}.
\end{aligned}$$

As $L\mathbf{1} = 0$, we have

$$\dot{x} \in -L(x - x_{ave}\mathbf{1}) + A\mathscr{K}(q(x) - x) \subseteq -L(x - x_{ave}\mathbf{1}) + A(\mathscr{K}q(x) - x).$$

For any $v \in \mathscr{K}q(x) - x$, it holds $\|v\| \leqslant \frac{\sqrt{N}}{2}$. Then, if $v \in \mathscr{K}q(x) - x$ is such that $\dot{y} = -L(x - x_{ave}\mathbf{1}) + Av$, we have

$$\begin{aligned}\nabla V(y)^\top \dot{y} =& (x - x_{ave}\mathbf{1})^\top[-L(x - x_{ave}\mathbf{1}) + Av]\\ =& -(x - x_{ave}\mathbf{1})^\top L(x - x_{ave}\mathbf{1}) + (x - x_{ave}\mathbf{1})^\top Av\\ \leqslant& -\lambda_* \|x - x_{ave}\mathbf{1}\|^2 + \|x - x_{ave}\mathbf{1}\| \|A\| \frac{\sqrt{N}}{2}\\ \leqslant& \|x - x_{ave}\mathbf{1}\| \left[-\lambda_* \|x - x_{ave}\mathbf{1}\| + \|A\| \frac{\sqrt{N}}{2} \right].\end{aligned}$$

We conclude that $dist(x(t), M) \to 0$ as $t \to +\infty$, because otherwise V would decrease unboundedly along solutions, which is forbidden by V being nonnegative. □

We remark that this result is tight in the following sense: on some graphs, the estimate on the limit set is asymptotically tight for large networks in the sense of the Euclidean distance from the consensus. More precisely, if the graph is a path with N nodes and weights are uniform, for all points in the attractor M it holds true that $\frac{1}{\sqrt{N}}\|x - x_{ave}\| = O(N^2)$ as $N \to \infty$. At the same time, the equilibrium x^* that was constructed in the proof of Proposition 14.6 is such that (for odd N)

$$\frac{1}{\sqrt{N}}\|x^* - x^*_{ave}\| = \frac{1}{\sqrt{120}} N^2 + o(N^2) \quad \text{as } N \to \infty.$$

Hence, the estimate of M cannot be improved in general in terms of distance from consensus. Details of these computations can be found in [16].

Even though not guaranteed in general, the consensus is achieved on some topologies. An example of such result is the following.

Theorem 14.8 (Sufficient conditions for consensus) *If the graph underlying system* (QB) *is either complete or complete bipartite and its adjacency matrix A has all non-null entries equal to* 1, *then all Krasovskii solutions of* (QB) *converge to a consensus point.*

The proof of this result, which can be found in [16], is based on showing that $\max_i x_i(t) - \min_i x_i(t)$ is decreasing and converges to zero.

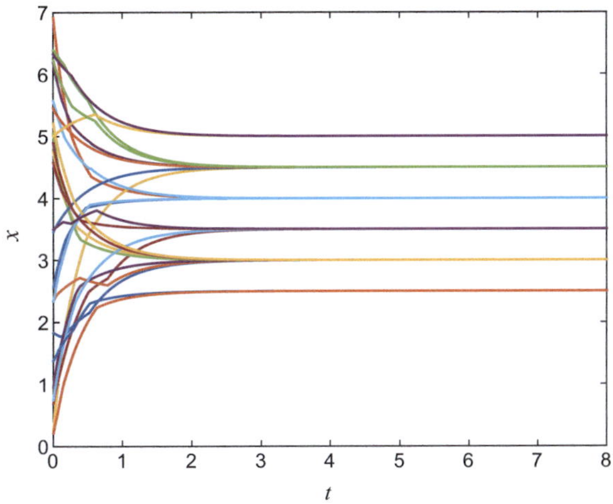

Fig. 14.8 Evolution of a solution of (QB), assuming the same initial conditions and graph as in Fig. 14.7

14.6 Discussion: The Origins of Disagreement in Opinion Dynamics

The dynamics analyzed in this chapter are meant to describe opinion dynamics in social networks. In this context, the nodes of the graph are individuals, an edge between two nodes means that they socially interact and the ith component of the state represents the value of the ith individual's opinion. This graph-based modeling approach has a strong support in mathematical sociology [27, 43] as well as in economics [36] and in the physics of complex systems [5, 11, 32].

The basic assumption in these models of opinion dynamics is that if an individual communicates with another, then his/her opinion is attracted by the others. If one translates this assumption into a set of differential equations, then one gets system (14.1), as already proposed in [1]. This dynamics asymptotically leads to consensus, i.e., agreement of the individuals on the same opinion, except in case there are different groups of individuals which do not communicate with each other, i.e., the communication graph has separated connected components. However, it has been noted that agreement is rare in societies [28], even if individuals do communicate: for this reason, more complex models have been elaborated with the aim of explaining agreement and disagreement at the same time.

In this chapter, we have focused on a group of models involving different kinds of threshold phenomena leading to discontinuities. Before going back to discuss their features, it is important to mention that these are not the only possible explanations for disagreement. In [29], disagreement is explained as the effect of obstinacy that is translated into the dependence of any individual's opinion on its initial value. Stubbornness as the source of disagreement is also considered in other models, such as [39, 42], also in connection with the occurrence of randomized asynchronous interactions [2, 25, 45]. Another explanation has been proposed to be the presence of contrarians [31] or of negative interactions, i.e., negative weights in the adjacency matrix [3]. Similar dynamics on "signed graphs" may also feature randomized interactions [46] or bounded confidence [18].

Going back to the models considered in this chapter, we now try to summarize their features in the opinion dynamics context, beginning with (BC). Bounded confidence dynamics allow for the existence of complete Carathéodory solutions for every initial condition and all Krasovskii solutions are proved to converge to an equilibrium. The structure of these equilibria is a set of separated clusters of individuals sharing the same opinion. In [8, 14], it is proved that, due to robustness issues, one can expect the opinion values of different clusters to be approximately twice the threshold apart. The most recent results on this matter are probably those in [49]. Actually, a fine understanding of how the final opinions depend on the initial ones is still missing. In this chapter, we have reported a sufficient condition for consensus, which asks for the initial opinions to be already quite close to each other. Other models that involve assumptions of bounded confidence include [9, 22, 26, 38, 52].

Consensus dynamics with quantization have first been studied with engineering motivations, while seeking controlled dynamics that could lead to (approximate) consensus despite the constraint of quantization [10, 41]. Proposed in this context by [17], the quantized states dynamics (QS) does not allow for global existence of Carathéodory solutions and thus requires to consider Krasovskii solutions: all Krasovskii solutions converge to equilibria such that the quantized opinions are equal. This is not exactly consensus, as individuals' opinions may slightly differ, but they agree on their quantized values. Consistently with its history, dynamics (QS) better fits engineering applications than social dynamics[5]: we believe that a better model of quantized social interactions is given by the quantized behavior dynamics (QB), which we proposed in [15, 16]. This model allows for the existence of complete Carathéodory solutions for every initial conditions, but Krasovskii solutions are preferred to avoid the pathology of solutions converging to nonequilibrium point. In general, a result of convergence to equilibria is missing, but a tight result of convergence to a set is available. Remarkably, there can be equilibria very far from consensus, in which the difference among different opinions of individuals is proportional to N^2.

Beyond the specific dynamics considered in this chapter, we believe that dynamical models that involve discontinuities can be useful in the study of social dynamics: we thus hope that the tools collected here can also be useful in the analysis of new and richer models.

[5]The discretization of the opinions in social systems has been observed by social scientists [30, Chap. 10] and addressed in several models including [20, 37, 48].

References

1. Abelson, R.P.: Mathematical models in social psychology. Adv. Exp. Soc. Psychol. **3**, 1–54 (1967)
2. Acemoğlu, D., Como, G., Fagnani, F., Ozdaglar, A.: Opinion fluctuations and disagreement in social networks. Math. Oper. Res. **38**(1), 1–27 (2013)
3. Altafini, C.: Consensus problems on networks with antagonistic interactions. IEEE Trans. Autom. Control **58**(4), 935–946 (2013)
4. Aubin, J.P., Cellina, A.: Differential Inclusions, Grundlehren der Mathematischen Wissenschaften, vol. 264. Springer, Berlin (1984)
5. Aydoğdu, A., Caponigro, M., McQuade, S., Piccoli, B., Pouradier Duteil, N., Rossi, F., Trélat, E.: Interaction Network, State Space, and Control in Social Dynamics. Springer, Berlin (2017). https://doi.org/10.1007/978-3-319-49996-3_3
6. Bacciotti, A.: Some remarks on generalized solutions of discontinuous differential equations. Int. J. Pure Appl. Math. **10**(3), 257–266 (2003)
7. Blondel, V.D., Hendrickx, J.M., Tsitsiklis, J.N.: Existence and uniqueness of solutions for a continuous-time opinion dynamics model with state-dependent connectivity (2009). http://www.mit.edu/jnt/Papers/BHT10-solutions-DA.pdf
8. Blondel, V.D., Hendrickx, J.M., Tsitsiklis, J.N.: Continuous-time average-preserving opinion dynamics with opinion-dependent communications. SIAM J. Control Optim. **48**(8), 5214–5240 (2010)
9. Canuto, C., Fagnani, F., Tilli, P.: An Eulerian approach to the analysis of Krause's consensus models. SIAM J. Control Optim. **50**(1), 243–265 (2012)
10. Carli, R., Fagnani, F., Speranzon, A., Zampieri, S.: Communication constraints in the average consensus problem. Automatica **44**(3), 671–684 (2008)
11. Castellano, C., Fortunato, S., Loreto, V.: Statistical physics of social dynamics. Rev. Mod. Phys. **81**(2), 591–646 (2009)
12. Ceragioli, F.: Discontinuous ordinary differential equations and stabilization. Ph.D. thesis, Università di Firenze (2000)
13. Ceragioli, F.: Finite valued feedback laws and piecewise classical solutions. Nonlinear Anal. Theory Methods Appl. **65**(5), 984–998 (2006)
14. Ceragioli, F., Frasca, P.: Continuous and discontinuous opinion dynamics with bounded confidence. Nonlinear Anal. Real World Appl. **13**(3), 1239–1251 (2012)
15. Ceragioli, F., Frasca, P.: Continuous-time consensus dynamics with quantized all-to-all communication. In: European Control Conference, pp. 1120–1125. Linz, Austria (2015)
16. Ceragioli, F., Frasca, P.: Consensus and disagreement: the role of quantized behaviours in opinion dynamics. SIAM Journal on Control and Optimization (2018). To appear. arXiv:1607.01482
17. Ceragioli, F., De Persis, C., Frasca, P.: Discontinuities and hysteresis in quantized average consensus. Automatica **47**(9), 1916–1928 (2011)
18. Ceragioli, F., Lindmark, G., Veibäck, C., Wahlström, N., Lindfors, M., Altafini, C.: A bounded confidence model that preserves the signs of the opinions. In: European Control Conference 2016, pp. 543–548. Aalborg, Denmark (2016)
19. Chazelle, B., Wang, C.: Inertial Hegselmann-Krause systems. IEEE Trans. Autom. Control **62**(8), 3905–3913 (2016)
20. Chowdhury, N.R., Morarescu, I.C., Martin, S., Srikant, S.: Continuous opinions and discrete actions in social networks: a multi-agent system approach. In: IEEE Conference on Decision and Control, pp. 1739–1744. Las Vegas, NV, USA (2016)
21. Cortés, J.: Discontinuous dynamical systems - a tutorial on solutions, nonsmooth analysis, and stability. IEEE Control Syst. Mag. **28**(3), 36–73 (2008)
22. Deffuant, G., Neau, D., Amblard, F., Weisbuch, G.: Mixing beliefs among interacting agents. Adv. Complex Syst. **3**(1–4), 87–98 (2000)
23. Fagnani, F., Frasca, P.: Introduction to Averaging Dynamics over Networks. Lecture Notes in Control and Information Sciences. Springer, Berlin (2017)

24. Filippov, A.: Differential Equations with Discontinuous Righthandside. Kluwer, Berlin (1988)
25. Frasca, P., Ravazzi, C., Tempo, R., Ishii, H.: Gossips and prejudices: Ergodic randomized dynamics in social networks. In: IFAC Workshop on Estimation and Control of Networked Systems, pp. 212–219. Koblenz, Germany (2013)
26. Frasca, P., Tarbouriech, S., Zaccarian, L.: A hybrid model of opinion dynamics with limited confidence. In: IFAC Symposium on Nonlinear Control Systems. Monterey, CA, USA (2016)
27. Friedkin, N.E.: A Structural Theory of Social Influence. Cambridge University Press, Cambridge (2006)
28. Friedkin, N.E.: The problem of social control and coordination of complex systems in sociology: a look at the community cleavage problem. IEEE Control Syst. **35**(3), 40–51 (2015)
29. Friedkin, N.E., Johnsen, E.C.: Social influence networks and opinion change. In: Lawler, E.J., Macy, M.W. (eds.) Advances in Group Processes, vol. 16, pp. 1–29. JAI Press (1999)
30. Friedkin, N.E., Johnsen, E.C.: Social Influence Network Theory: A Sociological Examination of Small Group Dynamics. Cambridge University Press, Cambridge (2011)
31. Galam, S.: Contrarian deterministic effects on opinion dynamics: "the hung elections scenario". Phys. A: Stat. Mech. Appl. **333**(Supplement C), 453–460 (2004)
32. Galam, S.: Sociophysics: a physicist's modeling of psycho-political phenomena. Springer Science & Business Media (2012)
33. Hájek, O.: Discontinuous differential equations I. J. Differ. Equ. **32**, 149–170 (1979)
34. Hegselmann, R., Krause, U.: Opinion dynamics and bounded confidence: models, analysis and simulation. J. Artif. Soc. Soc. Simul. **5**(3), 1–33 (2002)
35. Jabin, P.E., Motsch, S.: Clustering and asymptotic behavior in opinion formation. J. Differ. Equ. **257**(11), 4165–4187 (2014)
36. Jackson, M.O.: Social and Economic Networks. Princeton University Press, Princeton (2010)
37. Martins, A.C.R.: Continuous opinions and discrete actions in opinion dynamics problems. Int. J. Mod. Phys. C **19**(04), 617–624 (2008)
38. Mirtabatabaei, A., Bullo, F.: Opinion dynamics in heterogeneous networks: convergence conjectures and theorems. SIAM J. Control Optim. **50**(5), 2763–2785 (2012)
39. Mobilia, M.: Does a single zealot affect an infinite group of voters? Phys. Rev. Lett. **91**(2), 028–701 (2003)
40. Motsch, S., Tadmor, E.: Heterophilious dynamics enhances consensus. SIAM Rev. **56**(4), 577–621 (2014)
41. Nedic, A., Olshevsky, A., Ozdaglar, A., Tsitsiklis, J.N.: On distributed averaging algorithms and quantization effects. IEEE Trans. Autom. Control **54**(11), 2506–2517 (2009)
42. Parsegov, S.E., Proskurnikov, A.V., Tempo, R., Friedkin, N.E.: Novel multidimensional models of opinion dynamics in social networks. IEEE Trans. Autom. Control **62**(5), 2270–2285 (2017)
43. Proskurnikov, A.V., Tempo, R.: A tutorial on modeling and analysis of dynamic social networks. Part I. Annu. Rev. Control **43**, 65–79 (2017)
44. Pucci, A.: Traiettorie di campi di vettori discontinui. Rendiconti dell'Istituto Matematico dell'Università degli Studi di Trieste **8**, 84–93 (1976)
45. Ravazzi, C., Frasca, P., Tempo, R., Ishii, H.: Ergodic randomized algorithms and dynamics over networks. IEEE Trans. Control Netw. Syst. **2**(1), 78–87 (2015)
46. Shi, G., Proutiere, A., Johansson, M., Baras, J.S., Johansson, K.H.: The evolution of beliefs over signed social networks. Oper. Res. **64**(3), 585–604 (2016)
47. Tangredi, D., Iervolino, R., Vasca, F.: Consensus stability in the Hegselmann–Krause model with coopetition and cooperosity. In: IFAC World Congress, pp. 12,426–12,431. Toulouse, France (2017)
48. Urbig, D.: Attitude dynamics with limited verbalisation capabilities. J. Artif. Soc. Soc. Simul. **6**(1), 1–23 (2003)
49. Wang, C., Li, Q., Weinan, E., Chazelle, B.: Noisy Hegselmann-Krause systems: phase transition and the 2r-conjecture. J. Stat. Phys. **166**(5), 1209–1225 (2017)
50. Wei, J., Yi, X., Sandberg, H., Johansson, K.: Nonlinear consensus protocols with applications to quantized systems. In: IFAC World Congress, pp. 16,010–16,015. Toulouse, France (2017)

51. Yang, Y., Dimarogonas, D.V., Hu, X.: Opinion consensus of modified Hegselmann-Krause models. Automatica **50**(2), 622–627 (2014)
52. Zhang, J., Hong, Y.: Opinion evolution analysis for short-range and long-range Deffuant-Weisbuch models. Phys. A **392**(21), 5289–5297 (2013)

Chapter 15
Information Constraints in Multiple Agent Problems with I.I.D. States

S. Lasaulce and S. Tarbouriech

Abstract In this chapter, we describe several recent results on the problem of coordination among agents when they have partial information about a state which affects their utility, payoff, or reward function. The state is not controlled and rather evolves according to an independent and identically distributed (i.i.d.) random process. This random process might represent various phenomena. In control, it may represent a perturbation or model uncertainty. In the context of smart grids, it may represent a forecasting noise (Beaude et al., 6th IEEE international conference on smart grid communications (SmartGridComm 2015), Miami, Florida, 2015, [3]). In wireless communications, it may represent the state of the global communication channel. The approach used is to exploit Shannon theory to characterize the achievable long-term utility region. Two scenarios are described. In the first scenario, the number of agents is arbitrary, and the agents have causal knowledge about the state. In the second scenario, there are only two agents, and the agents have some knowledge about the future of the state, making its knowledge noncausal.

Chapter Overview

This chapter concerns the problem of coordination among agents. Technically, the problem is as follows. We consider a set of $K \geqslant 2$ agents. Agent k has a utility, payoff, or reward function $u_k(x_0, x_1, \ldots, x_K)$ where $x_k, k \geqslant 1$, is the action of Agent k while x_0 is the action of an agent called Nature. The Nature's actions correspond to the system state and are assumed to be noncontrolled; more precisely, Nature corresponds to an independent and identically distributed (i.i.d.) random process. The problem studied in this chapter is to characterize the long-term utility region under various assumptions in terms of observation at the agents. By long-term utility for Agent k we mean the following quantities:

S. Lasaulce (✉)
L2S, 3 Rue Joliot Curie, 91191 Gif-sur-Yvette, France
e-mail: lasaulce@l2s.centralesupelec.fr

S. Tarbouriech
LAAS-CNRS, Université de Toulouse, Toulouse, France
e-mail: sophie.tarbouriech@laas.fr

S. Tarbouriech et al. (eds.), *Control Subject to Computational and Communication Constraints*, Lecture Notes in Control and Information Sciences 475, https://doi.org/10.1007/978-3-319-78449-6_15

$$U_k(\sigma_1, \ldots, \sigma_K) = \lim_{T \to +\infty} \frac{1}{T} \mathbb{E}\left[\sum_{t=1}^{T} u_i(X_0(t), \ldots, X_K(t))\right], \tag{15.1}$$

where $\sigma_k = (\sigma_{k,t})_{t \geqslant 1}$ is a sequence of functions which represent the strategy of Agent k, $x_k(t)$ is the action chosen by Agent k at time or stage $t \geqslant 1$, t being the time or stage index; concerning notation, as far as random variables are concerned, capital letters will stand for random variables whereas, small letters will stand for realizations. Note that, implicitly, we assume sufficient conditions (such as utility boundedness) under which the above limit exists. The functions $\sigma_{k,t}$, $k \in \{1, \ldots, K\}$, map the available knowledge to the action of the considered agent. The available knowledge depends on the information assumptions made (e.g., the knowledge of the state can be causal or noncausal). We will distinguish between **two scenarios**. In the first scenario, agents are assumed to have some causal knowledge (in the wide sense) about the state whereas, in the second scenario noncausal knowledge (i.e., some knowledge about the future) about the state is assumed. The second scenario is definitely the most difficult one technically, which is why only two agents will be assumed.

Remarkably, the long-term utility region, whenever available, can be characterized in terms of elegant information constraints. For instance, in the scenario of noncausal state information, determining the long-term utility region amounts to solving a convex optimization problem whose nontrivial constraints are the derived information-theoretic constraints.

15.1 Introduction

An important example, which illustrates well how the results reported in this chapter can be used, is given by the problem of power control in wireless networks (see Fig. 15.1). Each transmitter has to adapt its transmit power not only to the fluctuations of the quality of the link (or channel gain) between itself and its respective receiver but also to the transmit power levels of the other transmitters that use the same radio resources (and therefore create interference). This problem is a multi-agent problem where the agents are the transmitters, the actions of the agents are their transmit power level, and the system state is given by the set of channel gains of the various links in presence; channel gains are typically noncontrolled variables (they do not depend on the transmit power levels) and evolve in a random manner; in practice, each transmitter has a partial and imperfect knowledge of the system state. Now, if the agents (namely, the transmitters in the considered example) have a certain performance criterion, which will be referred to as a utility function for the general setup considered in the chapter, the important problem of knowing the best achievable utilities appears. For instance, a transmitter might be designed to maximize its communication rate. The best data rate of a given transmitter would be obtained if all the other transmitters would be silent (i.e., when they do not transmit) and when the transmitter perfectly adapts its power to the channel gain

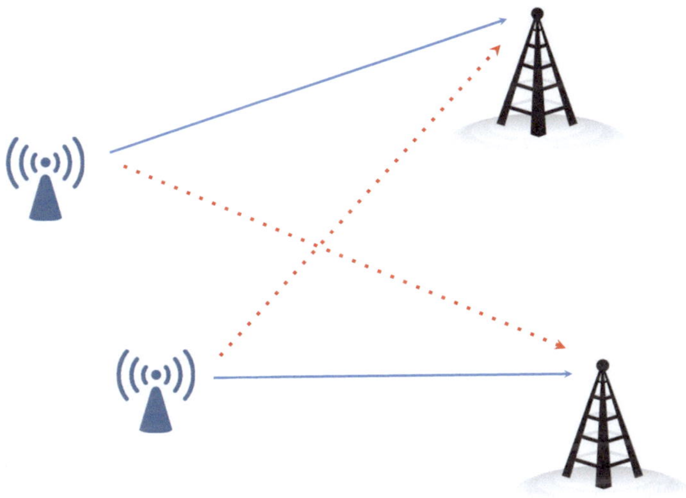

Fig. 15.1 The problem of power control in wireless networks is a typical application for the results provided in this chapter. The agents are the transmitters, the agents' actions are given by the transmit power level, and the agent utility function may be its communication rate with its intended receiver

fluctuations of the link between itself and its intended receiver. Obviously, in the real life, several transmitters will transmit at the same time, hence the need to coordinate as well as possible, which leads to the problem of characterizing the best performance possible in terms of coordination. This precisely corresponds to the problem of characterizing the *long-term utility region*, i.e., the set of possible achievable points $(U_1, U_2, \dots, U_K)$ for a given definition for the strategies. In Sects. 15.3 and 15.4, we will consider two different definitions for the strategies, each of them corresponding to a given observation structure that is, to some given information assumptions.

15.2 General Problem Formulation

This chapter aims at describing a few special instances of a general problem which has been addressed in several recent works [2, 5–9, 11].

We consider $K \geqslant 2$ agents, where Agent $k \in \{1, \dots, K\}$ produces time-t action $x_k(t) \in \mathcal{X}_k$[1] for $t \in \{1, \dots, T\}$, $T \geqslant 1$, the set $\mathcal{X}_k$ representing the set of actions for Agent k. Each agent has access to some observations associated with the chosen actions and the realization of a random process $\{X_{0,t}\}_{t=1}^{T} = \{X_{0,1}, \dots, X_{0,T}\} \in \mathcal{X}_0^T$. In the motivating example described in the introduction, the random process was given by the global wireless channel state, i.e., the set of qualities of all the links in presence. In a control problem, the random process may represent a noncontrolled

[1]Throughout the chapter, we assume that all the alphabets such as $\mathcal{X}_k$ are finite.

perturbation or some uncertainty. All agents' actions and the random process also affect the agents' individual *stage or instantaneous utility functions* $u_1, \dots, u_K$ where for all $k \in \{1, \dots, K\}$ the function u_k writes

$$u_k : \begin{array}{ccc} \mathscr{X}_0 \times \mathscr{X}_1 \times \cdots \times \mathscr{X}_K & \to & \mathbb{R} \\ (x_0, x_1, \dots, x_K) & \mapsto & u_k(x_0, x_1, \dots, x_K) \end{array}. \tag{15.2}$$

One of the main goals of the chapter is to explain how to determine the set of feasible *expected long-term utilities*:

$$U_k^{(T)} = \mathbb{E}\left[\frac{1}{T}\sum_{t=1}^{T} u_k(X_{0,t}, X_{1,t}, \dots, X_{K,t})\right], \tag{15.3}$$

that are reachable by some strategies for the agents. The set of feasible utilities is fully characterized by the set of feasible averaged joint probability distributions on the $(K+1)-$tuple $\{(X_{0,t}, X_{1,t}, \dots, X_{K,t})\}_{t=1}^{T}$. Indeed, denoting by $P_{X_{0,t}X_{1,t}\dots X_{K,t}}$ the joint probability distribution of the time $(K+1)-$tuple $(X_{0,t}, X_{1,t}, \dots, X_{K,t})$, we have

$$\begin{aligned} U_k^{(T)} &= \frac{1}{T}\sum_{t=1}^{T} \mathbb{E}\left[u_k(X_{0,t}, X_{1,t}, \dots, X_{K,t})\right] \\ &= \frac{1}{T}\sum_{t=1}^{T} \sum_{x_0,\dots,x_K} P_{X_{0,t}X_{1,t}\dots X_{K,t}}(x_0, x_1, \dots, x_K) u_k(x_0, x_1, \dots, x_K) \\ &= \sum_{x_0,\dots,x_K} u_k(x_0, x_1, \dots, x_K) \frac{1}{T}\sum_{t=1}^{T} P_{X_{0,t}X_{1,t}\dots X_{K,t}}(x_0, x_1, \dots, x_K). \end{aligned}$$

Therefore, the problem of characterizing the long-term utility region amounts to determining the set of averaged distributions

$$P^{(T)}(x_0, x_1, \dots, x_K) = \frac{1}{T}\sum_{t=1}^{T} P_{X_{0,t}X_{1,t}\dots X_{K,t}}(x_0, x_1, \dots, x_K) \tag{15.4}$$

that can be induced by the agents' strategies. For simplicity, and in order to obtain closed-form expressions, we shall focus on the case where $T \to \infty$ [4, 5].

We consider two types of scenarios with two different observation structures. In the first scenario, referred to as the *noncausal state information* scenario, the agents observe the system states noncausally. That means, at each stage $t \in \{1, \dots, T\}$ they have some knowledge about the entire state sequence $X_0^T = (X_{0,1}, \dots, X_{0,T})$. In the second scenario, referred to as the *causal state information* scenario, the agents learn the states only causally and therefore, at any stage t, the agents have some knowledge about the sequence $X_0^t = (X_{0,1}, \ldots, X_{0,t})$, where throughout the chapter, we use the shorthand notation A^m and a^m for the tuples $(A_1, \ldots, A_m)$ and $(a_1, \ldots, a_m)$, when m is a positive integer.

15.3 Coordination Among Agents Having Causal State Information

15.3.1 Limiting Performance Characterization

First, we define the information structure under consideration. At every instant or stage t, Agent k is assumed to have an image or a partial observation $S_{k,t} \in \mathcal{S}_k$ of the nature state $X_{0,t}$ with respect to which all agents are coordinating. In the case of the wireless power control example described in the introduction, this might be the knowledge of local channel state information, e.g., a noisy estimate of the direct channel between the transmitter and the associated receiver. The observations $S_{k,t}$ are assumed to be generated by a memoryless channel. By memoryless it is meant that the joint conditional probability on sequences of realizations factorizes the product of individual conditional probabilities. Denoting by $\daleth_k$ the transition probability for the observation structure of Agent k, the memoryless condition can be written as

$$P(s_K^T|x_0^T) = \prod_{t=1}^{T} \daleth_k(s_k(t)|x_0(t)). \tag{15.5}$$

The strategy or the sequence of decision functions for Agent k, $\sigma_{k,t}$, is defined by

$$\sigma_{k,t} \quad : \quad \mathcal{S}_k^t \longrightarrow \mathcal{X}_k \tag{15.6}$$

$$(s_k(1), s_k(2), \ldots, s_k(t)) \longmapsto x_k(t) \tag{15.7}$$

where $\mathcal{S}_k$ is observation alphabet for Agent k.

As mentioned in Sect. 15.2, the problem of characterizing the long-term utility region amounts to determining the achievable correlations measured in terms of joint distribution, hence the notion of implementability for a distribution.

Definition 15.1 (*Implementability*) The probability distribution $Q(x_0, x_1, \ldots, x_N)$ is implementable if there exist strategies $(\sigma_{1,t})_{t\geqslant 1}, \ldots, (\sigma_{K,t})_{t\geqslant 1}$ such that as $T \to +\infty$, we have for all $x \in \mathcal{X}$,

$$\frac{1}{T}\sum_{t=1}^{T} P_{X_{0,t}\ldots X_{K,t}}(x_0, \ldots, x_K) \longrightarrow Q(x_0, \ldots, x_K), \tag{15.8}$$

where $P_{X_{0,t}\ldots X_{K,t}}$ is the joint distribution induced by the strategies at stage t.

The following theorem is precisely based on the notion of implementability and characterizes the achievable long-term utilities that are implementable under the information structure (15.6); for this, we first define the *weighted utility function* w as a convex combination of the individual utilities u_k:

$$w = \sum_{k=1}^{K} \lambda_k u_k. \tag{15.9}$$

Theorem 15.1 [7] *Assume the random process $X_{0,t}$ to be i.i.d. following a probability distribution ρ and the available information to the transmitters $S_{k,t}$ to be the output of a discrete memoryless channel obtained by marginalizing the joint conditional probability $\urcorner$. An expected payoff $\overline{w}$ is achievable in the limit $T \to \infty$ if and only if it can be written as*

$$\begin{aligned} \overline{w} = & \sum_{\substack{x_0, x_1, \dots x_N, \\ u, s_1, \dots s_N}} \rho(x_0) P_U(u) \urcorner(s_1, \dots, s_K | x_0) \times \\ & \left(\prod_{k=1}^{K} P_{X_k|S_k,U}(x_k|s_k, u)\right) w(x_0, x_1, \dots, x_K), \end{aligned} \tag{15.10}$$

where U is an auxiliary variable, which can be optimized, and $P_{X_k|S_k,U}(x_k|s_k, u)$ is the probability that Agent k, chooses action x_k after observing s_k, u.

The auxiliary variable U is an external lottery known to the agents beforehand, which can be used to achieve better coordination, e.g., in presence of individual constraints or at equilibrium. Theorem 15.1 allows us to find all the achievable utility vectors $(U_1, \dots, U_K)$. Indeed, the long-term utility region being convex (this readily follows from a time-sharing argument), its Pareto boundary can be found by maximizing the weighted utility w. Of course, remains the problem of determining the strategies allowing to operate at a given arbitrary point of the utility region. Since this problem is nontrivial and there does not exist any methodology for this, we provide an algorithm which allows one to find a suboptimal strategy. Indeed, the associated multilinear optimization problem is too complex to be solved and to overcome this we resort to an iterative technique which is much less complex but is suboptimal.

15.3.2 An Algorithm to Determine Suboptimal Strategies

One of the merits of Theorem 15.1 is to provide the best performance achievable in terms of long-term utilities when agents have an arbitrary observation structure. However, Theorem 15.1 does not provide practical strategies which would allow a given utility vector to be reached. Finding "optimal" strategies consists in finding good sequences of functions as defined per (15.6), which is an open and promising direction to be explored. More pragmatically, the authors of [2] proposed to restrict to stationary strategies which are merely functions of the form $f_k : \mathcal{S}_k \to \mathcal{X}_k$. This choice is motivated by practical considerations such as computational complexity and it is also coherent with the current state of the literature. The water-filling solution is a special instance of this class of strategies. To find good decision functions, the idea, which is proposed in [2], is to exploit Theorem 15.1. This is precisely the purpose of this section.

The first observation we make is that the best performance only depends on the vector of conditional probabilities $P_{X_1|S_1,U}, \ldots, P_{X_K|S_K,U}$ and the auxiliary variable probability distribution P_U, the other quantities being fixed. It is therefore relevant to try to find an optimum vector of lotteries for every action possible and use it to make decisions. Since this task is typically computationally demanding, a possible and generally suboptimal approach consists in applying a distributed algorithm to maximize the expected weighted utility. The procedure proposed in [2] is to use the sequential best response dynamics (see, e.g., [10]). The idea is to fix all the variables (that are probability distributions here) expect one and maximize the expected weighted utility with respect to the only possible degree of freedom. This operation is then repeated by considering another variable. The key observation to be made is then to see that when the distributions of the other agents are fixed, the best distribution for Agent k boils down to a function of s_k, giving us a candidate for a decision function which can be used in practice.

To describe the algorithm of [2] (see also Fig. 15.2), we first rewrite the expected weighted utility in the following manner:

$$\overline{W} = \sum_{x_0,x_1,\ldots x_K,u,s_1,\ldots s_K} \rho(x_0) P_U(u) \times \tag{15.11}$$

$$\Gamma(s_1, \ldots, s_K | x_0, x_1, \ldots, x_K) \times \tag{15.12}$$

$$\left(\prod_{k=1}^{K} P_{X_k|S_k,U}(x_k|s_k, u) \right) w(x_0, x_1, \ldots, x_K)$$

$$= \sum_{i_k, j_k, u} \delta_{i_k, j_k, u} P_{X_k|S_k,U}(x_k|s_k, u), \tag{15.13}$$

where i_k, j_k, u are the respective indices of x_k, s_k, u and

$$\delta_{i_k, j_k, u} = \left[\sum_{i_0} \rho(x_{i_0}) \Gamma_k(s_k|x_{i_0}) \sum_{i_{-k}} u_k(x_{i_0}, x_{i_1}, \ldots x_{i_K}) \times \right.$$

$$\left. \sum_{j_{-k}} \prod_{k' \neq k} \Gamma_{k'}(s_{j_{k'}}|x_0) \prod_{k' \neq k} P_{X_k|S_k,U}(x_k|s_k, u) \right] P_U(u), \tag{15.14}$$

where i_{-k}, j_{-k} are the indices which represent i_k, j_k being constant, while all the other indices are summed over. To make the description of the algorithm clearer, we have also assumed the independence of the observation channels as

Algorithm 1: A possible decentralized algorithm for finding decision functions for the transmitters

inputs : $\mathcal{X}_k \quad \forall k \in \{0, ..., K\}, w(x_0, x_1, ...x_K) \quad \forall \underline{x}$,
$\rho(x_0), \Gamma_{\underline{S}|\underline{X}_0}(\underline{s}|\underline{x}_0) \quad \forall x_0, f_k^{init} \quad \forall k \in \{1...K\}, \varepsilon$
outputs: $f_k^*(s_k) \quad \forall k \in \{1...K\}$

Initialization: $\overline{f}_k^0 = \overline{f}_k^{\text{init}}$, iter $= 0$

while $\forall k \; \| f_k^{(iter-1)} - f_k^{iter} \|^2 \leq \varepsilon$ *OR* iter $= 0$ **do**
  iter = iter + 1;
  foreach $k \in \{1, \ldots, K\}$ **do**
    foreach $s_k \in \mathcal{S}_k$ **do**
      $\overline{f}_k^{\text{iter}}(s_k) \in \arg\max_{i_k} \delta_{i_k, j_k, u}$ using (15.14);
    end
  end
end
Final update: $\forall k \in \{1, ..., K\}$, $\overline{f}_k^{\star} = \overline{f}_k^{\text{iter}}$

Fig. 15.2 Pseudo-code of the algorithm proposed in [2] to find suboptimal strategies

well as independence of the signal with the strategies chosen by the agents, i.e., $\Gamma(s_1, \ldots, s_K|x_0, x_1, \ldots, x_K) = \Gamma_1(s_1|x_0) \times \ldots \times \Gamma_K(s_K|x_0)$. Written under this form, for every agent, optimizing the expected weighted utility in a distributed manner implies giving a probability 1 for the optimal coefficient $\delta_{i_k, j_k, u}$, and every player does that turn by turn.

To conclude, note that the above algorithm always converges. This can be proved, e.g., by induction or by calling for an exact potential game property (see, e.g., [10, 12]).

15.4 Coordination Between Two Agents Having Noncausal State Information

15.4.1 Limiting Performance Characterization

As explained previously, the problem of characterizing the utility region in the case where the state is known noncausally to the agents is much more involved technically. Even in the case of two agents, one may have to face with an open problem, depending on the observation structure assumed for the agents. Here, we consider an important case for which the problem can be solved, as shown in [7]. Therein, the authors consider an asymmetric observation structure. In the case of noncausal state information, agents' strategies are sequences of functions that are defined as follows. For Agent 1, the strategy is defined by

$$\sigma_{1,t} \quad : \quad \mathcal{S}_1^T \times \mathcal{Y}_1^{t-1} \longrightarrow \mathcal{X}_1 \tag{15.15}$$

$$(s_1(1), \ldots, s_K(T), y_1(1), \ldots, y_1(t-1)) \longmapsto x_1(t) \tag{15.16}$$

and for Agent 2, the strategy is defined by

$$\sigma_{2,t} \quad : \quad \mathcal{S}_2^T \times \mathcal{Y}_2^{t-1} \longrightarrow \mathcal{X}_2 \tag{15.17}$$

$$(s_2(1), \ldots, s_2(t), y_2(1), \ldots, y_2(t-1)) \longmapsto x_2(t), \tag{15.18}$$

where $y_k(t) \in \mathcal{Y}_k$ is the observation Agent k has about the triplet $(x_0(t), x_1(t), x_2(t))$ whereas $s_k(t) \in \mathcal{S}_k$ is the observation Agent k has about the state $x_0(t)$. Note that distinguishing between the two observations s_k and y_k is instrumental. Indeed, it does not make any sense physically speaking to assume that an agent might have some future knowledge about the actions of the other agents, which is why the feedback signal is strictly causal. On the other hand, assuming some knowledge about the future of the noncontrolled state x_0 perfectly makes sense, as motivated the chapter abstract and the works quoted in the list of references. More precisely the observation is assumed to be the output of a memoryless channel whose transition law is denoted by Γ:

$$\Pr\Big[Y_1(t) = y_1(t), Y_2(t) = y_2(t) \big| \boldsymbol{X}_0^t = \boldsymbol{s}_0^t, \boldsymbol{X}_1^t = \boldsymbol{x}_1^t, \boldsymbol{X}_2^t = \boldsymbol{x}_2^t, \boldsymbol{Y}_1^{t-1} = \boldsymbol{y}_1^{t-1}, \boldsymbol{Y}_2^{t-1} = \boldsymbol{y}_2^{t-1}\Big]$$
$$= \Gamma(y_1(t), y_2(t) | s x_0(t), x_1(t), x_2(t)). \tag{15.19}$$

We now provide the characterization of the set of implementable probability distributions both for the considered noncausal strategies.

Theorem 15.2 [7] *The distribution Q is implementable if and only if it satisfies the following condition*[2]

$$I_Q(S_1; U) \leq I_Q(V; Y_2|U) - I_Q(V; S_1|U). \tag{15.20}$$

where U and V are auxiliary random variables and Q is any joint distribution that factorizes as

$$Q(x_0, s_1, s_2, u, v, x_1, x_2, y_1, y_2) =$$
$$\rho(x_0) \urcorner(s_1, s_2|x_0) P_{UVX_1|S_1}(u, v, x_1|s_1) P_{X_2|US_2}(x_2|u, s_2) \Gamma(y_1, y_2|x_0, x_1, x_2). \tag{15.21}$$

In practice, to plot the utility region, one typically has to solve a convex optimization problem. To be illustrative, we consider the special case of [5], namely, $\mathcal{Y} = \mathcal{X}_1$. Denoting by H the entropy function, the problem of finding the Pareto-frontier of the utility region exactly corresponds to solving the following optimization problem:

---

[2]The notation $I_Q(A; B)$ indicates that the mutual information should be computed with respect to the probability distribution Q.

$$\begin{array}{ll}\text{minimize} & -\sum_{x_0,x_1,x_2} Q(x_0,x_1,x_2)w(x_0,x_1,x_2) \\ \text{subject to} & H_Q(X_0)+H_Q(X_2)-H_Q(X_0,X_1,X_2)\leq 0 \\ & -Q(x_0,x_1,x_2)\leq 0 \\ & -1+\sum_{x_0,x_1,x_2} Q(x_0,x_1,x_2)=0 \\ & -\rho(x_0)+\sum_{x_1,x_2} Q(x_0,x_1,x_2)=0\end{array}.$$

The above problem can be shown to be convex (see [5]). In the next section, we exploit this result to assess the performance gain brought by implementing coordination for distributed power control in wireless networks.

15.4.2 Application to Distributed Power Control

Here, we apply the results of the previous section to the wireless power control problem.

A flat-fading interference channel (IC) with two transmitter–receiver pairs is considered. Transmissions are assumed to be time-slotted and synchronized; the time-slot or stage index is denoted by $t \in \mathbb{N}^*$. For $k \in \{1,2\}$ and "$\ell = -k$" ($-k$ stands for the terminal other than k), the signal-to-noise plus interference ratio (SINR) at Receiver k on a given stage writes as $\text{SINR}_k = \frac{g_{kk}x_k}{\sigma^2+g_{\ell k}x_{-k}}$ where $x_k \in \mathscr{X}_i^{\text{IC}} = \{0, P_{\max}\}$ is the power level chosen by Transmitter k, $g_{k\ell}$ represents the channel gain of link $k\ell$, and σ^2 the noise variance. If Transmitter 1 is fully informed of $x_0 = (g_{11}, g_{12}, g_{21}, g_{22})$ for the next stage and Transmitter 2 has no transmit CSI while both transmitters want to maximize the average of a common stage payoff which is $w^{\text{IC}}(x_0, x_1, x_2) = \sum_{k=1}^{2} f(\text{SINR}_k(x_0, x_1, x_2))$, there may be an incentive for Transmitter 1 to inform Transmitter 2 what to do for the next stage; a typical choice for f is $f(a) = \log(1+a)$. Since Transmitter 1 knows the optimal pair of power levels to be chosen on the next stage, say $(x_1^*, x_2^*) \in \arg\max_{(x_1,x_2)} w(x_0, x_1, x_2)$, a simple coded power control (CPC) policy for Transmitter 1 consists in transmitting on stage t at the level Transmitter 2 should transmit on stage $t+1$. Therefore, if Transmitter 2 is able to observe the actions of Transmitter 1, power levels will be optimally tuned half of the time. Such a simple policy, which will be referred to as semi-coordinated PC (SPC), may outperform (in terms of average payoff) pragmatical PC policies such as the one for which the maximum power level is always chosen by both Transmitters ($(x_1, x_2) = (P_{\max}, P_{\max})$ is the Nash equilibrium of the static game whose individual utilities are $u_k = f(\text{SINR}_k)$).

The channel gain of the link between Transmitter k and Receiver ℓ is assumed to be Bernoulli distributed: $g_{k\ell} \in \{g_{\min}, g_{\max}\}$ is i.i.d. and Bernoulli distributed $g_{k\ell} \sim \mathscr{B}(p_{k\ell})$ with $P(g_{k\ell} = g_{\min}) = p_{k\ell}$. The utility function is either $f(a) = \log(1+a)$ or $f(a) = a$. We define $\text{SNR[dB]} = 10\log_{10}\frac{P_{\max}}{\sigma^2}$ and set $g_{\min} = 0.1$, $g_{\max} = 1.9$,

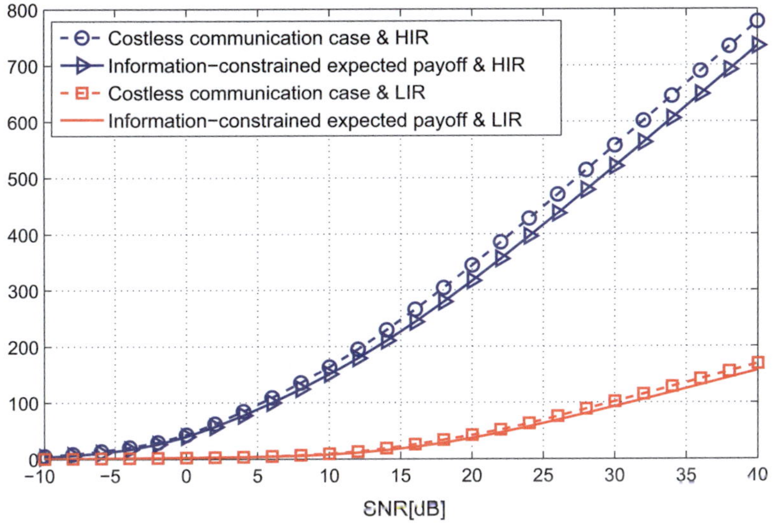

Fig. 15.3 Relative gain in terms of expected payoff ("CPC/FPC - 1" in [%]) vs. SNR[dB] obtained with CPC (with and without communication cost) when the reference power control policy is to transmit at full power (FPC)

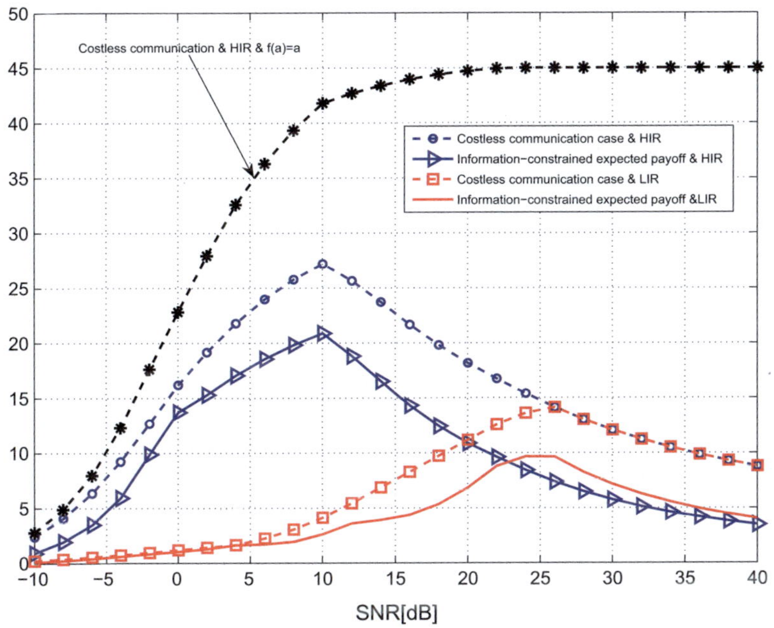

Fig. 15.4 The difference with Fig. 15.3 is that the reference power control policy is the semi-coordinated power control policy (SPC), which is already a CPC policy. Additionally, the top curve is obtained with $f(a) = a$

$\sigma^2 = 1$. The low and high interference regimes (LIR for low interference regime, HIR and for high interference regime) are, respectively, defined by $(p_{11}, p_{12}, p_{21}, p_{22}) = (0.5, 0.9, 0.9, 0.5)$ and $(p_{11}, p_{12}, p_{21}, p_{22}) = (0.5, 0.1, 0.1, 0.5)$. At last, $Y \equiv X_1$ and we define two reference PC policies: full power control (FPC) policy $x_k = P_{\max}$ for every stage; the semi-coordinated PC (SPC) policy $x_2 = P_{\max}$, $x_1^\dagger \in \arg\max_{x_1} w^{\text{IC}}(x_0, x_1, P_{\max})$. Figures 15.3 and 15.4 depict the relative gain in % in terms of average payoff versus SNR[dB] which is obtained by costless optimal coordination and information-constrained coordination. Compared to FPC, gains are very significant whatever the interference regime and provided the SNR has realistic values. Compared to SPC, the gain is of course less impressive since SPC is precisely a coordinated PC scheme but, in the HIR and when the communication cost is negligible, gains as high as 25% can be obtained with $f(a) = \log(1 + a)$ and 45% with $f(a) = a$.

15.5 Conclusion

In this chapter, we have described an information-theoretic framework to characterize the limiting performance of a multiple agent problem. More precisely, the theoretical performance analysis has been conducted in terms of long-term utility region. We have seen that the problem amounts to finding the set of implementable joint distribution over the system state and actions. Both in the scenarios of causal and noncausal state information, auxiliary random variables appear in the characterization of implementable joint distribution. To be able to assess numerically the limiting performance for given utility functions, an optimization problem has to be solved. In the causal state information scenario, the problem is multilinear and the challenge is due to the dimension of the vectors involved. In the noncausal state information scenario, the problem to be solved is a convex problem; more precisely, the information constraint function which translates the agent capabilities in terms of coordination is a convex function of the joint distribution. Note that although the state is not controlled and evolves randomly, the general problem of characterizing the utility region for any number of agents is not trivial. Of course, the problem is even more difficult in the case of controlled states, which therefore constitutes one possible nontrivial extension of the results reported in this chapter. Another interesting research direction would be to consider the case where the state and actions are continuous. A first attempt to this has been made in [1]. Interestingly, the corresponding problem can be shown to be strongly connected to the famous Witsenhausen problem [13, 14], which is a typical decentralized control problem where control and communication intervene in an intricate manner.

References

1. Agrawal, A., Danard, F., Larrousse, B., Lasaulce, S.: Implicit coordination in two-agent team problems. In: European Conference on Control (ECC), Linz, Austria, July 2015
2. Agrawal, A., Lasaulce, S., Beaude, O., Visoz, R.: A framework for decentralized power control with partial channel state information. In: IEEE Proceedings of the Fifth International Conference on Communications and Networking (ComNet 2015), Hammamet, Tunisia, 4–7 Nov 2015
3. Beaude, O., Agrawal, A., Lasaulce, S.: A framework for computing power consumption scheduling functions under uncertainty. In: 6th IEEE International Conference on Smart Grid Communications (SmartGridComm 2015), Miami, Florida, USA, Nov 2015
4. Gossner, O., Hernandez, P., Neyman, A.: Optimal use of communication resources. Econometrica **74**(6), 16031636 (2006)
5. Larrousse, B., Lasaulce, S.: Coded power control: performance analysis. In: IEEE International Symposium on Information Theory (ISIT), Istanbul, Turkey, 2013
6. Larrousse, B., Agrawal, A., Lasaulce, S.: Implicit coordination in two-agent team problems. Application to distributed power allocation. In: IEEE 12th International Symposium on Modeling and Optimization in Mobile, Ad Hoc, and Wireless Networks (WiOpt), Hammamet, Tunisia, May 2014
7. Larrousse, B., Lasaulce, S., Wigger, M.: Coordination in state-dependent distributed networks. In: IEEE Proceedings of the Information Theory Workshop (ITW), Jerusalem, Israel, April–May 2015
8. Larrousse, B., Lasaulce, S., Wigger, M.: Coordination in state-dependent distributed networks: the two-agent case. In: IEEE International Symposium on Information Theory (ISIT), Hong Kong, June 2015
9. Larrousse, B., Lasaulce, S., Bloch, M.: Coordination in distributed networks via coded actions with application to power control. In: IEEE Transactions on Information Theory. https://arxiv.org/pdf/1501.03685.pdf
10. Lasaulce, S., Tembine, H.: Game Theory and Learning for Wireless Networks: Fundamentals and Applications, pp. 1–336. Academic Press, Elsevier (2011). ISBN 978-0123846983
11. Le Treust, M.: Joint empirical coordination of source and channel. In: IEEE Transactions on Information Theory. https://arxiv.org/pdf/1406.4077.pdf
12. Mondrere, D.: Potential games. Games Econ. Behav. **14**, 124–143 (1996)
13. Sahai, A., Grover, P.: Demystifying the Witsenhausen counterexample. IEEE Control Syst. **30**, 20–24 (2010)
14. Witsenhausen, H.S.: A counterexample in stochastic optimum control. SIAM J. Control **6**, 131–147 (1968)

Chapter 16
Networked Hybrid Dynamical Systems: Models, Specifications, and Tools

R. G. Sanfelice

Abstract Models, specifications, and tools for networked hybrid dynamical systems are presented. The proposed modeling framework allows the agent, the network, and the algorithms to have hybrid dynamics. Notions that properly capture key specifications for networked systems, namely, formation, synchronization, safety, and security, are provided. Tools for analysis of the closed-loop hybrid system and for the design of distributed hybrid algorithms are presented. Applications of the methods to estimation, consensus, and synchronization over complex networks are presented throughout the chapter.

16.1 Introduction

The objective of this paper is to present mathematical models, specifications, notions, and tools for the design of algorithms for networked hybrid dynamical systems. A network of such systems is defined as multiple agents running algorithms that are allowed to share information over a network so as to fulfill a given design specification. The mathematical models of the agents, the algorithms, and the network are all given in terms of hybrid inclusions. In the autonomous case, a hybrid inclusion is given by

$$\begin{aligned} \dot{x} &\in F(x) \qquad x \in C \\ x^+ &\in G(x) \qquad x \in D, \end{aligned} \tag{16.1}$$

where x is the state. This model allows the state to change continuously according to the constrained differential inclusion in (16.1) during flows and, at jumps, change discretely according to the constrained difference inclusion in (16.1). With such a general model, the agents may have states that evolve continuously and discretely,

R. G. Sanfelice (✉)
Department of Computer Engineering, University of California,
Santa Cruz, CA 95064, USA
e-mail: ricardo@ucsc.edu

S. Tarbouriech et al. (eds.), *Control Subject to Computational and Communication Constraints*, Lecture Notes in Control and Information Sciences 475, https://doi.org/10.1007/978-3-319-78449-6_16

the models of the algorithms can have logic statements and conditions under which their response changes, and the models of the network may capture the conditions triggering communication events for agents to exchange information over the network. Due to the combination of heterogeneous continuous and discrete dynamics, the analysis of the resulting system as well as the design of algorithms and system parameters to satisfy a particular design specification cannot be carried out with tools for purely continuous-time or discrete-time systems.

Several unique features of networked hybrid dynamical systems make their analysis and design challenging. These features include the unavoidable effect of the network, which, in most cases, does not allow continuous exchange of information, perturbations, and the inherently hybrid dynamics of the agents. More precisely:

1. *Distributed agents with hybrid dynamics:* the intervals of time over which the state of the agents changes continuously may be different among agents. The time instants at which the state of the agents change discretely may also not be the same. In fact, the assumption that all of the agents flow and jump at the same time might be too restrictive.
2. *Asynchronous communication events at unknown times:* the time instances at which agents exchange information may not be synchronized, meaning that each agent may receive information at different time instances. Furthermore, as in the previous item, the amount of ordinary time elapsed between communication events for each agent might be different; for instance, an agent can receive information at a much faster rate than others. In addition, the exact times at which information is exchanged may not be known a priori.
3. *Lack of full information at the same time:* The information about the states of the neighboring agents may not be available at the same time. In fact, most realistic models of networks would not provide information continuously, but rather, at isolated time instances. To meet certain design specifications, such a constraint may require algorithms that can cope with limited information, both in terms of its value and the time information is received.
4. *Perturbations in the dynamics, parameters, and measurements:* the lack of knowledge of the actual models of each component of a network of hybrid systems would prevent one from compensating for their effect at the design stage. Designs that are robust to perturbations such as measurement noise, unmodeled dynamics, and delay are mandatory.

Section 16.2 of this paper pertains to modeling of networked hybrid dynamical systems. A mathematical model of each of the agents is introduced first. Each agent is modeled as a hybrid system, similar to (16.1). Such a model is general enough to allow for nonlinear, nonautonomous, set-valued, and heterogeneous dynamics with solutions that evolve continuously and, at times, jump. Hybrid dynamical models capturing the mechanisms behind the networks connecting the agents are presented. The hybrid dynamics in these models allows to capture the discrete nature of communication events in digital networks. These models are also modular to permit their use in the definition of interconnections between the agents in the network, where its topology is defined by a graph. Finally, a general model of hybrid algorithms for the control of the agents is given in terms of hybrid inclusions as in (16.1) as well.

The interconnection between the agents, the networks, and the algorithms defines a closed-loop system that, after appropriate design, is to meet certain given specifications. With such a model at hand, Sect. 16.3 introduces specifications that are of typical interest in networked systems problems. These specifications are given in terms of the dynamical properties of the resulting closed-loop system. The property of all agents converging to a desired relative configuration, typically referred to as *formation*, is introduced as the property that solutions converge (in the limit or in finite time) to the set of points defining the formation. Synchronization is defined as the property that all solutions (or some of its components) converge to each other, property that we define as *asymptotic synchronization*, potentially with stability, which we refer to as *stable synchronization*. In addition, specifications that capture safety and security are also presented.

With the specifications introduced in Sect. 16.3, notions and tools that can be used to satisfy the given specifications are introduced in Sect. 16.4. The notions include asymptotic stability, finite time convergence, forward invariance, and robustness. Due to space constraints, we provide pointers to the literature of hybrid dynamical systems where formal statements and further applications of these notions and tools can be found. These methods have been recently used to solve problems pertaining to certain classes of networked hybrid dynamical systems, specifically, to solve state estimation [28, 29], consensus [53], synchronization [48, 49, 51, 52], and security [54] problems over networks. Section 16.5 provides a summary of some of these applications.

16.2 Networked Hybrid Dynamical Systems

In this section, we introduce a general model of N networked hybrid systems. A graph defines the network structure, in particular, the nodes and the communication links between them. Each node in the graph corresponds to an agent with general hybrid dynamics. The exchange of information between the agents is also modeled as a hybrid system, in particular, to capture the events at which communication events occur. Each agent is controlled by an algorithm that may also be hybrid.

16.2.1 Agents

For each $i \in \mathcal{V} := \{1, 2, \dots, N\}$, the ith agent is modeled as a hybrid system $\mathcal{H}_i^a$ with data $(C_i^a, F_i^a, D_i^a, G_i^a, E_i^a, H_i^a)$ and given by the hybrid inclusion with inputs and outputs

$$\begin{aligned} \dot{z}_i &\in F_i^a(z_i, u_i) & (z_i, u_i) &\in C_i^a \\ z_i^+ &\in G_i^a(z_i, u_i) & (z_i, u_i) &\in D_i^a \\ y_i &\in H_i^a(z_i, u_i) & (z_i, u_i) &\in E_i^a, \end{aligned} \tag{16.2}$$

where $z_i \in \mathbb{R}^{n_i^a}$ is the state, $u_i \in \mathbb{R}^{m_i^a}$ the input, and $y_i \in \mathbb{R}^{p_i^a}$ the output of the ith agent. The set-valued map F_i^a is the flow map capturing the continuous dynamics, and C_i^a defines the flow set on which flows are allowed. The set-valued map G_i^a defines the jump map and models the discrete behavior, and D_i^a defines the jump set, which is where jumps are allowed. The set E_i^a defines the output set. A solution[1] to $\mathcal{H}_i^a$ is given by a pair (ϕ_i, u_i) parametrized by $(t, j) \in \mathbb{R}_{\geqslant 0} \times \mathbb{N}$, where t denotes ordinary time and j denotes jump time. The domain $\text{dom}(\phi_i, u_i) \subset \mathbb{R}_{\geqslant 0} \times \mathbb{N}$ is a hybrid time domain if for every $(T, J) \in \text{dom}(\phi_i, u_i)$, the set $\text{dom}(\phi_i, u_i) \cap ([0, T] \times \{0, 1, \ldots, J\})$ can be written as $\cup_{j=0}^{J}(I_j \times \{j\})$, where $I_j := [t_j, t_{j+1}]$ for a time sequence $0 = t_0 \leqslant t_1 \leqslant t_2 \leqslant \cdots \leqslant t_J \leqslant t_{J+1}$. The t_j's with $j > 0$ define the time instants when the state of the hybrid system jumps and j counts the number of jumps. The set $\mathcal{S}_{\mathcal{H}_i^a}$ contains all maximal solutions to $\mathcal{H}_i^a$, and the set $\mathcal{S}_{\mathcal{H}_i^a}(\xi)$ contains all maximal solutions to $\mathcal{H}_i^a$ with initial condition ξ.

Example 16.1 A widely studied problem in the literature of multi-agent systems is the problem of controlling the state of point-mass systems over a network to reach consensus. In such a case, the dynamics of the agents are simply $\dot{z}_i = u_i$ for each $i \in \mathcal{V}$, where $z_i, u_i \in \mathbb{R}^{n_i^a}$ for some $n_i^a = m_i^a$. Certainly, such dynamics can be modeled as shown in (16.2) by choosing $F_i^a(z_i, u_i) := u_i$, $G_i^a(z_i, u_i)$ arbitrary, $C_i^a = \mathbb{R}^{n_i^a} \times \mathbb{R}^{n_i^a}$, and D_i^a empty. The model in (16.2) also allows to include constraints in the state and the input of each agent. For instance, if the input of the agent is constrained to $|u_i| \leqslant \overline{u}$ for some $\overline{u} > 0$ then the flow set can be defined as $C_i^a = \mathbb{R}^{n_i^a} \times \left\{u_i \in \mathbb{R}^{m_i^a} \ : \ |u_i| \leqslant \overline{u}\right\}$. More interestingly, the model in (16.2) permits capturing agents with point-mass hybrid dynamics, such as

$$\dot{z}_i = u_{i,1} =: F_i^a(z_i, u_i)$$

during flows and

$$z_i^+ = u_{i,2} =: G_i^a(z_i, u_i)$$

at jumps, where $u_i = (u_{i,1}, u_{i,2})$. In such a model, the conditions on the state and the input imposed by the flow and jump sets would determine when the input $u_{i,1}$ affecting the flows is active, and when the input $u_{i,2}$ assigning the state after jumps is active. ∎

[1]A solution to $\mathcal{H}_i^a$ is called maximal if it cannot be extended, i.e., it is not a (proper) truncated version of another solution. It is called complete if its domain is unbounded. A solution is Zeno if it is complete and its domain is bounded in the t direction. A solution is precompact if it is complete and bounded.

Example 16.2 Synchronization of the state of nonlinear continuous-time systems of the form $\dot{z}_i = f_i(z_i, u_i)$ emerges in many problems in science and engineering. Such an agent model is captured by defining $F_i^a(z_i, u_i) := f_i(z_i, u_i)$, $G_i^a(z_i, u_i)$ arbitrary, $C_i^a = \mathbb{R}^{n_i^a} \times \mathbb{R}^{m_i^a}$, and D_i^a empty. More interestingly, the model in (16.2) allows for jumps in the state that can emerge due to hybrid dynamics in the agents themselves. The mathematical models of impulse-coupled oscillators used in the literature to capture the dynamics of populations of fireflies and neurons exhibit such dynamics; see, e.g., [40]. For instance, one such a model consists of a scalar state z_i of each oscillator taking values in the compact set $[0, T]$, where $T > 0$ is a parameter, and that, during flows, increases monotonically toward T. During this regime, the change of z_i is governed by the autonomous system $\dot{z}_i = f_i(z_i)$, and the state z_i is constrained to $[0, T]$. Upon reaching a threshold T, the state z_i self-resets to zero. Furthermore, when agents that are neighbors to the ith agent self-reset their states to zero, they trigger a reset of the state z_i to a value that may depend on the state of the ith agent and of its neighbors. Letting u_i be the input to the ith agent, which is to be assigned to a function of the state of the neighbors so as to externally reset z_i as just described, the change of z_i at Self-Triggered jumps is $z_i^+ = 0$ and at externally triggered jumps as $z_i^+ = g_i(z_i, u_i)$ An agent model as in (16.2) is given by

$$\dot{z}_i = f_i(z_i) =: F_i^a(z_i, u_i) \qquad (z_i, u_i) \in [0, T] \times \mathbb{R}^{m_i^a} =: C_i^a,$$

$$z_i^+ \in G_i^a(z_i, u_i) := \begin{cases} 0 & \text{if } z_i = T,\ u_i \notin D_i^e \\ g_i(z_i, u_i) & \text{if } z_i \in [0, T),\ u_i \in D_i^e \\ \{0, g_i(z_i, u_i)\} & \text{if } z_i = T,\ u_i \in D_i^e \end{cases}$$

$$(z_i, u_i) \in (\{T\} \times \mathbb{R}^{m_i^a}) \cup ([0, T] \times D_i^e) =: D_i^a,$$

$$y_i = z_i =: H_i^a(z_i, u_i) \qquad (z_i, u_i) \in [0, T] \times \mathbb{R}^{m_i^a} =: E_i^a.$$

In this model, the events are triggered when $z_i = T$ or u_i is equal to a value, or more generally, belong to an appropriately defined set describing the conditions that externally reset z_i. The latter set is denoted as D_i^e in the model above. Note that when both reset conditions occur simultaneously, the jump map of the agent is set valued, meaning that either one of the two possible resets is possible. ■

16.2.2 Networks

A directed graph (digraph) is defined as $\Gamma = (\mathcal{V}, \mathcal{E}, \mathcal{G})$. The set of nodes of the digraph are indexed by the elements of $\mathcal{V}$ and the edges are pairs in the set $\mathcal{E} \subset \mathcal{V} \times \mathcal{V}$. Each edge directly links two different nodes, i.e., an edge from i to k, denoted by (i, k), implies that agent i can send information to agent k. The adjacency matrix of the digraph Γ is denoted by $\mathcal{G} \in \mathbb{R}^{N \times N}$, whose entries g_{ik} take values on $\{0, 1\}$ according to the connectivity map: $g_{ik} = 1$ if $(i, k) \in \mathcal{E}$, and $g_{ik} = 0$ otherwise. The set of indices corresponding to the neighbors that can send information to the ith agent is denoted by $\mathcal{N}(i) := \{k \in \mathcal{V} : (k, i) \in \mathcal{E}\}$. The in-degree and out-degree of agent i are defined by $d_i^{\text{in}} = \sum_{k=1}^{N} g_{ki}$ and $d_i^{\text{out}} = \sum_{k=1}^{N} g_{ik}$. The in-degree matrix $\mathcal{D}$ is the diagonal matrix with entries $D_{ii} = d_i^{\text{in}}$ for all $i \in \mathcal{V}$. The Laplacian matrix of the digraph Γ, denoted by $\mathcal{L} \in \mathbb{R}^{N \times N}$, is defined as $\mathcal{L} = \mathcal{D} - \mathcal{G}$. A digraph is said to be

- *weight balanced* if, at each node $i \in \mathcal{V}$, the out-degree and in-degree are equal; i.e., for each $i \in \mathcal{V}$, $d_i^{\text{out}} = d_i^{\text{in}}$;
- *completely connected* if every pair of distinct vertices is connected by a unique edge; that is, $g_{ik} = 1$ for each $i, k \in \mathcal{V}, i \neq k$;
- *strongly connected* if and only if any two different nodes of the digraph can be connected via a path that traverses the directed edges of the digraph.

In most applications involving networks, the transfer of information between neighboring agents is driven by events. The events triggering communication between neighboring agents may depend on the state, the input, output information, or on a local quantity. The following general hybrid system model, denoted $\mathcal{H}_{ik}^{\text{net}}$, is used to trigger such events for each $(i, k) \in \mathcal{E}$:

$$\begin{aligned} \dot{\mu}_{ik} &\in F_{ik}^{\text{net}}(\mu_{ik}, \omega_{ik}) && (\mu_{ik}, \omega_{ik}) \in C_{ik}^{\text{net}} \\ \mu_{ik}^{+} &\in G_{ik}^{\text{net}}(\mu_{ik}, \omega_{ik}) && (\mu_{ik}, \omega_{ik}) \in D_{ik}^{\text{net}} \\ \chi_{ik} &\in H_{ik}^{\text{net}}(\mu_{ik}, \omega_{ik}) && (\mu_{ik}, \omega_{ik}) \in E_{ik}^{\text{net}}, \end{aligned} \tag{16.3}$$

where $\mu_{ik} \in \mathbb{R}^{n_{ik}^{\text{net}}}$ is a state variable associated to the communication of information from agent i to agent k, $\omega_{ik} \in \mathbb{R}^{m_{ik}^{\text{net}}}$ is its input, which might be assigned to information that agent i has to transmit to agent k as well as state variables in agent i that determine whether μ_{ik} should evolve continuously or discretely, and $\chi_{ik} \in \mathbb{R}^{p_{ik}^{\text{net}}}$ is its output, which includes the information that is transmitted from agent i to agent k. The hybrid model $\mathcal{H}_{ik}^{\text{net}}$ is general enough to capture most communication mechanisms or protocols in the literature. The following sample-and-hold mechanism defines perhaps the simplest version of a model to trigger communication of information from agent i to agent k.

Example 16.3 (*Periodic communication events with memory*) The simplest event-driven communication protocol is perhaps one that collects information and transmits it periodically. Let $T > 0$ denote the period for the events. A model that, after the first event, updates the information provided by the network after every T seconds have elapsed can be modeled as (16.3) for each $(i, k) \in \mathcal{E}$. Let τ_{ik} denote a timer state that triggers the communication events and let ℓ_{ik} be a memory state that stores the information at those events. Then, defining the state of (16.3) as $\mu_{ik} = (\tau_{ik}, \ell_{ik})$, the following model captures the network described above:

$$\begin{aligned}\dot{\mu}_{ik} &= \begin{bmatrix}\dot{\tau}_{ik}\\ \dot{\ell}_{ik}\end{bmatrix} = \begin{bmatrix}1\\0\end{bmatrix} \qquad \text{when } \tau_{ik} \in [0, T]\\ \mu_{ik}^{+} &= \begin{bmatrix}\tau_{ik}^{+}\\ \ell_{ik}^{+}\end{bmatrix} = \begin{bmatrix}0\\ \omega_{ik}\end{bmatrix} \qquad \text{when } \tau_{ik} = T,\end{aligned} \tag{16.4}$$

where ω_{ik} is the input to the network, which has the information to communicate, and the output is $\chi_{ik} = \ell_{ik}$. Then, the data of $\mathscr{H}_{ik}^{\text{net}}$ is given by

$$\begin{aligned}F_{ik}^{\text{net}}(\mu_{ik}, \omega_{ik}) &:= \begin{bmatrix}1\\0\end{bmatrix}\\ C_{ik}^{\text{net}} &:= \{(\mu_{ik}, \omega_{ik}) \ : \ \tau_{ik} \in [0, T]\}\\ G_{ik}^{\text{net}}(\mu_{ik}, \omega_{ik}) &:= \begin{bmatrix}0\\ \omega_{ik}\end{bmatrix}\\ D_{ik}^{\text{net}} &:= \{(\mu_{ik}, \omega_{ik}) \ : \ \tau_{ik} = T\}\\ H_{ik}^{\text{net}}(\mu_{ik}, \omega_{ik}) &:= \ell_{ik}\\ E_{ik}^{\text{net}} &:= \{(\mu_{ik}, \omega_{ik}) \ : \ \tau_{ik} \in [0, T]\}.\end{aligned}$$

A network model in which collection and transmission of information do not occur simultaneously can be obtained by adding a timer and a memory state to the model above. In such a model, one of the timers, denoted as $\tau_{ik,1}$, triggers the events every T_1 seconds, at which events the input ω_{ik} is stored in a memory state, denoted as $\ell_{ik,1}$. The other timer, denoted as $\tau_{ik,2}$, triggers the events every T_2 seconds updating the memory state assigning the output, denoted $\ell_{ik,2}$, to the recorded value of ω_{ik} in $\ell_{ik,2}$. A model as in (16.3) capturing such mechanism has state $\mu_{ik} = (\tau_{ik,1}, \ell_{ik,1}, \tau_{ik,2}, \ell_{ik,2})$ and data

$$\begin{aligned}F_{ik}^{\text{net}}(\mu_{ik}, \omega_{ik}) &:= (1, 0, 1, 0)\\ C_{ik}^{\text{net}} &:= \left\{(\mu_{ik}, \omega_{ik}) \ : \ \tau_{ik,1} \in [0, T_1], \tau_{ik,2} \in [0, T_2]\right\}\\ G_{ik}^{\text{net}}(\mu_{ik}, \omega_{ik}) &:= \begin{cases}(0, \omega_{ik}, \tau_{ik,2}, \ell_{ik,2}) & \text{if } \tau_{ik,1} = T_1, \tau_{ik,2} \in [0, T_2),\\ (\tau_{ik,1}, \ell_{ik,1}, 0, \ell_{ik,1}) & \text{if } \tau_{ik,1} \in [0, T_1), \tau_{ik,2} = T_2,\\ \left\{(0, \omega_{ik}, \tau_{ik,2}, \ell_{ik,2}), (\tau_{ik,1}, \ell_{ik,1}, 0, \ell_{ik,1})\right\} & \\ & \text{if } \tau_{ik,1} = T_1, \tau_{ik,2} = T_2,\end{cases}\\ D_{ik}^{\text{net}} &:= \left\{(\mu_{ik}, \omega_{ik}) \ : \ \tau_{ik,1} = T_1\right\} \cup \left\{(\mu_{ik}, \omega_{ik}) \ : \ \tau_{ik,2} = T_2\right\}\\ H_{ik}^{\text{net}}(\mu_{ik}, \omega_{ik}) &:= \ell_{ik}\\ E_{ik}^{\text{net}} &:= \left\{(\mu_{ik}, \omega_{ik}) \ : \ \tau_{ik,1} \in [0, T_1], \tau_{ik,2} \in [0, T_2]\right\}.\end{aligned}$$

The jump set D_{ik}^{net} of this hybrid system captures the two possible events, which are when $\tau_{ik,1} = T_1$ or when $\tau_{ik,2} = T_2$, and the jump map G_{ik}^{net} resets the state variables according to which even has occurred. Note that when both events happen simultaneously, the jump map is set valued. In general, the parameters T_1 and T_2 in the models above may depend on each agent, in which case they will be denoted as T_1^i and T_2^i. ■

While the models in Example 16.3 capture the key property that information transmitted over networks is typically only available at isolated time instances, they make the assumption that transmissions occur periodically. The following model relaxes that assumption by allowing consecutive communication events to occur within a window of finite length. For simplicity, this extension is carried out for the model in (16.4) and without a memory state. An extension for the case with memory states and two timers follows similarly.

Example 16.4 (*Aperiodic communication events*) The first model in Example 16.3 guarantees that every solution has a hybrid time domain defined by a sequence $t_0 = 0 \leqslant t_1 < t_2 < t_3 < \ldots$ satisfying

$$t_{j+1} - t_j = T$$

for all $j > 0$ such that (t, j) is in the domain of the solution. When the time in between consecutive events is not constant, but rather known to occur no later than T_2 seconds and no sooner than T_1 seconds after every event, the sequence of times $\{t_j\}$ would satisfy

$$t_{j+1} - t_j \in [T_1, T_2] \tag{16.5}$$

for all $j > 0$ such that (t, j) is in the domain of the solution. The parameters T_1 and T_2 are such that $T_2 \geqslant T_1 > 0$. In principle, the event times t_j can be thought of being determined by a random variable taking values in the interval $[T_1, T_2]$. The following model generates solutions satisfying (16.5) by exploiting nondeterministic behavior due to an overlap between the flow and jump sets:

$$\dot{\tau}_{ik} = 1 \qquad \tau_{ik} \in [0, T_2] \tag{16.6a}$$

$$\tau_{ik}^{+} = 0 \qquad \tau_{ik} \in [T_1, T_2], \tag{16.6b}$$

In fact, whenever the timer state τ_{ik} is in $[T_1, T_2)$, both flows and jumps are possible, meaning that there exist solutions that jump or that flow when τ_{ik} is equal to any point in that set. A model as in (16.3) capturing such mechanism has state $\mu_{ik} = \tau_{ik}$, and flow and jump maps/sets given by

$$F_{ik}^{\text{net}}(\mu_{ik}, \omega_{ik}) := 1$$

$$C_{ik}^{\text{net}} := \{(\mu_{ik}, \omega_{ik}) \ : \ \tau_{ik} \in [0, T_2]\}$$

$$\begin{aligned} G_{ik}^{\mathrm{net}}(\mu_{ik}, \omega_{ik}) &:= 0 \\ D_{ik}^{\mathrm{net}} &:= \{(\mu_{ik}, \omega_{ik}) \ : \ \tau_{ik} \in [T_1, T_2]\}. \end{aligned}$$

An alternative model that generates solutions satisfying (16.5) but, instead, through a set-valued jump map is given by

$$\dot{\tau}_{ik} = -1 \qquad \tau_{ik} \in [0, T_2] \tag{16.7a}$$

$$\tau_{ik}^{+} \in [T_1, T_2] \qquad \tau_{ik} = 0. \tag{16.7b}$$

In this model, the communication events are triggered by a timer τ_{ik} that decreases and upon reaching zero, it is reset to a point in $[T_1, T_2]$. The hybrid system (16.7) can be captured by (16.3) by choosing the state as $\mu_{ik} = \tau_{ik}$, and flow and jump maps/sets given by

$$\begin{aligned} F_{ik}^{\mathrm{net}}(\mu_{ik}, \omega_{ik}) &:= -1 \\ C_{ik}^{\mathrm{net}} &:= \{(\mu_{ik}, \omega_{ik}) \ : \ \tau_{ik} \in [0, T_2]\} \\ G_{ik}^{\mathrm{net}}(\mu_{ik}, \omega_{ik}) &:= [T_1, T_2] \\ D_{ik}^{\mathrm{net}} &:= \{(\mu_{ik}, \omega_{ik}) \ : \ \tau_{ik} = 0\}. \end{aligned}$$

■

Network mechanisms and protocols that employ timers, memory states, and logic can be fit in the network model (16.3), for instance, TCP/IP [7], wireless Ethernet, and Bluetooth protocols [64] can be modeled with such a hybrid model.

16.2.3 Algorithms

For each $i \in \mathcal{V}$, the algorithm associated to the ith agent is modeled as a hybrid system $\mathcal{H}_i^K$ with data $(C_i^K, F_i^K, D_i^K, G_i^K, E_i^K, H_i^K)$ and given by the hybrid inclusion with inputs and outputs

$$\begin{aligned} \dot{\eta}_i &\in F_i^K(\eta_i, v_i) & (\eta_i, v_i) &\in C_i^K \\ \eta_i^{+} &\in G_i^K(\eta_i, v_i) & (\eta_i, v_i) &\in D_i^K \\ \zeta_i &\in H_i^K(\eta_i, v_i) & (\eta_i, v_i) &\in E_i^K, \end{aligned} \tag{16.8}$$

where $\eta_i \in \mathbb{R}^{n_i^K}$ is the state, $v_i \in \mathbb{R}^{m_i^K}$ the input, and $\zeta_i \in \mathbb{R}^{p_i^K}$ the output of the ith agent. As for $\mathcal{H}_i^a$ in (16.2), the set-valued map F_i^K is the flow map capturing the continuous dynamics and C_i^K defines the flow set on which flows are allowed. The set-valued map G_i^K defines the jump map and models the discrete behavior, and D_i^K

defines the jump set which is where jumps are allowed. The set E_i^K defines the output set. A solution to the algorithm $\mathcal{H}_i^K$ can also be defined, as done for $\mathcal{H}_i^a$.

Example 16.5 Algorithms that, at isolated time instants, measure the received data and compute a feedback control law that is to be applied to the agent can be modeled as the algorithm in (16.8). An algorithm with sampling events triggered when one of its inputs reaches a particular value, at which event uses the information in another of its inputs to compute the control law and stores it in a memory state that is to be applied to the agent is given as follows. Let $v_i = (v_{i,1}, v_{i,2})$ be the input to the algorithm, where $v_{i,1}$ is the input triggering the events and $v_{i,2}$ is the input with the information needed to compute the control law. Suppose that $v_{i,1}$ reaching zero triggers the computation events. Let ℓ_i be a state variable that, at the events, stores the value of the feedback control law, which is given by the function κ, and in between events remains constant. The discrete dynamics of the algorithm are

$$\ell_i^+ = \kappa(v_{i,2})$$

which are active when $v_{i,1} = 0$. The continuous dynamics of the algorithm are simply

$$\dot{\ell}_i = 0$$

which, in principle,[2] are active when $v_{i,1} > 0$. In this way, the state ℓ_i operates as a memory state. This algorithm is given by $\mathcal{H}_i^K$ as in (16.8) with state $\eta_i = \ell_i$, input $v_i = (v_{i,1}, v_{i,2})$, and data given by

$$\begin{aligned} F_i^K(\eta_i, v_i) &= 0 \\ C_i^K &= \{(\eta_i, v_i) \ : \ v_{i,1} > 0\} \\ G_i^K(\eta_i, v_i) &= \kappa(v_{i,2}) \\ D_i^K &= \{(\eta_i, v_i) \ : \ v_{i,1} = 0\} \\ H_i^K(\eta_i, v_i) &= \ell_i \end{aligned}$$

and E_i^K the entire state and input space of $\mathcal{H}_i^K$.

The model for $\mathcal{H}_i^K$ also allows to capture the dynamics behind an algorithm that does not trigger computations synchronously with the arrival of information. The computation events in such algorithm could be triggered by an internal state, at which events the last piece of information received is used to compute the feedback law. Such a mechanism can be modeled using a memory state that stores the information received, a memory state that stores the computed feedback law, and a state that

[2]In Sect. 16.5.1, we present a model for which the continuous dynamics are active when $v_{i,1} \geqslant 0$ due to such input being connected to a strictly decreasing timer, in which case, flows with $v_{i,1}$ identically zero are not possible. In such a case, the set C_i^K is closed.

triggers the events. Denote these state variables as $\ell_{i,1}$, $\ell_{i,2}$, and τ_i, respectively, which define the state of the algorithm $\eta_i = (\ell_{i,1}, \ell_{i,2}, \tau_i)$. As in the model with a single memory state given above, the memory state $\ell_{i,1}$ stores the information in $v_{i,2}$ when the input $v_{i,1}$ reaches zero. Let $\widetilde{\gamma}$ be a function that, when zero, triggers the computation of the feedback control law, which is denoted as κ. Then, the discrete dynamics of τ_i are active when

$$\widetilde{\gamma}(\eta_i) = 0.$$

At each jump, the discrete dynamics update τ_i according to

$$\tau_i^+ = \rho_i^d(\eta_i),$$

where ρ_i^d is a function to be defined, and $\ell_{i,2}$ according to

$$\ell_{i,2}^+ = \kappa(\ell_{i,1}).$$

We assume that flows of τ_i are active when

$$\widetilde{\gamma}(\eta_i) \geqslant 0$$

and that are governed by

$$\dot{\tau}_i = \rho_i^c(\eta_i).$$

The function ρ_i^c is assumed to not allow flows that remain in $\widetilde{\gamma}(\eta_i) = 0$. The memory state $\ell_{i,2}$ remains constant during flows. This algorithm is given by $\mathscr{H}_i^K$ as in (16.8) with state $\eta_i = (\ell_{i,1}, \ell_{i,2}, \tau_i)$, input $v_i = (v_{i,1}, v_{i,2})$, and data given by

$$\begin{aligned}
F_i^K(\eta_i, v_i) &= (0, 0, \rho_i^c(\eta_i)) \\
C_i^K &= \big\{(\eta_i, v_i) \,:\, v_{i,1} > 0, \widetilde{\gamma}(\eta_i) \geqslant 0\big\} \\
G_i^K(\eta_i, v_i) &= \begin{cases} (v_{i,1}, \ell_{i,2}, \tau_i) & \text{if } v_{i,2} = 0, \widetilde{\gamma}(\eta_i) > 0 \\ (\ell_{i,1}, \kappa(\ell_{i,1}), \rho_i^d(\eta_i)) & \text{if } v_{i,1} > 0, \widetilde{\gamma}(\eta_i) = 0 \\ \{(v_{i,2}, \ell_{i,2}, \tau_i), (\ell_{i,1}, \kappa(\ell_{i,1}), \rho_i^d(\eta_i))\} & \text{if } v_{i,1} = 0, \widetilde{\gamma}(\eta_i) = 0 \end{cases} \\
D_i^K &= \big\{(\eta_i, v_i) \,:\, v_{i,1} = 0\big\} \cup \{(\eta_i, v_i) \,:\, \widetilde{\gamma}(\eta_i) = 0\} \\
H_i^K(\eta_i, v_i) &= \ell_{i,2}
\end{aligned}$$

and E_i^K the entire state and input space of $\mathscr{H}_i^K$. ■

The model in $\mathscr{H}_i^K$ is general enough to allow for multimode, Event-Triggered, and predictive-based algorithms.

16.2.4 Closed-Loop System

Given a digraph Γ, the interconnection between the agents, algorithms, and network models results in a hybrid system. Assuming that, for each $i \in \mathcal{V}$, the input u_i of the ith agent is assigned to the output ζ_i of the ith algorithm, that the input v_i of the ith algorithm is assigned to a function of y_i and of the output of the networks connected to it, namely, $\{\chi_{ki}\}_{k\in\mathcal{N}(i)}$, and that, for each $k \in \mathcal{N}(i)$, the input ω_{ik} is assigned to y_i, the interconnection between these hybrid systems lead to an autonomous hybrid system $\mathcal{H}$ of the form

$$\begin{aligned} \dot{x} &\in F(x) \qquad x \in C \\ x^+ &\in G(x) \qquad x \in D, \end{aligned} \tag{16.9}$$

where

$$x = (x_1, x_2, \ldots, x_N) \in \mathbb{R}^n$$

is the state with $n = \sum_{i\in\mathcal{V}} \left(n_i^a + n_i^K + d_i^{\text{in}} n_{ki}^{\text{net}}\right)$, where x_i collects the states components of the agent, algorithm, and networks associated to the ith agent. The data (C, F, D, G) is constructed using the data of the individual systems. In Sect. 16.5, we provide numerous examples of such construction.

16.3 Design Specifications

In this section, we formulate specific properties of interest in the design of networked systems. The network-specific properties introduced include the situation when the states of the individual systems reach a particular set that depends on the local variables, which we call *formation*, that the states of all systems converge to each other, which is referred to as *synchronization*, that the entire interconnected hybrid system is safe, called *safety*, and that exogenous signals injected at specific agents are detectable, which we refer to as *security*. These properties are given in terms of the variables and inputs of the individual agents.

16.3.1 Formation

A property that is of interest in network system problems is when, for each $i \in \mathcal{V}$, the state x_i converges to a particular relative configuration. For the closed-loop system $\mathcal{H}$, the set of interest is given as

$$\mathcal{A} := \bigcap_{i\in\mathcal{V}} \mathcal{A}_i, \tag{16.10}$$

where

$$\mathscr{A}_i := \left\{x \in \mathbb{R}^n \; : \; \rho_i(x) = 0\right\}$$

and, for each $i \in \{1, 2, \ldots, N\}$, the function ρ_i defines the relative formation between the agent i and the other agents. Convergence of solutions to this set can be interpreted as the network reaching a formation, in particular, when components of x_i are related to physical quantities, such as position or angles. To formulate this property, denote the distance from x to $\mathscr{A}$ as $|x|_{\mathscr{A}}$, namely,

$$|x|_{\mathscr{A}} = \inf_{x' \in \mathscr{A}} |x - x'|.$$

Then, the goal is to design an algorithm such that every maximal solution ϕ to $\mathscr{H}$ converges to $\mathscr{A}$ in finite time or asymptotically, that is, in the limit as "hybrid" time gets large:

- For some $(t^*, j^*) \in \operatorname{dom} \phi$

$$\lim_{(t,j)\in \operatorname{dom} \phi,\; t+j \searrow t^*+j^*} |\phi(t, j)|_{\mathscr{A}} = 0.$$

- If ϕ is complete, then

$$\lim_{(t,j)\in \operatorname{dom} \phi,\; t+j \to \infty} |\phi(t, j)|_{\mathscr{A}} = 0.$$

Note that while in some network systems problems converging to the set $\mathscr{A}$ might be possible without exchanging information between agents, there are numerous problems where transmission of information between agents and algorithms is mandatory. One such a case for $N = 2$ is when the algorithm that controls agent $\mathscr{H}_1^a$ is $\mathscr{H}_2^K$, and the algorithm that controls agent $\mathscr{H}_2^a$ is $\mathscr{H}_1^K$.

Certainly, the construction of the set $\mathscr{A}$ in (16.10) covers the situation when state components of the algorithm for each agent are to converge to a common point, say z^*. In such a case, the definition of the sets $\mathscr{A}_i$ will include the condition $z_i = z^*$ for each $i \in \mathscr{V}$. It also covers the setting when the algorithm reconstructs the state z_i from measurements of y_i, namely, the algorithm includes an observer. In such a case, the set $\mathscr{A}_i$ will include a condition of the form $z_i = \hat{z}_i$, where $\hat{z}_i$ is the component of η_i that provides an estimate of z_i.

16.3.2 Synchronization

Another dynamical property of interest in many network systems problems is when particular components of the solutions to each agent converge to each other, rather than to a particular set or point. For the closed-loop system $\mathscr{H}$, this property is stated as follows. Let $x = (x_1, x_2, \ldots, x_N)$ be partitioned as $x_i = (p_i, q_i)$. The closed-loop system $\mathscr{H}$ is said to have

- *stable synchronization with respect to p* if for every $\varepsilon > 0$, there exists $\delta > 0$ such that every maximal solution $\phi = (\phi_1, \phi_2, \ldots, \phi_N)$, where $\phi_i = (\phi_{i,p}, \phi_{i,q})$, to $\mathscr{H}$ such that
$$|\phi_i(0,0) - \phi_k(0,0)| \leqslant \delta$$
for each $i, k \in \mathscr{V}$ implies
$$|\phi_{i,p}(t,j) - \phi_{k,p}(t,j)| \leqslant \varepsilon$$
for all $i, k \in \mathscr{V}$ and $(t, j) \in \operatorname{dom} \phi$.
- *globally attractive synchronization with respect to p* if every maximal solution is complete, and for each $i, k \in \mathscr{V}$
$$\lim_{\substack{(t,j)\in \operatorname{dom}\phi \\ t+j\to\infty}} |\phi_{i,p}(t,j) - \phi_{k,p}(t,j)| = 0.$$
- *global asymptotic synchronization with respect to p* if it has both stable synchronization and global attractive synchronization with respect to p.

In general, this is a partial state synchronization notion, but if $x_i = p_i$ for each $i \in \mathscr{V}$, then this notion can be considered to be a full-state synchronization notion. Note that stable synchronization with respect to p requires solutions ϕ_i for each $i \in \mathscr{V}$ to start close to each other, while only the components $\phi_{i,p}, i \in \mathscr{V}$ remain close to each other over their domain of definition. Similarly, global attractive synchronization with respect to p only requires that the Euclidean distance between each ϕ_i approaches zero, while the other components are left unconstrained. Also, note that boundedness of the solutions is not required.

16.3.3 Safety

Safety is a property of interest in the design of most algorithms for dynamical systems. Safety is typically characterized by conditions on the system variables, called *safety conditions*, that guarantee system operation within limits and away from undesired configurations. A system is said to be safe when its solutions are such that they remain within the set of points where the safety conditions are satisfied. For each $i \in \mathscr{V}$, let K_i denote the set of points defining the safety conditions for the variables of the ith agent and the set
$$K := K_1 \times K_2 \times \ldots \times K_N$$

be the set that captures all safety conditions for the closed-loop system $\mathcal{H}$. Then, a particular safety goal is to design the algorithms and the networks such that every solution ϕ to $\mathcal{H}$ with initial condition

$$\phi(0,0) \in K$$

is such that

$$\phi(t,j) \in K \qquad \forall (t,j) \in \operatorname{dom} \phi.$$

Note that this property enforces all solutions that start from K to remain in K, even if they are not complete. At times, one might be interested in the property that solutions starting from a potentially smaller set than K, stay in K. More precisely, let K_0 denote the set of allowed initial conditions. Then, such a safety property is as follows: design the algorithms and the networks such that every solution ϕ to $\mathcal{H}$ with initial condition

$$\phi(0,0) \in K_0$$

is such that

$$\phi(t,j) \in K \qquad \forall (t,j) \in \operatorname{dom} \phi,$$

where, in most cases, the set K_0 would be strictly contained in the K.

16.3.4 Security

The general closed-loop system $\mathcal{H}$ allows to model the dynamics of the physical components, such as sensors and actuators, the cyber components, which include digital devices and computing, as well as their interfaces. These interfaces can be exploited by adversaries to, for example, deny access or corrupt the information transmitted among agents. The characterization of which attacks are detectable and the design of algorithms to detect them are of great importance. Modeling the attacks as exogenous signals w_c and w_d affecting the continuous and discrete dynamics of $\mathcal{H}$, respectively, the closed loop under the effect of attacks is given by

$$\begin{aligned} \dot{x} &\in F(x + w_{c,1}) + w_{c,2} \qquad & x + w_{c,3} \in C \\ x^+ &\in G(x + w_{d,1}) + w_{d,2} \qquad & x + w_{d,3} \in D, \end{aligned}$$

where $w_c = (w_{c,1}, w_{c,2}, w_{c,3})$ and $w_d = (w_{d,1}, w_{d,2}, w_{d,3})$. We refer to this closed-loop system as $\mathcal{H}_w$. In this context, the security problem consists of detecting when the exogenous signal $w := (w_c, w_d)$ is nonzero. One way to accomplish that is to design a function that, when evaluated along solutions only, is nonzero if the attacker's input w is nonzero. For instance, one would be interested in designing a function r such that for every solution pair (ϕ, w) to $\mathcal{H}_w$

$$\exists (t, j) \in \mathrm{dom}(\phi, w) \ : \ |w(t, j)| > 0 \qquad \Rightarrow \qquad |r(\phi(t, j))| > 0.$$

Note that when the input w is nonzero over an interval, it might suffice to have a function r that becomes nonzero at some time over that interval, within some reasonable amount of time since the attack started.

16.4 Notions and Design Tools

In this section, we present dynamical properties that are suitable to certify the network-specific properties given in Sect. 16.3. These properties are stated for general hybrid systems given as $\mathscr{H}$ in (16.9) and later, in Sect. 16.5, specialized to networked systems problems. The presentation of these properties and related results is informal, and pointers to the literature with formal statements are given.

16.4.1 Asymptotic Stability

Given a subset of the state space of a dynamical system, asymptotic stability captures the property that solutions starting close to the set stay close to it, and that solutions that are complete converge to it asymptotically. For a hybrid system $\mathscr{H}$ as in (16.9) with state space $\mathbb{R}^n$, a closed set $\mathscr{A} \subset \mathbb{R}^n$ is said to be

- *stable* for $\mathscr{H}$ if for each $\varepsilon > 0$ there exists $\delta > 0$ such that each solution ϕ to $\mathscr{H}$ with initial condition such that

$$|\phi(0, 0)|_{\mathscr{A}} \leqslant \delta$$

  satisfies

$$|\phi(t, j)|_{\mathscr{A}} \leqslant \varepsilon \qquad \forall (t, j) \in \mathrm{dom}\, \phi.$$

- *globally asymptotically attractive* for $\mathscr{H}$ if every maximal solution ϕ to $\mathscr{H}$ is complete[3] and satisfies

$$\lim_{(t,j)\in \mathrm{dom}\, \phi,\ t+j\to\infty} |\phi(t, j)|_{\mathscr{A}} = 0.$$

[3]This attractivity notion enforces that every maximal solution to $\mathscr{H}$ is complete, which is a property that is not for free. Sufficient conditions guaranteeing that maximal solutions are complete are given in [23, Propositions 2.10 and 6.10]. An attractivity notion that does not require every maximal solution to be complete is given in [23, Definitions 3.6 and 7.1], which, to emphasize the potential lack of completeness, has the prefix "pre."

- *globally asymptotically stable* for $\mathcal{H}$ if it is stable and globally asymptotically attractive.

Algorithms $\mathcal{H}_i^K$ for $\mathcal{H}_i^a$ that, under the effect of the networks $\mathcal{H}_{ik}^{\text{net}}$, guarantee asymptotic stability of the set $\mathcal{A}$ can be designed using the Lyapunov stability analysis tools in [23, Chaps. 3 and 7]. In particular, asymptotic stability is of interest to networked systems problems as it can be employed to guarantee formation and synchronization. In fact, the problem of guaranteeing that the network $\mathcal{H}$ asymptotically reaches a formation can be solved by showing that the set $\mathcal{A}$ in (16.10) is asymptotically attractive. The problem of designing algorithms $\mathcal{H}_i^K$ that guarantee full-state asymptotic synchronization of $\mathcal{H}$ can be recast as the problem of asymptotically stabilizing the closed set

$$\mathcal{A} := \left\{ x \in \mathbb{R}^n \ : x_1 = x_2 = \cdots = x_N \right\}.$$

Sufficient conditions for asymptotic stability in terms of Lyapunov functions can be found in [23, Chaps. 3, 6, and 7]; see Sect. 16.5 for illustrations.

16.4.2 Finite Time Convergence

At times, convergence to the set of points of interest in finite time is desired. For instance, in a network system, one might be interested in assuring that the state of the individual systems converge to a particular formation, and after that, accomplish a different task. For a hybrid system $\mathcal{H}$ on $\mathbb{R}^n$, given a closed set $\mathcal{A} \subset \mathbb{R}^n$, an open neighborhood $\mathcal{N}$ of $\mathcal{A}$, and a function $\mathcal{T} : \mathcal{N} \to [0, \infty)$ called the settling-time function, the closed set $\mathcal{A}$ is said to be

- *finite time attractive* for $\mathcal{H}$ if each solution ϕ to $\mathcal{H}$ with initial condition such that
$$\phi(0,0) \in \mathcal{N}$$
satisfies
$$\sup_{(t,j)\in \operatorname{dom}\phi} t + j \geqslant \mathcal{T}(\phi(0,0)) \tag{16.11}$$
and
$$\lim_{(t,j)\in \operatorname{dom}\phi : t+j \nearrow \mathcal{T}(\phi(0,0))} |\phi(t,j)|_{\mathcal{A}} = 0. \tag{16.12}$$

This property becomes global when $\mathcal{N}$ can be picked such that $\overline{C} \cup D \subset \mathcal{N}$. Condition (16.11) assures that convergence occurs at a point in dom ϕ, in turn guaranteeing that the solution actually converges to $\mathcal{A}$.

The design of networked systems with such finite time convergence properties can be designed using the tools in [26], in particular, to guarantee network formation in finite time.

16.4.3 Forward Invariance

A set K is said to be forward invariant for a dynamical system if every solution to the system from K stays in K for all future time. Also referred in the literature as flow-invariance and positively invariance, this property assures the key property of interest in dynamical systems that solutions remain in a desired region of the state space. For a hybrid system $\mathcal{H}$ on $\mathbb{R}^n$, a given set

$$K \subset \overline{C} \cup D$$

is said to be *forward pre-invariant* for $\mathcal{H}$ if for each $x \in K$, each solution ϕ to $\mathcal{H}$ with initial condition $\phi(0, 0) = x$ is such that

$$\phi(t, j) \in K \qquad \forall (t, j) \in \operatorname{dom} \phi.$$

The condition on K belonging to $\overline{C} \cup D$ is so that a set being pre-invariant is such that a solution exists from each point in it.[4] The prefix "pre" indicates that the notion does not enforce that maximal solutions are complete, and when every maximal solution from K is complete, then the notion reduces to forward invariance as defined in [11] – see therein also "weak" notions of forward invariance.

Sufficient conditions guaranteeing forward pre-invariance of sets are given in [11–13] for general hybrid systems modeled as $\mathcal{H}$ in (16.9). In particular, these conditions can be used as design tools to certify safety in a networked system.

16.4.4 Robustness

In real-world settings, networked systems are affected by a variety of perturbations that may compromise the satisfaction of the properties that they were designed for. Unmodeled dynamics in the models used for the agents $\mathcal{H}_i^a$ leads to perturbations of the data $(C_i^a, F_i^a, D_i^a, G_i^a, E_i^a, H_i^a)$. In particular, additive (in the general set-valued sense) perturbations to the flow map F_i^a and the jump map G_i^a can be used to capture terms that were omitted at the modeling stage, potentially with the intention of providing a simplified agent model that would enable analysis and design. Deflations and inflations of the sets C_i^a and D_i^a can be defined to model perturbations in the

[4] The solution might be *trivial* though, in the sense that its domain might be just one point – otherwise, points that are neither in $\overline{C}$ nor in D would satisfy the invariance notion vacuously.

conditions allowing flows and jumps. Similar perturbations may appear in the models of the networks and algorithms. When such perturbations are not known at the design stage, one typically performs the design in nominal conditions, with the expectation that when the perturbations are present and have small size, then the established properties will hold, practically and semiglobally.

A semiglobal and practical (on the size of the perturbation) version of the asymptotic stability property of a set $\mathcal{A}$ defined in Sect. 16.4.1 would guarantee, for every compact set $M \subset \mathbb{R}^n$ and every level of closeness $\varepsilon > 0$, the existence of a maximum allowed perturbation size $\delta^* > 0$ such that every complete solution $\widetilde{\phi}$ to $\mathcal{H}$ under perturbations with size smaller that δ^* that has initial condition $\widetilde{\phi}(0, 0) \in M$ is such that, in the limit as t or j grow unbounded, the distance from $\widetilde{\phi}$ to $\mathcal{A}$ is less than or equal to ε. When $\mathcal{A}$ is an asymptotically stable compact set for $\mathcal{H}$, this property is guaranteed to hold under mild conditions on the data of $\mathcal{H}$. Such a result can be found in Chap. 7 of [23]; see Definition 7.18, Lemma 7.19, and Theorem 7.21 therein.

The property outlined above can be formally written in terms of a $\mathcal{KL}$ bound. First, when $\mathcal{H}$ is nominally well posed and a compact set $\mathcal{A}$ is asymptotically stable, then there exists a class-$\mathcal{KL}$ function β such that every solution ϕ to $\mathcal{H}$ satisfies

$$|\phi(t, j)|_{\mathcal{A}} \leqslant \beta(|\phi(0, 0)|_{\mathcal{A}}, t + j) \qquad \forall (t, j) \in \operatorname{dom} \phi. \tag{16.13}$$

See [23, Theorem 7.12]. Then, [23, Theorem 7.21] implies that for each compact set $M \subset \mathbb{R}^n$ and every level of closeness $\varepsilon > 0$, there exists $\delta^* > 0$ such that

$$|\widetilde{\phi}(t, j)|_{\mathcal{A}} \leqslant \beta(|\phi(0, 0)|_{\mathcal{A}}, t + j) + \varepsilon \qquad \forall (t, j) \in \operatorname{dom} \widetilde{\phi} \tag{16.14}$$

for every solution $\widetilde{\phi}$ that starts from M and that is under the effect of perturbations with size smaller than δ^*. A similar result for the case of finite time convergence is in Theorem 4.1 in [26].

Another typical perturbation in networked systems is the presence of noise in the quantities measured and transmitted by the network, and the values that finally arrive to the agents. When such noise is small, a semiglobal property that is practical on the size of the noise can be established using the tools mentioned above. When the noise is large, one is typically interested in characterizing the effect of the noise on the nominal asymptotic stability property, namely, on the distance to the set $\mathcal{A}$. The notion of input-to-state stability (ISS) is one way to characterize the effect of large noise. Denote by $\widetilde{\mathcal{H}}$ as the hybrid system under the effect of an exogenous disturbance d. The hybrid system $\widetilde{\mathcal{H}}$ is input-to-state stable with respect to $\mathcal{A}$ if there exist $\beta \in \mathcal{KL}$ and $\kappa \in \mathcal{K}$ such that each solution $\widetilde{\phi}$ to $\widetilde{\mathcal{H}}$ with associated disturbance d satisfies

$$|\widetilde{\phi}(t, j)|_{\mathcal{A}} \leqslant \max\{\beta(|\phi(0, 0)|_{\mathcal{A}}, t + j), \kappa(\|d\|_{(t,j)})\} \tag{16.15}$$

for each[5] $(t, j) \in \text{dom}\,\widetilde{\phi}$. Several characterizations and Lyapunov-based tools to certify ISS in hybrid systems are given in [8]. The set $\mathcal{KL}$ is the set of $\mathcal{KL}$ functions and $\mathcal{K}$ is the set of $\mathcal{K}$ functions; see [23, Sect. 3.5].

A direct approach to design algorithms conferring robustness is to perform the design task using a model that explicitly includes a model of the perturbations. Such an approach allows for a variety of perturbations, as long as they can be modeled and a certificate guaranteeing the desired properties can be found. When robust asymptotic stability is of interest, robust control Lyapunov functions for hybrid systems can be employed to guarantee robust asymptotic stability of sets; see [56]. Forward invariance of sets with robustness to perturbations can be certified for hybrid dynamical systems when the model includes the perturbations. Tools for the design of algorithms conferring robust forward invariance to general sets and, in particular, to sets given by sublevel sets of Lyapunov functions are available in [13].

Delay is a perturbation that is of particular interest in networked systems as it is unavoidable in real-world settings. Compared to the tools to deal with the sources of perturbations mentioned above, design methods to guarantee robustness to delays are much less developed. Works pertaining to systems with hybrid dynamics and delays have focused on guaranteeing pre-asymptotic stability through the use of Razumikhin functions [37, 67] and Lyapunov functionals for retarded functional impulse differential equations [63]. Results for switched systems with delays are also available in [14, 33, 62, 66]. Results for linear reset systems with delays developed using passivity appeared in [4, 5]. Tools for the study of delays in hybrid systems modeled as in (16.9) have recently appeared in the sequence of articles [30–32, 34], which provide tools to study the effects of general delays. Along a different vein, in [2], we have recently proposed a way to exploit well posedness and a $\mathcal{KL}$ bound as in (16.13) to handle the sole effect of delays on events in hybrid systems.

16.5 Applications

The models and tools presented in the previous sections have recently been used to solve problems pertaining to certain classes of networked hybrid dynamical systems. In [28, 29], a distributed hybrid observer to estimate the state of a linear time-invariant system was designed to guarantee asymptotic stability of a set on which the state estimation error is zero. In [53], a solution to the control problem of steering the state of the agents with point-mass dynamics to the same value over a network that only allows exchange of information at isolated, aperiodic time instances is proposed. The

[5]A pair $(\widetilde{\phi}, d)$ defines a solution to $\widetilde{\mathcal{H}}$ if it satisfies its dynamics. Given a hybrid arc d, its sup norm at $(t, j) \in \text{dom}\, d$ is

$$\|d\|_{(t,j)} := \max\left\{ ess\ sup_{(s,k)\in \text{dom}\, d\setminus\Gamma(d), s+k\leqslant t+j}|d(s,k)|, \sup_{(s,k)\in\Gamma(d), s+k\leqslant t+j}|d(s,k)| \right\}$$

where $\Gamma(d)$ denotes the set of all $(t, j) \in \text{dom}\, d$ such that $(t, j + 1) \in \text{dom}\, d$.

algorithm discretely updates the input to the point-mass system at communication events and, in between events, changes continuously according to a linear differential equation. A hybrid control algorithm for the synchronization of multiple systems with general linear time-invariant dynamics over a similar communication network appeared in [50, 51]. The remainder of this section presents a summary of these results.

16.5.1 Distributed Estimation

State estimation in networked systems has seen increased attention recently. These include continuous-time algorithms for distributed estimation of the state of a plant in [27] with robustness guarantees and in [25, 65], both when information is exchanged continuously. Algorithms for which information arrives at common discrete time instances include the network of local observers proposed in [47] for linear time-invariant plants and the optimal estimators in [68], for time-varying networked systems, both in discrete time and with information shared at each discrete time instant. Approaches that keep the continuous dynamics of the plant and treat the communication events as impulsive events include the observer-based controller [42] for network control systems modeled as time-varying hybrid systems, the observer-protocol pair in [16] to asymptotically reconstruct the state of a linear time-invariant plant using periodic measurements from multiple sensors, the distributed observer in [19] designed by partitioning the dynamics into disjoint areas and attaching an algorithm to each area that updates the estimates over time windows with common length, and the robust continuous-time observer for estimation in network control systems in [55] designed via an emulation-like approach and exploiting trajectory-based and small-gain arguments. Other approaches that mix continuous and discrete dynamics have appeared in the nonlinear and stochastic systems literature; see [1, 10, 15, 18, 20, 39, 43, 59, 60].

In this section, we consider the problem of estimating the state of a dynamical system from intermittent measurements of functions of its output over a network with N' nodes, each running a decentralized state estimator. The communication events occur according to one of the models in Example 16.4. Under nominal conditions, the model governing the dynamics of the system to estimate the state of is given by a linear time-invariant system. The algorithm we propose builds from the hybrid observer in [22], which is shown to guarantee global exponential stability of the zero-estimation error under sporadic measurements. Without loss of generality, following the model in (16.2) and defining $N = N' + 1$, we assume that the first agent corresponds to this dynamical system, while the dynamics of agents with $i \in \mathcal{V}' := \{2, 3, \ldots, N' + 1\}$ implement the decentralized state estimators. In this way, the dynamics of the first agent are given by

$$\dot{z}_1 = A z_1, \tag{16.16}$$

where $z_1 \in \mathbb{R}^{n_1^a}$ denotes its state and $A \in \mathbb{R}^{n_1^a \times n_1^a}$ is the system matrix. For each $i \in \mathscr{V}'$, the ith agent running a state estimator receives the measurement

$$y_i = H_i z_1 \tag{16.17}$$

and the outputs y_k of its neighbors, that is, for each $k \in \mathscr{N}(i)$, at time instances t_j^i satisfying

$$t_{j+1}^i - t_j^i \in [T_1^i, T_2^i], \qquad j > 0, \tag{16.18}$$

where $H_i \in \mathbb{R}^{p_i^a \times n_i^a}$ is the local output matrix of the ith agent and $T_2^i \geqslant T_1^i > 0$ are parameters that, as T_2 and T_1 in (16.5), determine the minimum and maximum amount of time to elapse between communication events for the ith agent. Following the network models proposed in Sect. 16.2.2, in particular, those in Example 16.4, we employ the model

$$\dot{\tau}_i = 1 \qquad \tau_i \in [0, T_2^i] \tag{16.19a}$$

$$\tau_i^+ \in [T_1^i, T_2^i] \qquad \tau_i = 0 \tag{16.19b}$$

to trigger the events at which the ith agent receives y_i and the y_k's. Since the information from all neighbors to agent i arrives simultaneously, we can employ a single state μ_i for each agent, rather than d_i^{in} states μ_{ik} for each agent, $i \in \mathscr{V}'$. A model as in (16.3) can be derived following the construction in Example 16.4, where the state μ_i would be given by τ_i and the input ω_i by the information to transmit, namely, y_i and the y_k's.

We propose a decentralized hybrid algorithm that, at each agent and by employing information received from the neighbors over a communication graph, generates a converging estimate of the state of the first agent. More precisely, at the ith agent, $i \in \mathscr{V}'$, the hybrid algorithm has a state with a variable $\hat{z}_i \in \mathbb{R}^{n_1^a}$ storing the estimate of the state z_1 and an information fusion state variable, denoted ℓ_i, storing the measurements received from its neighbors. These state variables are continuously updated by differential equations

$$\dot{\hat{z}}_i = A\hat{z}_i + \ell_i \tag{16.20a}$$

$$\dot{\ell}_i = h_i \ell_i \tag{16.20b}$$

when no information is received, while when information is received, the states $\hat{z}_i$ and ℓ_i are updated according to

$$\hat{z}_i^+ = \hat{z}_i \tag{16.21a}$$

$$\ell_i^+ = \sum_{k \in \mathscr{N}(i)} G_{oi}^k(\hat{z}_i, \hat{z}_k, y_i, y_k) \tag{16.21b}$$

with

$$G^k_{oi}(\hat{z}_i, \hat{z}_k, y_i, y_k) = \frac{1}{d^{\text{in}_i}} K_{ii} y^e_i + K_{ik} y^e_k + \gamma(\hat{z}_i - \hat{z}_k), \tag{16.22}$$

where, for each $i, k \in \mathcal{V}'$, $y^e_i = H_i \hat{z}_i - y_i$ is the output estimation error; the scalars h_i and γ, and the matrix K_{ik} define the parameters of the algorithm. The constants g_{ik} in (16.21) and d^{in}_i in (16.22) are associated with the communication graph, which is assumed to be given. The map G^k_{oi} defines the impulsive update law when new information is collected from the first agent and the kth neighbor for agent i. The information fusion state ℓ_i is injected into the continuous dynamics of the local estimate $\hat{z}_i$ and, at communication events, injects new information impulsively – the right-hand side of (16.21) is the "innovation term" of the proposed observer. The specific update law in (16.22) is such that the second term in (16.22) uses the output error of each kth agent that is a neighborhood of the ith agent, and the third term in (16.22) uses the difference between the estimates $\hat{z}_i$ and $\hat{z}_k$. These are the quantities that are transmitted (instantaneously) at communication events only.

The continuous and discrete dynamics in (16.20) and (16.21) can be modeled as a hybrid algorithm $\mathcal{H}^K_i$ as in (16.8) with state $\eta_i = (\hat{z}_i, \ell_i)$, input $v_i = (y_i, \{(\hat{z}_k, y_k)\}_{k \in \mathcal{N}(i)}, \mu_i)$, and data given by

$$F^K_i(\eta_i, v_i) := \begin{bmatrix} A\hat{z}_i + \ell_i \\ h_i \ell_i \end{bmatrix}$$

$$C^K_i := \{(\eta_i, v_i) \ : \ \mu_i \in [0, T_2]\}$$

$$G^K_i(\eta_i, v_i) := \begin{bmatrix} \hat{z}_i \\ \sum_{k \in \mathcal{N}(i)} \frac{1}{d^{\text{in}}_i} K_{ii}(H_i \hat{z}_i - y_i) + K_{ik}(H_k \hat{z}_k - y_k) + \gamma(\hat{z}_i - \hat{z}_k) \end{bmatrix}$$

$$D^K_i := \{(\eta_i, v_i) \ : \ \mu_i = 0\}$$

$$H^K_i(\eta_i, v_i) := \hat{z}_i$$

and E^K_i the entire state and input space of $\mathcal{H}^K_i$. Note that the input to the algorithm includes the output $\mu_i = x_i$ of the network model in (16.7). Due to μ_i triggering the jumps in $\mathcal{H}^K_i$, for each $i \in \mathcal{V}'$, jumps of the network and the hybrid algorithm for the ith agent occur simultaneously.

The goal of the algorithm is to guarantee that, for each $i \in \mathcal{V}'$, the estimate $\hat{z}_i$ converges to the state z_1. When the estimates are equal to z_1, the update law maps ℓ_i to zero. Noting that the timers τ_i $(= \mu_i)$ in the model of the network remain within the set $[0, T^i_2]$, the goal of the algorithm is to render the set

$$\mathcal{A} := \left\{x \ : \ z_1 = \hat{z}_i, \mu_i \in [0, T^i_2], \ell_i = 0 \ \ \forall i \in \mathcal{V}\right\} \tag{16.23}$$

for the resulting closed-loop system with state x, which is given by the stack of the state variables of the first agent (z_1), each algorithm (η_i's), and each network (μ_i's).

A result for the design of the parameters of the proposed hybrid algorithm can be found in [29] to guarantee that the set $\mathscr{A}$ in (16.23) is globally exponentially stable for the closed-loop hybrid system $\mathscr{H}$. Given the network parameters $0 < T_1^i \leqslant T_2^i$ for each $i \in \mathscr{V}'$, it is assumed that the N' agents are connected via a digraph $\Gamma = (\mathscr{V}, \mathscr{E}, \mathscr{G})$ that is such that there exist a constant $\delta > 0$ and matrices K_g, $P = P^\top > 0$, $Q_i = Q_i^\top > 0$ satisfying[6]

$$\mathscr{M}(\tau) := \begin{bmatrix} \mathrm{He}(A_\theta, P) & -P + \widetilde{A}_\theta^\top \mathscr{K}^\top \widetilde{Q}(\tau) \\ \star & -\delta \widetilde{Q}(\tau) - \mathrm{He}(\widetilde{\mathscr{K}}, \widetilde{Q}(\tau)) \end{bmatrix} < 0 \quad \forall \tau \in \mathscr{T}, \tag{16.24}$$

where $\tau = (\tau_2, \tau_3, \ldots, \tau_N)$, $\mathscr{T} := [0, T_2^2] \times [0, T_2^3] \times \cdots \times [0, T_2^N]$,

$$\begin{aligned} A_\theta &= I_N \otimes A + \mathscr{K} \\ \mathscr{K} &= (K_g H_g) * (I_N + \mathscr{G}) + \gamma \mathscr{L} \otimes I_n \\ H_g &= \mathrm{diag}(H_2, H_3, \ldots, H_N) \\ \widetilde{A}_\theta &= A_\theta - \widetilde{H} \\ \widetilde{\mathscr{K}} &= \mathscr{K} - \widetilde{H} \\ \widetilde{H} &= \mathrm{diag}(h_2 I_n, h_3 I_n, \ldots, h_N I_n) \\ \widetilde{Q}(\tau) &= \mathrm{diag}\left(\widetilde{Q}_2(\tau_2), \widetilde{Q}_3(\tau_3), \ldots, \widetilde{Q}_N(\tau_N)\right) \\ \widetilde{Q}_i(\tau_i) &= \exp(\delta \tau_i) Q_i. \end{aligned}$$

These design conditions are obtained using sufficient conditions for asymptotic stability in [23] (specifically, Proposition 3.29 therein), which for the current data turns out to be exponential, and a convenient change of coordinates. The Lyapunov function used to show global exponential stability of the set $\mathscr{A}$ in (16.23) is given by

$$V(x) := e^\top P e + \theta^\top \widetilde{Q}(\tau)\theta,$$

where $e = (e_2, e_3, \ldots, e_N), e_i = \hat{z}_i - z_1, \tau = (\tau_2, \tau_3, \ldots, \tau_N), \theta = (\theta_2, \theta_3, \ldots, \theta_N)$, and

$$\theta_i = K_{ii} y_i^e + \sum_{k \in \mathscr{N}(i)} K_{ik} y_k^e + \gamma \sum_{k \in \mathscr{N}(i)} (\hat{z}_i - \hat{z}_k) - \ell_i \tag{16.25}$$

for each $i \in \mathscr{V}'$, with P and $\widetilde{Q}$ as defined above. Note that $V(x) = 0$ for each $x \in \mathscr{A}$, while for any $x \notin \mathscr{A}$, $V(x)$ is positive. More importantly, intuitively, regardless of

[6]Given matrices A and B, $\mathrm{He}(A, B) = A^\top B + B^\top A$, $A \otimes B$ defines the Kronecker product, and $A * B$ the Khatri–Rao product. The matrix I_n is the $n \times n$ identity matrix.

which timer triggers a jump, this function satisfies the useful property that $V(x^+) - V(x)$ is upper bounded by a nonpositive function of θ_i for all x in the jump set. Such a property is possible due to the convenient choice of the update law of the observer used at jumps, which, in the coordinates in (16.25), leads to e being mapped by the identity and θ_i to zero. The injection of ℓ_i in the flows of the local estimate in (16.20) and the continuous dynamics of ℓ_i further permit a decrease of V during flows, which conveniently uses exponential functions in the definition of $\widetilde{Q}$. These properties are exploited to arrive to the result above. The interested reader is referred to [29], where in addition to several other results pertaining to design, nominal and ISS-type robustness of the above algorithm, several examples are provided.

16.5.2 Distributed Synchronization

Synchronization is a property of interest in many problems emerging in science and engineering, such as spiking neurons [41, 48], formation control and flocking [21, 44], distributed sensor networks [45], and satellite constellation formation [57], among others. The literature about synchronization is quite rich, with numerous contributions employing a variety of techniques, such as Lyapunov functions [6, 24], convergence [46, 58], contraction theory [61], and incremental input-to-state stability [3, 9]. Synchronization for continuous-time systems where communication coupling occurs at discrete events is an emergent area of study. In [9], the authors study a case of synchronization where agents have nonlinear continuous-time dynamics with continuous coupling and impulsive perturbations. In [38], the authors use Lyapunov-like analysis to derive sufficient conditions for the synchronization of continuously coupled nonlinear systems with impulsive resets on the difference between neighboring agents. In [36], a distributed Event-Triggered control strategy was developed to drive the outputs of the agents in a network to synchronization. Using a sample-and-hold Self-Triggered controller policy, a practical synchronization result was established in [17] for the case of first-order integrator dynamics. On the other hand, methods for the design of algorithms that guarantee synchronization of multi-agent systems with information arriving at impulsive, asynchronous time instances are not available.

In this section, we consider the problem of synchronizing the state of N networked agents from intermittent measurements of the state (or of a function of it) over a digraph. Each agent runs a decentralized hybrid algorithm that uses information received from its neighbors. The nominal model of the agents is given as follows: for each $i \in \mathcal{V}$,

$$\dot{z}_i = Az_i + Bu_i, \tag{16.26}$$

where A is the nominal system matrix and B is the input matrix. The ith agent in the network measures its local output, denoted y_i, and the information received from its neighbors, denoted y_k, at the communication events, where

$$y_i = Hz_i \tag{16.27}$$

with H being the output matrix. Following the network models proposed in Sect. 16.2.2, in particular, those in Example 16.4, we employ the model in (16.7) to trigger the events at which the ith agent receives the y_k's.

To globally synchronize the states of the N agents, we propose the following decentralized hybrid algorithm for each $i \in \mathscr{V}$: the algorithm has a memory state, denoted ℓ_i, that when information arrives, is updated to the relative error between the output of the ith agent and those received from its neighbors, namely,

$$\ell_i^+ = K \sum_{k \in \mathscr{N}(i)} (y_i - y_k) = KH \sum_{k \in \mathscr{N}(i)} (z_i - z_k), \tag{16.28}$$

where K is a constant matrix to be designed, and in between communication events is continuously updated according to

$$\dot{\ell}_i = M\ell_i, \tag{16.29}$$

where M is a constant matrix to be designed. Following the construction of the hybrid algorithms in Sect. 16.5.1, this algorithm can be modeled as $\mathscr{H}_i^K$ in (16.8) with state $\eta_i = \ell_i$, input $v_i = (y_i, \{y_k\}_{k \in \mathscr{N}(i)}, \mu_i)$, and data given by

$$F_i^K(\eta_i, v_i) := M\ell_i \tag{16.30}$$

$$C_i^K := \left\{(\eta_i, v_i) \ : \ \mu_i \in [0, T_2^i]\right\} \tag{16.31}$$

$$G_i^K(\eta_i, v_i) := KH \sum_{k \in \mathscr{N}(i)} (z_i - z_k) \tag{16.32}$$

$$D_i^K = \{(\eta_i, v_i) \ : \ \mu_i = 0\} \tag{16.33}$$

$$H_i^K(\eta_i, v_i) := \ell_i \tag{16.34}$$

and E_i^K the entire state and input space of $\mathscr{H}_i^K$. Also, note that the input to the algorithm includes the output μ_i of the network model in (16.7), leading to jumps of the network and the hybrid algorithm for the ith agent occurring simultaneously.

The goal of the synchronization algorithm introduced above is to guarantee that, for each $i, k \in \mathscr{V}$, the error between z_i and z_k converges to zero, with stability. These requirements correspond to the notions of stable and attractive synchronization introduced in Sect. 16.3.2. When the estimates are equal to z_1, the update law maps ℓ_i to zero. Noting that when the states of all of the agents coincide we have that the ℓ_i's are reset to zero and that the timers τ_i in the model of the network remain within the set $[0, T_2^i]$, the goal of the algorithm is to render the set

$$\mathscr{A} := \left\{x \ : \ z_i = z_k \ \ \forall i, k \in \mathscr{V}, \mu_i \in [0, T_2^i], \ell_i = 0 \ \ \forall i \in \mathscr{V}\right\} \tag{16.35}$$

globally asymptotically stable for the resulting closed-loop system $\mathscr{H}$ with state x, which is given by the stack of the state variables of each agent (z_i), each algorithm (η_i's), and each network (μ_i's).

Results for the design of the parameters M and K of the proposed hybrid algorithm can be found in [50] to guarantee that the set $\mathscr{A}$ in (16.35) is globally exponentially stable for $\mathscr{H}$, and hence, global exponential synchronization is achieved. Given the network parameters $0 < T_1^i \leqslant T_2^i$ for each $i \in \mathscr{V}$ and a undirected graph Γ, the set $\mathscr{A}$ in (16.35) is globally exponentially stable for the hybrid closed-loop system $\mathscr{H}$ resulting from controlling the agents in (16.26) with hybrid algorithms as in (16.30)–(16.34) over a network modeled as in (16.7) if there exist scalars $\sigma > 0$, $\varepsilon \in (0, 1)$, matrices K and M, and positive definite symmetric matrices P_i, Q_i for each $i \in \mathscr{V}'$, satisfying

$$\mathscr{M}(\nu) := \begin{bmatrix} \mathrm{He}(P, \bar{A}) & -P\bar{B} + \exp(\sigma\nu)(\bar{K}\bar{A} - \bar{M}\bar{K})^\top Q \\ \star & \mathrm{He}(\exp(\sigma\nu)Q, \bar{M} - \bar{K}\bar{B} - \frac{\sigma}{2}I) \end{bmatrix} < 0 \qquad \forall \nu \in [0, \overline{T}], \tag{16.36}$$

where $\bar{A} = I \otimes A + \Lambda \otimes BKH$, $\bar{B} = I \otimes B$, $\bar{M} = I \otimes M$, $\bar{K} = \Lambda \otimes KH$, $\Lambda = \mathrm{diag}(\lambda_2, \lambda_3, \ldots, \lambda_N)$ where λ_i are the nonzero eigenvalues of $\mathscr{L}$, and

$$(1 - \varepsilon)\underline{T} - \frac{\alpha_2 \sigma \overline{T}}{\beta} > 0, \tag{16.37}$$

where $\underline{T} := \min_{i\in\mathscr{V}} T_1^i$, $\overline{T} := \max_{i\in\mathscr{V}} T_2^i$,

$$\beta = -\max_{\nu\in[0,\overline{T}]} \bar{\lambda}(\mathscr{M}(\nu))$$

$$\alpha_2 = \max\{\bar{\lambda}(P), \bar{\lambda}(Q)\exp(\sigma\overline{T})\}.$$

Moreover, every maximal solution ϕ to the closed-loop system satisfies

$$|\phi(t, j)|_{\mathscr{A}} \leqslant \kappa \exp\left(-r(t + j)\right) |\phi(0, 0)|_{\mathscr{A}} \qquad \forall (t, j) \in \mathrm{dom}\,\phi, \tag{16.38}$$

where $\kappa = \sqrt{\frac{\alpha_2}{\alpha_1}} \exp\left(\frac{\beta(1-\varepsilon)\underline{T}}{2\alpha_2}\right)$ and $r = \frac{\beta}{2\alpha_2 N} \min\left\{\varepsilon N, (1 - \varepsilon)\underline{T} - \frac{\alpha_2\sigma\overline{T}}{\beta}\right\}$, and $\alpha_1 = \min\{\underline{\lambda}(P), \underline{\lambda}(Q)\}$.

To arrive at these design conditions, we employed the property that

$$\theta_i = KH \sum_{k\in\mathscr{N}(i)} (z_i - z_k) - \ell_i \tag{16.39}$$

is reset to zero at jumps due to the timer τ_i expiring alone. It follows that the quantity

$$V(x) = \begin{bmatrix} z \\ \theta \end{bmatrix}^\top \bar{\Psi} R(\tau) \bar{\Psi}^\top \begin{bmatrix} z \\ \theta \end{bmatrix}, \tag{16.40}$$

with[7] $\bar{\Psi} = \text{diag}(\widetilde{\Psi} \otimes I_n, \widetilde{\Psi} \otimes I_p)$, where $\widetilde{\Psi} = (\psi_2, \psi_3, \ldots, \psi_N) \in \mathbb{R}^{N \times N-1}$, $\psi_i = (\psi_{i1}, \psi_{i2}, \ldots, \psi_{iN})$ being the orthonormal eigenvector corresponding to the nonzero eigenvalue λ_i of $\mathscr{L}$, $i \in \mathscr{V}$ (furthermore, $\sum_{k=1}^{N} \psi_{ik} = 0$), $R(\tau) = \text{diag}(P, Q \exp(\sigma \bar{\tau}))$, $\bar{\tau} = \frac{1}{N} \sum_{i=1}^{N} \tau_i$, $P = \text{diag}(P_2, P_3, \ldots, P_N)$, and $Q = \text{diag}(Q_2, Q_3, \ldots, Q_N)$, decreases during flows due to (16.36), while at jumps, its potential growth can be dominated by imposing (16.37); cf. the construction of the Lyapunov function in Sect. 16.5.1, where such a Lyapunov function decreases during flows and has a nonpositive change at jumps. To guarantee exponential stability of the synchronization set, the result [23, Proposition 3.29], which uses a balancing condition between jumps and flows to guarantee that solutions converge to the desired set, exploited.

16.6 Final Remarks and Acknowledgments

Hybrid systems models, along with their associated notions and tools, lead to powerful methods for the design of algorithms conferring desired dynamical properties in complex networks. The methods summarized in this book chapter are suitable for settings in which the combination of continuous and discrete behavior is unavoidable, digital networks govern the exchange of information between the agents, information is limited and with uncertainty, and the algorithms are distributed. The proposed networked hybrid systems framework allows for hybrid models at the agent, network, and algorithm level. The applications of the notions and tools to estimation, consensus, and synchronization over networks are just examples of the power of the hybrid systems framework, being the hope that they will inspire the formulation of new notions and tools suitable for networked hybrid systems as well as the solution to challenging applications.

I would like to acknowledge and thank my collaborators who have contributed to the ideas presented in this book chapter. Part of the work presented here was done in collaboration with my Ph.D. students Yuchun Li and Sean Phillips, who, respectively, have lead our research on distributed estimation and distributed synchronization using hybrid systems methods. The distributed estimation strategy and the nondeterministic network model using timers were developed with Francesco Ferrante, Frederic Gouaisbaut, and Sophie Tarbouriech. The formulation of the safety notion was inspired by work with my Ph.D. student Jun Chai. The formulation of the security notion follows our recent work with Sean Phillips, Alessandra Duz, and Fabio Pasqualetti. Part of the work presented here has recently appeared in conference venues and journal publications, and associated papers are available at

[7] A digraph is undirected if and only if the Laplacian is symmetric. The construction of $\widetilde{\Psi}$ is inspired by [35].

https://hybrid.soe.ucsc.edu. I would also like to thank the support received to fund part of this work from the National Science Foundation under CAREER Grant No. ECS-1450484 and Grant No. CNS-1544396, the Air Force Office of Scientific Research under Grant No. FA9550-16-1-0015, as well as CITRIS and the Banatao Institute at the University of California.

References

1. Ahmed-Ali, T., Karafyllis, I., Lamnabhi-Lagarrigue, F.: Global exponential sampled-data observers for nonlinear systems with delayed measurements. Syst. Control Lett. **62**(7), 539–549 (2013)
2. Altın, B., Sanfelice, R.G.: On Robustness of Pre-Asymptotic Stability to Delayed Jumps in Hybrid Systems. In: Proceedings of the American Control Conference, (2018)
3. Angeli, D.: A Lyapunov approach to incremental stability properties. IEEE Trans. Autom. Control **47**(3), 410–421 (2002)
4. Banos, A., Barreiro, A.: Delay-independent stability of reset systems. IEEE Trans. Autom. Control **54**(2), 341–346 (2009)
5. Barreiro, A., Baos, A.: Delay-dependent stability of reset systems. Automatica **46**(1), 216–221 (2010)
6. Belykh, I., Belykh, V., Hasler, M.: Generalized connection graph method for synchronization in asymmetrical networks. Phys. D Nonlinear Phenom. **224**(1–2), 42–51 (2006)
7. Bohacek, S., Hespanha, J., Lee, J., Obraczka, K.: Modeling communication networks using hybrid systems. IEEE/ACM Trans. Netw. **15**(6), 663–672 (2007)
8. Cai, C., Teel, A.R.: Characterizations of input-to-state stability for hybrid systems. Syst. Control Lett. **58**, 47–53 (2009)
9. Cai, S., Zhou, P., Liu, Z.: Synchronization analysis of hybrid-coupled delayed dynamical networks with impulsive effects: a unified synchronization criterion. J. Franklin Inst. **352**(5), 2065–2089 (2015)
10. Cattivelli, F.S., Sayed, A.H.: Diffusion strategies for distributed Kalman filtering and smoothing. IEEE Trans. Autom. Control **55**(9), 2069–2084 (2010)
11. Chai, J., Sanfelice, R.G.: On notions and sufficient conditions for forward invariance of sets for hybrid dynamical systems. In: Proceedings of the 54th IEEE Conference on Decision and Control, pp. 2869–2874 (2015)
12. Chai, J., Sanfelice, R.G.: Results on feedback design for forward invariance of sets in hybrid dynamical systems. In: Proceedings of the 55th IEEE Conference on Decision and Control, pp. 622–627 (2016)
13. Chai, J., Sanfelice, R.G.: On robust forward invariance of sets for hybrid dynamical systems. In: Proceedings of the American Control Conference, pp. 1199–1204 (2017)
14. Chen, W.-H., Zheng, W.X.: Input-to-state stability and integral input-to-state stability of nonlinear impulsive systems with delays. Automatica **45**(6), 1481–1488 (2009)
15. Cortes, J.: Distributed Kriged Kalman filter for spatial estimation. IEEE Trans. Autom. Control **54**(12), 2816–2827 (2009)
16. Dačić, D., Nešić, D.: Observer design for wired linear networked control systems using matrix inequalities. Automatica **44**(11), 2840–2848 (2008)
17. De Persis, C., Frasca, P.: Self-triggered coordination with ternary controllers. IFAC Proc. Vol. **45**(26), 43–48 (2012)
18. Dinh, T.N., Andrieu, V., Nadri, M., Serres, U.: Continuous-discrete time observer design for Lipschitz systems with sampled measurements. IEEE Trans. Autom. Control **60**(3), 787–792 (2015)

19. Dörfler, F., Pasqualetti, F., Bullo, F.: Continuous-time distributed observers with discrete communication. IEEE J. Sel. Top. Signal Process. **7**(2), 296–304 (2013)
20. Farza, M., M'Saad, M., Fall, M.L., Pigeon, E., Gehan, O., Busawon, K.: Continuous-discrete time observers for a class of mimo nonlinear systems. IEEE Trans. Autom. Control **59**(4), 1060–1065 (2014)
21. Fax, J.A., Murray, R.M.: Information flow and cooperative control of vehicle formations. IEEE Trans. Autom. Control **49**(9), 1465–1476 (2004)
22. Ferrante, F., Gouaisbaut, F., Sanfelice, R.G., Tarbouriech, S.: State estimation of linear systems in the presence of sporadic measurements. Automatica **73**, 101–109 (2016)
23. Goebel, R., Sanfelice, R.G., Teel, A.R.: Hybrid Dynamical Systems: Modeling, Stability, and Robustness. Princeton University Press, New Jersey (2012)
24. Hui, Q., Haddad, W.M., Bhat, S.P.: Finite-time semistability theory with applications to consensus protocols in dynamical networks. In: Proceedings of the American Control Conference, pp. 2411–2416 (2007)
25. Kamal, A.T., Farrell, J.A., Roy-Chowdhury, A.K.: Information weighted consensus filters and their application in distributed camera networks. IEEE Trans. Autom. Control **58**(12), 3112–3125 (2013)
26. Li, Y., Sanfelice, R.G.: Results on finite time stability for a class of hybrid systems. In: Proceedings of the American Control Conference, pp. 4263–4268 (2016)
27. Li, Y., Sanfelice, R.G.: Interconnected observers for robust decentralized estimation with performance guarantees and optimized connectivity graph. IEEE Trans. Control Netw. Syst. **3**(1), 1–11 (2016)
28. Li, Y., Phillips, S., Sanfelice, R.G.: On distributed observers for linear time-invariant systems under intermittent information constraints. In: Proceedings of 10th IFAC Symposium on Nonlinear Control Systems, pp. 654–659 (2016)
29. Li, Y., Phillips, S., Sanfelice, R.G.: Robust distributed estimation for linear systems under intermittent information. To appear in IEEE Trans. Autom. Control (2018)
30. Liu, J., Teel, A.R.: Lyapunov-based sufficient conditions for stability of hybrid systems with memory. IEEE Trans. Autom. Control **61**(4), 1057–1062 (2016)
31. Liu, J., Teel, A.R.: Hybrid Dynamical Systems with Finite Memory, pp. 261–273. Springer International Publishing, New York (2016)
32. Liu, J., Teel, A.R.: Invariance principles for hybrid systems with memory. Nonlinear Anal. Hybrid Syst. **21**, 130–138 (2016)
33. Liu, J., Liu, X., Xie, W.-C.: Input-to-state stability of impulsive and switching hybrid systems with time-delay. Automatica **47**(5), 899–908 (2011)
34. Liu, K.-Z., Sun, X.-M.: Razumikhin-type theorems for hybrid system with memory. Automatica **71**, 72–77 (2016)
35. Liu, T., Hill, D.J., Liu, B.: Synchronization of dynamical networks with distributed event-based communication. In: 2012 IEEE 51st IEEE Conference on Decision and Control, pp. 7199–7204 (2012)
36. Liu, T., Cao, M., De Persis, C., Hendrickx, J.M.: Distributed event-triggered control for synchronization of dynamical networks with estimators. IFAC Proc. Vol. **46**(27), 116–121 (2013)
37. Liu, X., Shen, J.: Stability theory of hybrid dynamical systems with time delay. IEEE Trans. Autom. Control **51**(4), 620–625 (2006)
38. Lu, J., Ho, D.W.C., Cao, J.: A unified synchronization criterion for impulsive dynamical networks. Automatica **46**(7), 1215–1221 (2010)
39. Mazenc, F., Andrieu, V., Malisoff, M.: Continuous-Discrete Observers for Time-Varying Nonlinear Systems: A Tutorial on Recent Results, Chapter 25, pp. 181–188 (2015)
40. Mirollo, R.E., Strogatz, S.H.: Synchronization of pulse-coupled biological oscillators. SIAM J. Appl. Math. **50**, 1645–1662 (1990)
41. Murthy, V.N., Fetz, E.E.: Synchronization of neurons during local field potential oscillations in sensorimotor cortex of awake monkeys. J. Neurophysiol. **76**(6), 3968–3982 (1996)
42. Naghshtabrizi, P., Hespanha, J.P.: Designing an observer-based controller for a network control system. In: Proceedings of the 44th IEEE Conference on Decision and Control, pp. 848–853 (2005)

43. Olfati-Saber, R.: Distributed Kalman filter with embedded consensus filters. In: Proceedings of the 44th IEEE Conference on Decision and Control and 2005 European Control Conference, pp. 8179–8184 (2005)
44. Olfati-Saber, R., Murray, R.M.: Graph rigidity and distributed formation stabilization of multi-vehicle systems. In: Proceedings of the 41st IEEE Conference on Decision and Control, 2002, vol. 3, pp. 2965–2971. IEEE (2002)
45. Olfati-Saber, R., Shamma, J.S.: Consensus filters for sensor networks and distributed sensor fusion. In: Proceedings of the Conference on Decision and Control and European Control Conference, pp. 6698–6703. IEEE (2005)
46. Olfati-Saber, R., Fax, J.A., Murray, R.M.: Consensus and cooperation in networked multi-agent systems. Proc. IEEE **95**(1), 215–233 (2007)
47. Park, S., Martins, N.C.: Design of distributed LTI observers for state omniscience. IEEE Trans. Autom. Control **62**(4), 1–16 (2017)
48. Phillips, S., Sanfelice, R.G.: A framework for modeling and analysis of robust stability for spiking neurons. In: Proceedings of the American Control Conference, pp. 1414–1419 (2014)
49. Phillips, S., Sanfelice, R.G.: Synchronization of two linear systems over intermittent communication networks with robustness. In: Proceedings of the IEEE Conference on Decision and Control, pp. 5569–5574 (2015)
50. Phillips, S., Sanfelice, R.G.: Robust synchronization of interconnected linear systems over intermittent communication networks. In: Proceedings of the American Control Conference, pp. 5575–5580 (2016)
51. Phillips, S., Sanfelice, R.G.: On asymptotic synchronization of interconnected hybrid systems with applications. In: Proceedings of the American Control Conference, pp. 2291–2296 (2017)
52. Phillips, S., Sanfelice, R.G., Erwin, R.S.: On the synchronization of two impulsive oscillators under communication constraints. In: Proceedings of the American Control Conference, pp. 2443–2448 (2012)
53. Phillips, S., Li, Y., Sanfelice, R.G.: On distributed intermittent consensus for first-order systems with robustness. In: Proceedings of 10th IFAC Symposium on Nonlinear Control Systems, pp. 146–151 (2016)
54. Phillips, S., Duz, A., Pasqualetti, F., Sanfelice, R.G.: Recurrent attacks in a class of cyber-physical systems: hybrid-control framework for modeling and detection. In: Proceedings of the IEEE Conference on Decision and Control, 1368–1373 (2017)
55. Postoyan, R., Nešić, D.: A framework for the observer design for networked control systems. IEEE Trans. Autom. Control **57**(5), 1309–1314 (2012)
56. Sanfelice, R.G.: Robust asymptotic stabilization of hybrid systems using control Lyapunov functions. In: Proceedings of the 19th International Conference on Hybrid Systems: Computation and Control, pp. 235–244 (2016)
57. Sarlette, A., Sepulchre, R., Leonard, N.: Cooperative attitude synchronization in satellite swarms: a consensus approach. In: Proceedings of the 17th IFAC Symposium on Automatic Control in Aerospace, pp. 223–228 (2007)
58. Scardovi, L., Sepulchre, R.: Synchronization in networks of identical linear systems. Automatica **45**(11), 2557–2562 (2009)
59. Schenato, L., Sinopoli, B., Franceschetti, M., Poolla, K., Sastry, S.S.: Foundations of control and estimation over lossy networks. Proc. IEEE **95**(1), 163–187 (2007)
60. Shi, L., Xie, L., Murray, R.M.: Kalman filtering over a packet-delaying network: a probabilistic approach. Automatica **45**(9), 2134–2140 (2009)
61. Slotine, J.J., Wang, W., El-Rifai, K.: Contraction analysis of synchronization in networks of nonlinearly coupled oscillators. In: Proceedings of the 16th International Symposium on Mathematical Theory of Networks and Systems (2004)
62. Sun, X.-M., Wang, W.: Integral input-to-state stability for hybrid delayed systems with unstable continuous dynamics. Automatica **48**(9), 2359–2364 (2012)
63. Sun, Y., Michel, A.N., Zhai, G.: Stability of discontinuous retarded functional differential equations with applications. IEEE Trans. Autom. Control **50**(8), 1090–1105 (2005)

64. Suri, A., Baillieul, J., Raghunathan, D.V.: Control using feedback over wireless ethernet and bluetooth. Handbook of Networked and Embedded Control Systems, pp. 677–697. Springer, Berlin (2005)
65. Wang, S., Ren, W.: On the consistency and confidence of distributed dynamic state estimation in wireless sensor networks. In: Proceedings of the 54th IEEE Conference on Decision and Control, Osaka, Japan, pp. 3069–3074 (2015)
66. Yan, P., Özbay, H.: Stability analysis of switched time delay systems. SIAM J. Control Optim. **47**(2), 936–949 (2008)
67. Yuan, R., Jing, Z., Chen, L.: Uniform asymptotic stability of hybrid dynamical systems with delay. IEEE Trans. Autom. Control **48**(2), 344–348 (2003)
68. Zhang, W., Yu, L., Feng, G.: Optimal linear estimation for networked systems with communication constraints. Automatica **47**(9), 1992–2000 (2011)

Chapter 17
Stabilization of Linear Hyperbolic Systems of Balance Laws with Measurement Errors

A. Tanwani, C. Prieur and S. Tarbouriech

Abstract This chapter considers the feedback stabilization of partial differential equations described by linear balance laws when the measurements are subjected to disturbances. Compared to our previous work on robust stabilization of linear hyperbolic systems, the presence of source terms in the system description complicates the analysis. We first consider the case of static controllers and provide conditions on system data and feedback gain which result in stability of the closed-loop system, and robustness with respect to measurement errors. Motivated by the applications where it is of interest to bound the maximum norm of the state trajectory, we also study feedback stabilization with dynamic controllers. Conditions in terms of matrix inequalities are proposed which lead to robust stability of the closed-loop system in the presence of measurement errors in the feedback. As an application, we study the problem of quantized control, where the quantization error plays the role of disturbance in the measurements. The simulations for an academic example are reported as an illustration of our theoretical results.

17.1 Introduction

Balance laws are used to describe the physical systems with certain conservative properties, and hyperbolic partial differential equations (PDEs) provide the mathematical framework to model systems governed by such laws. Stability and stabilization of this class of systems are indeed relevant from several applications viewpoint, and several tools are now available for analyzing such properties of hyperbolic systems.

A. Tanwani (✉) · S. Tarbouriech
LAAS – CNRS, University of Toulouse, CNRS, Toulouse, France
e-mail: tanwani@laas.fr

S. Tarbouriech
e-mail: tarbour@laas.fr

C. Prieur
Grenoble INP, GIPSA-lab, Univ. Grenoble Alpes, CNRS, 38000 Grenoble, France
e-mail: christophe.prieur@gipsa-lab.fr

S. Tarbouriech et al. (eds.), *Control Subject to Computational and Communication Constraints*, Lecture Notes in Control and Information Sciences 475, https://doi.org/10.1007/978-3-319-78449-6_17

We encourage the reader to consult [1] for physical examples of hyperbolic PDEs, and an overview of tools used for studying solutions and stability of this system class.

This chapter concerns the problem of feedback stabilization for a class of boundary controlled hyperbolic systems, which model linear balance laws. We are particularly focused on studying a notion of robust stability when the measurements used for feedback control are subjected to unknown disturbances. In the literature on ordinary differential equations (ODEs), the property of *input-to-state stability* (ISS), coined in [15], captures the desired robust behavior that we want to study here, while regarding the disturbances as exogenous inputs in the closed-loop system. The Lyapunov function-based techniques available for verifying ISS are thus generalized in the context of hyperbolic PDEs in this chapter. Our previous work on robust stabilization of hyperbolic PDEs [18, 19] only considers systems with conservation laws and no source term in the dynamics. Whereas, in this chapter, we generalize our results to linear balance laws by including a source term in the PDE. This results in novel stability conditions and calculations.

One finds the Lyapunov stability criteria with $\mathscr{L}^2$-norm and dissipative boundary conditions in [2]. Lyapunov stability in $\mathscr{H}^2$-norm for nonlinear systems is treated in [3]. Thus, the construction of Lyapunov functions in $\mathscr{H}^2$-norm for the hyperbolic PDEs with static control laws can be found in the literature. In the literature, one finds various instances where the ISS-related tools are used for stability analysis of interconnected systems. For infinite-dimensional systems, the problem of ISS has attracted attention recently but most of the existing works treat the problem with respect to uncertainties in the dynamics. See, for example, [11], where a class of linear and bilinear systems is studied. See also [4] where a linearization principle is applied for a class of infinite-dimensional systems in a Banach space. When focusing on parabolic partial differential equations, some works to compute ISS Lyapunov functions have also appeared, such as [9, 10]. For time-varying hyperbolic PDEs, construction of ISS Lyapunov functions has also been addressed in [13]. The recent work reported in [6, 7] derives ISS bounds for 1-D parabolic systems in the presence of boundary disturbances but without the use of Lyapunov-based techniques.

For systems of conservation laws, when seeking robust stabilization with measurement errors, one could see that the results in [5] provide robust stability of $X(\cdot, t)$ in $\mathscr{L}^2((0, 1); \mathbb{R}^n)$ space by using static controllers. The use of dynamic controller for stabilization of systems of conservation laws with ISS estimates in $\mathscr{H}^1$-norm and maximum norm is studied in our previous works [18] and [19], respectively. Inspired by the applications of such notions in finite-dimensional systems [16, 17], these works also discuss the applications of ISS notion in the context of sampled-data and quantized control of hyperbolic systems, which require stability in a functional space equipped with maximum norm. An intermediate exposition of such applications in finite and infinite dimensions appears in [14].

In this chapter, we build on our works [18, 19] dealing with input-to-state stabilization in maximum norm and using dynamic feedbacks. The novelty here appears due to the presence of source terms as we migrate from conservation laws to balance laws. The presence of source terms induces some changes in the criterion for achieving ISS. The system class and the stability notions of our interest are discussed in Sect. 17.2. To clearly highlight the role of the source term, we first deal with the static feedback case and ISS estimates in $\mathscr{L}^2$-norm in Sect. 17.3. We then develop ISS estimates in supremum norm in Sect. 17.4 and carry out the design of a dynamic feedback to achieve that purpose. The applications of these notions in the context of quantized control are discussed in Sect. 17.5. Section 17.6 illustrates the main results of the chapter through an academic example. Finally, some concluding remarks end the chapter in Sect. 17.7.

17.2 System Class and Stability Notions

The problem of interest for us is to address feedback control for the class of linear hyperbolic balance laws described by the equation

$$\frac{\partial X}{\partial t}(z,t) + \Lambda\frac{\partial X}{\partial z}(z,t) = SX(z,t), \tag{17.1a}$$

where $z \in [0,1]$, and $t \in [0,\infty)$. The matrix Λ is assumed to be diagonal and positive definite. The expression $SX(z,t)$ in (17.1a) denotes the source terms. We call $X : [0,1] \times \mathbb{R}_+ \to \mathbb{R}^n$ the state trajectory, and the initial condition is defined as

$$X(z,0) = X^0(z), \quad z \in (0,1) \tag{17.1b}$$

for some function $X^0 : (0,1) \to \mathbb{R}^n$. The value of the state X is controlled at the boundary $z = 0$ through some input $u : \mathbb{R}_+ \to \mathbb{R}^m$ so that

$$X(0,t) = HX(1,t) + Bu(t), \tag{17.2}$$

where $H \in \mathbb{R}^{n\times n}$ and $B \in \mathbb{R}^{n\times m}$ are constant matrices. We consider the case when only the measurement of the state X at the boundary point $z = 1$ is available for each $t \geqslant 0$. We thus denote the output of the system by

$$y(t) = X(1,t) + d(t), \tag{17.3}$$

where $d \in \mathscr{L}^\infty([0,\infty), \mathbb{R}^n)$ is seen as the perturbation in the measurement of the state trajectory at the boundary point.

We are interested in designing a control law u as a function of the output measurement y, which stabilizes the system in some appropriate sense. In case there are no perturbations, that is, $d \equiv 0$, one typically chooses $u(t) = Ky(t)$. Following this recipe with uncertain measurements, we obtain the closed-loop boundary condition

$$X(0,t) = (H + BK)X(1,t) + BKd(t), \tag{17.4}$$

where K is chosen such that $(H + BK)$ satisfies a certain dissipative condition. We are interested in studying the stability of system (17.1)–(17.4) with respect to the measurement disturbance d.

Definition 17.1 System (17.1)–(17.4) is said to be input-to-state stable in $\mathscr{L}^2$-norm ($\mathscr{L}^2$-ISS) with respect to the disturbance d if there exist constants $c, a > 0$ and a class $\mathscr{K}$ function γ such that

$$\|X(z,t)\|_{\mathscr{L}^2((0,1);\mathbb{R}^n)} \leqslant c\,\mathrm{e}^{-at}\|X^0\|_{\mathscr{L}^2((0,1);\mathbb{R}^n)} + \gamma\left(\|d_{[0,t]}\|_\infty\right). \tag{17.5}$$

In Sect. 17.3, we treat this case and propose conditions for choosing feedback gain K, which results in aforementioned stability estimate for the closed-loop system with static control.

Stabilization in $\mathscr{L}^2$-norm does not necessarily guarantee convergence of the maximum norm of $X(\cdot, t)$ over the spatial domain [0, 1]. To do that, we have to consider the stability of $X(\cdot, t)$ in $\mathscr{H}^1$-norm, which is defined as

$$\|X\|_{\mathscr{H}^1((0,1);\mathbb{R}^n)} := (\|X\|^2_{\mathscr{L}^2((0,1);\mathbb{R}^n)} + \|\partial X\|^2_{\mathscr{L}^2((0,1);\mathbb{R}^n)})^{1/2}.$$

The following proposition, proved in [19], allows us to make the connection between $\mathscr{C}^0$-norm and $\mathscr{H}^1$-norm.

Proposition 17.1 *Given any function $X : [0,1] \to \mathbb{R}^n$ such that $X \in \mathscr{C}^0([0,1];\mathbb{R}^n) \cap \mathscr{H}^1((0,1);\mathbb{R}^n)$. It holds that, for every $z \in [0,1]$,*

$$\max_{z\in[0,1]} |X(z)|^2 \leqslant |X(0)|^2 + \|X\|^2_{\mathscr{H}^1((0,1);\mathbb{R}^n)}. \tag{17.6}$$

By definition, functions with finite $\mathscr{H}^1$-norm must be differentiable Lebesgue almost everywhere, and since $d \in \mathscr{L}^\infty$, we can no longer use static feedbacks. The use of dynamic controller allows us to circumvent this problem, see [19] for details. The dynamic controller driven by the output y that we choose for our purposes is described by the following equations:

$$\dot{\eta}(t) = -\alpha(\eta(t) - y(t)) = -\alpha\,\eta(t) + \alpha X(1,t) + \alpha d(t) \tag{17.7a}$$

$$\eta(0) = \eta^0 \tag{17.7b}$$

$$u(t) = K\eta(t), \tag{17.7c}$$

where $\eta^0 \in \mathbb{R}^n$ is the initial condition for the controller dynamics.

Definition 17.2 System (17.1), (17.2), and (17.7) is said to be input-to-state stable in $\mathscr{C}^0$-norm ($\mathscr{C}^0$-ISS) with respect to the disturbance d if there exist constants $c, a > 0$ and a class $\mathscr{K}$ function γ such that

$$\max_{z\in[0,1]} |X(z,t)| \leqslant c\,\mathrm{e}^{-at} M_{X^0,\eta^0} + \gamma\left(\|d_{[0,t]}\|_\infty\right). \tag{17.8}$$

In (17.8), M_{X^0,η^0} is a constant that depends on some norm associated with the function X^0 and the initial state η^0 chosen for the dynamic compensator η. We are interested in designing α, K such that the closed-loop system (17.1), (17.2), and (17.7) is stable in the sense of Definition 17.2.

In terms of analysis, the addition of dynamic controller introduces a coupling of ODEs and PDEs in the closed loop, which makes the analysis more challenging. The derivations of the main results are also more involved compared to [19] due to the presence of source terms. We use Lyapunov function-based analysis to synthesize the controller and guarantee ISS with respect to the perturbation d. After designing controllers which achieve the desired ISS estimates, we study an application of these notions in the context of quantized control: We establish the practical stability of the system and derive the ultimate bounds on the state trajectory in terms of the quantization error. The problem of quantized control has mostly been studied in finite-dimensional systems so far [8, 12, 14, 18], and this chapter extends this problem setting to the case of hyperbolic balance laws.

17.3 Static Control and $\mathscr{L}^2$-Estimates

We first address the problem of finding conditions for the system to be $\mathscr{L}^2$-ISS as formulated in Definition 17.1. In the following, $\mathscr{D}_+^n$ denotes the set of diagonal positive definite matrices.

Theorem 17.1 *If there exist scalars $\kappa \in (0, 1)$, $c < \lambda_{\min}(\Lambda)$, a matrix $D \in \mathscr{D}_+^n$, and a matrix $K \in \mathbb{R}^{m\times n}$, such that*

$$(H + BK)^\top \Lambda D(H + BK) \leqslant \kappa \Lambda D \tag{17.9a}$$

$$S^\top D + DS \leqslant c \log\left(\frac{1}{\kappa}\right) D, \tag{17.9b}$$

then system (17.1) *and* (17.2) *with $u = Ky$ is ISS with respect to the disturbance d.*

Proof The proof is based on introducing a Lyapunov function and analyzes its derivative with respect to time. As a candidate, we choose $V : \mathscr{L}^2((0, 1); \mathbb{R}^n) \to \mathbb{R}_+$ given by

$$V(X) := \int_0^1 X^\top(z) D X(z) e^{-\mu z}\, dz,$$

where D is a diagonal positive definite matrix satisfying (17.9). The constant $\mu > 0$ is chosen such that

$$\frac{c}{\lambda_{\min}(\Lambda)} \log\left(\frac{1}{\kappa}\right) < \mu < \log\left(\frac{1}{\kappa}\right), \tag{17.10}$$

which is possible because $c < \lambda_{\min}(\Lambda)$.

Using an integration by parts, along the solutions to (17.1) and (17.2) with $u = Ky$, the time derivative of V yields

$$\begin{aligned}\dot{V} &= \int_0^1 (\partial_t X^\top DX + X^\top D\partial_t X)\mathrm{e}^{-\mu z}\,dz \\ &= -\int_0^1 (\partial_z X^\top \Lambda DX + X^\top D\Lambda\partial_z X)e^{-\mu z}\,dz + \int_0^1 X^\top (S^\top D + DS)Xe^{-\mu z}\,dz \\ &\leqslant -[X^\top \Lambda DXe^{-\mu z}]_0^1 - \mu\int_0^1 X^\top D\Lambda Xe^{-\mu z}dz + c\log\left(\frac{1}{\kappa}\right)\int_0^1 X^\top DXe^{-\mu z}dz \\ &\leqslant -e^{-\mu}\, X(1,t)^\top \Lambda DX(1,t) + X(0,t)^\top \Lambda DX(0,t) - \sigma V,\end{aligned}$$

where $\sigma := (\mu\lambda_{\min}(\Lambda) - c\log(1/\kappa)) > 0$ due to the first inequality in (17.10). We now substitute the expression for boundary control to get

$$X(0,t) = (H + BK)X(1,t) + BKd(t).$$

Using (17.9a), we get

$$\begin{aligned}\dot{V}(t) &\leqslant -\sigma V(t) - (\mathrm{e}^{-\mu} - \kappa)X^\top(1,t)\Lambda DX(1,t) + d(t)^\top \Lambda Dd(t) \\ &\leqslant -\sigma V(t) + \chi d(t)^\top d(t).\end{aligned}$$

where $\chi = \lambda_{\max}(\Lambda D)$, and $\mathrm{e}^{-\mu} > \kappa$ due to the second inequality in (17.10). The ISS estimate now follows from the last inequality. □

17.4 Dynamic Control and $\mathscr{C}^0$-Estimates

In this section, we are interested in analyzing the closed-loop system (17.1), (17.2), and (17.7). Since we are interested in computing estimates on the $\mathscr{H}^1$-norm of the state X, we need to look at the evolution of its derivative X_z. We recall that, for the closed-loop system with dynamic controller, the state trajectory X satisfies

$$X_t(z,t) + \Lambda X_z(z,t) = SX(z,t), \tag{17.11a}$$

$$X(z,0) = X^0(z),\ \forall z \in [0,1], \tag{17.11b}$$

$$X(0,t) = HX(1,t) + BK\eta(t). \tag{17.11c}$$

For what follows, we are also interested in analyzing the dynamics of $X_z := \partial_z X$ which are derived as follows:

$$\frac{\partial X_z}{\partial t}(z,t) + \Lambda\frac{\partial X_z}{\partial z}(z,t) = SX_z(z,t). \tag{17.12}$$

To obtain the boundary condition for X_z, from (17.11c), we have

$$X_t(0,t) = HX_t(1,t) + BK\dot{\eta}(t).$$

Substituting $X_t(z,t) = -\Lambda X_z(z,t) + SX(z,t)$ for each $z \in [0,1]$, we get

$$\begin{aligned}
\Lambda X_z(0,t) &= H\Lambda X_z(1,t) + (SH - HS)X(1,t) + SBK\eta(t) - BK\,\dot{\eta}(t) \\
&= H\Lambda X_z(1,t) + S(H+BK)X(1,t) + SBK(\eta(t) - X(1,t)) - HSX(1,t) \\
&\quad - BK(\dot{\eta}(t) - X_t(1,t)) + BK(\Lambda X_z(1,t) - SX(1,t)) \\
&= (H+BK)\Lambda X_z(1,t) + \big[S(H+BK) - (H+BK)S\big]X(1,t) \\
&\quad + SBK(\eta(t) - X(1,t)) - BK(\dot{\eta}(t) - X_t(1,t)) \\
&= -(H+BK)X_t(1,t) - BK(\dot{\eta}(t) - X_t(1,t)) + S(H+BK)X(1,t) \\
&\quad + SBK(\eta(t) - X(1,t)),
\end{aligned} \tag{17.13}$$

using (17.11a) for the last equality. We now use these equations as the system description and proceed to develop the next result.

17.4.1 Stability Result

The second main contribution of this chapter is to present conditions on the controller dynamics (17.7) which result in stability of system (17.1) and (17.2), and robustness with respect to the measurement disturbances d in the sense of Definition 17.2. To state the result, we introduce some notation. For scalars $\mu > 0$ and $0 < \kappa < 1$, let $\rho := e^{-\mu} - \kappa$; let $F := BK$, and $Q := F^\top \Lambda DF$ for $D \in \mathscr{D}_+^n$; and finally, let $G := H^\top \Lambda DF$. We introduce the matrices

$$\Omega_1 := \begin{bmatrix} \rho\beta_1\Lambda D & -\beta_1(G+Q) & 0 \\ * & 2\alpha\beta_3 I - (\beta_1 + \alpha^2\beta_2)Q & \beta_3 I + \alpha\beta_2 G \\ * & * & \beta_2(\rho\Lambda D + Q + G + G^\top) \end{bmatrix}$$

and with $\widetilde{S} := S(H+F)$,

$$\Omega_2 := \beta_2 \begin{bmatrix} -\widetilde{S}^\top \Lambda D\widetilde{S} & -\widetilde{S}^\top \Lambda DSF & H^\top \Lambda D\widetilde{S} \\ * & -F^\top(S^\top \Lambda DS + \Lambda DS + S^\top \Lambda D)F & H^\top \Lambda DSF \\ * & * & 0 \end{bmatrix}$$

in which $\alpha, \beta_1, \beta_2, \beta_3$ are some positive constants, and $*$ denotes the transposed matrix block. It must be noted that the matrix $\Omega_2 = 0$ if there is no source term, that is, $S = 0$. In the following statement, we denote the induced Euclidean norm of a matrix M by $\|M\|_2$.

Theorem 17.2 *Assume that there exist scalars* $\mu \in \mathbb{R}_+$, $\kappa \in (0, 1)$, *a matrix* $D \in \mathscr{D}_+^n$, *the gain matrix* K, *and the positive constants* $\alpha, \beta_1, \beta_2, \beta_3$ *in the definitions of* Ω_1 *and* Ω_2 *such that*

$$(H + BK)^\top \Lambda D(H + BK) \leqslant \kappa \Lambda D \tag{17.14a}$$

$$S^\top D + DS < \mu \lambda_{\min}(\Lambda) D, \tag{17.14b}$$

$$S^\top \Lambda^2 D + \Lambda^2 DS \leqslant \mu \lambda_{\min}(\Lambda) \Lambda^2 D, \tag{17.14c}$$

$$\Omega_1 + \Omega_2 > 0. \tag{17.14d}$$

Then, the closed-loop system satisfies the ISS estimate (17.8) *with*

$$M_{X^0,\eta^0} := \|X^0\|^2_{\mathscr{H}^1((0,1);\mathbb{R}^n)} + |\eta^0 - X(1, 0)|^2. \tag{17.15}$$

Remark 17.2 Condition (17.14) provides a generalization of stability conditions that were studied earlier in [18, 19] in the sense that (17.14a) and (17.14d) (with $\Omega_2 = 0$) were already proposed there. Conditions (17.14b) and (17.14c) appear because of the nonzero source term SX. If S is symmetric, then (17.14b) and (17.14c) are equivalent.

The proof of Theorem 17.2 is based on constructing a Lyapunov function for the closed-loop system (17.1), (17.2), and (17.7). Within the remainder of this section, we provide this construction, and the analysis involving the computation of the derivative of this function. The required ISS estimate then follows from the condition (17.14) and the bound on the derivative of the Lyapunov function constructed.

17.4.2 Construction of the Lyapunov Function

As a candidate, we choose $V : \mathscr{H}^1((0, 1); \mathbb{R}^n) \times \mathbb{R}^n \to \mathbb{R}_+$ given by

$$V := V_1 + V_2 + V_3, \tag{17.16}$$

where $V_1 : \mathscr{H}^1((0, 1); \mathbb{R}^n) \to \mathbb{R}_+$ is defined as

$$V_1(X) := \int_0^1 X(z)^\top P_1 X(z) e^{-\mu z}\, dz,$$

where we choose $P_1 := \beta_1 D$, and β_1, D satisfy (17.14). Similarly, $V_2 : \mathscr{H}^1((0, 1); \mathbb{R}^n) \to \mathbb{R}_+$ is given by

$$V_2(X) := \int_0^1 \partial X(z)^\top P_2 \partial X(z) e^{-\mu z}\, dz,$$

with $P_2 = \beta_2 \Lambda^2 D$, and finally $V_3 : \mathscr{H}^1((0,1);\mathbb{R}^n) \times \mathbb{R}^n \to \mathbb{R}_+$ is given by

$$V_3(X, \eta) = (\eta - X(1))^\top P_3(\eta - X(1)),$$

with $P_3 = \beta_3 I$. With this choice of V, we can now introduce the constants $\underline{c}_P := e^{-\mu} \min_{i=1,2,3}\{\lambda_{\min}(P_i)\}$, $\overline{c}_P := \max_{i=1,2,3}\{\lambda_{\max}(P_i)\}$ such that, for all $\eta \in \mathbb{R}^n$, and $X \in \mathscr{H}^1((0,1);\mathbb{R}^n)$,

$$\begin{aligned} \underline{c}_P(\|X\|^2_{\mathscr{H}^1((0,1);\mathbb{R}^n)} + |\eta - X(1)|^2) \leqslant V(X,\eta) \leqslant \\ \overline{c}_P(\|X\|^2_{\mathscr{H}^1((0,1);\mathbb{R}^n)} + |\eta - X(1)|^2). \quad (17.17) \end{aligned}$$

17.4.3 Lyapunov Dissipation Inequality

We now derive the bound on $\dot{V}$ that was used in Sect. 17.4.2 to obtain the desired ISS estimate. This is done by analyzing the time derivative of each of the three functions in the definition of the Lyapunov function.

Analyzing V_1:

Using an integration by parts and recalling that $P_1 = \beta_1 D$ is a diagonal positive definite matrix, the time derivative of V_1 yields

$$\begin{aligned} \dot{V}_1 &= \beta_1 \int_0^1 (\partial_t X^\top D X + X^\top D \partial_t X) e^{-\mu z}\, dz \\ &= -\beta_1 \int_0^1 (\partial_z X^\top \Lambda D X + X^\top D \Lambda \partial_z X) e^{-\mu z}\, dz + \beta_1 \int_0^1 X^\top (S^\top D + DS) X e^{-\mu z}\, dz \\ &\leqslant -\beta_1 [X^\top \Lambda D X e^{-\mu z}]_0^1 - \beta_1 \mu \int_0^1 X^\top \Lambda D X e^{-\mu z} dz + \beta_1 \nu \int_0^1 X^\top D X e^{-\mu z} dz \\ &\leqslant -\beta_1 e^{-\mu} X(1,t)^\top D \Lambda X(1,t) + \beta_1 X(0,t)^\top D \Lambda X(0,t) - \sigma_1 V_1, \qquad (17.18) \end{aligned}$$

where $\nu < \lambda_{\min}(\Lambda)$ due to (17.14b) which results in $\sigma_1 := \beta_1(\mu\lambda_{\min}(\Lambda) - \nu)$ strictly positive. We now impose the boundary conditions by substituting the value of control u given in (17.11c) to get

$$X(0,t) = (H + BK)X(1,t) + BK(\eta - X(1,t))$$

which results in

$$\begin{aligned} \dot{V}_1 \leqslant\; & -\sigma_1 V_1 - \beta_1 e^{-\mu} X(1,t)^\top \Lambda D X(1,t) \\ & + \beta_1 X(1,t)^\top (H + BK)^\top \Lambda D (H + BK) X(1,t) \\ & + 2\beta_1 X(1,t)^\top (H + BK)^\top \Lambda D B K(\eta - X(1,t)) \\ & + \beta_1 (\eta - X(1,t))^\top K^\top B^\top \Lambda D B K (\eta - X(1,t)). \end{aligned}$$

With (17.14a), we thus get

$$\dot{V}_1 \leqslant -\sigma_1 V_1 - \beta_1(e^{-\mu} - \kappa)X(1,t)^\top \Lambda D X(1,t) \\ + 2\beta_1 X(1,t)^\top (H + BK)^\top \Lambda D B K(\eta - X(1,t)) \\ + \beta_1(\eta - X(1,t))^\top K^\top B^\top \Lambda D B K(\eta - X(1,t)). \quad (17.19)$$

Analyzing V_2:

Using (17.12) and repeating the same calculations as in the case of $\dot{V}_1$ when obtaining (17.18), we get

$$\dot{V}_2 \leqslant -\beta_2 e^{-\mu} X_z(1,t)^\top \Lambda^2 D \Lambda X_z(1,t) + \beta_2 X_z(0,t)^\top \Lambda^2 D \Lambda X_z(0,t) - \sigma_2 V_2 \,, \quad (17.20)$$

where $\sigma_2 := \beta_2(\mu\lambda_{\min}(\Lambda) - \nu) > 0$ by our choice of ν. Substitute the value of $\Lambda X_z(0,t)$ from (17.13) results in

$$\dot{V}_2 \leqslant -\sigma_2 V_2 - \beta_2 e^{-\mu} X_z(1,t)^\top \Lambda^2 D \Lambda X_z(1,t) + \beta_2(T_1 + T_2 + T_3),$$

where

$$T_1 := [(H+F)X_t(1,t) + F(\dot{\eta} - X_t(1,t))]^\top \Lambda D[(H+F)X_t(1,t) + F(\dot{\eta} - X_t(1,t))]$$
$$T_2 := [\widetilde{S}X(1,t) + SF(\eta(t) - X(1,t))]^\top \Lambda D[\widetilde{S}X(1,t) + SF(\eta(t) - X(1,t))]$$
$$T_3 := 2[-(H+F)X_t(1,t) - F(\dot{\eta} - X_t(1,t))]^\top \Lambda D[\widetilde{S}X(1,t) + SF(\eta(t) - X(1,t))] \\ = -2[(H+F)X_t(1,t) + F(\dot{\eta} - X_t(1,t))]^\top \Lambda D[\widetilde{S}X(1,t) + SF(\eta(t) - X(1,t))]$$

and we used the notation $\widetilde{S} = S(H+F)$ and $F = BK$. The term T_1 is already analyzed in a manner similar to our paper [19], whereas the terms T_2 and T_3 appear only because of the source term S, which is considered in this paper. After substituting η-dynamics in (17.7a), straightforward calculations yield

$$T_1 = X_t(1,t)^\top [(H+F)^\top \Lambda D(H+F) - H^\top \Lambda D F - F^\top \Lambda D H - F^\top \Lambda D F] X_t(1,t) \\ + \alpha^2(\eta(t) - X(1,t))^\top F^\top \Lambda D F(\eta(t) - X(1,t)) \\ - 2\alpha X_t(1,t)^\top H^\top \Lambda D F(\eta - X(1,t)) - 2\alpha^2(\eta(t) - X(1,t))^\top F^\top \Lambda D F d(t) \\ + 2\alpha X_t(1,t)^\top H^\top \Lambda D F d(t) + \alpha^2 d^\top F^\top \Lambda D F d(t).$$

Similarly, for T_2, we get

$$T_2 = X(1,t)^\top \widetilde{S}^\top \Lambda D \widetilde{S} X(1,t) + 2X(1,t)^\top \widetilde{S}^\top \Lambda D S F(\eta(t) - X(1,t)) \\ + (\eta(t) - X(1,t))^\top F^\top S^\top \Lambda D S F(\eta(t) - X(1,t)),$$

and finally, T_3 yields

$$
\begin{aligned}
T_3 = & 2\alpha(\eta(t) - X(1,t))^\top F^\top \Lambda D S F(\eta(t) - X(1,t)) \\
& - 2 X_t(1,t)^\top H^\top \Lambda D \widetilde{S} X(1,t) - 2 X_t(1,t)^\top H^\top \Lambda D S F(\eta(t) - X(1,t)) \\
& + 2\alpha(\eta(t) - X(1,t))^\top F^\top \Lambda D \widetilde{S} X(1,t) \\
& - 2\alpha X(1,t)^\top \widetilde{S}^\top \Lambda D F d(t) + 2\alpha(\eta(t) - X(1,t))^\top F^\top S^\top \Lambda D F d(t).
\end{aligned}
$$

One can use the Young's inequality to bound the terms with disturbances d, so that, for every $\overline{\zeta} > 0$, the disturbance terms in $\beta_2 T_1$ are bounded as

$$
|2\alpha^2\beta_2(\eta - X(1,t))^\top F^\top \Lambda D F d(t)| \leqslant \frac{\overline{\zeta}}{4}|\eta - X(1,t)|^2 + 4(\alpha^2\beta_2)^2 \frac{\|F^\top \Lambda D F\|_2^2}{\overline{\zeta}} |d(t)|^2
$$

$$
|2\alpha\beta_2 X_t(1,t)^\top H^\top \Lambda D F d(t)| \leqslant \overline{\zeta}|X_t(1,t)|^2 + (\alpha\beta_2)^2 \frac{\|H^\top \Lambda D F\|_2^2}{\overline{\zeta}} |d(t)|^2
$$

and the terms in $\beta_2 T_3$ can be bounded as

$$
|2\alpha\beta_2 X(1,t)^\top \widetilde{S}^\top \Lambda D F d(t)| \leqslant \overline{\zeta}|X(1,t)|^2 + (\alpha\beta_2)^2 \frac{\|\widetilde{S}^\top \Lambda D F\|_2^2}{\overline{\zeta}} |d(t)|^2,
$$

$$
\begin{aligned}
|2\alpha\beta_2(\eta(t) - X(1,t))^\top F^\top S^\top \Lambda D F d(t)| \leqslant & \frac{\overline{\zeta}}{4}|\eta - X(1,t)|^2 \\
& + 4(\alpha\beta_2)^2 \frac{\|F^\top S^\top \Lambda D F\|_2^2}{\overline{\zeta}} |d(t)|^2 .
\end{aligned}
$$

With (17.20), using (17.9a) and (17.13), we get

$$
\begin{aligned}
\dot{V}_2 \leqslant & -\sigma V_2 - \beta_2 X_t(1,t)^\top \big[(\mathrm{e}^{-\mu} - \kappa)\Lambda D + H^\top \Lambda D F + F^\top \Lambda D H + F^\top \Lambda D F\big] X_t(1,t) \\
& + \beta_2(\eta(t) - X(1,t))^\top F^\top \big[\alpha^2 \Lambda D + S^\top \Lambda D S + \Lambda D S + S^\top \Lambda D\big] F(\eta(t) - X(1,t)) \\
& + \beta_2 X(1,t)^\top \widetilde{S}^\top \Lambda D \widetilde{S} X(1,t) - 2\beta_2 X_t(1,t)^\top H^\top \Lambda D \widetilde{S} X(1,t) \\
& - 2\beta_2 X_t(1,t)^\top \big[\alpha(H+F)^\top \Lambda D F + H^\top \Lambda D S F\big](\eta - X(1,t)) \\
& + 2\beta_2 X(1,t)^\top \big[\widetilde{S}^\top \Lambda D S F + \alpha F^\top \Lambda D \widetilde{S}\big](\eta(t) - X(1,t)) \\
& + \overline{\zeta}|X(1,t)|^2 + \overline{\zeta}|X_t(1,t)|^2 + \frac{\overline{\zeta}}{2}|\eta(t) - X(1,t)|^2 + \chi_1|d(t)|^2,
\end{aligned}
$$

where

$$\chi_1 := \alpha^2 \, \|F^\top \Lambda D F\|_2^2 + \, 4\alpha^4 \beta_2^2 \frac{\|F^\top \Lambda D F\|_2^2}{\overline{\zeta}} + (\alpha\beta_2)^2 \frac{\|H^\top \Lambda D F\|_2^2}{\overline{\zeta}}$$
$$+ 4(\alpha\beta_2)^2 \frac{\|F^\top S^\top \Lambda D F\|_2^2}{\overline{\zeta}} + (\alpha\beta_2)^2 \frac{\|\widetilde{S}^\top \Lambda D F\|_2^2}{\overline{\zeta}}. \tag{17.21}$$

Analyzing V_3:

Once again, substituting the dynamics of η from (17.7a) in the expression of $\dot{V}_3$ to get

$$\begin{aligned}\dot{V}_3 &= 2\,\beta_3(\eta(t) - X(1,t))^\top(\dot{\eta}(t) - X_t(1,t)) \\ &= -2\alpha\beta_3\,|(\eta(t) - X(1,t))|^2 - 2\beta_3(\eta(t) - X(1,t))^\top X_t(1,t) \\ &\quad + 2\alpha\beta_3(\eta(t) - X(1,t))^\top d(t).\end{aligned}$$

Once again, Young's inequality is used to obtain, $\forall\,\overline{\zeta} > 0$

$$2\alpha\beta_3(\eta(t) - X(1,t))^\top d(t) \leqslant \frac{\overline{\zeta}}{4}|\eta(t) - X(1,t)|^2 + \frac{4(\alpha\beta_3)^2}{\overline{\zeta}}|d(t)|^2,$$

which further yields

$$\begin{aligned}\dot{V}_3 \leqslant &-\left(2\alpha\beta_3 + \frac{\overline{\zeta}}{4}\right)|(\eta(t) - X(1,t))|^2 - 2\beta_3(\eta(t) - X(1,t))^\top X_t(1,t) \\ &+ \frac{\overline{\zeta}}{2}|(\eta(t) - X(1,t))|^2 + \frac{4(\alpha\beta_3)^2}{\overline{\zeta}}|d(t)|^2.\end{aligned}$$

Combining $\dot{V}_1$, $\dot{V}_2$, $\dot{V}_3$:

By introducing the vector w as

$$w(t) := (X(1,t)^\top, (\eta(t) - X(1,t))^\top, X_t^\top(1,t))^\top,$$

one can manage the terms in the expressions for $\dot{V}_i$, $i = 1, 2, 3$ to get

$$\dot{V} \leqslant -\sigma_1 V_1 - \sigma_2 V_2 - \frac{\overline{\zeta}}{4} V_3 - w^\top(\Omega_1 + \Omega_2)w + \overline{\zeta} w^\top w + \chi|d(t)|^2,$$

where the constant χ is given by

$$\chi := \chi_1 + \frac{2(\alpha\beta_3)^2}{\overline{\zeta}}. \tag{17.22}$$

From (17.14d), there exists $\zeta > 0$ such that $\Omega_1 + \Omega_2 > \zeta\, I$, and hence by choosing $\overline{\zeta} = \zeta$, we obtain

$$\dot{V}(X(t), \eta(t)) \leqslant -\sigma V(X(t), \eta(t)) + \chi |d(t)|^2 \tag{17.23}$$

with $\sigma := \min\left\{\sigma_1, \sigma_2, \frac{\zeta}{4}\right\}$.

17.5 Quantized Control

We are interested in studying stabilization of the system (17.1), (17.2), and (17.7) when the output $X(1, \cdot)$ is quantized using a set of finite alphabets, and cannot be transmitted to the control precisely. In this case, the role of disturbance d is played by the quantization error. To define a quantizer, we first specify a set of finite alphabets $\mathcal{Q} := \{q_1, q_2, \ldots, q_N\}$. A quantizer with sensitivity $\Delta_q > 0$, and range $M_q > 0$, is then a function $q : \mathbb{R}^n \to \mathcal{Q}$ having the property that

$$|q(x) - x| \leqslant \Delta_q \quad \textbf{if} \quad |x| \leqslant M_q \tag{17.24}$$

and

$$|q(x)| \geqslant M_q - \Delta_q \quad \textbf{if} \quad |x| > M_q. \tag{17.25}$$

In other words, within the space $\mathbb{R}^n$, where the measurements of $X(1, \cdot)$ take values, we take a ball of radius M_q and partition it into N regions. Each of these regions is identified with a symbol q_i from the set $\mathcal{Q}$. If $|X(1, t)| \leqslant M_q$, the controller receives a valid symbol and knows the variable $X(1, t)$, modulo the error due to sensitivity of the quantizer. When the measurements are out of the range of the quantizer, then the quantizer just sends an out of bounds flag and no upper bound on the error between $X(1, t)$ and its quantized value can be obtained in that case. For this paper, we limit ourselves to the case of static quantizers, that is, the parameters of the quantizer are assumed to be fixed which introduces a bounded measurement error determined by the sensitivity of the quantizer.

The ratio between the range and the sensitivity of the quantizer M_q/Δ_q determines the rate at which the information is communicated by the quantizer on average. The basic idea of the quantized control in finite-dimensional systems is to show that the state of the system converges to a certain ball around the origin if this rate is sufficiently large (to dominate the most unstable mode) [12]. In the same spirit, we derive a lower bound on the ratio M_q/Δ_q which is required to achieve practical stability in the presence of quantization errors.

With quantized measurements, the controller (17.7) takes the form

$$\dot{\eta}(t) = -\alpha\, \eta(t) + \alpha\, q(X(1, t)) \tag{17.26a}$$

$$u(t) = K\eta(t). \tag{17.26b}$$

By writing $q(X(1,t)) = X(1,t) + q(X(1,t)) - X(1,t)$, and letting $d_q(t) := q(X(1,t)) - X(1,t)$, we are indeed in the same setup as earlier with $y(t) = q(X(1,t))$. Here, d_q is such that

$$|d_q| \leqslant \sqrt{n}\,|d_q|_\infty \leqslant \sqrt{n}\,\Delta_q, \quad \text{if } |X(1,t)|_\infty \leqslant M_q.$$

To state this result, we need the following lemma which relates $|X(1,t)|$ with the value of the Lyapunov function V considered in the previous section and defined in (17.16).

Lemma 17.1 *There exists a constant $C > 0$ such that*

$$|X(1,t)|^2 \leqslant C\,V(X(\cdot,t),\eta(t)), \quad \forall t \geqslant 0, \tag{17.27}$$

for the Lyapunov function V defined in (17.16).

The proof of Lemma 17.1 is omitted but we emphasize that the value of C in (17.27) can be computed directly in terms of closed-loop system data, see [19] for details.

Theorem 17.3 *Consider system* (17.1), (17.2), *and* (17.26), *and assume that the conditions of Theorem 17.2 hold, and the initial condition X^0 and η^0 satisfy*

$$C\,V(X^0,\eta^0) \leqslant M_q^2, \tag{17.28}$$

where the constant C is obtained from (17.27). *With the constants σ, χ appearing in* (17.23), *if the quantizer is designed such that*

$$\frac{M_q^2}{\Delta_q^2} > \frac{Cn\chi}{\sigma}, \tag{17.29}$$

then the following items hold:

- *The output $X(1,t)$ remains within the range of the quantizer for all $t \geqslant 0$, that is,*

$$|X(1,t)| \leqslant M_q, \qquad \forall t \geqslant 0.$$

- *The state of the system remains ultimately bounded in $\mathcal{H}^1$-norm, that is, there exists T such that for all $t \geqslant T$*

$$V(X(\cdot,t),\eta(t)) \leqslant \gamma_q(\Delta_q),$$

  where $\gamma_q(s) = \frac{nC\chi}{\sigma}s^2(1+\varepsilon)$, for some sufficiently small $\varepsilon > 0$, is a class $\mathcal{K}$ function.

We provide below a sketch of the proof of Theorem 17.3 and suggest the reader to consult [19] for more details.

In the light of condition (17.29), fix $\varepsilon > 0$ such that

$$\frac{n\chi}{\sigma}\Delta_q^2(1+\varepsilon) \leqslant \frac{M_q^2}{C}.$$

When the controller uses quantized measurements of $X(1,t)$, the derivative of the Lyapunov function in (17.23) satisfies, along the solutions to (17.1), (17.2), and (17.26),

$$\dot{V}(X(t),\eta(t)) \leqslant -\sigma V(X(t),\eta(t)) + \chi|q(X(1,t)) - X(1,t)|^2.$$

Thus, for the chosen $\varepsilon > 0$, if

$$\frac{n\chi}{\sigma}\Delta_q^2(1+\varepsilon) \leqslant V(X(t),\eta(t)) \leqslant \frac{M_q^2}{C}$$

then, using (17.27), $|X(1,t)| \leqslant M_q$ implying, with (17.24), that $|q(X(1,t)) - X(1,t)| \leqslant \sqrt{n}\Delta_q$, and hence

$$\dot{V}(X(t),\eta(t)) \leqslant -\varepsilon n\chi\Delta_q^2.$$

From the constraints imposed on the initial condition of the system, it readily follows from the above inequality that

$$|X(1,t)|^2 \leqslant CV(X(t),\eta(t)) \leqslant M_q^2, \quad \forall t \geqslant 0,$$

and hence, the quantization error is always upper bounded by Δ_q. The uniform decrease in the value of V also guarantees that

$$V(X(t),\eta(t)) \leqslant \frac{n\chi}{\sigma}\Delta_q^2(1+\varepsilon)$$

for sufficiently large t, and for initial condition satisfying $|X(1,0)| \leqslant M_q$. This concludes the sketch of proof of Theorem 17.3.

17.6 Example

To illustrate the effectiveness of the controller (17.7), we provide simulation results for the case of a 2×2 hyperbolic system. The system we simulate is of the form (17.1) with

$$\Lambda := \begin{bmatrix} 1 & 0 \\ 0 & 2 \end{bmatrix}, \quad S := \begin{bmatrix} 0.01 & 0 \\ 0 & 0.03 \end{bmatrix},$$

and the boundary condition (17.2) is described by

$$H = \begin{bmatrix} 0.25 & -1 \\ 0 & 0.25 \end{bmatrix}, \quad B = \begin{bmatrix} 1 & 0 \\ 0 & 1 \end{bmatrix}.$$

We first check conditions for Theorem 17.2. Selecting the matrix $K = \begin{bmatrix} 0 & 0.5 \\ -0.25 & -0.5 \end{bmatrix}$, it could be checked that the conditions of Theorem 17.2 are satisfied with

$$\mu = 0.1, \quad \kappa = 0.2, \quad D = \begin{bmatrix} 1 & 0 \\ 0 & 1 \end{bmatrix}, \quad \alpha = 90, \quad \beta_1 = 1, \quad \beta_2 = 1, \quad \beta_3 = 75.$$

Thus, the ISS estimate (17.8) holds for (17.11a) and (17.26) with the closed-loop boundary condition (17.11c). One can, for example, select the following initial condition, which satisfies the first-order compatibility condition for the existence of solutions in $\mathscr{H}^1((0, 1); \mathbb{R}^n)$:

$$X_1(z, 0) = \cos(4\pi z) - 1\ , \quad X_2(z, 0) = \cos(2\pi z) - 1,$$

with $z \in [0, 1]$.

Now to illustrate Theorem 17.3, let us consider the quantizer centered at the origin, and given by

$$q(x) = \lfloor \ell x + 0.5 \rfloor / \ell$$

with the parameter $\ell = 5$. The error due to quantization in this case is $\Delta_q = 1/\ell$, and for the sake of simplicity, we take a sufficiently large range M_q to bound the initial condition.

The time evolution of the solutions for the first and second components of X, as well as the state of the dynamic controller η, is plotted in Figs. 17.1 and 17.2, respectively. It could be seen that the solution to (17.11) converges to a neighborhood

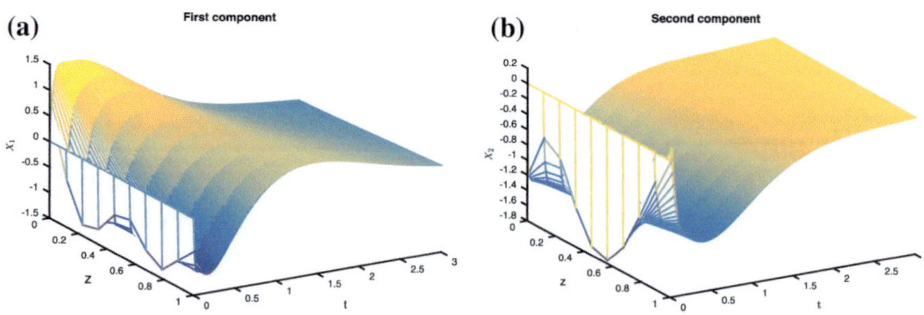

Fig. 17.1 Time and space evolution of state X: **a** X_1; **b** X_2

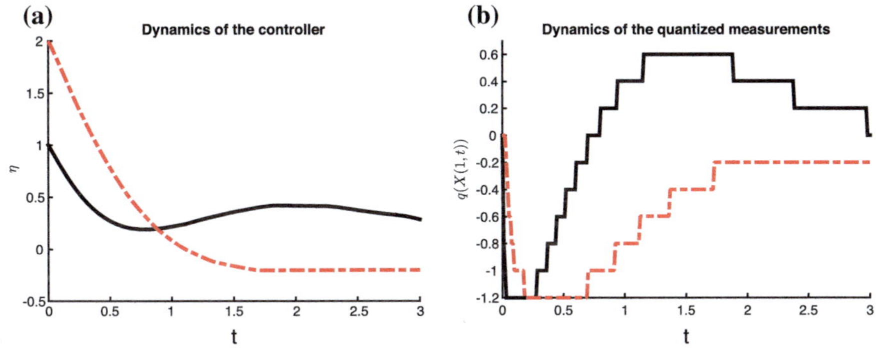

Fig. 17.2 Time evolution of controller state η and quantized measurements $q(X(1,t))$: **a** time evolution of η. Black: η_1, and red: η_2; **b** time evolution of $q(X(1,t))$. Black: $q(X_1(1,t))$, and red: $q(X_2(1,t))$

of the origin as the time increases. This simulation is thus in agreement with the result reported in Theorem 17.3. Figure 17.2b also gives the time evolution of the quantized measurements.

17.7 Conclusion

In this chapter, the stabilization of partial differential equations described by linear balance laws when the measurements are subjected to disturbances has been addressed via the use of a dynamic controller. This chapter can then be complementary to our previous work on robust stabilization of linear hyperbolic systems, because the presence of source terms in the system description has been taken into account. As a first step, the case of static controllers has been tacked allowing to provide conditions on stability of the closed-loop system and robustness with respect to measurement errors. In a second step, the case of dynamic controllers is considered. Conditions in terms of matrix inequalities have been then proposed to bound the maximum norm of the state trajectory, leading to robust stability of the closed-loop system in the presence of measurement errors in the feedback. The case of the quantized control, where the quantization error plays the role of disturbance in the measurement, have been also considered.

This work opens the door for studying other problems. For example, it could be interesting to study the design of a more general dynamical controller ensuring the robustness properties of the closed loop.

References

1. Bastin, G., Coron, J.-M.: Stability and Boundary Stabilization of 1-D Hyperbolic Systems. Subseries in Control: Progress in Nonlinear Differential Equations and Their Applications. Birkhauser, Basel (2016)
2. Bastin, G., Coron, J.-M., d'Andréa-Novel, B.: Using hyperbolic systems of balance laws for modeling, control and stability analysis of physical networks. In: Lecture Notes for the Pre-Congress Workshop on Complex Embedded and Networked Control Systems, Seoul, Korea (2008). 17th IFAC World Congress
3. Coron, J.-M., Bastin, G., d'Andréa, B.: Dissipative boundary conditions for one-dimensional nonlinear hyperbolic systems. SIAM J. Control Optim. **47**(3), 1460–1498 (2008)
4. Dashkovskiy, S., Mironchenko, A.: Input-to-state stability of infinite-dimensional control systems. Math. Control Signals Syst. **25**(1), 1–35 (2013)
5. Espitia, N., Girard, A., Marchand, N., Prieur, C.: Event-based control of linear hyperbolic systems of conservation laws. Automatica **70**, 275–287 (2016)
6. Karafyllis, I., Krstic, M.: ISS with respect to boundary disturbances for 1-D parabolic PDEs. IEEE Trans. Autom. Control **61**(12), 3712–3724 (2016)
7. Karafyllis, I., Krstic, M.: ISS in different norms for 1-D parabolic PDEs with boundary disturbances. SIAM J. Control Optim. **55**(3), 1716–1751 (2017)
8. Liberzon, D.: Hybrid feedback stabilization of of systems with quantized signals. Automatica **39**(9), 1543–1554 (2003)
9. Mazenc, F., Prieur, C.: Strict Lyapunov functions for semilinear parabolic partial differential equations. Math. Control Relat. Fields **1**(2), 231–250 (2011)
10. Mironchenko, A., Ito, H.: Construction of Lyapunov functions for interconnected parabolic systems: an iISS approach. SIAM J. Control Optim. **53**(6), 3364–3382 (2015)
11. Mironchenko, A., Wirth, F.: A note on input-to-state stability of linear and bilinear infinite-dimensional systems. In: 54th Conference on Decision and Control, pp. 495–500 (2015)
12. Nair, G., Fagnani, F., Zampieri, S., Evans, R.J.: Feedback control under data rate constraints: an overview. Proc. IEEE **95**(1), 108–137 (2007)
13. Prieur, C., Mazenc, F.: ISS-Lyapunov functions for time-varying hyperbolic systems of balance laws. Math. Control Signals Syst. **24**(1), 111–134 (2012)
14. Prieur, C., Tanwani, A.: Asymptotic stabilization of some finite and infinite dimensional systems by means of dynamic event-triggered output feedbacks. In: Feedback Stabilization of Controlled Dynamical Systems, vol. 473. Lecture Notes in Control and Information Sciences, pp. 201–230. Springer (2017)
15. Sontag, E.D.: Smooth stabilization implies coprime factorization. IEEE Trans. Autom. Control **34**(4), 435–443 (1989)
16. Tanwani, A., Teel, A.R., Prieur, C.: On using norm estimators for event-triggered control with dynamic output feedback. In: Proceedings of 54th IEEE Conf. on Decision and Control, pp. 5500–5505 (2015)
17. Tanwani, A., Prieur, C., Fiacchini, M.: Observer-based feedback stabilization of linear systems with event-triggered sampling and dynamic quantization. Syst. Control Lett. **94**, 46–56 (2016)
18. Tanwani, A., Prieur, C., Tarbouriech, S.: Input-to-state stabilization in H^1-norm for boundary controlled linear hyperbolic pdes with application to quantized control. In: 2016 IEEE 55th Conference on Decision and Control (CDC), pp. 3112–3117 (2016)
19. Tanwani, A., Prieur, C., Tarbouriech, S.: Disturbance-to-state stabilization and quantized control for linear hyperbolic systems. In: IEEE Transactions on Automatic Control (2017). arXiv:1703.00302

Index

S. Tarbouriech et al. (eds.), *Control Subject to Computational and Communication Constraints*, Lecture Notes in Control and Information Sciences 475, https://doi.org/10.1007/978-3-319-78449-6

MIX
Papier aus verantwortungsvollen Quellen
Paper from responsible sources
FSC® C105338

If you have any concerns about our products,
you can contact us on
ProductSafety@springernature.com

In case Publisher is established outside the EU,
the EU authorized representative is:
Springer Nature Customer Service Center GmbH
Europaplatz 3, 69115 Heidelberg, Germany

Printed by Libri Plureos GmbH
in Hamburg, Germany